BIOLOGY OF THE LAND CRABS

BIOLOGY OF THE LAND CRABS

EDITED BY

WARREN W. BURGGREN
DEPARTMENT OF ZOOLOGY
UNIVERSITY OF MASSACHUSETTS
AMHERST, MASSACHUSETTS
UNITED STATES

BRIAN R. McMAHON
DEPARTMENT OF BIOLOGY
UNIVERSITY OF CALGARY
CALGARY, ALBERTA
CANADA

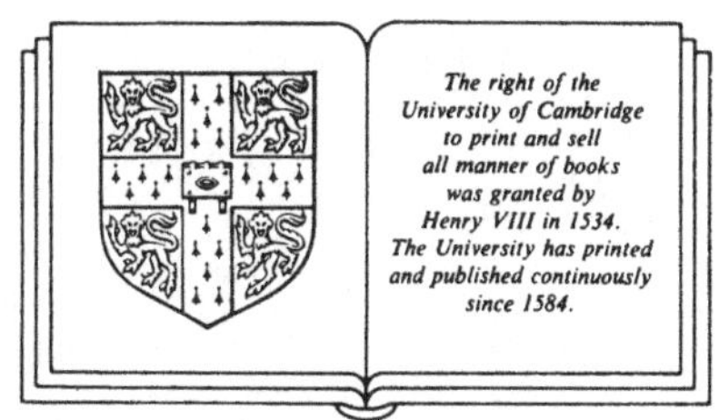

CAMBRIDGE UNIVERSITY PRESS
CAMBRIDGE
NEW YORK • NEW ROCHELLE • MELBOURNE • SYDNEY

CAMBRIDGE UNIVERSITY PRESS
Cambridge, New York, Melbourne, Madrid, Cape Town, Singapore, São Paulo, Delhi

Cambridge University Press
The Edinburgh Building, Cambridge CB2 8RU, UK

Published in the United States of America by Cambridge University Press, New York

www.cambridge.org
Information on this title: www.cambridge.org/9780521112925

First published 1988
This digitally printed version 2009

A catalogue record for this publication is available from the British Library

Library of Congress Cataloguing in Publication data
Biology of the land crabs / edited by Warren W. Burggren and Brian R. McMahon
p. cm.
Bibliography: p.
Includes index.
ISBN 0–521–30690–6
1. Grapsidae. 2. Gecarcinidae. 3. Coenobotidae. 4. Crabs.
I. Burggren, Warren W. II. McMahon, Brian R. (Brian Robert), 1936–.
QL444.M33B564 1988 87–20734
595.3′842—dc19 CIP

ISBN 978-0-521-30690-4 hardback
ISBN 978-0-521-11292-5 paperback

Contents

Contributors

Adiyodi, Rita G., Vatsyayana Center of Invertebrate Reproduction, Department of Zoology, University of Calicut, Kerala 673 635, India

Burggren, Warren W., Department of Zoology, University of Massachusetts, Amherst, Massachusetts 01003–0027, United States

Dunham, David W., Department of Zoology, University of Toronto, Toronto, Ontario M5S 1A1, Canada

Full, Robert J., Department of Zoology, University of California at Berkeley, Berkeley, California 94720, United States

Gilchrist, Sandra L., Department of Biology, New College, University of South Florida, Sarasota, Florida 33580, United States

Greenaway, Peter, School of Zoology, University of New South Wales, Kensington, N.S.W. 2033, Australia

Hartnoll, Richard G., Department of Marine Biology, University of Liverpool, Port Erin, Isle of Man, United Kingdom

Herreid II, Clyde F., Department of Biology, State University of New York, Buffalo, New York 14260, United States

McMahon, Brian R., Department of Biology, University of Calgary, Calgary, Alberta T2V 1N4, Canada

Wolcott, Thomas G., Department of Marine, Earth and Atmospheric Sciences, North Carolina State University, Box 8208, Raleigh, North Carolina 27695–8208, United States

Preface

This volume, detailing the biology of what are essentially subtropical and tropical animals, had its genesis during the winter of 1983–1984, during which time our home institutions in Massachusetts and Alberta were blanketed in snow. In spite of this unlikely environment (or perhaps because of it!), we turned our thoughts and research efforts once more to collaborative studies of cardiorespiratory physiology in land crabs. Although we felt we knew where to look for information on that particular aspect of their physiology, we frequently lamented the fact that there was no single source of information where we could learn about the other fields of biological research on land crabs that potentially impinged upon our physiological studies.

Having recognized the need for a scholarly volume on many different aspects of land crab biology, we then set about what we thought would be the straightforward task of assembling a group of experts to contribute to a "small" monograph on the biology of land crabs. After arbitrating (and contributing to) pitched battles among contributors about which animals were actually to be considered "land crabs" and how they should be classified, and, following the generation of several hundred pages of information and the collection of approximately one thousand references, there is now this single source of information on the biology of land crabs. Rather than suggest that this volume be regarded as the authoritative source, however, we must emphasize that contributors were constrained by book length to report only on selected aspects of their assigned topic. In fact, far more is known about land crabs than we had ever imagined at the outset of this project. Yet, paradoxically, far less is known about them than we had hoped for, and as most authors freely admit, there are many aspects of the biology of land crabs about which we know very little.

In retrospect (despite this being a preface), perhaps the single most notable achievement of this joint effort by ten authors has been to define clearly not only what we currently know about land crabs but, as importantly, to also indicate what we still need to find out. Land crabs represent a fascinating evolutionary chapter in the colonization of land by aquatic animals, and as such will continue to be a focal point for biological research.

Warren W. Burggren
Brian R. McMahon

1: Biology of the land crabs: an introduction

WARREN W. BURGGREN and
BRIAN R. McMAHON

CONTENTS

1. Why study land crabs?

The Decapoda represent a highly diverse and successful crustacean order, with over 8,500 described species. Although the great majority of decapods are strictly aquatic, inhabiting either marine or freshwater habitats, some of the most fascinating (and certainly most readily observable) of the decapods are those that have become adapted in varying degrees to life on land. Various aspects of the biology of land crabs have been treated in a number of reviews and research articles (for examples see reviews by Bliss, 1968; Powers and Bliss, 1983), but to date there has been no systematic attempt to assemble the available material into a single, cohesive account. Thus, the purpose of this book is to provide a broadly based, yet unified, presentation of the biology of the land crabs.

Biologists have been documenting species and describing the natural history, morphology, and ecology of the semiterrestrial and terrestrial crabs for more than a century. Yet, the scientific interest in these decapods has burgeoned in recent years. For example, of the approximately 1,000 references that have been cited in this book, more than 40% have been published just in the last decade. How can we account for this recently expanding interest in land crabs?

First, those decapods adapted for life on land present fascinating case studies in many different disciplines. For example, the ecologist is presented with an unusual situation in many tropical islands, where the top of the energy pyramid may be occupied not by a vertebrate but rather by a land crab. The physiologist encounters a very complex arrangement where gas exchange may be achieved simultaneously by "lungs," gills, and other body surfaces rather than just by gills. The behaviorist observes elaborate behaviors that depend on extreme visual acuity, permitted by the relatively transparent nature of air compared to water. These are but a few examples of the striking differences that are frequently encountered when comparing semiterrestrial or terrestrial crabs with their aquatic counterparts.

Land crabs are also studied because of their interactions with humans. For example, land crabs play a significant role as a food source in several tropical island communities (see Chapter 3). The impact of these terrestrial decapods may even be quite unexpected and amusing, as in the chronic disruption of golf games on Christmas Island (in the Indian Ocean) by mass migrations of red crabs (*Gecarcoidea natalis*) across the green (Hicks, Rumpff, and Yorkston, 1984).

Perhaps the most compelling reason for studying land crabs, however, relates to the evolutionary transitions that semiterrestrial and terrestrial crabs have experienced as they emerge from "ancient" marine or freshwater environments. As will become evident in the ensuing chapters, biologists from many diverse disciplines have found the land crabs a highly useful paradigm for studying the evolutionary transition from water to land. By understanding the adjustments in, for example, behavior, morphology, and physiology that have allowed the land crabs to successfully exploit a variety of terrestrial environments, we will come to a better understanding of opportunistic colonization of unexploited habitats in all animals.

2. Defining a "land" crab

It is important at the outset to indicate what is meant by "land crab," and thereby to rationalize the inclusion or exclusion of particular species in a detailed treatment in the various areas covered in this book. This is no simple task. As Bliss (1968) emphasizes, "It is apparent that one can draw no sharp line between terrestrial and semi-terrestrial decapod crustaceans nor between semi-terrestrial and aquatic forms." Yet, in the past many authors have explicitly or implicitly adopted "schemes" that categorize the various degrees of adaptation for terrestrial existence that are shown by land crabs. Can any one such scheme be universally adopted? As will become apparent in several chapters in this book, the degree of apparent terrestriality shown by any given species can vary greatly as a function of season, reproductive state, molting condition, etc. Consequently, many species defy rigid categorization. An additional complication with classification schemes for semiterrestrial and terrestrial crabs is that a scheme that might be internally consistent for one discipline might be constantly at odds with that appropriate for another discipline. A classic example is the hermit crab *Coenobita clypeatus*. An ecologist might reasonably classify this anomuran as a highly terrestrial species, since it can be found many kilometers from the shoreline and does not require periodic immersion in water (De Wilde, 1973). From a morphological perspective, however, *C. clypeatus* is not nearly as highly modified from the aquatic anomuran condition as is the closely related coenobitid crab *Birgus latro*. Consequently, a morphologist might generate a scheme in which *C. clypeatus* does not occupy a position as a fully adapted land crab. Finally, from the physiological perspective of respiratory gas and ion exchange, *C. clypeatus* is but little modified from its marine ancestors, since it carries the aquatic medium with it in the gastropod shell it inhabits. The physiologist thus might also be quite unwilling to categorize this species as highly terrestrial.

Clearly, then, no single scheme can be mutually acceptable to the many different disciplines represented in the following chapters. This became increasingly evident to the editors as we witnessed the passionate exchanges on this subject between the various contributors to this book! Thus, although contributors may have developed their own methods of classification to organize their material more expeditiously, none of these is intended to serve as a global scheme of classification.

What, then, constitutes a "land crab"? In popular accounts "land crab" is used to refer exclusively to highly terrestrial brachyuran or "true" crabs, for example, *Gecarcinus*. The basis for this classification is somewhat arbitrary, since from many viewpoints it could be argued that the terrestrial anomuran decapods deserve comparable inclusion (see Chapter 2 for further discussion of the phylogenetic relationships between the infraorders Anomura and Brachyura). To serve as a basis for discussion and to define the scope of this book, we broadly define land crabs as crabs that show significant behavioral, morphological, physiological, or biochemical adaptation permitting extended activity out of permanent water. In this regard the emphasis is on "activity" and not mere "survival," since many marine and aquatic decapods can survive hours or days of exposure to humid air.

Such a broad definition would include not only the "obvious" land crabs such as *Cardisoma, Gecarcinus,* or *Birgus,* but clearly would also include many upper intertidal crabs such as *Pachygrapsus, Uca,* or *Paguristes*. It should be emphasized that inclusion or omission of a particular intertidal species from this book has not been decided strictly on the basis of our definition of land crab. In virtually every discipline represented, it has been necessary to describe the condition in the aquatic and strictly intertidal crabs to allow the unusual and often unique adaptations of land crabs to be put in proper perspective. Clearly, the degree of emphasis on terrestrial versus intertidal and aquatic forms will vary from chapter to chapter. For example, in Chapter 10, Herreid and Full indicate that relatively little is known about many aspects of digestion and assimilation in land crabs, but that a thorough understanding of these processes in aquatic crabs can allow informed speculation on the modifications shown by land crabs. Comparatively more work, however, has been done on water and ion relationships in land crabs (Chapter 7), and consequently in this chapter Greenaway is able to concentrate more intensively on data from highly terrestrial species. Clearly, arbitrarily to exclude discussion of intertidal species not intuitively considered terrestrial would not promote understanding of land crabs and how they evolved.

3. The scope and coverage of *Biology of the land crabs*

Chapter 2 initiates the detailed discussion of land crabs by describing their taxonomy, evolution, and geographic distribution. This chapter addresses in greater detail which decapods should be considered land crabs and the phylogenetic relationships between them. Chapter 3 continues with consideration of the ecology of land crabs and describes how opportunistic they have been in exploiting a wide range of niches and habitats. The behavior of land crabs, particularly their courtship displays, is discussed in Chapter 4. Chapter 5 explores both the special problems in reproduction and development faced by land crabs, and the special solutions that have evolved. Growth and molting and the particular obstacles to these processes presented by the terrestrial habitat are dealt with in Chapter 6. Land crabs have particularly crucial problems in water and ion conservation, and their many varied mechanisms for maintaining hydromineral balance are discussed in Chapter 7. Chapters 8 and 9 discuss the modifications of the respiratory and circulatory systems that have enabled land crabs to exploit successfully an aerial existence. Chapter 10 extends these considerations of life on land, dealing with energetics and locomotion in land crabs. The concluding chapter, Chapter 11, reexamines the questions posed in the introductory chapters, in light of the ensuing multifaceted discussion of the biology of land crabs. Finally, the Appendix presents in tabular form the natural history of many of the more common species of land crabs.

The coverage of this book, though broad, can be neither balanced nor all-inclusive. The chapters dealing with cardiovascular, respiratory, and osmoregulatory physiology of land crabs, for example, are rather substantial, since these subjects have been the focus of a considerable degree of recent experimental investigation. On the other hand, many important and interesting aspects of the biology of land crabs have not been allocated entire chapters. In some instances, these aspects are effectively treated within individual chapters. For example, anatomy and functional morphology are discussed individually in each chapter within the context of that discipline. Other areas only poorly represented are the nervous and endocrine systems, and the roles they play in regulatory processes. Unfortunately, little information on these systems exists for land crabs, and the subjects could not reasonably stand on their own as discrete chapters. Again, where relevant information does exist, it has been incorporated into the individual chapters. Of course, the hope of all of the book's contributors is that by defining what is not known, as well as what is, research on the biology of land crabs will be further stimulated, especially in the many areas that are but poorly represented in *Biology of the Land Crabs*.

2: Evolution, systematics, and geographical distribution

RICHARD G. HARTNOLL

CONTENTS

1. Introduction

At the outset of this chapter two points must be clarified in order to define the limits of coverage: first, which taxa are to be accepted as "crabs," and second, which characteristics will qualify members of those taxa for consideration as "land" crabs.

Here, in discussing the taxonomic limits, and in later sections of this chapter, I have followed the classification of the Crustacea developed by Bowman and Abele (1982) in the most recent major review of crustacean biology. This will be convenient, although later in this chapter it will become clear that this classification is certainly not ideal in all respects (see section 4). However, in the currently uncertain state of brachyuran phylogeny an additional different scheme might only increase confusion. The relevant parts of Bowman and Abele's classification are reproduced in Table 2.1. The eleven families of the Potamoidea are listed, but since the subdivision of that superfamily is still to some extent uncertain, the families are not mentioned further in this chapter. A case could be made for restricting coverage to the Brachyura or "true crabs," but this would exclude the land hermit crabs: These show terrestrial adaptations very similar to the true crabs, and a comprehensive treatment demands their inclusion. The simplest solution is to extend consideration to the infraorders Anomura and Brachyura. The Anomura is used in a rather more restricted sense by Bowman and Abele (1982) than in some previous classifications. Included are six families of hermit crabs (Coenobitidae, Diogenidae, Lomisidae, Pomatochelidae, Paguridae, and Parapaguridae), the stone or king crabs (Lithodidae), the porcellain crabs (Porcellanidae), and the mole crabs (Albuneidae, Hippidae): All of these could arguably be included in a broad definition of "crabs." Also in the Anomura are the squat lobsters (Aeglidae, Chirostylidae, and Galatheidae), whose inclusion as crabs would be more debatable. However, since none of them show any measure of terrestrial adaptation, their systematic position is academic in the present context.

The second problem is to decide on the characteristics of habit and behavior that characterize a crab as a land crab. As mentioned in Chapter 1, Bliss (1968) has previously emphasized the difficulty of discriminating between terrestrial, semiterrestrial, and aquatic decapods. An added complication is that species may be highly specialized

Table 2.1. *Classification of the Anomura and Brachyura, after Bowman and Abele (1982)*

	Grades of terrestrial adaptation
Infraorder Anomura H. Milne Edwards, 1832	
Superfamily Coenobitoidea Dana, 1851	
Family **Coenobitidae** Dana, 1851	$(T_3–T_4)$
Diogenidae Ortmann, 1892	$(T_1–T_2)$
Lomisidae Bouvier, 1895	
Pomatochelidae Miers, 1879	
Superfamily Paguroidea Latreille, 1803	
Superfamily Galatheoidea Samouelle, 1819	
Family Aeglidae Dana, 1852	
Chirostylidae Ortmann, 1892	
Galatheidae Samouelle, 1819	
Porcellanidae Haworth, 1825	$(T_1–T_2)$
Superfamily Hippoidea Latreille, 1825	
Infraorder Brachyura Latreille, 1803	
Section Dromiacea De Haan, 1833	
Superfamily Dromioidea De Haan, 1833	
Section Archaeobrachyura Guinot, 1977	
Superfamily Tymoloidea Alcock, 1896	
Superfamily Homoloidea De Haan, 1839	
Superfamily Raninoidea De Haan, 1839	
Section Oxystomata H. Milne Edwards, 1834	
Superfamily Dorippoidea MacLeay, 1838	
Superfamily Leucosioidea Samouelle, 1819	
Section Oxyrhyncha Latreille, 1803	
Superfamily Majoidea Samouelle, 1819	
Superfamily Hymenosomatoidea MacLeay, 1838	
Superfamily Mimilambroidea Williams, 1979	
Superfamily Parthenopoidea MacLeay, 1838	
Section Cancridea Latreille, 1803	
Superfamily Cancroidea Latreille, 1803	
Section Brachyrhyncha Borradaile, 1907	
Superfamily Portunoidea Rafinesque, 1815	
Family Geryonidae Colosi, 1923	
Portunidae Rafinesque, 1815	(T_1)
Superfamily Bythograeoidea Williams, 1980	
Superfamily Xanthoidea MacLeay, 1838	
Family Goneplacidae MacLeay, 1838	
Hexapodidae Miers, 1886	
Platyxanthidae Guinot, 1977	
Xanthidae MacLeay, 1838	$(T_1–T_2)$
Superfamily Bellioidea Dana, 1852	
Superfamily Grapsidoidea MacLeay, 1838	
Family **Gecarcinidae** MacLeay, 1838	$(T_3–T_4)$
Grapsidae MacLeay, 1838	$(T_1–T_5)$
Mictyridae Dana, 1851	(T_2)
Superfamily Pinnotheroidea De Haan, 1833	
Superfamily Potamoidea Ortmann, 1896	$(T_2–T_5)$
Family **Deckeniidae** Bott, 1970	
Gecarcinucidae Rathbun, 1904	
Isolapotamidae Bott, 1970	

Table 2.1. (*cont.*)

	Grades of terrestrial adaptation
Parathelphusidae Alcock, 1910	
Potamidae Ortmann, 1896	
Potamocarcinidae Ortmann, 1899	
Potamonautidae Bott, 1970	
Pseudothelphusidae Ortmann, 1893	
Sinopotamidae Bott, 1970	
Sundathelphusidae Bott, 1969	
Trichodactylidae H. Milne Edwards, 1853	
Superfamily Ocypodoidea Rafinesque, 1815	
Family **Ocypodidae** Rafinesque, 1815	**(T_2–T_5)**
Palicidae Rathbun, 1898	
Retroplumidae Gill, 1894	
Superfamily Hapalocarcinoidea Calman, 1900	

Note: The scheme is presented to the superfamily level and elaborated to the family level only in those superfamilies containing terrestrially adapted families. These are shown in boldface, and the range of terrestrial adaptation found in each is indicated.

for terrestrial life during part of their life cycle, yet very dependent on water at other phases. Since there is no simple discrimination, a series of degrees of terrestrial adaptation will be discussed in the next section.

2. Levels of terrestrial adaptation

The series of grades of terrestrial adaptation described will generally imply increasing levels of independence from the aquatic environment, although they may not correlate closely with actual tolerance of desiccation. Compared to insects, for example, even the most terrestrial of land crabs have poor resistance to water loss, and it is only by appropriate behavior patterns that they are able to survive out of water (see Chapter 7). Water loss is minimized by residing in cryptic and damp microenvironments, and by being active predominantly at night or during damp weather. Water uptake is maintained by regular visits to water, by drinking dew or casual water, or by the uptake of capillary water from the substrate (Wolcott, 1976; see also Chapters 3 and 7).

Any crab that spends part of its time out of water could be regarded as showing a degree of terrestrial adaptation. Possibly this is too broad a view, and a more realistic limit is to consider only those that are *active* while out of water. This would exclude the many intertidal crabs that shelter in pools, beneath rocks, or buried in the substrate at low tide, and are active only when covered by the incoming tide, for example, many of the Portunidae and Xanthidae. However, intertidal crabs active in air would be included. Five grades of terrestrial adaptation, T_1 to T_5, are specified in the following description, but the borders between them are not well defined. Moreover, the detailed information that would enable a species to be firmly categorized is not always available.

Grade T_1. Species resident in fresh or salt water, active in the water, but displaying intermittent activity in the air, usually for brief periods. Examples include the more aquatic of the potamonids and some of the grapsids such as *Percnon*.

Grade T_2. Intertidal species, which remain secreted in crevices or under boulders, or buried, while covered by the tide. They are active only in air when the tide has receded, either diurnally, nocturnally, or both. Ocypodids such as *Uca* or *Dotilla* typify this type.

Grade T_3. Species resident supratidally or out of fresh water, and active in air, most usually by night. Generally burrowing or otherwise cryptic in habit. Requiring regular access to water in which they can immerse themselves, either by visits to the sea or other sources, or to groundwater in the base of the burrow. Dependent on water for the pelagic larval stages. Examples are species of *Cardisoma* and *Ocypode*.

Grade T_4. Species as in T_3, but not requiring to immerse themselves regularly in water. Can obtain water from food, by drinking dew or casual water, or by capillary or osmotic uptake from damp substrates (see Wolcott, 1976, 1984). Dependent on water for the pelagic larvae. Examples are *Gecarcinus* and species of *Geograpsus*.

Grade T_5. Species as in T_4, but not requiring access to water bodies for the release of larvae, since development is abbreviated. Some species of *Geosesarma*, and certain potamonids, are examples of this type, which represents the most advanced level of terrestrial adaptation.

The preceding classification is far from perfect. Although it is possible to select clear-cut examples from each category, there are also many "borderline" species. The need for access to surface water is particularly hard to answer specifically, since it may vary seasonally and may be a requirement for some activities such as ecdysis, though crabs may otherwise survive without water for extended periods. The need for water for larval development is also problematic. There are a number of examples of abbreviated development – all of the potamonids, and some grapsids and ocypodids (see Chapter 5). This certainly gives a freedom from the need for access to the sea, but there may still be a requirement for water by the abbreviated larvae or young stages. A more complex scheme of classification would probably be self-defeating, however, especially since the detailed information to implement it is often lacking. The grades of terrestrial adaptation are listed for each family containing "land crabs" in Table 2.1. The information given during the systematic analysis in the next section should make it clear, as far as possible, how the various examples of land crabs relate to this outline scheme.

3. Systematic account

A. Introduction

In this section the crabs that can be regarded as land crabs according to the criteria detailed earlier will be reviewed on a family-by-family basis. As previously discussed, the taxonomic coverage will embrace the infraorders Anomura and Brachyura in the classification used by Bowman and Abele (1982). The relevant part of this is reproduced to the superfamily level in Table 2.1, but elaborated to the family level only for superfamilies containing terrestrially adapted forms: those families that contain members showing terrestrial adaptation are indicated in boldface, and the range of terrestrial adaptation found in each family is also indicated.

The attention given to families will vary substantially with the level of terrestrial adaptation. Those consisting entirely of highly adapted terrestrial species, such as the Coenobitidae and the Gecarcinidae, will be examined in detail. The species will be listed, their geographical distribution summarized, and their degree of terrestrial adaptation outlined. Families including many species showing only limited terrestrial adaptation will generally be treated by reviewing the situation and illustrating the level of adaptation by reference to selected examples. Within those families, species showing higher levels of adaptation will be discussed in detail.

This chapter is not intended as a formal taxonomic treatment, nor does it attempt to enable species to be identified. The morphological features of the various taxa will be introduced only where they seem particularly relevant to a point under discussion. Many of the special features related to terrestrial life are best discussed in detail in later chapters devoted to their various functions, and only passing reference will be made here. Certain of the more obvious features of land crabs will be described, though in general they display few external characteristics that distinguish them from their aquatic relatives. The more common external modifications include the following. The branchial regions of the carapace are enlarged to facilitate aerial respiration, and the flagellae of the maxillipeds are reduced or lost. There are tufts of setae at various sites on the ventral surface that enable moisture from the substrate to be passed to the branchial chamber. The antennules are reduced, or modified in form, to act as sense organs in air rather than in water.

One distinction that is relatively easily made is between the land hermit crabs (Coenobitidae and Diogenidae) and the other land crabs. The hermit crabs have a large, uncalcified abdomen that is normally kept concealed within an empty gastropod shell, and the last two pairs of walking legs are reduced (Fig. 2.1). *Birgus*, as described later (see Fig. 2.3), has departed somewhat from this pattern but is still recognizable as an anomuran. In contrast, the brachyuran land crabs have a flattened, calcified abdomen that is folded beneath the body of the crab, and all four pairs of walking legs are fully developed.

B. Coenobitidae

This family contains only two genera, *Birgus* and *Coenobita*, both well adapted for terrestrial environments. *Coenobita* comprises a number of species, the typical land hermit crabs. *Birgus* consists of only the one species, *B. latro* (L.) the robber crab, which in the adult stage has a somewhat calcified abdomen and does not inhabit a gastropod shell.

i. Coenobita. There has not been a comprehensive revision of this genus since that of Alcock (1905), and so the following list of species and their distribution must be regarded as tentative. The literature is very confused as a result of synonymy and misidentification, but the general consensus from several works (Fize and Sèrene, 1955; Gordan, 1956, in particular) is that there are probably 11 valid species, as described in the following list. No full synonymy is attempted, but one or two points of particular confusion are mentioned. Outlines of the recorded distributions are given, but some of these may be far from complete.

C. cavipes Stimpson. A widely distributed species from the mainland coast of East Africa to the East Indies and the Ryu Kyu Islands (Alcock, 1905; Barnard, 1950; Lewinsohn, 1982).

C. clypeatus (Herbst). This must not be confused with *C. clypeata* (Latreille), which is a synonym of *C. brevimanus*. From southern Florida to Venezuela, and the West Indian islands (Chace and Hobbs, 1969). Also Bermuda (Bright, 1966). The only species in the Western Atlantic.

C. compressus H. Milne Edwards. West coast of America from Mexico (Lower California) to Chile (Bright, 1966; Ball, 1972). Only species definitely known from American West Coast (Holthuis, 1954; Ball and Haig, 1974), and restricted to this coast. Records from the Indo-West Pacific are misidentifications (Holthuis, 1954).

C. perlatus H. Milne Edwards. From Aldabra, Mauritius, and Seychelles through the Indo-Pacific to Samoa (Alcock, 1905; Holthuis,

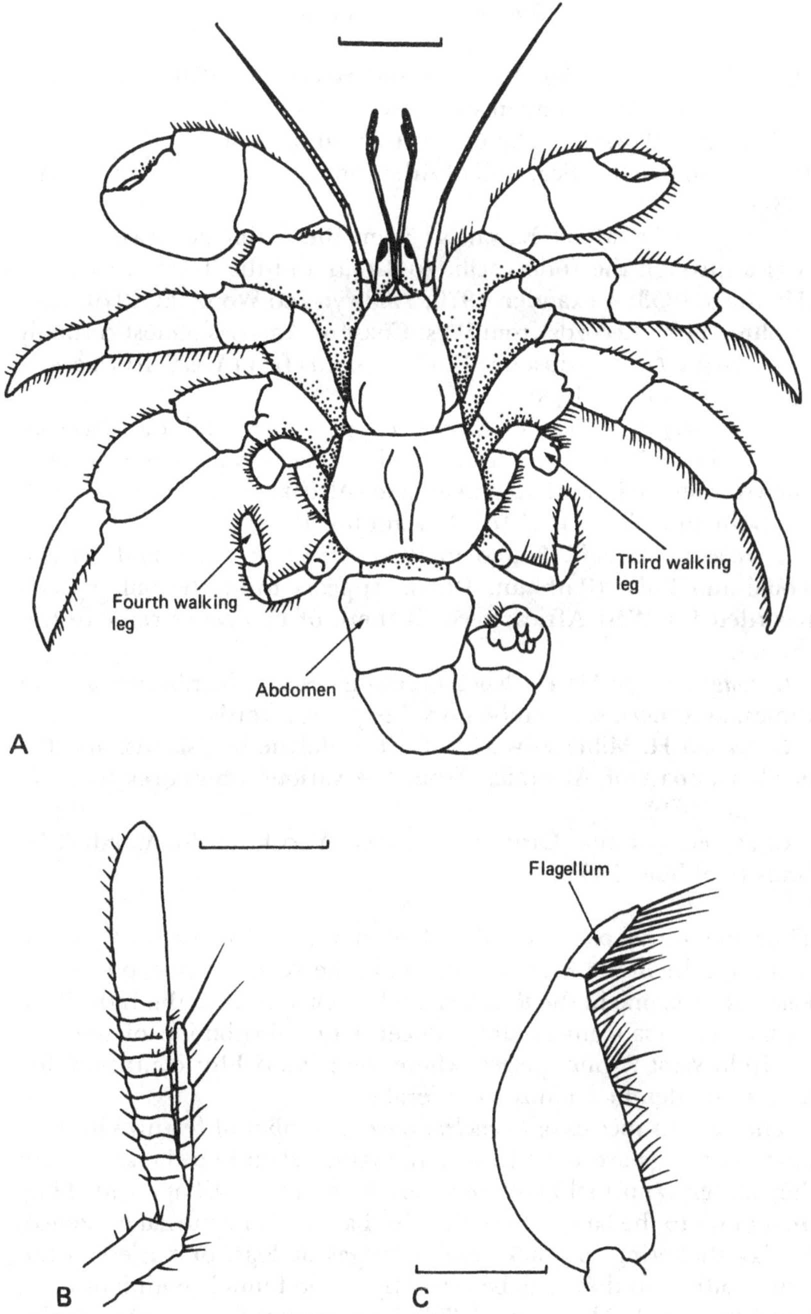

Fig. 2.1. A. Specimen of the land hermit crab, *Coenobita perlatus*, removed from its shell. Scale bar = 10 mm. (Redrawn after Alcock, 1905.) B. Terminal part of right antennule of *Coenobita clypeatus*. Scale bar = 2 mm. C. Expodite of right third maxilliped of *C. clypeatus* to show the reduced flagellum. Scale bar = 2 mm.

1953; Alexander, 1976; Yaldwyn and Wodzicki, 1979). Not known from the East African mainland (Lewinsohn, 1982).

C. scaevola (Forskal). The only species in the Red Sea (Lewinsohn, 1969). Found in Red Sea, Gulf of Aden, Somalia, Pakistan (Lewinsohn, 1982).

C. rugosus H. Milne Edwards. From the mainland coast of East Africa through the Indo-Pacific to Tahiti and the Tuamotu Islands (Holthuis, 1953; Alexander, 1976; Yaldwyn and Wodzicki, 1979; Lewinsohn, 1982). Records from West Coast of America almost certainly *C. compressus* (see Lewinsohn, 1982). Records from West Africa are *C. rubescens* (Rathbun, 1900).

C. brevimanus Dana (= *C. clypeata* (Latreille) and *C. hilgendorfi* Terao). From Zanzibar to Tahiti and the Tuamotu Islands, but not known from the East African mainland (Alcock, 1905; Holthuis, 1953; Yaldwyn and Wodzicki, 1979; Lewinsohn, 1982).

C. rubescens Greeff. Found in West Africa on the islands of São Tomé and Rolas (Rathbun, 1900). Appears to be the only species recorded for West Africa; records there of *C. rugosus* refer to this species.

C. longitarsis de Man. Alcock (1905) gives the distribution as East Indies, and there seem to be no subsequent records.

C. spinosa H. Milne Edwards. Various Polynesian islands, and the northern coast of Australia (from the various references listed in Gordan, 1956).

C. carnescens Dana. Eastern Polynesia (Alcock, 1905), Marshall Islands (Holthuis, 1953).

Thus, the overall pattern of distribution of the 11 species of *Coenobita* is one species in the East Pacific, one in the West Atlantic, one in the East Atlantic, one in the Red Sea, and seven species in the Indo-West Pacific. The main uncertainty concerns the distribution of some of the Indo-West Pacific species, where the potential for confusion due to past misidentification is considerable.

The various species of *Coenobita* have a number of features in common. The eggs are hatched and the larvae shed into the sea where they undergo normal planktonic development (see Chapter 5). They then move to the land, where they lead a mainly nocturnal existence, by day sheltering in cracks under ledges or logs, or buried in the sand – although they may be active by day in humid conditions or in rain (Ball, 1972; Alexander, 1976). The general features that distinguish *Coenobita* (Fig. 2.1) from typical hermit crabs are relatively minor: morphological and physiological changes in the osmoregulatory mechanisms; development of compressed and truncate antennular flagella; reduction of the flagella of the maxillipeds (McLaughlin, 1983b). The species vary in size, from medium-sized ones such as *C. perlatus* and *C. compressus*, to large species like *C. rugosus* and *C. clypeatus*, which can weigh up to 200 g. There are also substantial interspecific differences in their water relations (also see Chapter 7).

Some species are always found close to the shore, and enter either the sea or brackish water regularly to replenish the water in their shells. Examples are *C. cavipes* (Holthuis, 1953; Gross et al., 1966; Alexander, 1976), *C. compressus* (Ball, 1972), *C. perlatus* (Alexander, 1976, Gross, 1964a), and *C. scaevola* (Volker, 1965). All of these species seem to need regular access to water of high salinity, which effectively binds them to the shore.

Coenobita rugosus is found near the shore but inland from the beach, though never more than about 100 m from high tide (Johnson, 1965; Alexander, 1976; Page and Willason, 1982). However, despite occurring near the sea, it is reluctant to enter salt water, and will in preference fill its shell with fresh water.

Finally, a number of species consistently occur far from the shore and normally return to the sea only for the eggs to hatch. *C. brevimanus* is found in the interior of Eniwetok away from the sea (Page and Willason, 1982), and Gross (1964a) records it in heavy vegetation, never in or near the sea. It will fill its shell with fresh or brackish water. *C. clypeatus* can occur some 3 km from the coast and to an altitude of 400 m on Dominica (Chace and Hobbs, 1969), and elsewhere in the Caribbean to an altitude of 900 m and up to 15 km from the coast (De Wilde, 1973). It prefers dry rather than humid areas, and whereas specimens near the shore will use seawater, inland specimens depend on freshwater, which they pass to the mouth with the aid of their chelae (De Wilde, 1973). *C. brevimanus* occurs on grassy areas away from the shore on Aldabra, and is the most terrestrially adapted of the four species of *Coenobita* found there (Alexander, 1976). *C. rubescens* occurs far inland (von Hagen, 1977). It will exchange its marine shells for terrestrial ones, and is found in shells that do not occur below 800 m. It is also found in the tests of sea urchins (Rathbun, 1900).

It is not clear to what extent the more terrestrial species of *Coenobita* are dependent on the availability of permanent water. Certainly they are probably able to use dew and small amounts of temporary water, since they can use the chelae to drink the water or fill the shell, and do not need to immerse themselves. They also have the ability to absorb water from damp substrates: *C. rugosus* can take up water from damp sand (Vannini, 1975); *C. brevimanus* can survive for over two weeks in a damp substrate (Gross, 1964a) (see also Chapter 5).

ii. Birgus. This genus contains only a single species, *Birgus latro* (L.). It is widespread in the tropical Indo-Pacific, and a detailed account of its distribution has been provided by Reyne (1939). Its westward limit is Zanzibar and other smaller islands off the coast of East Africa. From here *B. latro* extends eastward, occurring on many islands in the tropical parts of the Indian and Pacific oceans (Fig. 2.2), reaching its eastward limit in the Gambier Islands (135° W). The only record from outside the tropics is from the Ryu Kyu Islands of Japan. Although widely distributed, almost all of these records are from small islands.

Birgus latro (Fig. 2.3) is colloquially known as the coconut or robber crab. These are apt names, given its habits of feeding on coconuts and of carrying off almost any portable article, edible or not (see Chapter 3). The comments of Alexander (1976) are typical: "Nothing could be left out at camps in safety; cooking utensils, cutlery, machetes; all were taken, and on one occasion a Primus stove was dragged off into the bush. On another occasion a *Birgus* was observed on the coastal strand line towing a whisky bottle behind it."

The basic life history of *Birgus* is similar to *Coenobita* in that the eggs are hatched in the sea, and there is a pelagic larval phase. The later stages are described in detail by Harms (1932, 1938). The glaucothoe larva returns to the land already inhabiting a small gastropod shell. It may leave this shortly, with only a slightly asymmetrical abdomen, and after a molt or two acquire the symmetrical abdomen of the adult. Alternatively they may retain the marine shell to the first molt, and at that molt transfer to a land shell, developing a markedly asymmetrical abdomen. Then after a few molts, and perhaps a transfer to a larger land shell, become free-living and develop the symmetry of the adult abdomen. This species grows to a very large size: Large males may span 75 cm or more across the legs, and weigh at least 2–3 kg (see Reyne, 1939). Harms (1932) suggests that this large size of *Birgus* may be linked to the fact that there is no longer any need to seek a larger shell on molting. This free-living habit gives access to nutritious food inland. Interestingly, Harms (1932) found that if he provided larger shells for the young crabs they grew to a greater-than-usual size before entering the free-living stage.

Birgus shows certain structural differences from *Coenobita*. These include changes in the eyestalks, antennules, and antennae; the gill chamber containing well-developed lungs (see Chapter 8); the rear portion of the carapace and the abdomen lightly calcified; the abdomen more or less symmetrical, and the uropods very reduced. Once the free-living habit of *Birgus* is assumed, the fourth pereiopods enlarge and redevelop an ambulatory function (Harms, 1932; Wolff, 1961).

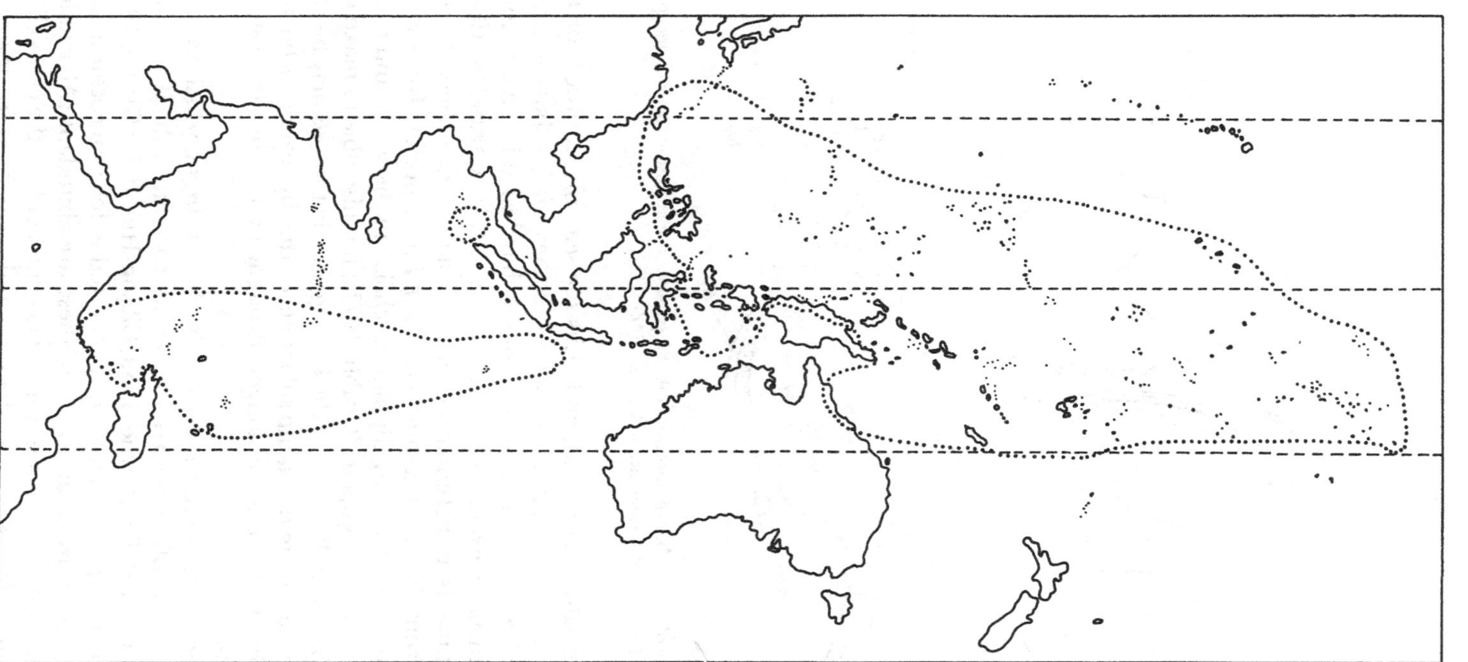

Fig. 2.2. The distribution of the robber or coconut crab, *Birgus latro*, delimited by the area enclosed within the dotted lines. (After Reyne, 1939.)

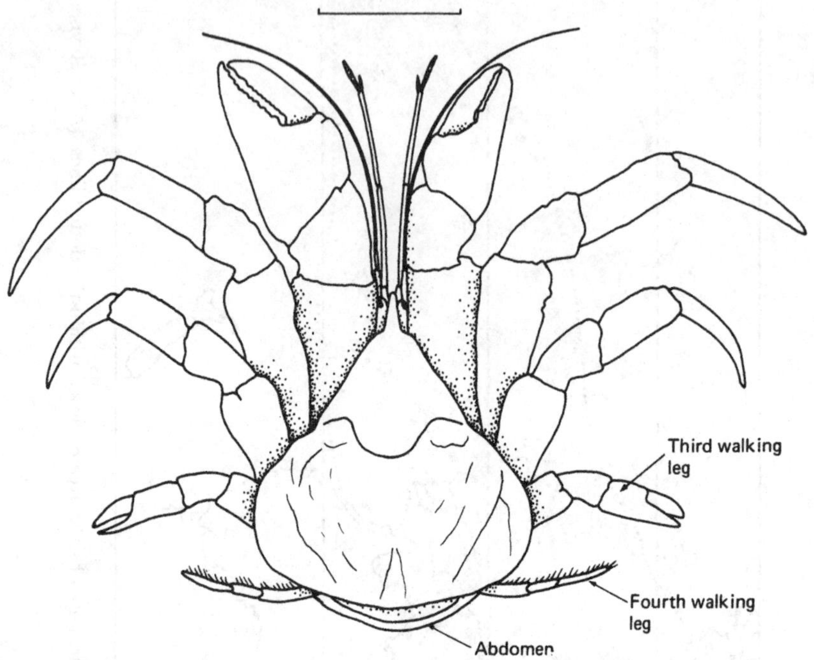

Fig. 2.3. Male of the robber or coconut crab, *Birgus latro*, the largest terrestrial decapod. Scale bar = 50 mm. (Redrawn after Alcock, 1905.)

Birgus usually lives inland, and visits the sea only in order for the eggs to hatch. It inhabits rock shelters or digs shallow burrows, and may be active by day, though on most inhabited islands it has become very nocturnal (Gibson-Hill, 1947; Grubb, 1971; Alexander, 1976). On Christmas Island (Indian Ocean), it is abundant inland at 6 km or more from the sea (Gibson-Hill, 1947). *Birgus* prefers fresh water and is able to drink this by dipping its chelae in the water and transferring it to the maxillipeds (Grubb, 1971); it is also able to make use of dew (Harms, 1932). Gross (1964a) suggests that *Birgus* may be able to absorb water from the damp substrate, or that by sealing its burrow may be able to tolerate prolonged drought by limiting the rate of desiccation.

A final point of interest relates to the diet of *Birgus*, which as mentioned earlier is often referred to as the coconut crab. It certainly consumes a varied diet (Reyne, 1939; Gibson-Hill, 1947; Grubb, 1971; Alexander, 1976), but is frequently reported to feed on coconuts and reaches a larger size in areas where these are abundant (Alexander, 1976). The literature contains varied anecdotes of how the crabs climb the coconut palms, tear down the nuts, and open them (reviewed by

Reyne, 1939). However, experimental observations suggest that in fact they are only capable of consuming nuts that are already damaged (Gibson-Hill, 1947; Alexander, 1976).

C. Diogenidae

Apart from the Coenobitidae, the hermit crabs show little terrestrial adaptation. The only ones to have developed a limited degree of terrestriality belong to the Diogenidae: Various species live in the upper intertidal, are active while uncovered, and on occasions forage above the high-tide level. Various species of *Clibanarius* are the most frequently recorded in this habit: *C. longitarsus* (de Haan) has been found in Indo-Pacific mangroves climbing above high-tide level (Johnson, 1965; Macnae, 1968); *C. tricolor* on rocky cliffs in Florida (Pearse, 1929); *C. vittatus* (Bosc) high on mud flats and rocky jetties in the eastern United States (Williams, 1965; Herreid, 1969a). Another genus is represented by *Calcinus laevimanus* (Randall), which climbs high on rocky cliffs in East Africa (Hartnoll, 1976), and is active in the supralittoral fringe in Hawaii, as is *Clibanarius zebra* (see Reese, 1969). These noncoenobitid hermit crabs may possess respiratory adaptations (see Chapter 8), but show no external features related to terrestrial life.

D. Porcellanidae

This family gains a somewhat marginal entry in respect of *Petrolisthes quadratus* Benedict. In the Caribbean this species lives beneath rocks above the high-tide mark, though still within the splash zone (Chace and Hobbs, 1969).

E. Portunidae

The Portunidae is a large family with many intertidal species, but virtually all of these are active only when covered by water. During low tide they either shelter beneath stones or remain buried in the substrate. However, there are several records of activity in air, for the common European shore crab, *Carcinus maenas* (L.) (see Flower, 1931), and in the laboratory it will survive in moist air for over 10 days (Crothers, 1968; Taylor and Butler, 1978).

F. Xanthidae

This very large family is represented by a great diversity of intertidal species, but like the portunids, almost all of them are active only when immersed, and are well hidden at low tide. There are some exceptions though, with the best documented being species of *Eriphia.* In East Africa *E. laevimanus* (Guerin) and *E. smithi* McLeay are common in holes or crevices on the lower part of the shore (Hartnoll, 1976). They are inactive while covered by the tide and also when uncovered by day, becoming active only during nocturnal low tides, when they show consistent homing to their refuges from over 50 m (M. Vannini, pers. comm). The structure of the branchial chamber in *E. smithi* is modified to act as a "lung" (Vuillemin, 1967). Also in East Africa the small *Lydia annulipes* (H. Milne Edwards) extends to above high-water springs, and is also mainly nocturnal (Hartnoll, 1976). In the Marshall Islands *E. sebana* is active at the supralittoral fringe, where it has been seen to prey on *Coenobita* (Reese, 1969).

G. Gecarcinidae

This family is generally referred to as the "land crabs," but although all of its members show distinct terrestrial adaptations, they do not necessarily include the most completely adapted species. For example, whatever the degree of terrestriality of the adult, the eggs of all species have to be hatched in the sea, where the larvae undergo a normal planktonic development. Thus independence from the sea is by no means complete. The family includes four genera: *Cardisoma, Epigrapsus, Gecarcinus,* and *Gecarcoidea. Ucides* has in the past been included in the Gecarcinidae, e.g., in Rathbun (1918), but has now been transferred to the Ocypodidae (Chace and Hobbs, 1969; Turkay, 1983).

Although the family is highly terrestrial, there is little external evidence of this adaptation. The lateral regions of the carapace are inflated (Fig. 2.4) to accommodate the highly modified respiratory structures (see Chapter 8). The flagellum of the exopodite of the third maxilliped is lost in *Gecarinus* and *Gecarcoidea* (Fig. 2.4B) and reduced in *Epigrapsus*. There are some tufts of setae related to water uptake – between the bases of the second and third legs in *Epigrapsus*, and between the base of the abdomen and the fifth leg in *Gecarcinus* and *Gecarcoidea*. Otherwise they appear typical brachyurans, and there are no striking features to differentiate the genera. They are medium- to large-sized crabs, and a large *Cardisoma* exceeds 100 mm in carapace width. A striking phenomenon in gecarcinids is that the breeding migration to the sea is highly synchronized on a lunar and seasonal basis, so that mass movements of the crabs occur. Accounts have been given for *Cardisoma crassum* (Garth, 1948), *C. carnifex* (Alexander, 1976), *Gecarcoidea natalis* (Gibson-Hill, 1947), *Gecarcinus lateralis* (Bliss, et al., 1978), *G. ruricola* (see Rathbun, 1918), and *G. planatus* (Niaussat and Ehrhardt, 1971). The most interesting are the observations over a 20-year period cited by Gibson-Hill (1947), which demonstrate the consistency of this behavior.

i. Cardisoma. This genus has a worldwide distribution in the tropics

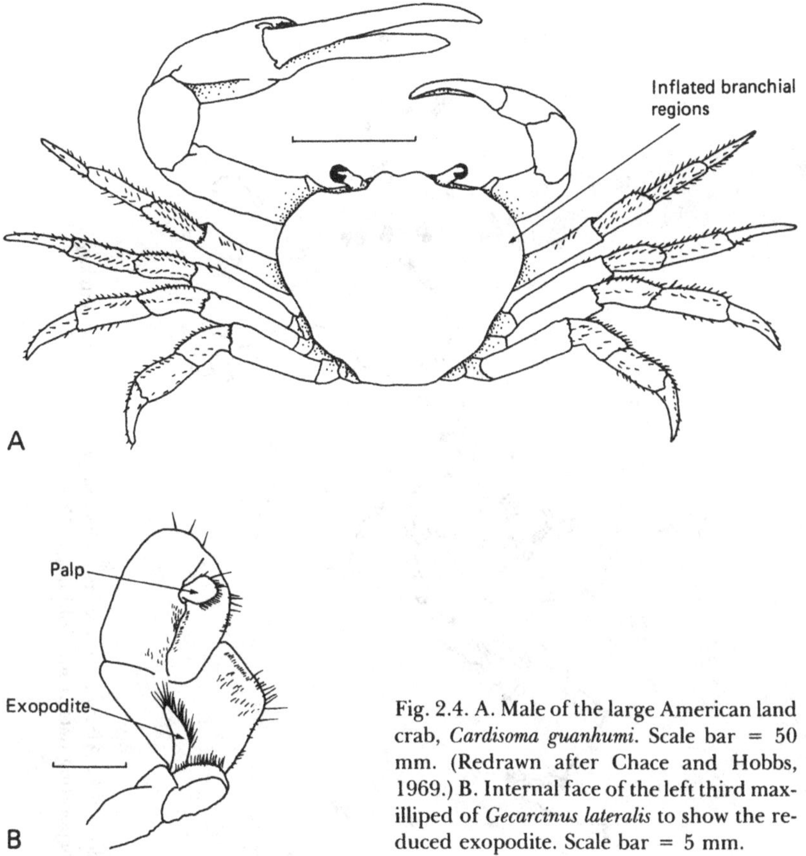

Fig. 2.4. A. Male of the large American land crab, *Cardisoma guanhumi*. Scale bar = 50 mm. (Redrawn after Chace and Hobbs, 1969.) B. Internal face of the left third maxilliped of *Gecarcinus lateralis* to show the reduced exopodite. Scale bar = 5 mm.

(Figs. 2.5 and 2.6) and comprises the following seven species, distributed as indicated.

C. armatum Herklots. West Africa, from Senegal to Angola.

C. carnifex (Herbst). The Red Sea (Holthuis, 1977), and generally in the tropical Indo-West Pacific from East Africa to the Pacific islands.

C. crassum Smith. The west coast of America, from the Gulf of California to Peru.

C. guanhumi Latreille. The east coast of America from Florida to Brazil; Bermuda, and the Caribbean islands.

C. hirtipes Dana. In the tropical Indo-West Pacific from the Andaman and Nicobar Islands eastward; confined to islands, not found on continental land masses.

C. longipes (A. Milne Edwards). Restricted to islands of the southern West Pacific, from New Caledonia eastward to the Paumotu Archipelago.

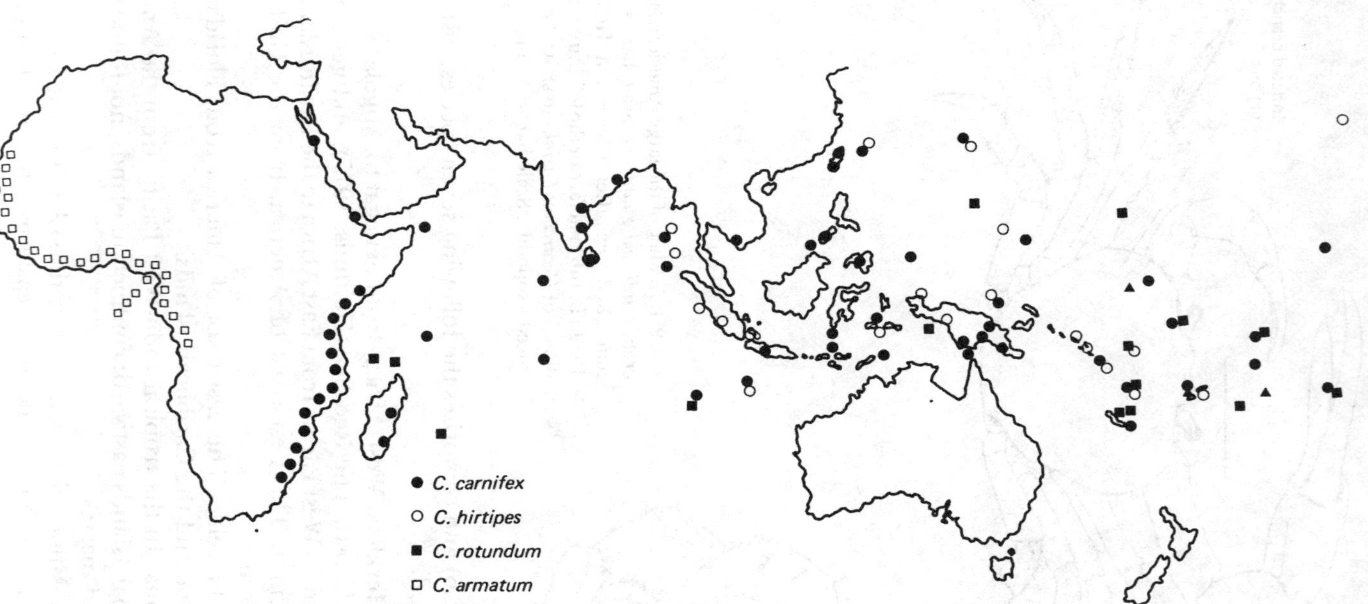

Fig. 2.5. The distribution of the African and Asian species of *Cardisoma*. (After Turkay, 1973b, 1974, and incorporating data from Holthuis, 1977.)

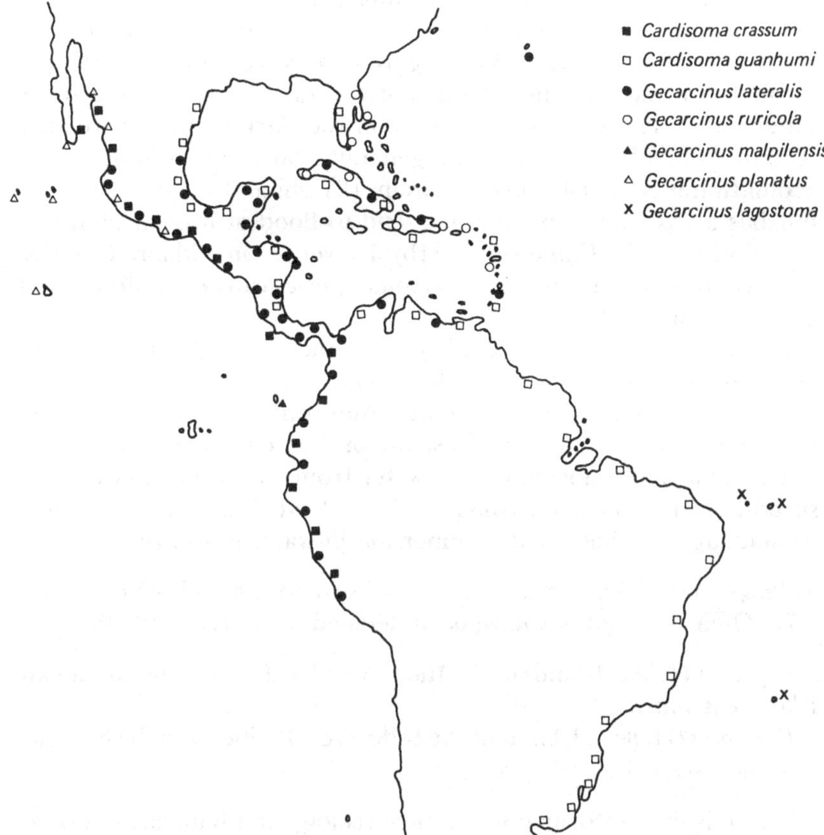

Fig. 2.6. The distribution of the American species of the Gecarcinidae. (After Turkay 1970, 1973a, b.)

C. rotundum (Quoy and Gaimard). Islands of the Indo-West Pacific from Mauritius eastward.

Complete details of the geographical distribution of these species, as well as detailed morphological descriptions, are to be found in Turkay (1970, 1973b, 1974), and in Turkay and Sakai (1976).

Of the various gecarcinids, *Cardisoma* (Fig. 2.4) seems the least tolerant of dry conditions. It may be found several kilometers from the sea – e.g., *C. guanhumi* in Jamaica (pers. obser.), *C. hirtipes* on Okinawa (Goshima, Ono, and Nakasone, 1978) – but only in damp environments. The habitat of *C. guanhumi* is well documented in a number of areas, and in each case the burrows are either adjacent to water of some form, or else extend to the groundwater (Gifford, 1962a; Bliss,

1963; Chace and Hobbs, 1969; Henning and Klaassen, 1973; Henning, 1975a). It has even been recorded some distance offshore in the sea (Gifford, 1962a; W. Burggren, pers. comm.). *C. hirtipes* is confined to wet areas near fresh water, and the burrows reach the water table: The crabs emerge only in the dark or in wet weather (Gibson-Hill, 1947). *C. carnifex* generally burrows in soils where groundwater is available in dry seasons (Bright and Hogue, 1972), or inhabits areas where the burrows tend to flood at least at high tide (Alexander, 1976; Cameron, 1981b). However, on Aldabra *C. rotundum* occurs in much drier areas, so some species can be less dependent on water (Alexander, 1976).

Despite the preceding records of specimens found offshore, *Cardisoma* usually enters the sea only to hatch the eggs (Goshima et al., 1978). Nevertheless, both *C. carnifex* (Alexander, 1976) and *C. guanhumi* (Gifford, 1962b) can withstand prolonged immersion in seawater. *Cardisoma* is able to extract water from damp substrates by the suction of capillary water (Bliss, 1968; Wolcott, 1984), but its normal habitat suggests that regular immersion in water is also required.

ii. Epigrapsus. Two species occur, confined to the Indo-Pacific (Fig. 2.7). Their distributions have been detailed by Turkay (1974).

E. politus Heller. Islands of the Indo-West Pacific from the Andaman Island eastward.

E. notatus (Heller). Islands of the Indo-West Pacific from the Nicobar Islands eastward.

There is little information on their ecology and habitats. *E. politus* appears to frequent stony areas close to the waterline on the Marshall Islands and on Tahiti (Holthuis, 1953). This would suggest a less terrestrial habitat than is typical of the rest of the family.

iii. Gecarcinus. This genus includes six species, found on the western coast of America and on both sides of the Atlantic (Figs. 2.6 and 2.7). The distributions are described by Turkay (1970, 1973a, b). There has recently been a revision of the status of various species: *G. quadratus* is now considered a synonym of *G. lateralis* (Turkay, 1973a); *G. weileri* has been distinguished from *G. lagostoma* (Turkay, 1973b). Thus earlier accounts of distribution are misleading. The genus is separated into two subgenera, *Gecarcinus* and *Johngarthia* (Turkay, 1970).

G. (Gecarcinus) ruricola (L.). Florida, Bahamas, the Antilles.

G. (Gecarcinus) lateralis (Freminville). Bermuda, Antilles, Florida, and the east coast of Central America to Guyana, west coast of America from Mexico to Peru.

G. (Johngarthia) lagostoma H. Milne Edwards. A limited distribution

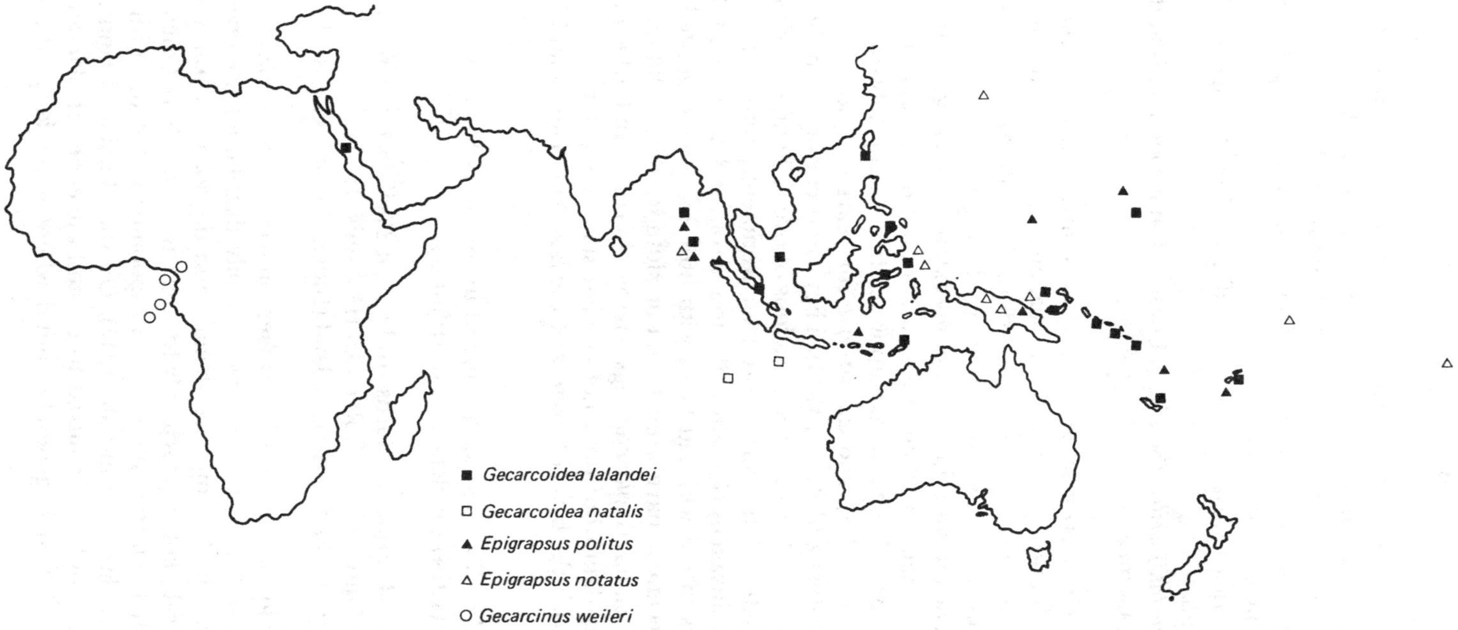

Fig. 2.7. The distribution of the African and Asian species of *Epigrapsus*, *Gecarcinus*, and *Gecarcoidea*. (After Turkay, 1973b, 1974, and incorporating data from Holthuis, 1977.)

in Atlantic islands, with records from Trinidad, Fernando de Noronha, and Ascension Island. A further record from Atol das Rocas (Fimpel, 1975). A questionable record from Australia.

G. (Johngarthia) malpilensis Faxon. Known only from Malpelo Island off the Pacific coast of Panama.

G. (Johngarthia) planatus. Gulf of California, Revilla Gigedo, Clipperton Island.

G. (Johngarthia) weileri (Sendler). Coast of Cameroon and islands of the Gulf of Guinea.

This is a highly terrestrial genus, and in comparison with *Cardisoma* is generally found farther from the sea, and in drier environments. *G. lateralis* has been particularly well studied. It burrows in dry areas and is mainly active nocturnally, except during wet periods (Bliss et al., 1978). The burrows do not generally extend down to groundwater (Bliss and Mantel, 1968), and the crabs are not dependent on access to surface water. They can take up water from damp substrates (see Chapters 3, 7, and 8), and can survive for years with access only to damp sand (Bliss and Boyer, 1964). They can occur far from the sea, and in Dominica they occur at elevations of over 300 m. *G. ruricola* is similar in habits, but prefers forest environments, rather than the sandy areas chosen by *G. lateralis* (Britton, Kroh, and Golightly, 1982), and shelters in crevices and holes with little burrowing needed. In Jamaica specimens have been found at altitudes over 1,000 m. On Clipperton Island, *G. planatus* digs burrows in coral sand, which usually remains damp but does not contain standing water; the crabs frequently visit the brackish waters of the lagoon (Niaussat and Ehrhardt, 1971).

iv. Gecarcoidea. Two species are found in the Indo-Pacific (Fig. 2.7); see Turkay (1974) for details of distribution.

G. lalandei H. Milne Edwards. Indo-West Pacific islands from the Andaman Islands eastward. Red Sea (Holthuis, 1977).

G. natalis (Pocock). Christmas Island (Indian Ocean), Cocos Islands.

The ecology of *Gecarcoidea* is rather similar to that of *Gecarcinus*. On Christmas Island, *G. natalis* is generally distributed wherever it can burrow. The burrows are shallow and do not reach the water table. The crabs enter the sea only for the eggs to hatch, but they will drink readily from standing water by "spooning up" drops with the tips of the chelae (Gibson-Hill, 1947). On the Marshall Islands, *G. lalandei* occurs in holes in mixed forest and the edges of plantations (Holthuis, 1953). It is generally found where water keeps the soil moist (Bright and Hogue, 1972), though on the islands of Palau it occurs in higher, drier areas (Cameron, 1981a).

H. Grapsidae

This is a large and diverse family. Almost all of its members show some degree of terrestrial adaptation, though in the great majority of species such adaptation is of limited extent. Examples of such species will be given but will be discussed only superficially. A limited number of genera occur in which some or all of the species show more substantial adaptations: These include *Geograpsus, Geosesarma, Metopaulias,* and *Sesarma*, which will be considered in detail. Some, such as *Geograpsus*, are still tied to the sea by having a normal pelagic larval development. However, others (e.g., *Geosesarma*) have abbreviated larval lives that considerably reduce dependence on water, and thus among these we find possibly the most fully terrestrial "land" crabs.

Despite the almost universal terrestrial adaptation in the Grapsidae, there is little external modification to indicate such specialization. The family can be readily distinguished from the Gecarcinidae by having a squarish carapace (Fig. 2.8; see also Fig. 2.9), rather than the smoothly rounded one (Fig. 2.4A), which characterizes the gecarcinids; otherwise there is considerable diversity in size and appearance. There are only a few relatively minor external features that are clearly related to terrestrial life. One of these is the occurrence of long tufts of setae between the bases of the walking legs, which function to transport water into the branchial chamber (see Chapter 7). In some genera there is one tuft between the second and third walking legs (e.g., *Geograpsus, Cyclograpsus*), whereas in *Sesarma* there are also tufts between the first and second walking legs (Fig. 2.8F). The development of these tufts is strongly related to the habit of a species. Thus in *Sesarma*, they are strongly developed in the more terrestrial species such as *S. ricordi* and *S. ortmanni*, but little evident in the more aquatic ones such as *S. bidentatum*. Interestingly, the setal tufts are not as well developed in the most terrestrial grapsids of all, such as *Geograpsus grayi, Sesarma cookei,* and species of *Geosesarma*. A second feature is related to the circulation of water from the gill chamber over the outer surface of the carapace in order to oxygenate it (see Chapter 9). This modification is limited to the subfamily Sesarminae (e.g., *Sesarma, Geosesarma, Aratus, Metopaulias*) and consists of a dense array of geniculate setae on the pterygostomial regions of the carapace. Finally, the flagellum of the third maxilliped is absent in some highly terrestrial grapsids such as *G. grayi* and in *Geosesarma*.

Grapsids at the lowest level of terrestrial adaptation are mostly those

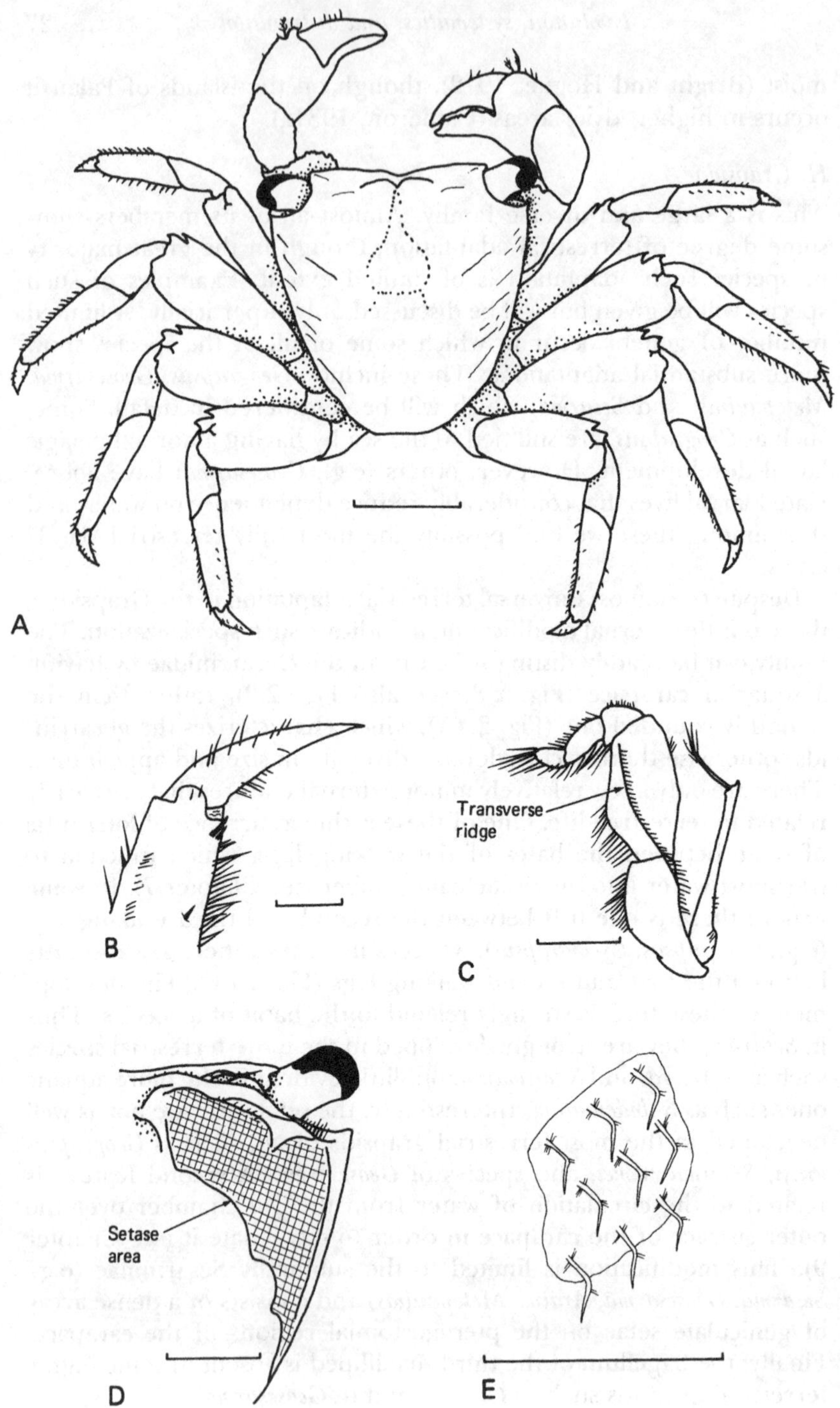

Fig 2.8 (Caption on the facing page.)

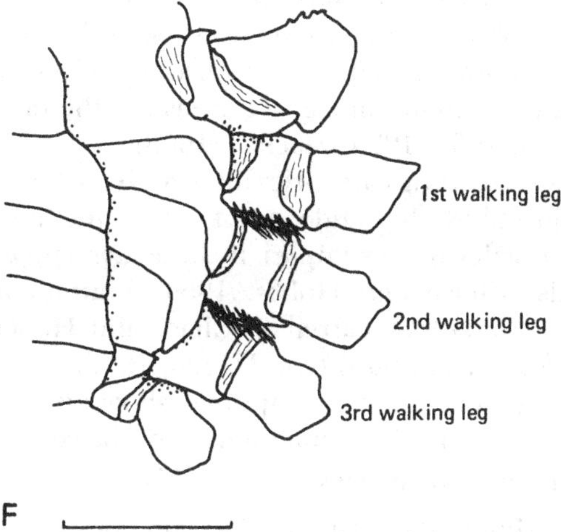

Fig. 2.8. A. Male of the mangrove tree crab, *Aratus pisonii*. Scale bar = 10 mm. (Redrawn after Chace and Hobbs, 1969.) B. Dactylus of the second right walking leg of *Aratus pisonii*. Scale bar = 1 mm. C. Left third maxilliped of *Aratus pisonii* to show transverse ridge. Scale bar = 2 mm. D. Pterygostomial region of *Aratus pisonii* to show area with reticulate pattern of geniculate setae. Scale bar = 5 mm. E. Geniculate setae from D. Scale bar = 1 mm. F. Ventral view of *Sesarma ricordi* to show setal tufts between bases of walking legs. Scale bar = 5 mm.

that are resident intertidally (or where tides are minimal, in the immediate subtidal) on hard substrata – the colloquial "rock crabs." Included are *Grapsus, Hemigrapsus, Pachygrapsus, Metopograpsus, Plagusia, Percnon,* and other genera. They are active both in and out of the water, and will often move up and down the shore with the tide. There is considerable variation in the time spent in the air. Thus species of *Percnon* are largely aquatic and make only brief forays out of the water. At the other extreme, *Grapsus* spends most of its time out of the water, though usually close to it, tending to enter water only when caught by waves or when alarmed. On Dominica, *G. grapsus* (L.) is found on wave-exposed rocks in the surf zone, or in the splash zone on rocky beaches (Chace and Hobbs, 1969); similarly in Jamaica (Hartnoll, 1965). On Aldabra, *G. tenuicrustatus* (Herbst) normally occurs out of water on wave-exposed rocks, and is indeed clearly reluctant to enter the water. Occasional specimens were found a short distance inland (Alexander, 1976; Grubb, 1971).

The second group showing limited terrestrial adaptation includes those that remain burrowed, or sheltered under rocks, during high tide, and become active only when uncovered by the receding tide. This includes some of the smaller and less active species on rocky shores, such as *Cyclograpsus* (Hartnoll, 1965) and *Nanosesarma* (Hartnoll, 1975). Primarily, however, it involves species burrowed on sandy, and particularly muddy, shores. The various species of *Sesarma* and related genera that inhabit mangrove areas are the most obvious examples (Hartnoll, 1965, 1975; Macnae, 1968).

Species in the preceding categories intergrade, of course, with those that live at or just above high tide; and the dividing line is often hard to designate. Borderline examples include *Cyclograpsus integer* H. Milne Edwards (Chace and Hobbs, 1969; Vannini and Valmori, 1981a), *Goniopsis cruentata* (Latreille) (Chace and Hobbs, 1969), and *Metasesarma rubripes* (Rathbun) (von Hagen, 1977).

The various genera of the Grapsidae showing substantially increased levels of terrestrial adaptation will now be considered in turn.

i. Aratus. There are two species.

A. pisonii (H. Milne Edwards) (Fig. 2.8). On both sides of the American continent: on the east from Florida to Brazil, the Bahamas, and the Antilles; on the west from Nicaragua to Peru (Rathbun, 1918; Chace and Hobbs, 1969).

A. elegans (Herklots). Only recently recognized as a species of *Aratus* (Green, 1986), previously referred to as *Sesarma elegans*. West Africa, from Sierra Leone to Angola (Monod, 1956).

The habits of *Aratus* are unusual in that it is predominantly arboreal, spending most of its time out of water on the trunks, branches, and prop roots of mangrove trees (Hartnoll, 1965; Warner, 1969; von Hagen, 1977; Green, 1986). The dactyli of the walking legs are specialized for this arboreal habit, being short but extremely sharp (Fig. 2.8B), thus giving the crab an exceptional grip on the trees. There are a variety of crabs that climb trees occasionally (von Hagen, 1977), but *Aratus* is the only one to do so on such a regular basis. The larvae have a normal planktonic existence in the sea. In *A. pisonii* the young stages frequent the lowest parts of the trees or hide in litter on the mangrove floor, but the larger crabs spend most of their time in the upper parts of the trees well clear of the water. Crabs are seen entering and leaving the water regularly, and when alarmed will leap into the water where they swim rapidly to the mangrove roots. However, it is not known just how frequently these visits to the sea, presumably to replenish water supplies, normally take place (Hartnoll, 1965). Interestingly, there is no comparable species among the very diverse grapsid fauna of the Indo-West Pacific mangroves.

ii. Geograpsus. As the name suggests, this genus contains species adapted to life on land, though some species are far better adapted than others. There are three Indo-Pacific species, revised by Banerjee (1960), and one Atlantic/East Pacific species.

G. lividus (H. Milne Edwards). West coast of Africa from Senegal to North Angola; Cape Verde Islands, Florida to Brazil, Bermuda, Antilles; west coast of America from Lower California to Chile (Rathbun, 1918; Monod, 1956; Chace and Hobbs, 1969). References to distribution outside these limits probably refer to the subspecies *stormi*, now regarded as a separate species.

G. stormi De Man. This is now considered to be a distinct species from *G. lividus* (see Banerjee, 1960). In the literature this species has also at times been confused with *G. crinipes* (see discussion in Hartnoll, 1975), so records must be looked at critically. Its basic distribution is from the East coast of Africa and the Arabian Sea, through the Indo-Pacific to Japan, the Pacific islands, and possibly Hawaii. From the comments on color and behavior it would appear that the records of *G. crinipes* from the Red Sea (Holthuis, 1977) could perhaps be *G. stormi*.

G. grayi (H. Milne Edwards). Zanzibar, Madagascar, Aldabra, and small Indian Ocean islands, but never from the mainland of East Africa (Banerjee, 1960; Hartnoll, 1975; Vannini and Valmori, 1981a). Throughout the Indo-West Pacific to Japan, Polynesia, and Australia (Banerjee, 1960).

G. crinipes (Dana). Only one record from the mainland of East Africa in Somalia (Vannini and Valmori, 1981a). However, common on small islands in the Indian Ocean and found through the Indo-West Pacific to Japan, Polynesia, and Hawaii (Banerjee, 1960). The only species recorded from the Red Sea (Holthuis, 1977), but as mentioned earlier confusion with *G. stormi* is possible.

Although one of the more terrestrial genera of the Grapsidae, the only external evidence of this is the reduction or loss of the flagellum of the exopodite of the third maxilliped, and the tufts of setae between the bases of second and third pairs of walking legs. They are medium to large crabs, ranging from *G. lividus*, which reaches 30 mm in carapace breadth, to *G. crinipes* and *G. grayi*, which can both exceed 60 mm.

The species of *Geograpsus* show considerable variation in the level of terrestrial adaptation, with *G. lividus* and *G. stormi* the least adapted. Neither occurs very far from the sea. In Jamaica *G. lividus* shelters by day just above the waterline in damp debris or under stones, emerging by night to forage along the waterline (Hartnoll, 1965). It occupies similar situations on Dominica, and is never more than 4 or 5 m from the surf (Chace and Hobbs, 1969). *G. stormi* occurs high on rocky shores, by day well concealed under overhangs or in caves or crevices in damp places (Hartnoll, 1975; Vannini and Valmori, 1981a). This species probably forages nocturnally and is never found far from the sea. Reports of *G. stormi* in more terrestrial habitats are due to confusion with *G. crinipes* (see Hartnoll, 1975).

The other two species are more terrestrial, though both have normal pelagic larvae that must be released into the sea. On Aldabra, *G. crinipes* does not extend far from the shore and is usually found sheltering under rocky overhangs. It is active both by day and by night, and forages into the coastal woodland (Alexander, 1976). Grubb (1971) recorded them a maximum of 300 m from the shore. In the Marshall Islands they are recorded from under rotting logs and coconut debris (Holthuis, 1953).

On the other hand, *G. grayi* is not restricted to coastal regions on Aldabra, but occurs all over the atoll. It is found in rock crevices, under leaves and debris, and in burrows made by *Birgus, Ocypode,* and *Cardisoma* (Alexander, 1976). Grubb (1971) recorded it over 1 km from the shore. On Christmas Island (Indian Ocean), Gibson-Hill (1947) recorded *G. grayi* from dryish habitats under rocks and boulders. Johnson (1965) records that *G. grayi* is reported "from great heights."

There is little information regarding the precise water requirements of the more terrestrial *Geograpsus,* which have had little attention in contrast to the coenobitids and gecarcinids. Alexander (1976) found that both *G. crinipes* and *G. grayi* "drowned" in seawater in less than 24 hours, so they are substantially adapted for terrestrial life. *G. crinipes* has been observed to "drink" from fresh water (Alexander, 1976).

iii. Sesarma sensu largo. The genus *Sesarma* Say *sensu largo* has been redefined and a much-reduced genus, *Sesarma* Say *sensu stricto,* instituted (Sèrene and Soh, 1970): The majority of species originally in *Sesarma* were redistributed into a number of other genera. This revision was only extended to the specific level by Sèrene and Soh (1970) for the Indo-Pacific species, and so the situation is currently rather confused for species from other areas. Even for the Indo-Pacific there are a number of inconsistencies (see Holthuis, 1978). Consequently, the revised classification has not been generally adopted (see, e.g., Vannini and Valmori, 1981a). To avoid confusion this conservative policy will be continued here, except that *Geosesarma,* as redefined by Sèrene and Soh (1970), will be treated as a separate genus.

Sesarma comprises a range of small and medium-sized species, the largest of which reach a carapace width of about 40 mm, though the more highly terrestrial species tend to be smaller. The external features include a transverse setose ridge on the third maxillipeds, and geniculate setae on the pterygostomial regions. In the more terrestrial species there are prominent tufts of setae between the bases of the first and second, and second and third walking legs.

As mentioned earlier there are a large number of intertidal species of *Sesarma*, mostly burrowing in sand or mud, and active when exposed to the air. However, a number of species show greater terrestrial adaptation. The first group comprises species that occur supratidally but that have normal planktonic larval stages: These must return to the sea for the eggs to hatch, and do not normally extend far from the water. Examples can be found in most areas, with different levels of terrestrial adaptation. On the east coast of America, *S. ricordi* H. Milne Edwards is the best example. On Dominica it occurs among rocks and debris on the tide line and up to 50 m inland (Chace and Hobbs, 1969). In Trinidad it hides above high water in strand-line debris by day, and forages on land at night (von Hagen, 1977). In Jamaica *S. ricordi* has similar habits, also shares *Uca* and *Cardisoma* burrows, and was found up to 100 m from the sea (Hartnoll, 1965). *Sesarma roberti* (H. Milne Edwards) is another West Indian species found some way from the sea – up to 300 m on Dominica – but always near fresh water and in damper situations than *S. ricordi* (Hartnoll, 1965; Chace and Hobbs, 1969; von Hagen, 1977). In Brazil, *S. angustipes* Dana occurs in the water retained in the bases of bromeliads (Abele, 1972; von Hagen, 1977); however, the bromeliads are growing near brackish streams, and it is assumed that it must return to the sea for the larvae to hatch (compare with *Metopaulias*, discussed later).

On East African shores *Sesarma obesus* (Dana) was found supratidally on rocky cliffs, and *S. eulimine* de Man, *S. ortmanni* Crosnier, and *S. meinerti* de Man burrow above highest tide level on sandy/muddy shores (Hartnoll, 1975). There are various other records. *S. trapezoidea* H. Milne Edwards is recorded at an altitude of 500 m in Tahiti (Forest and Guinot, 1961). In Indonesia, *S. taeniolata* White occupies a zone above highest tides, and will climb mangroves to avoid the rising tide (Verwey, 1930). In the Ryu Kyu Islands, *S. dehaani* is semiterrestrial on river banks and active out of the water (Goshima et al., 1978).

The preceding list is not intended to be complete, and there are doubtless various other species of *Sesarma* living supratidally – e.g., *S. rotundatum* Hess in the Marshall Islands (Holthuis, 1953) and *S. gardineri* Borradaile in the Tokelau Islands (Yaldwyn and Wodzicki, 1979). *Sesarma haematocheir* (de Haan) is an unusual species. It burrows well above high-tide mark and has normal pelagic larvae. However, at least some of the females, rather than returning to the sea, enter freshwater streams and ditches in order to hatch the larvae, which are then carried to the sea (Hashimoto, 1965).

Finally there are those species of *Sesarma* that have reduced larval development, with suppression of the planktonic phase, so that these crabs do not have to return to the sea to breed. Such species have restricted geographical distributions, which is not surprising in view of their lack of a pelagic dispersal stage. For some of these species, although it is clear that they no longer have any dependence on the sea, it is not clear how reliant they are on fresh water, with which they are often closely associated.

These "landlocked" species form two groups in widely separated areas. The first are cave dwellers in the East Indies. *Sesarma jacobsoni* Ihle occurs in Java (Ihle, 1912), and *S. cerberus* Holthuis in Amboina, in the Moluccas (Holthuis, 1964). Both species are adapted to cave life by having smaller eyes (though the corneas are still pigmented) and elongated legs. It is clear that they are both confined to the caves, but it is not known how long they spend in fresh water, nor are any details of their development known. Sèrene and Soh (1970) have included both species in a new genus, *Sesarmoides*. Details on the ecology of other species included by them in *Sesarmoides* are few, but *Sesarma kraussi* de Man is an intertidal form in Malaysian mangroves (Tweedie, 1954), so terrestrial-freshwater adaptation is not a universal feature of the group.

The second "landlocked" group comprises a series of species endemic to Jamaica. These were originally in the subgenus *Sesarma* (see Rathbun, 1918), but their position in the revised classification of Sèrene and Soh (1970) is not totally clear. Like the first group, they would probably be included in *Sesarmoides* (see Abele and Means, 1977), and when the relationships within *Sesarma* are better known it will be very interesting to reassess the affinities of these Jamaican species with the East Indian species.

The Jamaican species, and their habits and development so far as either are known, have been described by Rathbun (1918), Hartnoll (1965, 1971), and Abele and Means (1977). There are four species, none of which ever approaches the sea. *Sesarma bidentatum* Benedict occurs in or near streams at altitudes between 400 and 1,500 m. By day some are active in the streams, and others shelter beneath stones on damp ground near the water. By night many crabs are active on land near the water. The larval stages are abbreviated and nonfeeding (Hartnoll, 1964), though they have not been observed in the wild, so it is not known where they develop. *Sesarma verleyi* Rathun occurs in caves, both in and near water. The eyes are reduced but functional, the legs elongated, and the integumental pigment reduced. Larval development is not known, but the eggs are large so it must be assumed that development is abbreviated (Hartnoll, 1964). *S. cookei* Hartnoll and *S. jarvisi* Rathbun are terrestrial, occurring under stones and rubble in damp and shaded sites, which need not be near water (Abele and Means, 1977). They both produce large eggs, and it is considered likely that the larvae are nonfeeding and develop rapidly in the interstitial water of the rubble during wet periods. This would render these two species effectively independent of permanent water at all stages, and so place them among the most fully terrestrial of crabs.

Another endemic terrestrial grapsid in Jamaica, *Metopaulias depressus*, is discussed later.

The final group of terrestrial sesarmids are the various species of *Geosesarma*. This is now generally accepted as a distinct genus and will be discussed separately.

iv. Geosesarma. The genus *Geosesarma* de Man has been redefined by Sèrene and Soh (1970), who list 23 species, though they suggest that the inclusion of some of these is uncertain. They are small crabs, with carapace widths generally not much over 10 mm. This, together with their cryptic nature, accounts for the paucity of data. All species except *G. maculatum* produce small numbers of large eggs, and most are amphibious or terrestrial, often living far from water (Lam, 1969). The genus is distributed in the region of Malaysia, Indonesia, the Philippines, New Guinea, the Solomon Islands, and the Hawaiian Islands, with most species having limited distributions. This is only to be expected where a pelagic dispersal phase is absent, as is presumably the case. The species given by Sèrene and Soh (1970) are listed in Table 2.2, with some details of their distributions and the sources of these data. An exhaustive search has not been made of the literature, since it would in any case probably be largely unprofitable in view of the substantial recent revision of the genus (see Sèrene, 1968).

There is information on the habitats of some species.

Geosesarma perracae is one of the more aquatic species. In Singapore it is found in and near streams, in burrows 5–20 cm deep, which are usually half filled by water (Lam, 1969). Equally aquatic is *G. noduliferum*, which is found in springs, runnels, waterfalls, and the water in leaf axils (Pesta, 1930). *Geograpsus gracillimum* occurs in damp forests up to 1,000 m (Holthuis, 1979). In Sarawak it occurs in periodically flooded forests, active by night, by day sheltering in flooded burrows up to 1 m deep (Collins, 1980). On the other hand, *G. ocypodum* is found in hilly wooded regions, which may be quite dry (Johnson, 1965; Lam, 1969). *Geosesarma foxi* occurs under stones, including those on mountain summits distant from permanent water (Johnson, 1965).

The eggs are large – 1.6 mm in *G. perracae* (Lam, 1969) and 2 mm in *G. noduliferum* (Pesta, 1930) – and few in number: 25–70 in *G. perracae* (Lam, 1969), 14–20 in *G. johnsoni* (Sèrene, 1968), and 26–40 in *G. noduliferum* (Pesta, 1930). Little is known of the larval development. In *G. perracae* the larvae hatch into the water in the base of the burrows, where they develop rapidly without feeding (Lam, 1969).

Table 2.2. *The species of* Geosesarma, *as listed by Sèrene and Soh (1970)*

Species	Distribution	Source
G. amphinome (de Man)	Borneo, Sumatra	Tweedie (1936)
G. angustifrons (A. Milne Edwards)	Hawaiian Islands	Milne-Edwards (1869)
G. araneum (Nobili)	Sarawak	Nobili (1900)
G. celebensis (Schenkel)	Sulawesi	Schenkel (1902)
G. clavicruris (Schenkel)	Sulawesi	Schenkel (1902)
G. foxi (Kemp)	Malaya, Langkawi Islands	Tweedie (1936)
G. gordonae Sèrene	New Guinea	Sèrene (1968)
G. gracillimum (de Man)	Sarawak, Malaya, Natuna Islands	Holthuis (1979)
G. johnsoni Sèrene	Penang	Sèrene (1968)
G. leprosum (Schenkel)	Sulawesi	Schenkel (1902)
G. maculatum (de Man)	Halmahera Island, Moluccas Solomon Islands, Sumba	Sèrene (1968) Holthuis (1978)
G. noduliferum (de Man)	Java	Pesta (1930)
G. ocypodum (Nobili)	Singapore, Malaya, Sumatra	Tweedie (1936)
G. penangensis (Tweedie)	Penang	Sèrene (1968)
G. perracae (Nobili)	Singapore	Soh (1969)
G. rathbunae Sèrene	Paway Island, Philippines	Sèrene (1968)
G. rouxi Sèrene	Java	Sèrene (1968)
G. sarawakensis Sèrene	Sarawak	Sèrene (1968)
G. solomonensis Sèrene	Guadalcanal, Solomon Islands	Sèrene (1968)
G. sylvicolum de Man	Java, Sumatra	Pesta (1930), Sèrene (1968)
G. ternatensis Sèrene	Ternate Island, Moluccas	Sèrene (1968)
G. thelxinoe (de Man)	Andaman Islands	Tweedie (1940)
G. vincentensis (Rathbun)	Luzon, Philippines	Rathbun (1914)

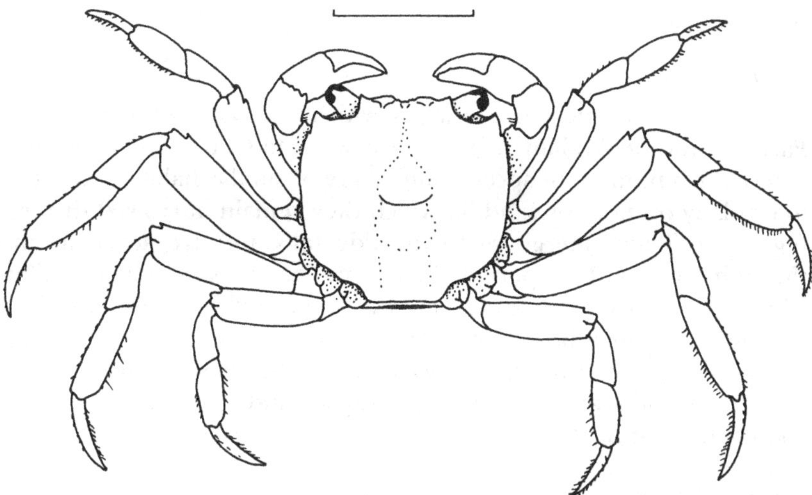

Fig. 2.9. The Jamaican bromeliad crab, *Metopaulias depressus*. Scale bar = 10 mm. (Redrawn after Chace and Hobbs, 1969.)

A female of *G. noduliferum* was found carrying 18 young crabs beneath the abdomen (Pesta, 1930). Thus it seems that at least some species could be independent of water for purposes of reproduction. Certainly the reproduction of *Geosesarma* merits further investigation.

v. Metopaulias. This genus is represented by only a single species, *M. depressus* Rathbun (Fig. 2.9), which is found only in Jamaica. It is a smallish crab, not exceeding 20 mm in carapace width. It occurs inland, usually above 300 m, in the pools of water collecting in the leaf bases of large bromeliads (Hartnoll, 1964). A variety of other crabs have been recorded from this habitat (see Abele, 1972), but they are casual visitors and must breed in the sea. *Metopaulias*, on the other hand, is found only in the bromeliads, though it presumably emerges at night to forage. The species is markedly flattened dorsoventrally, which helps it to move between the bromeliad leaves. Movement must also occur between bromeliads for mating, since mature males and females are not normally found together in the same plant (Hartnoll, 1964).

The eggs are large, 1.5 mm, and only between 60 and 100 are carried. They hatch in the pools of water generally found in the bromeliads. These pools occur throughout the year, but are more copious in the wet seasons when the crabs breed: A large plant may contain a liter of water. The larvae undergo a rapid nonfeeding development within this water (Hartnoll, 1964). This crab is highly terrestrial in the sense that it is free from dependence on large bodies of water, but it is restricted by its specialized adaptation and need for the water in the bromeliad tanks.

I. Mictyridae

There is only a single genus, *Mictyris*, which is distributed in the Indo-Pacific from the Andaman Islands eastward to China and Australia. There are a number of species, but all are of similar habits. They live intertidally on mud or sand, in which they remain burrowed during high tide. They emerge when the tide falls and are active in air through the low tide period, for example, *M. platycheles* Latreille (Kraus and Tautz, 1981) and *M. longicarpus* Latreille (Cameron, 1966). *M. longicarpus* may be active in air for over two hours, and during that time can take up capillary water from the substrate using rows of setae on the rear border of the carapace and first segment of the abdomen (Quinn, 1980).

J. Potamoidea

The Potamoidea have been divided into a number of families (see Bott, 1955, 1970), but the relationships are not finally settled by any means. For this reason the 11 families are not considered separately – there being no obvious correlation between the families and the degree of terrestrial adaptation – and a general account of the Potamoidea is given instead. They are medium to large crabs, with the larger species approaching 100 mm in carapace width. The carapace is smoothly rounded (Fig. 2.10) and may be somewhat inflated laterally, and in general appearance they resemble the Gecarcinidae (Fig. 2.4A). The potamonids and gecarcinids can be quite easily discriminated, however, by the position of the palp of the third maxilliped (Fig. 2.10). This arises laterally in the Gecarcinidae but medially in the Potamoidea. There are no obvious external adaptations to terrestrial life in the potamonids.

The Potamoidea are popularly known as "freshwater crabs." They do not occur in the sea, and as the name suggests they are typically inhabitants of streams, rivers, and lakes. They are specialized for this life-style by having what is effectively direct development. Small numbers of large eggs are produced, the larval stage is reduced and essentially passed within the egg, and hatching produces young crabs (Pace, Harris, and Jaccarini, 1976). Even after hatching these are protected and carried for some time under the abdomen of the female. Their habits and reproductive methods predispose them for terrestrial adaptation, and various degrees of this are found within the group. Probably the majority of species make occasional excursions on land, but they are normally restricted to the close proximity of water. These species are so numerous, and the accounts of precise

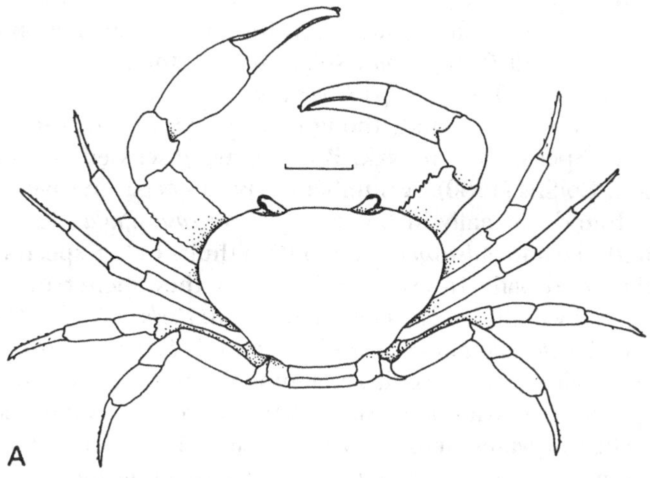

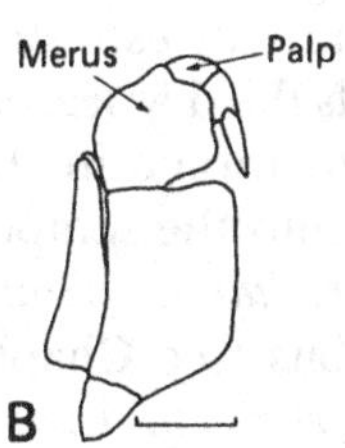

Fig. 2.10. A. *Guinotia dentata*, a West Indian freshwater crab. Scale bar = 20 mm. (Redrawn after Chace and Hobbs, 1969.) B. Right third maxilliped of *Potamonautes niloticus*, an African potamonid, to show palp inserted near the median edge of the merus. Scale bar = 5 mm.

habits often so vague, that no attempt will be made to compile a comprehensive review. However, examples will be discussed and some cases of very marked terrestrial adaptation considered in more detail.

There are many accounts of potamonids that are basically aquatic but that spend some time on land. In Europe *Potamon fluviatilis* (Herbst) is found in water, or burrowed adjacent to it, and does not travel far from it, whereas *P. potamios* Olivier is often found substantial distances from water (M. Vannini, pers. comm.). Chace and Hobbs (1969) mention a number of West Indian potamonids. For most, the habitat is noted only as "in and near fresh water," but more detail is given for *Guinotia dentata* (Latreille) (Fig. 2.10). This species occurs in streams and ponds, also burrows in seepage areas, and can be see on land; however, it does not extend into the rather drier areas occupied by *Cardisoma*. Fernando (1960) discusses seven potamonids in Ceylon, and notes that they are all found either in the water or in burrows close to water. Holthuis (1974) provides details for the habitats of the potamonids of New Guinea, which vary in terrestrial habit. Some species were found only in water, usually sheltering under rocks, e.g.,

Rouxana roushdyi Bott, *Geelvinkia ambaiana* Bott, and *Holthuisana festiva* (Roux). *Rouxana ingrami* (Calman) was found in water-filled burrows in river banks, and *R. papuana* (Nobili) under stones on river banks. The most terrestrial is perhaps *Holthuisana subconvexa* (Roux) found under moss on the ground, though the proximity to water is not mentioned. Species in Sarawak, Borneo, are discussed by Holthuis (1979) and Collins (1980). A number of species occur in caves, where they are found normally in water, e.g., *Cerberusa tipula* Holthuis, *C. caeca* Holthuis, and *Adeleana chapmani* Holthuis. Other species occur in alluvial flood-plain forests, where by day they shelter in flooded burrows, but by night feed actively on the land: *Perbrinckia loxophthalma* (de Man), *Thelphusula granosa* Holthuis, and *T. baramensis* (de Man).

Obviously all of the preceding potamonids, although in some cases showing regular activity on land, have ready access to water and normally shelter in water or in very damp sites. However, there are a few accounts of potamonids that are able to spend prolonged periods out of contact with water. This is usually during dry seasons when the crabs remain sheltered, more or less in a state of estivation. In India, *Parathelphusa guerini* Milne Edwards spends the dry season burrowed, out of contact with water. It emerges to forage in the wet season, and to shed its already hatched young into the temporarily available water (McCann, 1938). The Australian *Holthuisana transversa* Martens, found in arid regions, has similar habits (see Chapter 7). During the long dry periods it remains in its burrow, and for many months its only source of water is that which condenses in the burrows by night; it emerges only after heavy rain (Bishop, 1963; Greenaway and MacMillen, 1978; Taylor and Greenaway, 1979). Experiments show that the crab is able to absorb droplets of condensing water (Greenaway and MacMillen, 1978). In Madagascar there are several potamonid species inhabiting limestone forests, which are dry at certain seasons. Such species, e.g., *Madagapotamon humberti* Bott and *Gecarcinautes antongilensis* (Rathbun), are active during the wet summer season, when water is available from the small pools that collect in irregularities in the eroded limestone. However, in the dry season such water is not available, and the crabs take shelter for a period of months in small caves in the limestone that provide a cool, damp environment (Vuillemin, 1970, 1972).

Thus, a variety of potamonids spend considerable time out of water, and some species live in areas where they are without access to water for long periods of the year. However, during such periods they tend to be inactive and sheltering. They make use of temporary water during the wet seasons, when their feeding and reproductive activity is concentrated. There is little information on their reproduction, and although there is direct development in all cases, the indication is that water is usually required by the vulnerable young stages when they are first released. More data on this aspect would be very interesting.

K. Ocypodidae

The Ocypodidae are a large family of mainly small and medium-sized crabs with a worldwide distribution in the tropical and subtropical areas. In habits the majority of species are similar to the Mictyridae, living intertidally on soft substrates, remaining within burrows while covered by the tide, but emerging to forage over the surface when the water recedes. However, some of the family are more or less fully aquatic, whereas others occur supratidally and are markedly terrestrial. The main external evidence of terrestrial adaptation is the presence of structures concerned with water uptake. Brushes of setae, as already described for the Grapsidae, occur between the bases of the walking legs; these are found between the second and third legs in *Scopimera, Ocypode,* and *Uca,* and also between the first and second pairs in *Dotilloplax* and *Tmethypocoelis.* In *Dotilla,* the water-uptake setae arise from the fourth segment of the abdomen (see Hartnoll, 1973). The family is divided into five subfamilies that have different degrees of terrestrial habit.

The Macrophthalminae contain the genera *Macrophthalmus, Cleistostoma, Tylodiplax, Paracleistostoma, Austroplax,* and *Euplax.* The distribution is in the Indo-West Pacific, except for *Euplax,* which is found on the Pacific coast of America. *Macrophthalmus* generally occurs low on the shore, and is usually active only in water – the burrows are often located in pools. Some, such as *Macrophthalmus telescopicus,* are subtidal (Sèrene, 1973). On the other hand, *Cleistostoma* and *Paracleistostoma* forage actively on drained substances (Jones and Clayton, 1983).

The Camptandriinae include the rather obscure genera *Camptandrium, Shenius,* and *Leiopocten* (see Sèrene and Umali, 1972). They are intertidal, but precise ecological data are lacking.

The Scopimerinae include *Dotilla, Scopimera, Dotillopsis,* and a number of other genera. These are typical intertidal ocypodids, remaining burrowed at high tide, and becoming active in air when the tide recedes. The habits of typical species are described for *Dotilla fenestrata* Hilgendorf (Hartnoll, 1973), *D. blandfordi* Alcock and *D. mictyroides* (H. Milne Edwards) (Altevogt, 1957b), and *Scopimera proxima* Kemp (Silas and Sankarankutty, 1967).

The Heloecinae (see Turkay, 1983) consist of *Heloecius* and *Ucides. Heloecius,* found in Australia, is an intertidal genus active at low water. *Ucides* is found on both coasts of the Americas and occurs supratidally; it used to be included in the Gecarcinidae, but is now considered an ocypodid (see Chace and Hobbs, 1969; Turkay, 1970).

The last subfamily is the Ocypodinae, which contains *Uca* and *Ocypode*, two genera with worldwide tropical/subtropical distributions. Most of the species of *Uca*, and some species of *Ocypode*, occur intertidally and are active only at low water. However, others occur supratidally and may show considerable terrestrial adaptation.

Those species occurring supratidally in the genera *Ucides, Uca,* and *Ocypode* will now be considered in more detail.

i. Ucides. Ucides occur on both coasts of the Americas. Rathbun (1918) considered there were separate species, *U. cordatus* (Linnaeus) on the Atlantic side and *U. occidentalis* (Ortmann) on the Pacific. More recently, Turkay (1970) gives them only subspecific rank as *U. cordatus cordatus* and *U. cordatus occidentalis*, with intermediates occurring in the Panama/Colombia area. *U. cordatus cordatus* occurs from Florida to Uraguay, and in the Caribbean islands. *U. cordatus occidentalis* occurs from the Gulf of California to southern Peru. *Ucides* is the largest of the Ocyodidae, approaching 100 mm in carapace width.

Its habitat in Dominica is described by Chace and Hobbs (1969). It burrows both intertidally and supratidally, though never very far from the water, and overlaps the range of both *Uca* and *Cardisoma*. The burrows always descend to the water table, so the degree of terrestrial adaptation is modest. In Jamaica it occurs at all levels in the mangroves, but becomes increasingly common toward land (Warner, 1969). On the Pacific coast *U. cordatus occidentalis* is recorded from mangrove swamps (Rathbun, 1918). Bright (1966) describes its habitat in Costa Rica as mangroves, open fields, and areas where there is some standing water.

ii. Uca. Uca is a common genus with some 60 species, and with a worldwide distribution in the tropics and subtropics (Crane, 1975). The distribution patterns are given in detail by Crane (1975), and the general pattern of the occurrence of species is shown in Table 2.3. The striking feature is the proliferation of species on both sides of the Americas, notably the Pacific coast.

Most species of *Uca* are small, with only a few species substantially exceeding 25 mm in carapace width; the largest, *Uca tangeri* does not reach 50 mm carapace width. The most notable feature of *Uca* is the dimorphism of the chelae. Females have two small equal chelae, both used for feeding. Males have one chela of the female type, but the other is enormously enlarged (Fig. 2.11), often brightly colored, and used for combat and display (see Chapter 4). It is the waving of this large chela that has given the genus its popular name of "fiddler crabs."

The majority of species are intertidal, and display the typical ocy-

Table 2.3. *The distribution of species in some of the genera of land crabs*

Genus	Number of species			
	East Atlantic	West Atlantic	East Pacific	Indo-West Pacific
Coenobita	1	1	1	8
Birgus	0	0	0	1
Cardisoma	1	1	1	4
Epigrapsus	0	0	0	2
Gecarcinus	1	3	3	0
Geograpsus	1	1	1	3
Uca	1	15	31	18
Ocypode	2	1	2	15+

podid habit of remaining burrowed at high tide, and emerging to forage only when uncovered. The habitats and behavior of such species are reviewed in substantial detail by Crane (1975) and will not be elaborated here. However, there are a limited number of species that occur very high on the shore, and they will be considered further.

In Jamaica *U. leptodactyla* Rathbun occurs on dry flats at extreme high tide level in the company of *Sesarma ricordi, Cardisoma guanhumi,* and *Coenobita clypeatus* (Warner, 1969). In East Africa, *U. inversa* (Hoffman) occupies a similar zone, in regions covered only by the highest tides (Macnae and Kalk, 1962; Hartnoll, 1975). Comparable species in other areas are *U. uraguayensis* Nobili in Brazil and *U. lactea* (de Haan) in Java (see Warner, 1969). On the eastern coast of the United States, *U. minax* (le Conte) is the highest-ranging species, and its burrows are often well above high tide, though they always descend to the water table (Williams, 1965). These high-shore species are often only intermittently active on the surface during periods of high spring tides.

Although these species may live supratidally, they must all shed their larvae into the sea where they undergo a normal planktonic development. *U. subcylindrica* Stimpson, however, has overcome this restriction. It occurs in the Gulf of Mexico, where it is found in arid environments subject to hypersaline conditions. *U. subcylindrica* lives supratidally and may be found near ephemeral ponds up to 35 km from tidal water (Thurman, 1984; Rabalais and Cameron, 1985a). It burrows down to the water table, with burrows up to 1.5 m deep, though the groundwater may be highly hypersaline (Thurman, 1984). The reason that this species can live so far from the sea is that they produce large eggs, which hatch into nonfeeding larvae with an abbreviated development. These are found in ephemeral pools produced by rainfall (Rabalais and Cameron, 1983, 1985b).

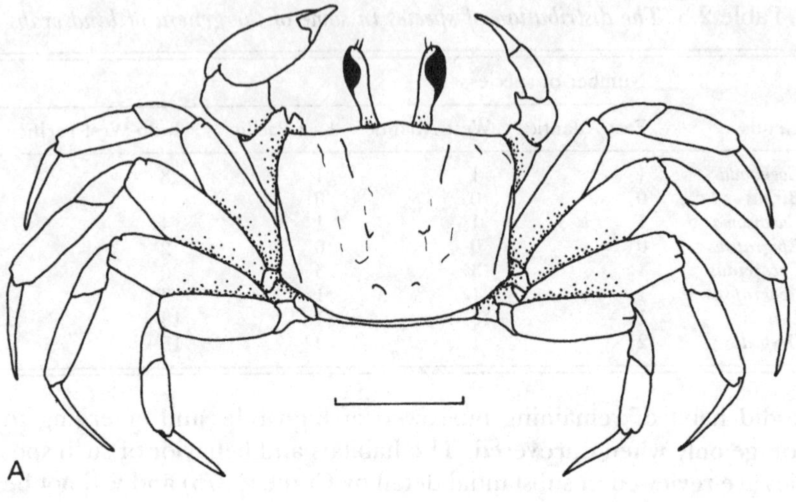

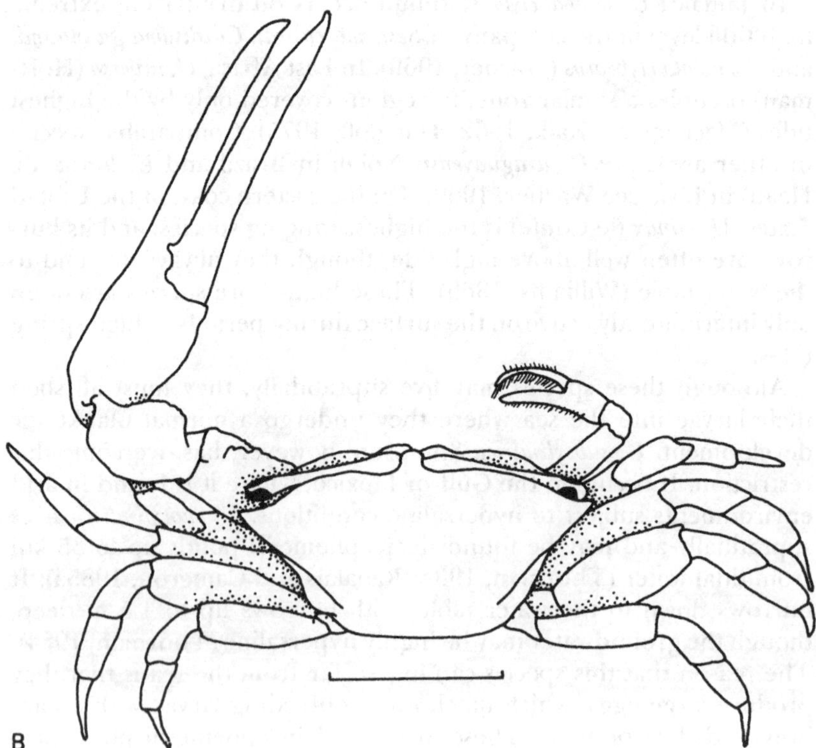

Fig. 2.11. A. Female of the ghost crab, *Ocypode africana*. Scale bar = 20 mm. (Redrawn after Monod, 1956.) B. Male of the fiddler crab, *Uca vocans*. Scale bar = 10 mm.

iii. Ocypode. This genus is rather more terrestrial than *Uca*, since all species occur in the upper intertidal or the supratidal, and a number of species venture some distance inland. They are medium-sized crabs, most species reaching a carapace width of 40–50 mm. The most striking features are the large eyes and long legs (Fig. 2.11), both presumably useful terrestrial adaptations. The exopodite of the third maxilliped lacks the flagellum. They are colloquially known as "ghost crabs" because of their pale color, and "racing crabs" because of their very rapid running on land: they are by far the fastest-moving terrestrial crabs. *Ocypode* occurs in all tropical and subtropical areas, and over 20 species are known. Of these, 2 occur in the East Pacific (*O. occidentalis* Stimpson and *O. gaudichaudii* Milne Edwards and Lucas), 1 in the West Atlantic (O. *quadrata* [Fabricius]), 2 in the East Atlantic (*O. cursor* [L.] and *O. africana* de Man), and the rest in the Indo-West Pacific.

All species are essentially active in air, and those that burrow intertidally remain in their plugged burrows while covered by the tide. However, they will enter the sea briefly, either spontaneously, or when chased. An example of one of the less terrestrial species is *O. ceratophthalmus* (Pallas), which in East Africa occupies burrows from below low-water neaps to just above high-water springs, with peak abundance just above mean tide level (Grubb, 1971; Jones, 1972; Hartnoll, 1975). The burrows are dug in such a position that they reach the water table at high tide (Hughes, 1966), and almost all activity takes place in the intertidal area. In East Africa, a second species, *O. ryderi* Kingsley, occupies a higher zone on the shore, with burrows from high-water neaps upward, and most abundant around high-water springs (Jones, 1972; Hartnoll, 1975). In Somalia, *O. ryderi* has yet another species, *O. cordimanus* Desmarest, above it; this species is essentially supralittoral and extends some way from the shore (Vannini, 1976b; Vannini and Valmori, 1981b). On Aldabra, the burrows of *O. cordimanus* occur over 150 m from the shore (Grubb, 1971; Alexander, 1976). Vannini (1976b) lists contrasting intertidal and supratidal species for a number of locations. Predominantly intertidal/littoral fringe species include *O. ceratophthalmus, O. urvillei* Guerin, *O. saratan* (Forskal), *O. macrocera* H. Milne Edwards, *O. platytarsis* H. Milne Edwards, *O. ryderi,* and *O. cursor*. Essentially supratidal species include *O. cordimanus, O. laevis* Dana, and *O. africana* (Fig. 2.11). Another species well adapted to terrestrial conditions is the western Atlantic *O. quadrata*. This burrows from high tide to 400 m inland, with the burrows seldom reaching the water table (Williams, 1965). Shuchman and Warburg (1978) list a further species, *O. aegyptiaca* Gerstaecker, as being essentially supratidal. There is some confusion in the literature as to whether some species are intertidal or supratidal, largely arising from different situations in areas with large tidal ranges on one hand, and minimal tides on the other.

It has been mentioned that in some intertidal species, such as *O. ceratophthalmus*, the burrows reach the water table and provide a ready source of water. This is not general, however, and the crabs usually forage near the waterline and will enter the sea regularly to moisten the gills – as in *O. quadrata* (Williams, 1965). *Ocypode* is also able to extract capillary moisture from damp sand, and this has been clearly demonstrated for *O. quadrata* (Wolcott, 1976, 1984; see also Chapters 3 and 7).

L. Conclusions on systematics and distribution

A few general comments are in order regarding, first, the level of terrestrial adaptation and, second, patterns of geographical distribution. No single group of crabs clearly emerges as being the most fully terrestrial. *Birgus* and *Gecarcinus* could be regarded as strong candidates, except that they must return to the sea where they retain a normal planktonic larval phase. For the crabs with an abbreviated development, there is a scarcity of precise data on the water requirements of, in some cases, the adults and, in nearly all cases, the larvae or juveniles. It seems likely that among them there are species that are effectively independent of standing water throughout their life cycle, but such species cannot yet be identified with certainty. There are candidates among the Jamaican sesarmids, the geosesarmids, and the potamonids.

With regard to the geographical distribution two features are noteworthy. One is that land crabs are essentially a tropical group, extending in smaller numbers into subtropical and warm temperate areas, but effectively excluded from cooler zones. This is paradoxical in that desiccation, a stress to which these crustaceans are poorly adapted, should pose less of a problem at lower temperatures. Second, there is the abundance of land crabs on islands, in contrast to their frequent scarcity on large land masses. This is almost certainly a result of reduced predation pressure on islands, the same factor that has permitted the development and survival of flightless birds, giant tortoises, and other specialized yet vulnerable forms.

Finally a brief analysis can be made of the regional occurrence of land crabs (see Table 2.3 for a summary of the occurrence of various genera in the major geographic zones). The most general pattern is of a relative paucity of species in the Atlantic and the East Pacific, contrasted with the more prolific Indo-West Pacific. This is in agreement with general zoogeographic patterns (see, e.g., Briggs, 1974) and is predictable on the basis of the extent of the various tropical shallow-water regions. The major deviant from this general pattern is *Uca*, with a great abundance of species in the East Pacific. The reason for this proliferation is not clear, but Crane (1975) cites a large tidal range and the absence of most other competing genera of the Ocypodidae as possible factors.

4. Phylogeny and evolution of land crabs

The fossil record of both the Anomura and the Brachyura is sparse, and the relationships of the major taxonomic groupings are still a matter of discussion and dispute (see later in this section). Consequently, the relationships of the various families containing land crabs are by no means settled and will not be considered at length. The evolution of the terrestrial habit within families is potentially more interesting, but in the absence of clear-cut evidence this also must be largely speculative.

A. Phylogenetic relationships of "land crab" families

The relationships between the Anomura and Brachyura are remote, and since the separation long preceded any tendency to terrestrial adaptation, it is hardly relevant. Thus, the Anomura and Brachyura will be considered in turn.

i. Anomuran phylogeny. This has recently been analyzed in some detail by McLaughlin (1983b), with particular reference to the relationships of the two superfamilies of hermit crabs – the Coenobitoidea and the Paguroidea. After a discussion of the various proposed relationships and an analysis of the similarities and differences between the taxa, McLaughlin (1983b) proposes the cladogram shown in Fig. 2.12. On the basis of these relationships she suggests a classification somewhat different from Bowman and Abele's (1982) presented in Table 2.1.

Superfamily Galatheoidea
Lomoidea
Hippoidea
Paguroidea: Family Pomatochelidae
Diogenidae
Coenobitidae
Paguridae
Parapaguridae
Lithodidae

The differences are the removal of the Lomisidae from the hermit crabs to a separate superfamily, and the inclusion of all hermit crabs in a single superfamily. McLaughlin (1983b) rejects the proposal that

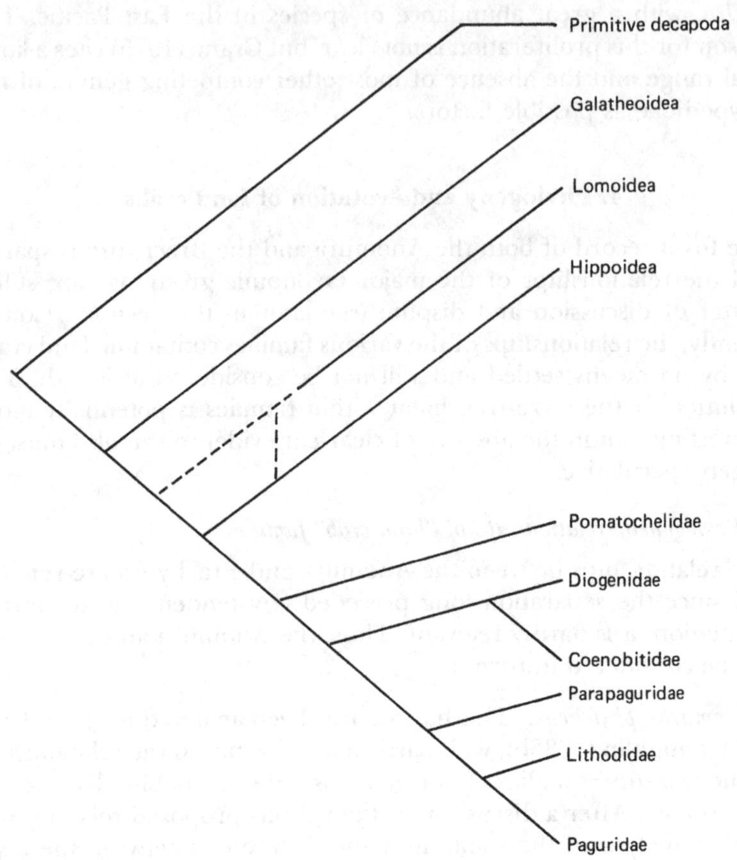

Fig. 2.12. Cladogram of the families of hermit crabs and related groups to indicate probable affinities. (After McLaughlin, 1983b.)

the Coenobitoidea and Paguroidea arose separately from ancestral anomurans (MacDonald, Pike, and Williamson, 1957). However, this point is not important to the present discussion. The main feature of interest in Fig. 2.12 is the close relationship of the Diogenidae and the Coenobitidae, a view expressed earlier by, among others, Wolff (1961), who suggested that the Coenobitidae arose from the Diogenidae. Thus, the only two families of Anomura with significant levels of terrestrial adaptation form a single phyletic unit, and it is reasonable to suggest that they both developed from a single stock with a terrestrial predisposition. Further speculation would not be profitable based on our present meager knowledge.

ii. Brachyuran phylogeny. The phylogeny of the Brachyura is by no means agreed to at the present, and the various arguments and schemes of classification (see, e.g., Borradaile, 1908; Bouvier, 1942; Glaessner, 1960; Guinot, 1978; de Saint-Laurent, 1980a, b; Rice, 1981) will not be elaborated here. Considering the classification from Bowman and Abele (1982) presented earlier in Table 2.1, the various groupings at the superfamily level are probably acceptable to most current workers, but at the higher levels there is still substantial disagreement. The sections Oxystomata and Oxyrhyncha, as demarcated by Bowman and Abele (1982), are open to objection on a variety of grounds (see, e.g., Guinot, 1978, and Rice, 1981). However, the section Brachyrhyncha, which includes all those brachyuran families showing any degree of terrestrial adaptation, has more general acceptance.

The Brachyrhyncha have been split in the past using various criteria. An early partitioning was Milne Edwards's (1834) subdivision into the Cyclometopa and Catometopa, a scheme followed with some variations by other workers such as Boas (1880) and Ortmann (1896). The Brachyrhyncha at that time comprised the sections Cancridea and Brachyrhyncha of the classification used here, and its division was based on shape of carapace and the position of the male genital openings. Without going into the rather tortuous details of these various classifications, which are now really only of historical interest, it is sufficient to state that the families with substantial terrestrial adaptation were partitioned as follows: the Potamoidea in the Cyclometopa, and the Gecarcinidae, Grapsidae, Mictyridae, and Ocypodidae in the Catometopa.

This partitioning of the Brachyrhyncha was criticized by Borradaile (1903a, 1908), largely on the grounds that the two groups intergraded, and has subsequently been little used. Recently, however, there has been renewed interest due to the very detailed morphological studies of Guinot (1977, 1978, 1979), who divides the Brachyura to the superfamily level according to the scheme given in Table 2.4. The basic division is based on the position of the sexual openings: In the Podotremata they are coxal in both sexes; in the Heterotremata they are coxal in the male, but sternal in the female; in the Thoracotremata they are sternal in both sexes. Guinot (1977, 1978, 1979) does not include the Potamoidea in her scheme, but provisionally it must be assigned to the Heterotremata on account of the coxal male and sternal female openings.

To a large degree the Thoracotremata corresponds with the earlier Catometopa, and includes all of the main groups of terrestrial Brachyura except for the potamonids. Guinot (1977, 1978) envisaged the Heterotremata and Thoracotremata as successive evolutionary levels, rather than as phylogenetic groups. However, de Saint-Laurent (1980a, b) considers the two groups as distinct phyletic units producing

Table 2.4. *Classification of the Brachyura to the superfamily level according to Guinot (1977, 1978)*

Section	Subsection	Superfamily
Podotremata	Dromiacea	Homolodromioidea
		Dromioidea
	Archaeobrachyura	Homoloidea
		Raninoidea
		Tymoloidea
Heterotremata		Dorippoidea
		Calappoidea
		Corystoidea
		Portunoidea
		Xanthoidea
		Potamoidea
		Majoidea
		Parthenopoidea
		Bellioidea
		Leucosioidea
Thoracotremata		**Gecarcinoidea**
		Grapsoidea
		Mictyroidea
		Pinnoteroidea
		Hexapodoidea
		Ocypodoidea
		Hymenosomatoidea

Note: Superfamilies containing a large component of terrestrially adapted species are indicated in boldface.

a phylogenetic relationship such as in Fig. 2.13. This would suggest that the potamonids separated from the other families with terrestrial adaptations a considerable time ago, and have independently evolved their terrestrial habits. On the other hand, it would give the Gecarcinidae, Grapsidae, Ocypodidae, and Mictyridae a more recent common ancestry in a group that seems predisposed to the terrestrial habit. Interestingly, de Saint-Laurent (1980b) suggests that the movement of the genital opening of both sexes to the sternum in the Thoracotremata, by freeing the legs from any direct involvement in reproduction, may have facilitated the development of the more sophisticated locomotory mechanisms presumably needed on land.

Borradaile (1908), though not accepting the Cyclometopa/Catometopa division, did consider that the families in the latter unit probably had a genetic affinity. Rice (1981) has examined brachyuran phylogeny on the basis of zoeal morphology, and considers that the Thoracotremata can be considered a discrete group on those grounds. He rather tentatively suggests that the Gecarcinidae separated earliest

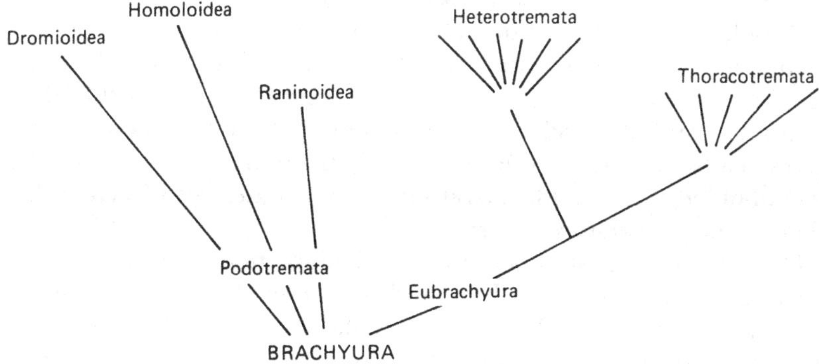

Fig. 2.13. The phylogenetic relationships of the main groups of the Brachyura according to the hypothesis of de Saint-Laurent. (After de Saint-Laurent, 1980b, redrawn from Rice, 1981.)

from the basic thoracotrematous stock, and that the Grapsidae could have given rise to the Ocypodidae.

There is little other evidence to supplement the morphological data on which the preceding considerations are based. The fossil record is not helpful: Most of the major groups were already recognizable in the Cretaceous (Glaessner, 1960), but there are no useful clues to their origins. Another possible line of investigation is to use biochemical techniques, such as the serological comparison employed by Leone (1950, 1951). His study showed that the Ocypodidae and Grapsidae were not closely related to the Portunidae, Xanthidae, and Cancridae, but that these last three had a distinct affinity with each other. This concurs with most of the morphological evidence, but rather surprisingly Leone's (1950, 1951) results suggested that the Grapsidae and Ocypodidae were not closely related, which conflicts with all other evidence. Nevertheless, such studies are promising, and it would be worthwhile to look at a wider range of crabs using the various more sophisticated techniques now available.

B. Evolution of the terrestrial habit in "land crab" families

In many cases the course of evolution of the terrestrial habit is relatively obvious, especially where the degree of adaptation is modest. Thus, intertidal or immediately supratidal crabs are clearly species that were originally fully marine but that have become adapted to spend increasing amounts of time in the air: They have migrated, or are in the course of migration, directly from sea to land.

Conversely, crabs that spend much time in or close to fresh water, but make excursions on land, have adapted to the terrestrial habit via the freshwater one. However, species that are effectively fully terrestrial could have arrived at that condition by either route, and various factors, including the situation of other members of the family, must be taken into account. The Diogenidae, Porcellanidae, Portunidae, Xanthidae, and Mictyridae need little comment; in these families any terrestrial habits have developed directly from the marine ones. The Coenobitidae, Gecarcinidae, Grapsidae, Potamoidea, and Ocypodidae will each be considered in turn.

The Coenobitidae show a gradation of adaptation from species that are always found near the shore and that enter the sea regularly, to those that penetrate well inland and make use of casual fresh water. However, there seems little doubt that the latter have migrated directly from the sea, since they still return to the sea to allow the eggs to hatch, and the juveniles migrate directly from the sea to the land.

The Gecarcinidae provide a less clear-cut situation, since all species are highly terrestrial. They all have marine larvae and must return to the sea for the eggs to hatch, but this alone is not evidence of a direct sea–land transition. The freshwater crab *Eriocheir*, which spends most of its adult life in the upper reaches of rivers, undertakes a similar breeding migration to the sea (Panning, 1939; Hoestlandt, 1948). Of the gecarcinids, *Cardisoma* is the least terrestrial, and is normally found in damp areas where it burrows to the water table, and is often in fresh water. Such a habit could have been reached by direct migration from the upper littoral, but a route via an estuarine/riverbank situation seems more tenable. Whether the more fully terrestrial *Gecarcinus* and *Gecarcoidea* followed the same route, or came direct from the sea, is a matter of conjecture. However, studies of ionic balance do not indicate a freshwater background for *Gecarcinus* (Lutz, 1969).

In the Grapsidae, the majority are intertidal/supratidal species with limited terrestrial adaptation, but four genera need some consideration: *Geograpsus, Sesarma, Metopaulias,* and *Geosesarma. Geograpsus* has marine planktonic larvae, and its species show a gradation from the immediately supratidal to the fully terrestrial – all indications are of a direct migration from the sea to land. The terrestrial species of *Sesarma*, the single species of *Metopaulias*, and probably all of the species of *Geosesarma*, have large eggs and abbreviated lecithotrophic larval development so that they no longer have to return to the sea to breed. These features tend to be characteristic of freshwater decapods, and indeed a number of them are found associated with freshwater or damp habitats. It is probable that all of these three genera have become terrestrial via the medium of freshwater penetration. Whether the abbreviated development and freshwater/terrestrial habit have evolved independently in each of the four groups (East Indian *Sesarma* spp., Jamaician *Sesarma* spp., *Metopaulias, Geosesarma*) is an unresolved point of considerable interest.

The potamonids are the typical freshwater crabs, with large eggs and direct development. It would seem obvious that all the potamonids that have developed terrestrial habits have acquired these through fresh water, but Lutz (1969) has suggested that this is not necessarily the case. On the basis of ionic ratio studies he concludes that *Sudanautes* has affinities with marine rather than freshwater crabs, in contrast to *Potamon*. He suggests that this indicates that although most potamonids have invaded freshwater/terrestrial environments via the estuarine route, *Sudanautes* at least has done so direct from the intertidal. This view is also held by Bott (1955). More extensive physiological investigation of other species may clarify the situation further. A possibility is that the evolution of direct development occurred initially, enabling the subsequent invasion of fresh water and land by these different routes.

Finally, we consider the Ocypodidae, which present a relatively straightforward case. Most species are intertidal, or immediately supratidal, have marine pelagic larvae, and have invaded the land direct from the sea. The unusual case is *Uca leptodactyla*, which has abbreviated larval development away from the sea; however, this species is characteristic of very hypersaline conditions, and the landward penetration is therefore more likely to have been direct from the supratidal, rather than through fresh water.

Thus, some highly terrestrial crabs have become terrestrial directly from the high intertidal, whereas others have done so after first adapting to a freshwater existence. The route is not certain in all cases, but it seems likely that most of the direct invaders still have to return to the sea where they have a pelagic larval stage. In contrast most (and perhaps all) of those that have reached land via fresh water have done so with the added advantage of abbreviated nonplanktonic larval development, which gives the potential of more complete independence from the aquatic environment. Only further ecological study of some of the more terrestrial of these will demonstrate just how total this independence has become.

5. Conclusions

"Land crabs" are defined as those members of the Anomura and Brachyura that are active in air, and it is seen that the extent of terrestrial adaptation varies substantially both within and between families. The most terrestrial crabs occur in the Coenobitidae, Gecarcinidae, Grapsidae, Potamoidea, and Ocypodidae. All of these include species where the adults are more or less independent of permanent water, whereas in a few of the Grapsidae and Potamoidea this independence extends to the larvae and juveniles. However, even in highly specialized species there is minimal external evidence of this terrestrial adaptation.

The geographical distribution of land crabs shows a clear concentration in tropical and subtropical areas, with the greatest diversity in the Indo-West Pacific region. They are particularly abundant on oceanic islands.

There is still uncertainty regarding the phylogenetic affinities of land crabs at the family level. Within the Anomura the affinity of the two families with terrestrial tendencies – the Diogenidae and the Coenobitidae – seems fairly clear. In the Brachyura, the Gecarcinidae, Grapsidae, and Ocypodidae almost certainly have a close relationship, with the Potamoidea having diverged rather earlier.

In most cases the terrestrial habit has arisen by a direct invasion of the land from the sea. However, in some of the Grapsidae, and possibly all of the Potamoidea, a route via fresh water is more probable.

3: Ecology

THOMAS G. WOLCOTT

CONTENTS

1. Introduction

This overview of the ecology of land crabs will emphasize species found in supratidal and upland habitats, including ghost crabs, gecarcinid land crabs, and land hermit crabs. It will not attempt to itemize the ecological observations made on all crabs that may emerge onto land. Information on other groups (e.g., the more intertidal species or the freshwater/land crabs) will be included where it illustrates the ecology of land crabs as a whole. The chapter will emphasize data published in the last 20 years. Bibliographies of earlier work appear in Bliss and Mantel's milestone 1968 symposium on land crabs (*American Zoologist*, vol. 8), and terrestrial adaptations of crustaceans in general are reviewed by Powers and Bliss (1983).

The robust, heavy-shelled gecarcinid genera *Gecarcinus, Gecarcoidea,* and *Cardisoma* are generally the subject when the term "land crab" is used

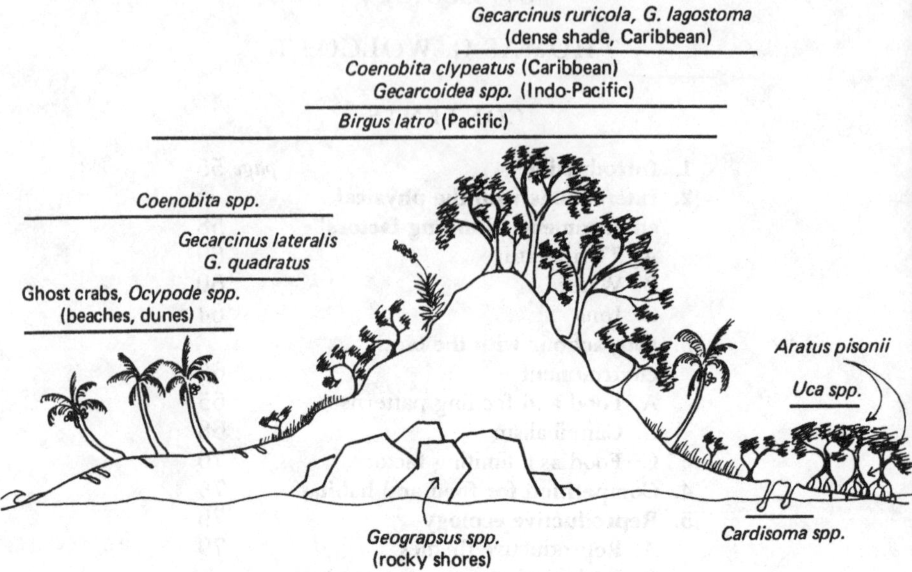

Fig. 3.1. Representative distributions of some better-known land crabs, superimposed on a section through a schematic subtropical island.

without modifiers. The first two occupy dry upland habitats (Fig. 3.1; also see the Appendix). Most is known about *Gecarcinus lateralis*, which occurs on Bermuda, in the Bahamas, and in the West Indies in sandy burrows within about 300 m of the sea (Bliss, 1979). Its larger congener, the black land crab or "mountain crab," *G. ruricola*, is common in the eastern Caribbean islands (Britton, Kroh, and Golightly, 1982), where it is a popular food item (Chace and Hobbs, 1969), yet it is known to science largely as museum specimens. It can live several kilometers inland, at hundreds of meters above sea level, in densely shaded locations (Chace and Hobbs, 1969; von Prahl, 1983). *Cardisoma* (Fig. 3.2) is a pantropical lowland genus, living in muddy soils near mangroves, swamps, and streams where its burrows can reach groundwater (Bruce-Chwatt and Fitz-John, 1951; Herreid and Gifford, 1963; Alexander, 1979; Cameron, 1981a; Cheng and Hogue, 1974; Henning, 1975a, b).

The ghost crabs (genus *Ocypode*, Fig. 3.3) are swift, lightly built, highly active predators, scavengers, and, in a few cases, deposit feeders (Koepcke and Koepcke, 1953; Hughes, 1966, 1973; Jones, 1972; Wolcott, 1978). They occur on semiprotected and exposed sandy beaches; some range hundreds of meters back into dunes behind beaches. The Coenobitidae, comprising land hermit crabs of the genus *Coenobita* and the coconut or robber crab *Birgus latro*, are the only Anomura to

Fig. 3.2. *Cardisoma guanhumi*, large adult male (approximately 9 cm carapace width), U.S. Virgin Islands National Park. (Photograph by T. G. Wolcott.)

invade land. *Coenobita* occupy adopted gastropod shells throughout life. Some species inhabit beaches; others range up into dry forests hundreds of meters above sea level (Chace and Hobbs, 1969; De Wilde, 1973; Page and Willason, 1982; Vannini, 1976a). *B. latro* inhabits gastropod shells only during its early postlarval life (Reese, 1968). The largest specimens known are too big for any snail shell; with adult body mass of up to 13 kg and a total leg span of nearly a meter (Vogel and Kent, 1971), they are the largest living land arthropods. *B. latro* inhabit jungles of Pacific and Indian Ocean islands, reaching elevations of several hundred meters and locations several kilometers from the sea (Gibson-Hill, 1947; Vogel and Kent, 1971; Helfman, 1973; Cameron, 1981a; Hicks, Rumpff, and Yorkston, 1984). The adult crabs have no predators other than fellow robber crabs and, most significantly, humans (Fig. 3.4).

Ecological information on other land crabs that presumably invaded from the sea (e.g., *Geograpsus* and other shore- and mangrove-dwelling grapsids) is fragmentary; only a few land crabs that have emerged from fresh water (e.g., *Holthuisana transversa*) have been studied in any detail (see Greenaway and MacMillen, 1978; see also Chapter 7).

It is now recognized that land crabs are behaviorally and ecologically plastic. Consequently, several levels of investigation will be needed to as-

Fig. 3.3. The North American Atlantic ghost crab, *Ocypode quadrata*, feeding at night on a mole crab (*Emerita talpoida*). (Photograph by T. G. Wolcott.)

semble a complete account of their ecology. The species need to be adequately described, and their distributions determined. Next, by extensive observations in the field, a qualitative picture of their natural history and environmental relationships should be formed. These sorts of data should be complemented by quantative experimental studies. Unfortunately, we have reached only the first of these steps in studying the ecology of most terrestrial crabs (see Chapter 2). Thus, the present treatment of the ecology of land crabs will necessarily be more descriptive in character than other chapters in this volume, due to the descriptive, often anecdotal, nature of existing ecological data. The needs and opportunities for additional research on the ecology of land crabs are obvious.

2. Interactions with the physical environment: "limiting factors"

The "limiting factors" that determine how far a species can penetrate into terrestrial habitats are poorly known for land crabs. Previous reports have often been based on observations of distributions in only

Fig. 3.4. *Birgus latro*, from Enewetak Atoll. Fishermen suspend crabs (destined for human consumption) individually by string to prevent cannibalism. (Courtesy of Gene Helfman.)

a few localities, and on preconceptions concerning the animals' physiological requirements. The more obvious potential "limiting factors" include temperature and water or salt shortage (see Chapter 7). More complex factors include the presence of habitat suitable for burrowing, presence of vegetation or shade, and distance from the sea with its attendant effects on energetic cost of reproductive migrations. Correlations between the distributions of crabs and macro-environmental measurements are usually of limited utility because crabs often select microenvironments inaccessible to conventional instruments or create modified microhabitats (burrows) that are more humid and thermally stable than the surface macroenvironment.

A. Temperature

Temperature clearly limits latitudinal distribution. Most land crabs are restricted to the tropics or subtropics. In warm climates, potentially lethal high temperatures can be avoided by moving into shade or into a burrow. The prolonged cold typical of temperate latitudes penetrates deep into the ground and cannot easily be escaped. At marginal latitudes, habitat differences take on extra importance. In the winter of 1976–7, the coldest since 1917 (a "fifty-year winter"), rangers of Fort Macon State Park (Carteret Co., North Carolina, U.S.A.; 35°42′ N, 76°42′ W) reported large numbers of ghost crabs emerging from their burrows on the beach and dying. However, no mortality was observed among crabs that had burrowed in the dunes. Burrows inland of the beach can extend below the lethal isotherm, about 6–8° C, without encountering the water table (T. G. Wolcott, unpub.). Increased survival in cold winters may explain the persistence of dune-burrowing behavior in *Ocypode quadrata* despite the apparent lack of food in dunes.

Limitation of land crab distributions by low temperatures need not involve mass mortality. Cold also limits the active season, and hence time available for feeding, and therefore the energy available for growth and reproduction. In temperate zones, the crabs for which information is available (fiddler crabs, ghost crabs, some gecarcinids) enter burrows and remain underground for the winter (Gifford, 1962a; Bliss et al., 1978; Shuchman and Warburg, 1978, Wolcott, 1978; Leber, 1982). *Gecarcinus lateralis* is active only above 18° C (Bliss et al., 1978). The active period is shorter at higher latitudes (Fig. 3.5), and extrapolation of this trend explains the lack of land crabs outside the subtropics or, in the case of ghost crabs, the warm temperate zones. As the length of the season suitable for surface activity declines, so does the probability of a crab population persisting.

B. Water

Availability of water is widely regarded as a "limiting factor" for crustaceans invading land. Obviously, crabs cannot live in terrestrial environments where essentially no water is available; hence, water is clearly an "ultimate" limiting factor. However, it is not obvious that any crabs press inland to the very limits of their ability to maintain

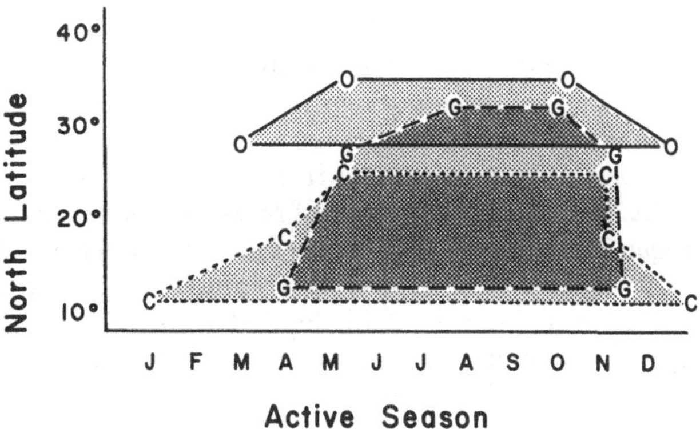

Fig. 3.5. Active seasons of land crabs decrease with increasing latitude, accounting for the restriction of these animals to tropical and subtropical climates. *O*, *Ocypode quadrata*; *G*, *Gecarcinus lateralis*; *C*, *Cardisoma guanhumi*.

water balance. In other words, it is not clear that availability of water directly determines the limits of ranges and thus serves as the "proximate" limiting factor in the penetration into terrestrial habitats. Water is acquired by various means (immersion, drinking with chelae, sucking moisture from soil, collecting condensation; see Chapter 7). Thus, the appropriate conditions (e.g., accessibility of the water table by burrowing, or amount and extractability of moisture in soil) must therefore be examined to determine if availability of water is indeed "limiting."

Both sides of the water budget, acquisition and loss, must be considered. Ghost crabs (*O. quadrata*) have high evaporation rates relative to other land crabs (see Chapter 7) and make up their losses by extracting soil moisture (Wolcott, 1984). Their water-uptake mechanism restricts them to sandy soils where the water table is close to the surface. For instance, in Brazil *O. quadrata* is found neither in impenetrable laterite soil nor in highly organic soil from which water cannot be readily extracted (Fimpel, 1975; Wolcott, 1984). Some ghost crabs appear to select beach zones on the basis of soil water content, moving up and down shores with monthly tidal regimes (Barrass, 1963; Hughes, 1966; Fimpel, 1975; Vannini, 1976b; Shuchman and Warburg, 1978), and select soils with preferred water content in the laboratory (Warburg and Shuchman, 1979). However, they are not found in all habitats where water availability is adequate, implying that something else is the proximate limiting factor. It may be that other processes (e.g., molting or defense against predation) place more stringent requirements on microhabitat selection. For instance, dense vegetation seems to exclude the crabs, perhaps because rapid running and escape become impossible (Henning and Klaassen, 1973; Fimpel, 1975).

Distribution of land hermit crabs is also often traceable to water availability. They usually seem to require access to standing water for drinking and to replenish the reserve carried in the shell, although at least some species can extract water from damp sand by an unknown mechanism (A. W. Pinder, University of Massachusetts, pers. comm.). *Coenobita rugosus* can only "drink" from sand that is nearly saturated with water (Vannini, 1975). *C. scaevola* lives along highly arid shores of the Sinai Peninsula, but is totally dependent on the sea for water and consequently limited to the nearshore area (Achituv and Ziskind, 1985). A well-fitting shell is essential for maintaining low evaporation rates and carrying ample water. On Curaçao in the Netherlands Antilles, an appropriately sized shell in good condition allows invasion of inland environments offering more shade, food, and fresh water for *C. clypeatus.* The hermit crabs with broken, ill-fitting shells are restricted to the coast, must rely on drinking seawater, and appear to be in relatively poor condition themselves (De Wilde, 1973). Rainfall determines the seasonal distribution patterns of *C. rugosus* in East Africa. During the dry season this crab returns each day to damp burrows at the foot of the coastal dune, after foraging during the night on the beach. During the rainy season it may leave the dune and spend weeks to months foraging in the "bush" a few hundred meters inland (Vannini, 1975). Vannini (1975) reviews a number of other coenobitid species (*C. brevimanus, C. clypeatus, C. diogenes, C. perlatus, C. scaevola*) that show either daily or seasonal migrations depending upon rainfall. *C. compressus* shows tidal rhythms that persist under constant conditions (Wheatly et al., 1985).

The genus *Cardisoma* requires burrows that can reach down to groundwater, but there are no studies showing that suitable habitat with access to groundwater is a limiting resource for any population. *C. guanhumi* can range far up tidal streams (Fimpel, 1975); it even occurs around pools along intermittent freshwater streams 200 meters above sea level in the Virgin Islands (T. G. Wolcott and D. L. Wolcott, unpublished).

The highly terrestrial land crabs, of the genera *Gecarcinus* and *Gecarcoidea*, are probably unable to draw on soil water under normal conditions (Wolcott, 1984). They rely on intermittent access to rain or dew, coupled with effective water conservation. On Bermuda, there is no evidence for gradients in availability of standing water, or for marked differences in conditions that would affect desiccation rates, with increasing distance from shore (Wolcott, 1984). In the absence of such gradients, water is unlikely to be the factor limiting inland penetration by Bermudian *G. lateralis*. No similar data are available for other species.

Land crabs of freshwater origin are found near streams, seeps, and swamps. No data are available on desiccation rates versus water availability in nature except for the Australian arid-zone crab *Holtuisana transversa*, the most remarkable example of this group. *H. transversa* occupies habitats where rainfall is low and unpredictable. It is active in water during infrequent flooding; during periods of months to years between floods, it remains inactive in mud burrows (Greenaway and MacMillen, 1978; also see Chapter 7). This species is not commonly found near permanent water or in wetter environments; there it is replaced by *H. valentula* and *H. agassizi* (Greenaway, 1984b), so the exact role of water in determining the species' distribution is not clear.

Water availability does seem to impose temporal limits on activity, with the probable energetic consequences mentioned earlier. Most terrestrial crabs show circadian activity rhythms (Palmer, 1971; Bliss, 1979); those foraging on beaches may also show a tidal component (Vannini, 1975, 1976a; Wheatly et al., 1985). Most crabs with limited access to free water are nocturnal, suggesting that they minimize water loss by foraging when evaporation rates will be lowest. All tend to become more active when rainfall dampens their habitat (e.g., Bliss et al., 1978; Hicks et al., 1984).

Ghost crabs are largely nocturnal but flexible in their behavior. Because they can readily replenish evaporative losses by sucking interstitial water from the sandy soil of their burrows (see Chapter 7), they can also forage by day (in areas undisturbed by humans) (Hughes, 1966; Vannini, 1976b; Wolcott, 1978). During the day crabs remain close to the refuge of the burrow, cleaning and enlarging it, and scavenging. At night they feed over much larger areas, and generally do not return the next morning to their burrow of the previous day. Juveniles tend to be more diurnal than adults (Fisher and Tevesz, 1979; T. G. Wolcott, unpub.). *Ocypode gaudichaudii* in Central America and *O. ryderi* in East Africa are unusual in being crepuscular or diurnal (Crane, 1941a; Vannini, 1976b).

Land hermit crabs usually hide in shallow burrows, crevices, or niches among branches or roots during the day (Vannini, 1975; 1976a; De Wilde 1973; Bliss et al., 1978). Like ghost crabs, they may replenish water from damp sand (Vannini, 1975; A. Pinder, University of Massachusetts, pers. comm.). They emerge during the night to feed and drink. Higher levels of activity, and activity outside normal active periods, can occur during rain or high humidity (De Wilde, 1973).

The coconut crab *B. latro* is nocturnal throughout its range, with limited diurnal activity on islands uninhabited by humans (Reyne, 1939; Vogel and Kent, 1971; Helfman, 1973; Hicks, 1985). An exception occurred on Christmas Island, where the crabs were diurnal when nocturnal rats were common on the island. This temporal shift presumably reduced predation by rats, and was made possible in part by the large size, low surface:volume ratio, and hence low evaporation rate of this crab. With rat control by humans *Birgus* has again become nocturnal (Gibson-Hill, 1947). During the day, crabs hide in shallow burrows, hollow logs, or perch 1–2 m up trees (Vogel and Kent, 1971; Cameron, 1981a).

Gecarcinus lateralis appears to rely on water conservation rather than having specialized mechanisms for water uptake. Its water turnover appears to be low. This species is always nocturnal on Nonsuch Island, Bermuda (Bliss et al., 1978; D. Wingate, Bermuda Agriculture and Fisheries, pers. comm.; T. G. Wolcott, unpub.). In contrast, *G. lateralis* is active only from a few hours before dawn until as late as noon at Spittal Pond, Bermuda, but hardly at all during the night (Wolcott and Wolcott, 1982a). In southern Florida, this species shows morning and evening peaks of activity extending longer into daylight hours where shade is dense (Bliss et al., 1978). The factors influencing the markedly different rhythms in two Bermudian populations remain unknown, though D. Wingate (Bermuda Agriculture and Fisheries, pers. comm.) suspects humidity differences may be involved. This behavioral plasticity of land crabs suggests that caution should be used in interpreting observations from a single habitat.

Cardisoma guanhumi tends to be diurnal or crepuscular for most feeding activities undertaken near the burrow in Colombia (Henning and Klaasen, 1973; Henning, 1975a) and the Virgin Islands (T. G. and D. L. Wolcott, unpub.), probably because the pool of water in the burrow makes high evaporative losses affordable. Extended migratory movements are largely nocturnal (Feliciano, 1962; Gifford, 1962a; Taissoun, 1974a; Henning, 1975b).

Gecarcoidea natalis, otherwise very similar in habits and habitat to *G. lateralis*, is almost exclusively diurnal in both foraging and migratory activity on Christmas Island (Indian Ocean), perhaps in response to predation pressure from nocturnal *B. latro* (Hicks, 1985).

C. Ions

Availability of ions (salts) may also be considered as a factor limiting the inland penetration of land crabs, particularly those of marine origin. Are any ions in short supply inland, and are there gradients that could account for the distributions of land crabs?

Even in species with high water turnover (e.g., ghost crabs) salt losses are much lower than would be predicted if isionic urine were produced (see Chapter 7). Ghost crabs (*O. quadrata*) resorb over 90% of Na^+ and Cl^- from urine before discarding it (T. G. Wolcott and D. L. Wolcott, 1985a), as do *G. lateralis* and *C. guanhumi* (T. G. Wolcott and D. L. Wolcott, 1982b, 1984b). Since some salt gain will come from food items, salt replacement may not be a severe limiting factor.

It would seem that there should be a gradient in availability of salt from the spray-washed seashore to several hundred meters inland. However, preferred forage plants of *G. lateralis* (St. Augustine grass, *Stenotaphrum secundatum*, and Australian "pine," *Casuarina equisetifolia*) in transects from Bermuda's spray zone to 200 m inland (60 m above sea level) showed no significant difference in salt content, even if spray-deposited salts were included (T. G. Wolcott and D. L. Wolcott, 1985b). Ion-depleted *G. lateralis* show recovery in hemolymph ion concentration when fed plants from beyond the inland limit of their normal range, suggesting that ample salts are available from food throughout the range (T. G. Wolcott, 1980; T. G. Wolcott and D. L. Wolcott, unpub.). Although budgets of individual ions require closer scrutiny, the low salt requirements of land crabs and the imperceptible gradients in availability of ions suggest that it will be difficult to establish a relationship between salt availability and limits of distributions.

In summary, several physical factors could potentially limit penetration of terrestrial habitats by land crabs. Nevertheless, studies to date have not demonstrated that any of these factors actually limit the distribution of a species by reaching, at the boundaries of the distribution, levels that exceed the physiological tolerances of the animals. Instead, new mechanisms have been discovered that confer physiological tolerance to physical conditions at or even outside the normal range limits. The search for a pervasive limiting factor has remained unrewarded, and the ecological barriers that exclude land crabs from more terrestrial habitats remain enigmatic (T. G. Wolcott and D. L. Wolcott, 1985b).

3. Interactions with the biotic environment

A. Food and feeding patterns

Land crabs have two basic foraging modes. The more lightly armored crabs that depend on speed and agility for defense tend to be active predators and facultative scavengers. Heavily armored crabs are usually more sluggish and tend to feed mostly on inactive items; since there are few sessile animals on land, these crabs are primarily herbivores or detritivores. On close inspection, most of these herbivores/detritivores also turn out to be predators or scavengers of carrion whenever possible. Preconceived notions of crabs as having relatively few stereotypic "hard-wired" action patterns have been challenged by observations of the great adaptability of land crab feeding behaviors (see Chapters 4 and 10).

Ghost crabs are, with few exceptions, highly adapted as predators. Various ghost crabs prey on sea turtle eggs and hatchlings on tropical and subtropical nesting beaches (Fimpel, 1975; Alexander, 1979). *O. quadrata* relies on beach macrofauna for 90% of its diet (T. G. Wolcott, 1978), but can also deposit feed on muddy sand, extracting diatoms almost at efficiently as fiddler crabs (Robertson and Pfeiffer, 1982). Other ghost crabs show similar flexibility. *O. cordimana*, a supratidal species in East Africa, takes many insects and ants; *O. ryderi* (misidentified *kuhli* in some papers) lives lower on the shore and subsists largely on cast-up algae, but also takes hippid crabs, insects, and fellow ghost crabs (Jones, 1972; Vannini, 1976b); it prefers animal to plant food when given a choice (S. M. Evans et al., 1976). *O. ceratophthalmus* also utilizes deposit feeding in addition to scavenging and preying on isopods and annelids (Braithwaite and Talbot, 1972; Jones, 1972). *O. gaudichaudii* is principally a deposit feeder but readily takes detritus and live prey (Koepcke and Koepcke, 1953).

Other agile, predatory crabs include the mangrove tree crab, *Aratus pisonii*, which feeds in the canopy on various insects as well as on mangrove leaves in Florida (Beever, Simberloff, and King, 1979). *A. pisonii* is preyed on in turn by the mangrove crab *Goniopsis cruentata* (Warner, 1967; Wilson, 1985). *Geograpsus crinipes*, a shore-dwelling grapsid on Aldabra Atoll (Indian Ocean, 9°30′ S, 46°0′ E), preys on the dominant land hermit crab, *Coenobita rugosus* (Alexander, 1979). On Christmas Island (Indian Ocean), *G. crinipes* preys on *Geograpsus grayi*; both prey on *Gecarcoidea natalis* (Hicks et al., 1984).

Land hermit crabs, with their heavy adopted shells, are the most cumbersome of land crabs. They feed on various forms of vegetation and detritus (Vannini, 1975), including the Caribbean manchineel (*Hippomane mancinella*, poisonous to most animals) and the feces of horses and cows (De Wilde, 1973), and of giant tortoises on Aldabra Atoll (Alexander, 1979). They also scavenge carrion so effectively on Aldabra and Pacific islands that they are thought to account for the low numbers of carrion-breeding flies there (Alexander, 1979; Page and Willason, 1982). On Curaçao, ". . . a dead donkey provided food for hundreds of [*C. clypeatus*] for weeks on end. Even when only an empty dried skin was left, animals were still entering it through the various holes. The dry skin functioned as a sounding box and the scuttling of all these animals could be heard over a great distance" (De Wilde, 1973). The land hermit crabs seem ungainly and ill equipped for predation, but large *C. rugosus*, like *Ocypode* spp., find and loot sea turtle nests (Alexander, 1979), probably by olfaction (Rittschof, Barlow, and Schmidt, 1985).

The coconut crab *B. latro* forages 30 m or more from its burrow (Vogel and Kent, 1971), feeding on various forms of vegetation, especially fruits of *Pandanus* (screw pine) and other plants (Helfman, 1973). Where coconuts (*Cocos nucifera*) are available, it eats them almost exclusively (Vogel and Kent, 1971). The average body mass of crabs on those Aldabran islets where the high-energy, high-water nuts are available is double that of crabs on other islets (Alexander, 1979). The crabs apparently do not climb to the crown of the coconut palm to harvest green nuts (although this is a persistent legend), but rely on fallen nuts. Reyne (1939) and Gibson-Hill (1947) mustered copious circumstantial evidence that the crabs could not open intact nuts without assistance. However, Vogel and Kent (1971), using night-vision equipment on Enewetak, witnessed a large *Birgus* hammering with its chelae at the "eyes" of a coconut until the shell was penetrated. The pointed first walking legs are also used to make the first hole (Hicks et al., 1984). The crabs then chip away the edges of the hole with the powerful claws, which are capable of exerting about 1.5 times the force of a human bite (Hicks et al., 1984) and shearing sticks up to 5 cm in diameter (Vogel and Kent, 1971). When the hole is large enough, claws and legs are inserted to extract the coconut meat. Possibly husks of ripe, fallen nuts must undergo some decomposition before they become manageable; in any case the whole process takes several days and may involve many crabs in succession (Hicks et al., 1984). Like most other herbivorous land crabs, *B. latro* is also a scavenger (Alexander, 1979). It reportedly requires animal protein in the laboratory (Harms, 1932) and is thought to prey on other land crabs in nature (Harms, 1932; Grubb, 1971; Hicks et al., 1984). Alexander (1979) relays anecdotal reports of *Birgus* predation on turtle hatchlings as well. The "robber crab" also removes all sorts of items (boots, pans, silverware, hats, wristwatches, etc.) from human habitations, abandoning them in the bush when it discovers they are inedible (Gibson-Hill, 1947).

The brachyuran land crab *C. guanhumi* feeds primarily on leaf litter in wooded areas inland of mangrove swamps (Herreid, 1963; Henning, 1975a). A wide variety of plants are consumed, including representatives of 35 families in southern Florida (Herreid, 1963). In dry weather, the crabs will even feed on underwater algae (Gifford, 1962a). The crabs also readily take animal protein obtained from various sources, particularly carrion (Herreid, 1963). Colombian *C. guanhumi* catch insects in laboratory tanks, and prefer animal to plant food in the laboratory. They show predatory behavior toward moving objects smaller than themselves in the field. In beach/dune habitat they have been seen charging into fiddler crab colonies and eating the claws autotomized by escaping fiddlers. These crabs also position themselves in trails of leafcutter ants, consuming both ants and leaf fragments. Juvenile crabs chase insects in the field, and adults readily approach, catch, and eat tethered lizards in field trials (Henning, 1975a). *C. carnifex* displays similar tastes on Aldabra (Alexander, 1979), as does *C. armatum* in West Africa (Bruce-Chwatt and Fitz-John, 1951).

In the omnivorous genus *Gecarcinus, G. lateralis* is by far the best studied. This species feeds on a wide variety of plant material on Bermuda (D. L. Wolcott and T. G. Wolcott, 1984). The preferred species in laboratory trials were St. Augustine grass and dry "needles" of the Australian pine, both of which are exotics. *Gecarcinus lagostoma*, a species restricted to southern Atlantic islands (R. B. Manning, U.S. National Museum of Natural History, pers. comm., contra Türkay, 1970) feeds largely on algae, mosses, lichens, and leaf litter (Fimpel, 1975). *G. quadratus*, the Pacific coast species, can be maintained on fruit in the laboratory (Abele, Robinson, and Robinson, 1973). *G. planatus* scavenges cast-up algae, and even enters the lagoon to eat submerged vegetation on Clipperton Atoll, where dense populations (up to 6 individuals$\cdot m^{-2}$; Ehrhardt, 1968) may exhaust stocks of preferred terrestrial plants (Ehrhardt and Niaussat, 1970).

Despite this usual herbivory, *G. lateralis* has been photographed by Bliss et al. (1978) feeding on a lizard tail; the species eats whistling frogs (*Eleutherodactylus* sp.) and dead birds and toads (Fig. 3.6) on Bermuda (T. G. Wolcott and D. L. Wolcott, unpub.). In the laboratory *G. lateralis*, normally a shy animal, becomes markedly more aggressive in the period following ecdysis and attempts to capture moving objects. Similarly, *G. lagostoma* on Trinidade (off the Brazilian coast; not Trinidad) takes carrion, bones, and feathers avidly, along with some more bizarre items like motor oil and soap remnants. It also preys on hatchling sea turtles (Fimpel, 1975). Swarms of *G. planatus* on Clipperton rapidly reduce to a skeleton any living or dead animal that falls into their clutches (Ehrhardt and Niaussat, 1970).

Gecarcoidea natalis of Christmas Island resembles *Gecarcinus* in form and habits. It feeds primarily on fallen leaves, flowers, and fruits in the rain forest, but Hicks et al. (1984) also observed predation on the introduced giant African land snail, *Achatina fulcicata*.

B. Cannibalism

Cannibalism is probably common in the predaceous ghost crabs. It has been observed in *Ocypode ceratophthalmus* (Hughes, 1966) and *O. ryderi* (Vannini, 1976b). Juvenile *O. quadrata* are susceptible to pre-

Fig. 3.6. *Gecarcinus lateralis* "mobbing" a dead toad. This is a typical response to carrion in "herbivorous" land crabs.

dation by the nocturnal adults; Fisher and Tevesz (1979) regard the diurnal activity of juveniles as a mechanism for avoiding nocturnally foraging adults. Conspecifics make up about 6% of the diet of *O. quadrata* (Wolcott, 1978). Since ghost crabs are larger than most other prey items, the percentage of dietary biomass attributable to cannibalism is presumably higher.

Cannibalism is also common among the nominally herbivorous crabs, as it is among other terrestrial herbivores (Fox, 1975). *A. pisonii* is occasionally cannibalistic (Beever et al., 1979). A large *B. latro* was observed feeding on a living smaller individual, which it had presumably caught and dismembered (Helfman, 1973, 1977a, 1979). Tethered *Birgus* are attacked and eaten by free-ranging robber crabs, and captive crabs must be kept separate or larger individuals invariably kill and eat the smaller ones (Helfman, 1979). Small *Birgus* tend to restrict their activity during the period when large individuals are foraging (Helfman, 1973). When other foods are extraordinarily abundant, however, *Birgus* may feed gregariously, with up to 340 animals in an area of 120 m^2 under two fruiting palms (*Arenga listeri*; Hicks et al., 1984).

Large *Gecarcinus lateralis* will sometimes eat their fellows in the field (Bliss et al., 1978), as do *G. quadratus* on Clipperton (Ehrhardt and Niaussat, 1970). The actual "kill" has rarely been seen, as might be expected for an event that occupies so little of the total foraging time. The possibility that many of these crabs were merely scavenging cannot be excluded. However, *G. lateralis* readily captures and eats conspecific juveniles in the laboratory. The tendency toward cannibalism is strongly affected by the quality of other food available to the crabs (D. L. Wolcott and T. G. Wolcott, 1984). Early crab stages of *G. lagostoma* live under rocks, where they are "protected from cannibalism by adult Gecarcinids" (Fimpel, 1975). *Gecarcoidea natalis* will not kill adult conspecifics, but they do avidly consume those killed on roads (Hicks et al., 1984). Gibson-Hill (1947) asserts that when baby crabs are migrating inland, adults "take up their position in the line of a stream of young crabs and scoop them into their mouths with both claws," although Hicks et al. (1984) did not corroborate this observation.

Cardisoma guanhumi in Florida and *C. armatum* in Africa fight with each other and scavenge dead conspecifics; by extrapolation it has been surmised that they also indulge in cannibalism (Herreid, 1963; Bruce-Chwatt and Fitz-John, 1951). In the laboratory, *C. guanhumi* readily consumes conspecific juveniles (Fig. 3.7) (D. L. Wolcott and T. G. Wolcott, 1985, 1987).

Cannibalism of juveniles, if common and particularly if modulated by food quality, has important ecological consequences (see the following section).

C. Food as a limiting factor

Food is probably a limiting resource for herbivorous land crabs, at least at some times and in some places. Dense populations of *C. guanhumi* consume all available leaf litter (Herreid, 1963), denuding the ground and clipping all leaves from trees and shrubs to 15–20 cm above ground level (Fig. 3.8A). Experimentally supplementing the food supply by scattering leaf clippings reduces foraging activity (Henning, 1975a). Henning (1975a) deduces from this that the crabs are usually looking intensively for scarce sustenance. In the Virgin Islands, leaf litter is absent in areas with many *C. guanhumi* burrows (Fig. 3.8A). It is abundant in areas where there are few burrows (Fig. 3.8B), suggesting that factors other than availability of food restrict habitat usage. Similarly, on Aldabran islets where *C. carnifex* is present, the ground is bare; where crabs are absent, 10 cm of leaf litter has accumulated (Alexander, 1979). On Christmas Island, where *G. natalis* is abundant, leaf litter reaches 50–80% coverage on the rain forest floor at the end of the dry season, when trees have been shedding

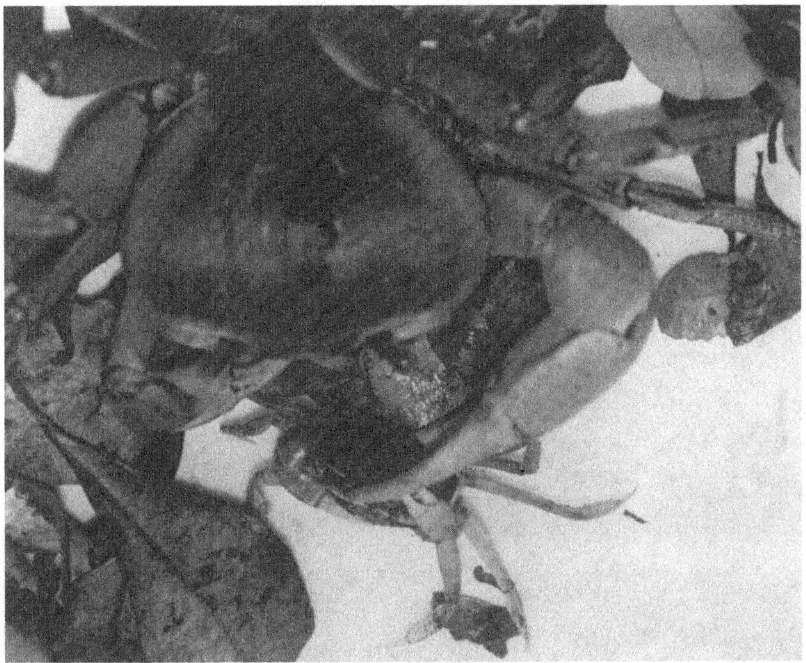

Fig. 3.7. Cannibalism by *Cardisoma guanhumi* under laboratory conditions. Juveniles were placed with an adult crab in a bucket, with leaf litter for food and cover. They were quickly captured and eaten in preference to the vegetable food. (Courtesy of D. L. Wolcott.)

leaves and the crabs have been relatively inactive. In the wet season, when crabs are most active, the forest floor is nearly bare (3–10% coverage), and the understory essentially absent (Hicks, 1983; Hicks et al., 1984). Hicks et al. (1984) quote an English visitor: "We can't get over the fact that one can walk so unhampered through the jungle, you have these useful little crabs keeping everything tidy!" In the Pacific, *Coenobita rugosus* is larger on islands with dense jungles. Page and Willason (1982) ascribe this to greater food availability, although greater protection from predation is also a possibility.

In other places and times, quantity (in contrast to quality) of food clearly is not limiting. *G. lateralis* on Bermuda, at least at Spittal Pond, has ample food; many burrows are hidden under thick stands of St. Augustine grass, a preferred forage plant (Fig. 3.9). Nonsuch Island, in contrast, has more closed canopy and the ground tends to be relatively clear (T. G. Wolcott, unpub.). Food was reported to be "virtually limitless" for the population of *G. lateralis* in Florida (Bliss et al., 1978). Britton et al. (1982) present no data, but also assume that

Fig. 3.8. Leaf litter availability in habitats supporting differing densities of herbivorous land crabs. A. Bare ground typical of areas densely populated by *Cardisoma guanhumi* (U.S. Virgin Islands National Park, St. John). Here availability of food is apparently a limiting factor. B. Leaf litter around an isolated burrow of *Cardisoma guanhumi* in an area of low crab density (U.S. Virgin Islands). Population density here is clearly not limited by availability of food. (Photograph by T. G. Wolcott.)

Fig. 3.9. Abundant food in habitat of *Gecarcinus lateralis* in Spittal Pond Nature Preserve, Bermuda. Grass in foreground and forest in background both provide preferred forage, and burrows are often concealed by lush grass. Availability of food does not appear to limit the density or distribution of this crab population. (D.L. Wolcott and T.G. Wolcott, 1984.)

food is normally not limiting for *G. lateralis* and *G. ruricola*. *B. latro* finds "abundant food" on Enewetak (Vogel and Kent, 1971), but this atoll may not be an entirely representative environment after suffering the ecological disturbances of nuclear explosions. *C. hirtipes* in Japan reportedly are active only about 2.3 hours each day (Goshima, Ono, and Nakasone, 1978); such limited foraging suggests ample food supplies.

Food quality may limit populations even where the quantity of food is ample. Availability of dietary nitrogen is increasingly being recognized as an important ecological factor. The nitrogen content of plant foods is low (typically 0.5–2% dry weight) relative to that of animals (7–14%) (Mattson, 1980). Limited gut volume and long gut cycling time (to 24 hours) limit the amount of food crabs can process, regardless of the quantity available (see Chapter 10). Given food with increased nitrogen content, *G. lateralis* can assimilate several times the maximum nitrogen that they could extract from an unlimited supply of their normal diet. The low availability of nitrogen in natural forage plants appears to limit growth rate, measured as accumulation of tissue (muscle + viscera) and total body nitrogen (D. L. Wolcott and T. G. Wolcott, 1984); the same appears true for *C. guanhumi* (D. L. Wolcott and T. G. Wolcott, 1985, 1987).

Herbivorous crabs appear to select high-nitrogen food under many circumstances. *G. natalis* prefers green leaves to dry (Hicks et al., 1984), and *C. carnifex* preferentially browses on the shoots of certain plants (Alexander, 1979). Shoots normally have a much higher proportion of nonstructural tissue, and thus a lower carbon:nitrogen ratio, than other plant parts (Mattson, 1980). Paradoxically, *C. guanhumi* in the Virgin Islands avoids green leaves of some plants (e.g., genips, *Melicocca bijugatus*) but readily consumes them as dry litter (D. L. Wolcott, unpub.). These plants may contain repellent allochemicals that are degraded as the leaves dry and begin to decompose. A preference for partially decomposed leaves may also be related to the increase of microbial nitrogen associated with "aged" detritus (e.g., Cundell et al., 1979). Insufficient data on the composition of items normally consumed by land crabs are available to test this hypothesis.

Vertebrate feces are another rich source of nitrogen. As mentioned earlier, *C. carnifex* on Aldabra eat bird droppings, and land hermit crabs quickly consume the fecal output of the atoll's giant tortoises (Alexander, 1979). On Bermuda dozens of *G. lateralis* have been observed clustered around human feces (T. G. Wolcott and D. L. Wolcott, unpub.), and Koepcke and Koepcke (1953) report finding bird feces in guts of *O. gaudichaudii.*

If nitrogen limitation is common among herbivorous land crabs, it may account for the frequent occurrence of cannibalism and its apparent modulation by the availability of other dietary nitrogen. Hypothetically, cannibalism of juveniles presents an appropriate feedback loop for population regulation in terrestrial animals with pelagic larvae. It operates at the level of recruitment rather than of reproductive output, which is particularly effective for broadcast spawners. Under conditions of locally poor food quality, restriction of recruitment will minimize competition between recruits and the reproductive population. Cannibalistic adults will not jeopardize their own reproductive output, since their larvae are likely to settle over a wide area. By the same token, juveniles that settle locally and are cannibalized are unlikely to be progeny of the cannibals. Cannibalism maximizes availability of nitrogen to the breeding population, including food materials acquired by the planktonic larvae in a habitat unavailable to adults. Regulation of land crab populations by such a mechanism could reduce intraspecific competition and the pressure to expand ranges inland. Helfman (1979) has advanced similar ideas based on his observation of cannibalism in *Birgus*. Young of *G. lateralis* and *C. guanhumi* have been found living in large burrows presumably occupied by adults (Feliciano, 1962; Klaassen, 1975), but we do not know the proportions of juveniles selecting (or avoiding) burrows of adults. In short, essentially no data are yet available to test theories about the ecological role of cannibalism.

4. Competition for food and habitat

In the literature on land crabs competition has not been precisely defined as "one crab denying a resource to another." Often it has been equated with agonistic displays or fighting. Such displays suggest that competition is occurring in many circumstances, but this has seldom been documented. Much of the following discussion deals with "aggression" rather than "competition" *sensu stricto.*

Lightly armored, agile terrestrial crabs tend to be much more aggressive than the heavily armored, slower species. Ghost crabs, for instance, readily fight and cannibalize each other in laboratory cages. They have developed various threat displays that actively maintain spacing and minimize damaging fights in nature (Evans et al., 1976; Fisher and Tevesz, 1979; Vannini, 1980a; see also Chapter 4). Gecarcinids and *Coenobita,* on the other hand, are relatively tolerant of each other. They may even feed gregariously when food is abundant (Chace and Hobbs, 1969; Grubb, 1971; De Wilde, 1973; Helfman, 1977a, 1979; Hicks et al., 1984) and can be kept in population cages (Gibson-Hill, 1947; Feliciano, 1962; Henning, 1975b). In the field, threat displays that serve to maintain individual spacing are often seen among gecarcinids, but fights are rare (Henning, 1975b), except among males contesting "copulation burrows" (Klaassen, 1975).

Interspecific interactions have been examined in very few species. In mangrove thickets, risk of predation is highest at the mud surface (Wilson, 1985). The habitat is partitioned vertically by a series of crabs, including the arboreal *A. pisonii,* on the basis of humidity gradients and differing physiological tolerances, but there is no evidence of competition for refuges from predators (Wilson, 1985). Semiterrestrial species of *Sesarma* with similar habitat preferences occur in pairs; where they are sympatric, habitat overlap is reduced. *Sesearma cinereum* occurs in dry supratidal locations in the subtropics, whereas *S. reticulatum* is found in adjacent wet microhabitats and has a more detritivorous diet. *S. ricordi* and *S. curacaoense* have a similar relationship in the tropics. These relationships imply resource partitioning and possibly competition, but concrete data are lacking (Abele, 1973).

On sandy shores, interspecific competition for food or habitat has been inferred by several investigators, but without quantitative data. Ghost crabs (*Ocypode cordimana, O. ryderi*) are asserted to compete "strongly" with other scavengers, particularly *C. rugosus,* on East African beaches (Vannini, 1975, 1976a, b). The hermit crabs are reportedly left with only the lower-quality food items. Vannini (1976a) attributes the inland migrations of *C. rugosus* to this competitive pressure. Competition between ghost crabs (*O. gaudichaudii*) and other scavengers (vultures, foxes, rats, occasionally *C. compressus*) in Peru has also been inferred by Koepcke and Koepcke (1953).

Both Vannini (1976b) and Braithwaite and Talbot (1972) comment on the common partitioning of tropical beaches between paired species of ghost crabs. One is predominantly extralittoral (terrestrial) and tends to be more carnivorous and aggressive; the other is predominantly supra/intertidal and relies more on detritus and deposit feeding. In the Indo-Pacific, for example, *O. cordimana* is typically found living above *O. ceratophthalmus*, *O. saratan*, *O. ryderi*, *O. macrocera*, or *O. platytarsis*. In West Africa, *O. africana* is found above *O. cursor*. In Central America, where *O. occidentalis* occupies exposed beaches, *O. gaudichaudii* is found principally on sheltered beaches and is diurnal (Crane, 1941a). In Peru, where *O. occidentalis* is absent, *O. gaudichaudii* is found in both habitats and is largely nocturnal (Koepcke and Koepcke, 1953). Koepcke and Koepke (1953) submit that the nocturnal, aggressive *O. occidentalis* displaces *O. gaudichaudii* both spatially and temporally. Similarly, Fellows (1975) suggests that *O. cordimana* displaces *O. laevis* on all islands other than Hawaii. There is no experimental evidence that these patterns of habitat partitioning are due to direct competition.

Another pair of potential congeneric competitors, *G. lateralis* and *G. ruricola*, are sympatric through much of the Caribbean. Toward the western part of the basin, *G. ruricola* becomes relatively rare and *G. lateralis* shows an increase in average size (Britton et al., 1982). Habitat suitable for burrowing is essential for *G. lateralis*. It is incidental for *G. ruricola*, which lives in forests with dense shade and usually does not burrow, retiring instead to crevices in rock or under logs, or to depressions in leaf litter. Britton et al. (1982) attribute the westward increase in size of *G. lateralis* to reduced competition with its congener for burrowing habitat, but present no data or observations to support this contention.

Intraspecific competition is also difficult to document, but aggressive interactions apparently related to resources are often taken as evidence that it occurs. Competition for shells may not be as intense among terrestrial hermit crabs as among their aquatic relatives. Abrams (1978) saw no fights for shells among *C. compressus* in the field in Panama, whereas aquatic hermit crabs frequently fight. Larger crabs tend to inhabit shells nearer their preferred size, i.e., shells that are greatly modified by previous hermit crab use. New shells have too little interior volume and are (very slowly) enlarged by hermit use.

Abrams (1978) suggests that rather than competition, terrestrial hermits show "shell facilitation"; that is, larger populations of crab generate, through wear, larger numbers of shells suitable for adult crabs. He accounts for limited adult populations by limited "recruitment" of entry-level shells, or possibly by predation on or food competition between adult hermits. On the other hand, De Wilde (1973) showed that coastal *C. clypeatus* on Curaçao were small, poorly housed, and thereby restricted to suboptimal habitat. Crabs with well-fitting, large shells were able to exploit inland environments. This suggests that large, intact shells are a limiting resource conferring adaptive value, and that competition for them may be occurring despite De Wilde's (1973) observation that fighting and shell exchanges never occurred on a large scale.

Burrows, as favorable microenvironments in a hostile world, are also a potentially limited resource. Competition for them may be expected if their construction requires substantial investment of energy, or if suitable sites are rare. Competition clearly occurs in ghost crabs (*O. quadrata*), which prefer taking an existing burrow to constructing a new one (Fimpel, 1975). Large individuals often enter a slightly undersized burrow, then back out and allow the smaller occupant to escape. The new resident proceeds to enlarge the burrow slightly, and the evictee goes and steals a burrow from another crab somewhat smaller than itself. This sequence can be repeated several times, with successive decreases in the size of the participants (T. G. Wolcott, unpub.). Spacing between burrows of adult *O. quadrata* is established by threat displays (Fisher and Tevesz, 1979); this is also true of *O. kuhli* (probably *ryderi*) (Evans et al., 1976; Vannini, 1980a). The concept of "territoriality" is sometimes attached to these interactions over burrows (e.g., Vannini, 1980a). This is not strictly appropriate, since little burrow fidelity is shown. Instead, there is an "individual space" that is associated with the currently occupied burrow. Monopoly of a resource (space) has been observed in one instance in which a male *O. ceratophthalmus* drove away another male constructing a copulation burrow/mound close by, and filled in the burrow (Lighter, 1974). Male *C. guanhumi*, like ghost crabs, defend the area surrounding their current burrows, but remove to another burrow frequently (Henning, 1975b).

Fighting over burrows is uncommon among *G. lateralis*, except among males near the shore during the breeding season (Klaassen, 1975; Wolcott and Wolcott, 1982a; Fig. 3.10). Klaassen (1975) refers to the nearshore zone as a "copulation zone" and asserts that males maintain the burrows as sites to which they can attract females for copulation. Inseminated females show sperm plugs (Bliss et al., 1978) and are unreceptive to further courtship (Klaassen, 1975), so this

Fig. 3.10. Combat by male *Gecarcinus lateralis* over burrows during the breeding season, near the shore of Bermuda. (Photograph by T. G. Wolcott.)

could be an effective mechanism of reproductive competition. However, the concept of females copulating immediately after shedding larvae is difficult to reconcile with the lengths of time between copulation, ovulation, and larval release reported by Klaassen (1975); the totals do not coincide with the length of the lunar breeding cycle. Copulation by *G. lateralis* occurs far inland but has not been observed in the "copulation zone" on Bermuda (T. G. Wolcott and D. L. Wolcott, unpub.) or in southern Florida (Bliss et al., 1978). Males of *Gecarcoidea natalis*, like those of *G. lateralis*, also excavate and defend burrows during breeding migrations (Hicks, 1985). Excavation of some "copulation burrows," and tracking of females after release of larvae, could elucidate the role of male–male combat over burrows in these species.

Among other land crabs, data regarding competition for burrows are largely negative. *B. latro* were not seen fighting over burrows (Vogel and Kent, 1971), nor do any authors mention competition in *Coenobita* spp. for their shallow holes or other refuges. In the only quantitative study of competition among semiterrestrial crabs, Wilson (1985) showed that mangrove crabs sustain higher risks of predation only after 25% of the burrows in a quadrat are obliterated, and argues that burrows are normally present in excess and therefore are not a limiting resource.

Competition for food is often assumed but seldom demonstrated. Vogel and Kent (1971) state that *B. latro* on Enewetak have "abundant food," yet Helfman (1977a) found that the crabs would engage in aggressive interactions over tethered coconuts. Ghost crabs tend to exclude each other from choice food items (T. G. Wolcott, unpub.), but it is impossible to distinguish this from the aggression that results in individual spacing and occasional cannibalism. It is not clear that food is limiting for ghost crabs nor that they compete for it in the classical sense of the word (Wolcott, 1978). Among gecarcinids, there is even less evidence for intraspecific competition for food. These crabs are remarkably placid in the presence of conspecifics. Up to 20 *C. guanhumi* have been observed pushing each other aside to feed on an opened coconut; in another instance 2 individuals continued feeding on the same twig, attempting to carry it off in different directions. No fights erupted in either case (Henning, 1975a). Very similar behavior occurs in Bermudian *G. lateralis* feeding on carrion (T. G. Wolcott, unpub.).

On the other hand, the apparent limitation of food for some populations of herbivorous crabs (*G. natalis, Cardisoma* spp.) implies that the crab that harvests a fallen leaf denies that food to other individuals, whether or not aggression is expressed. Most competition for food among land crabs, if it occurs, is "exploitation competition" (appropriating a resource before the other animal can do so) rather than "interference competition" (reserving a resource by interfering with the other animal's access to it).

5. Reproductive ecology

Reproduction in terrestrial land crabs has a number of common themes (also see Chapter 5). Discussion of these themes will be followed by specific examples of the better-known species.

A. Reproductive themes

Copulation occurs while both individuals are in intermolt (hard-shelled), often in or near burrows of males in *B. latro* and *C. perlatus* (Helfman, 1977b), *A. pisonii* (Warner, 1967), *G. lateralis* (Bliss, 1968; Klaassen, 1975; Bliss et al., 1978; T. G. Wolcott and D. L. Wolcott, punpub.), *C. guanhumi* (Henning, 1975b), *G. natalis* (Hicks et al., 1984; Hicks, 1985), and *O. quadrata* (Hughes, 1973; T. G. Wolcott, unpub.). Since the female need not be in a soft-shelled postmolt stage, mating is no longer dependent on a supportive aqueous medium.

Sex ratios in natural populations have been examined in only a few cases. In East African ghost crabs sex ratios are 1:1 (Vannini, 1976b), but disproportionate numbers of females have been reported for some populations of fiddler crabs (Frith and Brunenmeister, 1980), sand bubblers (*Scopimera globosa*) (Yamaguchi and Tanaka, 1974), and ghost crabs (*Ocypode quadrata*) (Leber, 1982). Estimation of sex ratios requires careful sampling; Hall (1982) found very different sex ratios in samples of ghost crabs collected several weeks apart at the same site, suggesting differences between sexes in temporal patterns of activity on the beach.

The larvae of terrestrial crabs of marine origin require high-salinity water for development (Costlow and Bookhout, 1968). This obligates at least the female adults, no matter how terrestrial their normal habitat, to migrate to permanent, high-salinity water to release larvae (Fig. 3.11). Breeding is typically cyclic, with lunar or semilunar rhythms peaking near the spring tides. Near the tropics reproduction may occur year-round (e.g., Jamaican mangrove crabs; Warner, 1967), but more typically it is seasonal, dependent on variation in rainfall. In temperate areas reproduction occurs during the warm months. Ovulation may occur several times between molts. Incubation of eggs requires about two weeks. Ovigerous females tend to be secretive; those of burrowing species usually remain underground. As hatching approaches, ovigerous crabs migrate to the shore. Larvae are often released near the highest tides of the lunar cycle. By some accounts, females copulate again immediately after releasing larvae.

Migrations of ovigerous females to the sea to release larvae can be as short as the distance from upper beach to surf (ghost crabs), or as long as 8 km (*Cardisoma guanhumi*). Frequently, migrations are directed toward only a few sites of the coast (Gibson-Hill, 1947; De Wilde, 1973). If these restricted localities are the only suitable settling sites, how do the late-stage larvae avail themselves of appropriate current regimes to return there? If they do not, and settling is widespread, are the animals settling at these sites the only ones to survive and spawn in succeeding years, or do crabs that settled elsewhere "learn" the traditional spawning migration routes? Answers to these questions await data on the distribution of settlement and juvenile survival, and on the ontogeny of migration behaviors.

The orientation or navigation mechanisms involved in the longer migrations have not been studied in detail. Potential cues include polarized light (Daumer, Jander, and Waterman, 1963), brightness

Fig. 3.11. Ovigerous *Cardisoma guanhumi* migrating to water to shed larvae, U.S. Virgin Islands National Park. (Photograph by T. G. Wolcott.)

of the horizon over the sea (Vannini, 1975; Bliss et al., 1978; T. G. Wolcott and D. L. Wolcott, 1982a), prevailing winds, and unspecified "celestial" cues (Vannini, 1975; Vannini and Chelazzi, 1981; also see Chapter 4). In many species, adult males also participate in migration; in the gecarcinids at least some mating occurs at the shore. The participation of subadult *C. guanhumi* in "nonbreeding" migrations (Gifford, 1962a) remains a puzzle.

Migration entails risks to both adults and larvae. Spawning females are liable to be swept into deep water and attacked by shore-feeding fishes (Gibson-Hill, 1947; Warner, 1967; Henning, 1975b; Alexander, 1979; Hicks et al., 1984). They may fall from sea cliffs and be crushed (Hicks, 1985) and, in developed areas, may be killed by road traffic, especially where their orientation mechanisms result in behavioral trapping under street lights (T. G. Wolcott and D. L. Wolcott, 1982a). Hicks et al. (1984) estimate the annual road kill of *G. natalis* on Christmas Island at 600,000 crabs, but this represents less than 0.07% of the total adult population! Eggs are susceptible to premature hatching when exposed to standing water or drenching rain (Klaassen, 1975). Ovigerous *G. lateralis* minimize such risks by migrating to the shore in stages, during the night (presumably using burrows as way stations), and by avoiding all standing water other than the sea itself (T. G. Wolcott and D. L. Wolcott, 1982a). Only *G. natalis* is reported to travel principally by day, perhaps because of a preponderance of nocturnal predators on Christmas Island (Hicks, 1985).

The terrestrial crabs derived of freshwater origin (Parathelphusidae, Pseudothelphusidae, Sundathelphusidae) tend to have abbreviated or direct development, in which highly developed larvae or miniature crabs emerge from eggs (Rabalais and Gore, 1985; also see Chapter 5). Little is known about their reproductive patterns; many live in areas where reproduction and surface activity are probably regulated by seasonal rainfall (e.g., *Sudanonautes africanus*: Lutz, 1969; Bertrand, 1979; *Potamon potamios*: Warburg, Goldenberg, and Rankevich, 1982; Warburg and Goldenberg, 1984). An extreme case is provided by the Australian arid-zone crab *Holthuisana transversa*, which may remain inactive in its burrow for years between sporadic floods (Greenaway and MacMillen, 1978; MacMillen and Greenaway, 1978; Greenaway, 1984b).

Only a few land crabs arising from marine progenitors have abbreviated development. *Sesarma jarvisi*, a grapsid inhabiting talus slopes and rock rubble between 300 and 900 m altitude on Jamaica, has exceptionally large eggs (> 1 mm), which hatch into highly developed zoeae. The larvae are thought to develop rapidly among moist litter and rocks (Abele and Means, 1977).

Most land crabs have the same sort of aquatic larval development as their marine ancestors, with several zoeal stages, and come ashore as megalops or first crab stages. One notable exception is *Uca subcylindrica*, a fiddler crab found up to 35 km inland along the courses of intermittent streams in Texas (Thurman, 1984; Rabalais and Cameron, 1985a). It releases larvae into ephemeral pools and has a markedly abbreviated development (to first crab stage within 8 days of hatching; Rabalais and Cameron, 1983) compared with other fiddlers. Development from hatching to settling typically requires four to six weeks in land crabs: e.g., one month for *A. pisonii* (Warner, 1967), 30–40 days for *C. guanhumi* (Costlow and Bookhout, 1968), 27 days for *G. natalis* (Hicks, 1985). Most mortality takes place during the planktonic larval phase of land crab life cycles. Warner (1967) estimates that of all *A. pisonii* larvae released into Jamaican mangrove swamps, only 0.04% survive to settlement, and (assuming stable populations) .007% to reproductive age. In other words, 99.97% of mortality occurs in the planktonic stages.

Repopulation of local stocks by planktonic larvae presents problems in areas of unidirectional ocean currents, especially isolated islands. Bliss et al. (1978) invoke eddies to account for larval recruitment in Florida, but the small-scale oceanographic studies necessary to address this question have not been done.

After the megalops come ashore and metamorphose into small land crabs, they are seldom seen, and little is known about their habitat or ecology. A few small *G. lateralis* and *G. ruricola* can be found under cobbles and supratidal stones (Bliss et al., 1978; Britton et al., 1982), and small *G. lagostoma* are widely distributed in crevices and under rocks (Fimpel, 1975).

Juvenile *A. pisonii* and several other mangrove crabs can be found in the burrows of other species or in "runways" under litter on the swamp floor (Warner, 1969). Juvenile *C. guanhumi* (< 1 cm carapace width) reportedly do not excavate their own burrows, and are found in those of the adults (Feliciano, 1962; Henning, 1975a). If this is their normal microenvironment, cannibalism cannot be a common occurrence. Juveniles in adult burrows may be exceptions and most may live elsewhere. In the Virgin Islands, only juveniles larger than 1.5 cm carapace width, occupying their own burrows, occur in the adult habitat (D. L. and T. G. Wolcott, unpub.). Juveniles can be collected by raking up leaf litter from low supratidal areas adjacent to mangroves near Mayaguez, Puerto Rico (H. Rojas, University of Puerto Rico, pers. comm.). Juveniles of *C. perlatus, C. rugosus,* and several species of *Ocypode* live closer to the sea than adults (Vannini, 1976a, b; Shuchman and Warburg, 1978; Page and Willason, 1982), presumably because of their high surface:volume ratios and consequently high susceptibility to evaporative water loss. However, no such correlation is evident in other species of *Ocypode* (Barrass, 1963; Jones, 1972; Vannini, 1976b; T. G. Wolcott, unpub.), and this hypothesis is difficult to reconcile with the greater diurnal activity, relative to adults, of juvenile *Ocypode* (Fisher and Tevesz, 1979) and *Coenobita* (Abrams, 1978).

In some species, juveniles and adults show marked ecological differences. *O. gaudichaudii* begins life as a predator and scavenger, with the pointed chelae characteristic of predators. As an adult, it develops truncated, shovel-shaped chelae and becomes a deposit feeder (Crane, 1941a). *S. africanum* burrows in marshy savannah as an adult, but crabs smaller than 27 mm carapace width were never collected from burrows and are thought to be aquatic (Bertrand, 1979).

B. Ecological aspects of reproductive behavior: case studies

The mangrove crab *A. pisonii* breeds year-round on Jamaica (Warner, 1967). Copulation occurs on mangrove roots. Females ovulate at least once after each postpubertal molt; molt interval averages 53 days. The eggs are incubated for 16 days, and go through a series of color changes. As hatching approaches, ovigerous females migrate toward the seaward fringe of the mangroves. Hatching begins in air, and the females then descend the mangrove roots to the water, immerse themselves, and fan the abdomen to shake off the larvae.

The red crab of Christmas Island (Indian Ocean), *G. natalis*, has been extensively observed (Hicks et al., 1984; Hicks, 1985). Like other large species with (presumably) long intermolt periods, *G. natalis* appears to ovulate more than once during intermolt. Multiple ovulations in intermolt are possible because mating in land crabs does not need to closely follow a molt. Breeding migrations from rain forests of the interior plateau begin with the onset of the rainy season. If the rains fail, little reproduction is seen in *G. natalis*. Up to three breeding cycles occur during a single season. In each cycle, the crabs migrate to the coast of the island en masse, and both sexes spend time "dipping" in the sea (perhaps to replenish salt stores). They then retire to the shore terrace where males excavate and compete for burrows. Copulation takes place in or near these burrows. Males "dip" again after the mating period, then return inland. Females stay in the shore terrace burrows for 12–13 days, ovulating and incubating eggs. Larvae are released at the turn of the high tide between the last quarter and new moon. The first crab stages come ashore some 27 days later, and take about 9–10 days to reach the rain forest on the island's interior plateau (Hicks, 1985).

Courtship and mating of *G. lateralis* in Colombia (Klaassen, 1975), Florida (Bliss et al., 1978), and Bermuda (T. G. Wolcott and D. L. Wolcott, 1982a) are similar to those of *Gecarcoidea lalandi*. Large adult males migrate to the shore in May, at the onset of the rainy season in Colombia, or July/August on Bermuda, where precipitation is not very seasonal. The male crabs construct burrows in the backshore ("*Ipomoea* zone" of Klaassen, 1975) or in the sand patches at the top of Bermuda's cliffs. Fighting over burrows in both areas is common. Copulation apparently can occur either in this zone (Klaassen, 1975) or inland (Bliss et al., 1978; T. G. Wolcott and D. L. Wolcott, 1982a). In Florida, following an undisturbed copulation in dunes back of the beach, the female crab went down a nearby burrow and the male moved away (Bliss et al., 1978), suggesting that a male's "copulation burrow" (Klaassen, 1975) is not always involved. Females bearing eggs appear up to several hundred meters inland in the days preceding full moons. Eggs on females far inland are usually brown (immature), whereas those near the shore are usually gray (near hatching), suggesting that the females have been incubating eggs in burrows well inland (T. G. Wolcott and D. L. Wolcott, 1982a), not exclusively in the "copulation zone" of Klaassen (1975).

Egg release by *G. lateralis* is highly synchronous, occurring within a few hours of midnight on the days following full moons (Klaassen, 1975; T. G. Wolcott and D. L. Wolcott, 1982a). Females move slowly down the beach or cliffs, approaching the water with apparent caution. On beaches they advance until shallow waves wash over them. Hatching of their 19,000–109,000 eggs (Klaassen, 1975) begins immediately, and the females rise up on their dactyls and violently vibrate their bodies, shaking off the hatching zoeae. On cliffs, females advance to wet areas, grip the rock tightly, and wait. If no splash comes to their position, they advance farther and wait again, repeating the sequence until shallow splash washes over them. At that moment they rise up, shake the body, and fan the abdomen as on sandy shores. After shedding larvae, all females move briskly upshore and inland (Klaassen, 1975; Bliss et al., 1978; T. G. Wolcott and D. L. Wolcott, 1982a).

Reproductive behavior in *G. lagostoma* differs from the preceding examples in that only the females migrate to the shore (Fimpel, 1975). *G. planatus* on Clipperton resemble *Gecarcoidea* in their behavior in that both sexes migrate to the shore, where mating occurs. Males then return inland, and females remain out of sight, presumably in burrows, until the eggs mature and larvae can be released into the sea (Niaussat and Ehrhardt, 1968).

Both sexes of *C. guanhumi* begin activity at the onset of the rainy season (late May in Puerto Rico). In the following months (June-Sept. in Puerto Rico) females establish a lunar breeding cycle with major migrations preceding full moons and minor ones preceding new moons. This pattern is maintained it until the weather grows dry or cold (Feliciano, 1962; Gifford, 1962a; Henning, 1975b). *C. hirtipes* and *C. carnifex* also breed at full moons in the rainy season (Gibson-Hill, 1947; Grubb, 1971). Copulation in *C. guanhumi* occurs outside burrows 1–2 days before full or new moon (Henning, 1975b; based on three occurrences). Females can copulate two (Henning, 1975b) to five times (Taissoun, 1974a) between molts, males many times. Ovulation follows copulation by 5–10 days (Taissoun, 1974a). Eggs number 20,000–1,200,000, depending on the size of the female (Gifford, 1962a; Taissoun, 1974a; Henning, 1975b). Females ovulate only after copulation (Henning, 1975b). Although Taissoun (1974a) found live sperm long after copulation and ovulation, his conclusion that a single mating can suffice for the reproductive lifetime of a female *C. guanhumi* appears unwarranted (also see Chapter 5). In the first four days after ovulation, eggs are susceptible to osmotic damage, and ovigerous females avoid water. Thereafter, they forage normally and even enter water, unlike the ovigerous females of other species. Development of fertilized eggs requires about 16 days (Henning, 1975b).

Cardisoma guanhumi makes exceptionally long migrations; the adults are found up to 8 km inland in lowlands along water courses (Gifford, 1962a; Fimpel, 1975). Single individuals occur near pools in beds of intermittent streams to about 200 m above sea level on St. John, U.S. Virgin Islands (T. G. and D. L. Wolcott, unpub.). Shedding of larvae is similar to that in *G. lateralis* (Taissoun, 1974a; Henning, 1975b). Larval development (in the laboratory) requires 30–40 days (Costlow and Bookhout, 1968).

Ghost crabs initiate oogenesis during winter inactivity, and female *O. quadrata* emerge with developed ovaries (Haley, 1972). Copulation occurs on the beach surface, and presumably within burrows, throughout the warm months (Hughes, 1973; T. G. Wolcott, unpub.). In several species, there is sexual zonation on beaches (reviewed in Vannini, 1980b), with females living higher than males. This probably facilitates mating, because receptive females must run the gauntlet of males to reach the lower beach and the water.

Ovigerous ghost crabs are usually very secretive, and most workers have seen only two or three specimens. Recent fieldwork (Dan Rittschof, Duke University, unpub.) has shed new light on this issue. Copulating females, tagged with string streamers in late July and early August, retired to burrows and allowed the openings to be obliterated by shifting sand. The burrows were not reopened until evening high tides on the few days near full or (to a lesser extent) new moon. When the tides were at their highest, about 30 minutes after sunset, numerous ovigerous females emerged and moved directly to the water. The egg masses dropped off instantly when the crabs became wet. These brief, highly synchronous pulses of spawning occupied less than an hour, which accounts for the paucity of previous observations. Larvae on ovigerous crabs carried in dry buckets to the laboratory were found to have egg membranes already ruptured. The larvae appeared desiccated, and were held in place by the capillary attraction, and perhaps viscosity, of the surrounding fluids. As soon as the adult female was dipped or shaken in seawater, the larvae fell off. They appeared to require hydration; at first they sank to the bottom of the vessel, but gradually became capable of swimming over the next 15–30 minutes. Ovigerous *O. ceratophthalmus* probably behave similarly; Fellows (1973) found them only in deep, plugged burrows. Ghost crabs in Texas may behave differently; Haley (1972) reports that the proportion of ovigerous females collected (by an unspecified method) on Padre Island (26–28° N, 97°20′ W) rose through the summer, to nearly 100% in August. Ovigerous *O. gaudichaudii* emerge only at dusk and go directly to the water to "drink"; Koepcke and Koepcke (1953) suggest that this is because distension of the abdomen with eggs prevents normal water uptake via the setal tufts (see Chapter 7).

Larval development of *O. quadrata* to the settling stages (megalops or first crab) requires about five weeks in the laboratory (Diaz and Costlow, 1972). From assumed molt intervals and increments, ghost crabs probably spend their first year as juveniles (Haley, 1969, 1973). Females may live up to two more years, producing two broods each year in Texas; there appears to be a major peak in ovulation during April and a less synchronous one in August (Haley, 1972). Population structure suggests two broods in North Carolina as well (T. G. Wolcott, 1978). African ghost crabs may have similar cycles; Vannini (1976b) found megalops in June (spring brood?), but observed copulation only in August (fall brood?).

Reproduction in coenobitids has received little attention in recent years. *B. latro* copulates on land quickly and without the ceremony common among smaller, less formidable aquatic hermit crabs (Helfman, 1977b). It carries unripe eggs in September (spring) on Enewetak (Vogel and Kent, 1971). Larval stages last 23–30 days, and the glaucothoes that come ashore use snail shells, like other hermit crabs, for the first three to four weeks of their lives. Settling in the laboratory is promoted by more complex, natural "beaches" including sand, coral rubble, coconut husks, and coconut meat (Reese, 1968).

Other land hermit crabs apparently go to the shore prior to mating, although copulation was not observed in De Wilde's (1973) extensive field study. Both sexes of *C. clypeatus* on Curaçao migrated between July 21 and August 14, moving 100–500 m per night toward a small number of coastal breeding sites (De Wilde, 1973). The breeding migration ("mamba") on Mona Island (between Hispaniola and Puerto Rico) is reported to be much more synchronous, with millions of crabs flowing down established tracks to the breeding beaches during about three days at the first quarter moon in August (Carpenter and Logan, 1945; M. Olson, Yale University, pers. comm.). On Curaçao, females carrying 1,000–50,000 eggs were first seen at the shore on August 14, and about 30 days later were moving toward the sea at low tide to drop or fling their eggs onto the wet rocks. Spawning females did not enter the water. Another two reproductive cycles followed, lasting through mid-December. Numbers of crabs at the shore gradually decreased as individuals straggled back inland (De Wilde, 1973). Larval development requires about 26 days to the glaucothoe and at least an additional month to first crab (the presumed settling stage; Provenzano, 1962). On Curaçao, *C. clypeatus* matures in the second year.

6. Growth and longevity

Estimation of growth rates and longevity requires knowledge of molt frequency and increment (also see Chapter 6). These data are lacking for most terrestrial crabs. Males often reach larger size than females; scanty information relates this to decreased growth of females due to energetic investment in reproduction in some species (e.g., sand bubblers *Scopimera globosa*: Yamaguchi and Tanaka, 1974). In others, molt interval and increment appear similar for both sexes, and smaller female size is attributed to reduced longevity due to the risks of breeding migrations (e.g., *C. guanhumi*: Herreid, 1967; Henning, 1975b).

Molting itself (natural or induced) has been observed in several species under laboratory conditions. *C. guanhumi*, and probably its congeners, require total immersion (Henning, 1975b) and maintain burrows sufficiently deep to hold a pool of water in the bottom. *G. lateralis* stores water (as increased hemolymph volume) in extraordinarily large pericardial sacs (Bliss, 1963; Mason, 1970) and is able to molt on dry land (see Chapter 7). This normally takes place underground; small crabs have been found with their recently cast exuviae in shallow burrows under rocks on Bermuda (T. G. Wolcott and D. L. Wolcott, unpub.).

Molting behavior and requirements of ghost crabs are poorly known. It seems unlikely that ghost crabs require immersion, given the impossibility of extending their sandy burrows below the water table and the risk of physical damage or predation if molting occurred in the surf. On the other hand, the normal mechanism for taking up water, presumably required in quantity for expanding the new exoskeleton, involves production of a partial vacuum (to 70 mmHg below ambient) in the branchial chambers (T. G. Wolcott, 1976, 1984). It is difficult to conceive how this could work when the exoskeleton is soft. Only two *O. quadrata* (of hundreds, maintained for months) ever attempted ecdysis on damp sand in our laboratory. Both failed to expand the new exoskeleton and eventually died. Perhaps water is acquired and stored in pericardial sacs preparatory to ecdysis, even though the sacs in *O. quadrata* are much smaller than in *G. lateralis* (Bliss, 1963). Pericardial sacs of the highly terrestrial *O. cordimana*, like those of *G. lateralis*, swell 5–10 days before molt, whereas those of the intertidal *O. platytarsis* and *O. macrocera* swell only during ecdysis (Rao, 1968, conducted with destalked crabs). The water storage is thought to be a terrestrial adaptation permitting *O. cordimana* to molt in a dry burrow. Several molting *O. cordimana* and *O. ceratophthalmus* were found in deep, plugged burrows by Fellows (1973), and an *O. quadrata* was found similarly engaged in a burrow by Henning and

Klaassen (1973) in Colombia. Molting without immersion seems to be the norm in ghost crabs, but the mechanisms need further study.

Data on relative growth of various body dimensions indicate that females mature at about 33 mm carapace width and males at 23–25 mm in *O. quadrata* (Haley, 1969) and females at 29–33 mm and males at about 27 mm in *O. ceratophthalmus* (Haley, 1973). Without molt staging, Haley (1972) was only able to estimate molt frequency and increment; from these he extrapolated a two- to three-year lifetime.

Growth rate and longevity have been estimated for *Aratus pisonii* by Warner (1967) in the only study using mass sampling and molt staging of land crabs in the field, coupled with measurements of molt increments in the laboratory. Crabs molt an average of every 53 days; molt frequency and increment decrease with increasing size. The mangrove tree crab lives a maximum of two to three years, but most crabs in the population are a year or less old.

Highly synchronous recruitment of *natalis* in one year allowed Hicks et al. (1984) to follow a cohort for three years; they found that carapace width in the first year class measured 12–20 mm; in the second 18–25 mm; and in the third, 28–33 mm. Females differentiate at 26–28 mm. Crabs smaller than 35 mm are seldom seen on the surface; adults 5 cm and larger must be several years older.

Growth and longevity data for the commercially important *C. guanhumi* are based on a few molt increments observed in the laboratory, and broad assumptions about molt frequency and trends in molt increment as a function of size. From the clean appearance of many crabs on spring emergence in Florida, Gifford (1962a) concludes that molting takes place during winter inactivity. Taissoun (1974a) also suggests that Venezuelan *C. guanhumi* larger than 3 cm molt annually in their winter burrows. In Puerto Rico, crabs begin to seal their burrows in November, and by January 90% have closed themselves in. Crabs dug from sealed burrows may have brittle, decalcified shells (Feliciano, 1962). Clean crabs emerge in April-May, and Feliciano (1962) also believes that molting occurs annually. He measured increments of 11 molts in the laboratory. Extrapolation, assuming an annual molt with size-specific molt increments from his figure, gives an estimated age of nearly 20 years for a 9 cm *C. guanhumi*.

Numerous molts were observed under more natural holding conditions in Colombia (Henning, 1975b). Molting can occur year-round at intervals of 20–200 days (depending on size of crab), and requires immersion in water within the sealed burrows. Molting is weakly correlated with lunar phases. A 4 cm crab spent 45 days sealed in its burrow, without feeding, prior to molt. Ecdysis required 3.5 hours. Over the next 1.5 days the crab hardened and ate the exuviae, which, along with some access to brackish water, is necessary to provide adequate calcium for full hardening of the new shell. Molt increment is highly variable. Henning (1975b) estimates that females at maturity (35–40 mm) are 3.75 years old, while a 98 mm female is 13 years old. These ages are based on an assumed two molts per year, with an average increment of 5%, but Henning (1975b) suspects they are underestimates.

From Henning's (1975b) data, A. H. Hines (Smithsonian Environmental Research Center, unpub.) has calculated that 40–45 molts would be required to reach maturity, and more than 60 to reach the maximum observed size in *C. guanhumi*. If molting is indeed annual, these are extremely long-lived crabs. In a large assortment of other crabs, the norm is 14–15 molts to maturity and about 19 to maximum size. Field studies of molt intervals and increments are needed to determine whether the herbivorous land crabs, with their poor-quality diet, truly have such anomalously low growth rates.

Land crabs store large quantities of lipids in the hepatopancreas, perhaps representing an adaptation to the variability of terrestrial environments. Unfortunately, few comparative data are available. Charles Darwin (cited in Reyne, 1939) remarked on the fact that over a liter of oil could be rendered from a large *B. latro*. The hepatopancreas of this animal contains up to 83% lipid (Lawrence, 1970; Storch, Janssen, and Cases, 1982), becoming particularly fat prior to molt (Wiens, 1962). Land crabs may rely heavily on a "lipid economy." Lipid biosynthesis increases markedly prior to ecdysis (O'Connor and Gilbert, 1968) concurrent with the degradation of muscle (particularly in chelae) that permits extracting the limbs through narrow joints in the old exoskeleton (Skinner, 1966b). Subsequent regeneration of muscle, and growth of new muscle tissue, will require nitrogen sources if based on stored lipids (see the previous discussion of dietary nitrogen).

7. Role of land crabs in food chains: predators and human exploitation

Anecdotal reports mention several potential predators on terrestrial crabs. Shore-feeding fishes take adult *C. carnifex* and *G. natalis* that enter the sea during breeding (Alexander, 1979; Hicks, 1985), and mangrove fishes take *A. pisonii* that leap from trees to avoid arboreal predators during high tide (Warner, 1967; Beever et al., 1979). Terrestrial crustaceans themselves are at least potential predators; cannibalism has been observed in several species (see section 3.B), and ghost crabs and *Geograpsus crinipes* eat small *C. rugosus* on Aldabra (Alexander, 1979). Hicks et al. (1984) suspect that *B. latro* is an important predator on *G. natalis*, biasing the behavior of the latter toward diurnal activity. Various terrestrial birds and mammals are mentioned as actual or potential predators. Land hermit crabs are important in the diet of Aldabran flightless rails (Alexander, 1979). *C. guanhumi* is preyed on by raccoons (*Procyon cancrivorus*), ocelots (*Felis pardalis*), foxes (*Cerdocyon thous*), "black hawks" (*Hipomorphus urubitinga*), caracaras (*Caracara plancus cheriway*), and various herons (Taissoun, 1974a; Henning, 1975a). Yellow-crowned night herons (*Nyctanassa violacea*) can eat 10 *G. lateralis* per day (D. Wingate, Bermuda Agriculture and Fisheries, pers. comm.). *O. ceratophthalmus* (juveniles?) are taken by turnstones and plovers on Aldabra (Alexander, 1979).

In Peru, *O. gaudichaudii* are taken as juveniles by various gulls and shorebirds, and by the iguanid lizard *Tropidurus peruvianus*. A nocturnal mustelid (*Conepatus* sp.) is a specific predator on adult *O. gaudichaudii*, digging into burrows and eating the crabs underground. Occasional predators on adults include large gulls, rats, domestic pigs, and foxes (*Canis [Dusicyon] sechurae*) (Koepcke and Koepcke, 1953). In Central America, heavy predation by another "black hawk" (*Buteogallus anthracinus*) forces *O. gaudichaudii* to "school" and forage on the backshore near their burrows rather than on the richer foreshore (Sherfy, 1984). *O. gaudichaudii* is the only brightly colored ghost crab (hence one common name, "painted ghost crab") and sports a "spectacle" pattern on the carapace. Koepcke and Koepcke (1953) speculate that this is warning coloration, but there is no evidence that the crab is either distasteful or better-armed than the cryptic ghost crabs.

Various land crabs (*G. lateralis*, *G. quadratus*, the freshwater crab *Potamocarcinus richmondi*) readily autotomize chelae after clamping them onto threatening objects; this "attack autotomy" is thought to be a deterrent against predators (Robinson, Abele, and Robinson, 1970).

Humans are among the most important predators for some species. There is excellent historical precedent; East African baboons dig up ghost crabs by hand, or by using cuttlebones as shovels (Messeri, 1978). Ghost crabs are occasionally used as food and bait in India and Pacific islands (Vannini, 1976b), and hermit crabs are used as bait on Curaçao (De Wilde, 1973) and Fiji (W. Burggren and A. Smits, unpub.). *B. latro*, by virtue of its large size, excellent meat, and fatty abdomen, is highly esteemed as food and consequently is rare on all inhabited islands. Some caution is required: If *Birgus* has been feeding on toxic plants, it may itself become poisonous (Reyne, 1939; Cameron, 1981a).

Gecarcinids are heavily exploited and economically important in some areas. *C. guanhumi* is under heavy pressure on Puerto Rico; in the 1950s it represented 7% of all Puerto Rican fisheries. Crabs are caught by hand or in traps during the active season. They are held

A

two days to a month and fed corn (maize) to "purge" and fatten them. If this is not done and they have been feeding on "manzanilla" (*Wedelia trilobata*), they are said to be bitter (Feliciano, 1962). They are sold in strings of 12 along roadsides (Fig. 3.12A) or to processors. Packing houses in Puerto Rico import additional crabs from Venezuela to meet demand for consumption on the island and for export to the Puerto Rican population of New York. In Venezuela, 800,000 *C. guanhumi* (185 metric tons) were taken, mostly during the September-August migrations, from the Golfo Triste area in 1973. The harvest brought U.S. $133,000 to crab hunters and up to U.S. $270,000 at retail (Taissoun, 1974b). Taissoun (1974a) estimates that these figures could be increased severalfold by exploiting other crab-producing areas (e.g., the Orinoco delta).

Gecarcinus ruricola ("black crab") and *C. guanhumi* ("white crab") are both harvested on Andros, Eleuthera, Grand Bahama, Dominica, and other Caribbean islands during breeding migrations. They are offered in the markets live, either intact or as "clipped crab" (all limbs auto-

B

Fig. 3.12. Exploitation of land crabs by humans. A. String of 12 "cangrejos" or"juey" (*Cardisoma guanhumi*). This is the standard unit of roadside sale in Puerto Rico. B. "Black crab" (*Gecarcinus ruricola*) offered for sale in the native market, Nassau, Bahamas. (Photograph by T. G. Wolcott.)

tomized). "Black crab" (*Gecarcinus* spp.; Fig. 3.12B) is preferred for traditional "crab and rice," "crab back" (stuffed crab), and calilou (spinach and crab stew) (Chace and Hobbs, 1969; Bahamian vendors, pers. comm.).

Habitat destruction also affects the ecology of land crabs. In the United States virtually all coastal hammock environments have been obliterated by "development," including the last known large population of *G. lateralis* on the U.S. Atlantic coast (the population described by Bliss et al., 1978). Driving off-road vehicles on beaches can drastically reduce ghost crab populations (T. G. Wolcott and D. L. Wolcott, 1984a). In southern Florida, replacement of mangroves by condominiums has greatly decreased habitat for *C. guanhumi*. At the same time this crab is coming under increasing pressure from foraging by Cuban and other Caribbean immigrants. This species is also under attack as an agricultural pest on Puerto Rico because it destroys young sugarcane and vegetable crops (Feliciano, 1962). In South America, conversion of mangroves to banana and coconut plantations is destroying habitat for *C. guanhumi* (Taissoun, 1974a, b). Even on remote Aldabra mankind exerts environmental pressure: Hermit crabs foraging on the beach often are fouled with oil (Alexander, 1979). On Pacific atolls, land crabs interfere with rat control by eating baits, and springing or stealing traps. The zinc phosphide, and possibly the anticoagulants, in the baits apparently kill crabs as well as rats (Yaldwyn and Wodzicki, 1979). Similar developments are likely elsewhere without regulation of environmental degradation and "fishing" pressure.

8. Effects of land crabs on environment

Land crabs can reach remarkable densities in the absence of heavy predation and, on islands with scarce mammalian and avian fauna, can be the dominant large animals. On Aldabra *C. carnifex* may reach densities of over 3,600 per hectare, with an average individual mass of 322 g (Alexander, 1979). *G. natalis*, the most abundant of the 16 species of semiterrestrial, terrestrial, and freshwater crustaceans on Christmas Island, reaches 11,900 adults, or about 0.8 metric ton, per hectare (Hicks, 1985). On Clipperton Atoll, *G. planatus* reaches the astonishing density of 60,000 individuals per hectare (Ehrhardt, 1968).

Effects on the physical environment are inferred in several cases. *Grapsus tenuicrustatus* may accelerate erosion of rock and its conversion to soil by rasping at encrusting algae; *C. rugosus* may turn over and enrich soil in the process of consuming and burying tortoise feces on Aldabra (Alexander, 1979). *Sudanonautes africanus* aerates savannah soils in the Ivory Coast (Bertrand, 1979), but with fewer than 0.01 burrows per square meter this seems unlikely to be an important process. On small, steep islets around Bermuda, burrowing activity of *G. lateralis* brings soil from crevices out onto surfaces where it can blow away or wash into the sea (T. G. Wolcott and D. L. Wolcott, unpub.); this may be an important route of soil loss.

Effects of land crabs on the biological composition and structure of their communities are sometimes more obvious. On Aldabra, abandoned hermit crab shells provide water catchments in otherwise arid areas (Alexander, 1979). On Christmas Island, Hicks et al. (1984) report, "Absent are the tangle of vines and profusion of shrubby understory which characterise Indo-Malayan rain forest and the forest floor is strangely bare – kept clean by scavenging land crabs." In addition to removing understory, land crabs may influence species composition by preferential feeding; or growth form and reproductive output of plants by selectively "pruning" buds, inflorescences, and fruit (Hicks, 1983; Hicks et al., 1984; Alexander, 1979), which typically contain the highest nitrogen and food value (Mattson, 1980). The gastric mill of land crabs ensures destruction of any ingested seeds (see Chapter 10). Thus, feeding on seeds, either directly or on those embedded in feces of birds and tortoises, may affect local species composition and represent an important barrier to invasion by exotics (Alexander, 1979). On the other hand, hard inedible seeds encased in edible pulp may be transported by crabs. *C. carnifex* feeding on the pulp of screw pine (*Pandanus tectorius*) fruits are important dispersers of the seeds, carrying them an average of 7.3 m from the parent plant (Lee, 1985).

Land crabs affect other terrestrial arthropods. The success of land crabs, particularly *Coenobita* spp., at finding and consuming carrion may account for the low densities of carrion-breeding flies on Aldabra and Pacific atolls (Alexander, 1979; Page and Willason, 1982). Burrowing crabs, particularly *Cardisoma* spp. with their pools of water, create habitats for a wide variety of symbiotic insects including flies that live on the crabs' bodies (Carson and Wheeler, 1973; Carson, 1974; Gomez, 1977) and those that breed in burrows. Mosquitoes are particularly common in the pools at the bottom of *Cardisoma* burrows (Bruce-Chwatt and Fitz-John, 1951; Bright and Hogue, 1972; Cheng and Hogue, 1974; Henning, 1975a). Some are of great epidemiological importance; the major vector of subperiodic filariasis in the South Pacific, *Aedes polynesiensis* Marks, breeds profusely in burrows of *C. carnifex* (Goettel et al., 1981). Various insects and arachnids also take shelter in freshwater land crab burrows (Bertrand, 1979).

Dense populations of land crabs can be important channels of energy in their ecosystems. Populations of *A. pisonii* transfer production from the mangrove canopy, where they attack 10–80% of the leaves, to the adjacent water through release of reproductive products and over 100 cc of finely divided fecal matter per crab each year (Beever et al., 1979. The thousands of eggs produced by other land crabs also represent a transfer of resources from the terrestrial to the marine system, but its importance has not been examined (Alexander, 1979).

9. Directions for future research in land crab ecology

Acquisition of food is one of the most basic ecological necessities, yet the field diets of many crabs are poorly known. In particular, the ecological role of nitrogen availability for herbivores needs further study. Cannibalism may be very important in this group, but quantitative data will be essential in determining its effects.

To understand competition among land crabs, we need a clearer understanding of "limiting resources." This will require perceptive field observation of crab behavior and biology to determine what environmental factors really represent "risks" and "resources." Controlled laboratory studies and field experiments will be needed to quantify differentials in requirements and exploitative abilities.

Land crab reproduction and life histories need further study in two fundamental areas. Juveniles of many species are seldom or never seen, and field work is needed to determine their habitat and ecological relationship to adults. Small-scale physical oceanography, coupled with studies of larval behavior (particularly vertical distribution in the water column) under natural conditions, are needed to elucidate mechanisms allowing replenishment of adult stocks by planktonic larvae in unidirectional current regimes, as on oceanic islands.

Field data on molt interval, molt increment, and growth rate on natural diets are almost totally lacking, even for the economically important species. Development of a permanent tagging method for land crabs would allow studies of growth rate as a function of food supply and food quality. These data, with assessments of food availability, would be useful in predicting potential "fishery" yields, and possibly in "terraculture" of these animals.

In short, land crab ecology needs observant workers who are willing to spend time with the animals in the field, finding out how the crabs interact with each other and their environments. From these data, we can be sure that the questions we ask under laboratory circumstances are relevant. If this chapter has kindled (or rekindled) interest in those objectives, it has served its primary purpose.

Acknowledgments

I thank John Hicks, Kim Wilson, Anson ("Tuck") Hines, David Wingate, and Linda Mantel for access to unpublished data or direction to obscure references; the Smithsonian Institution libraries for copies of the latter; Dr. Ted Rice and the NOAA Beaufort Laboratory for providing a field base for some of our North Carolina work; and especially Donna Wolcott for unpublished data, merciless criticism of the manuscript, scientific collaboration, and basic life support. Portions of work by T. G. Wolcott and D. L. Wolcott discussed in this chapter were supported by the U.S. National Science Foundation, the U.S. National Park Service, and corporate donors including Arco Solar, ACR Electronics, and Uniroyal.

4: Behavior

DAVID W. DUNHAM and SANDRA L. GILCHRIST

CONTENTS

1. Introduction

This chapter presents information on the wide variety of behavioral adaptations associated with terrestrial and semiterrestrial crab genera. Because of the fragmentary nature of much of the available literature, the following is by no means an exhaustive review. It is our intention to provide a comprehensive basis for those interested in continuing research on the fascinating array of behavioral adaptations observed in these crab genera.

Among the many "land crabs" discussed in this volume, the behaviors of the fiddler crabs (*Uca* spp.) and the ghost crabs (*Ocypode* spp.) are best known. These and their near relatives constitute the family

Ocypodidae Ortmann 1884. Some reference will be made to lesser-known ocypodid genera, such as *Dotilla, Scopimera* (sand-bubbler crabs), *Heloecius* (semaphore crabs), *Macrophthalmus* (sentinel crabs), and *Mictyris* (soldier crabs; Mictyridae), where information is available. Unfortunately, most of these genera are relatively poorly studied.

The classic work on ocypodid behavior is Crane's very comprehensive volume, *Fiddler Crabs of the World* (1975), in which she devotes two full chapters, and sections of others, to behavior. Since her review covers the literature through 1970 (with a few more recent references), and is well documented, the reader is referred to that volume for detailed sources of the older literature. Major aspects of behavior will be summarized here, with particular attention given to more recent studies.

Fiddler crabs and ghost crabs occur in open habitats, on sandy beaches and mudflats, where they are conspicuous to humans. They also actively move about in large numbers during periods of low tide (also see Chapter 3). Most species are brightly colored, especially the enlarged chela of male fiddler crabs, and their conspicuousness is even more enhanced by visual displays involving rhythmic movements of the chelipeds. Even in heavily settled areas where contact with humans seems to have encouraged more nocturnal than diurnal activity (Wolcott, 1978; cf. Robertson and Pfeiffer, 1982), they do not go unnoticed.

The popularity of ocypodids for behavior study is understandable when one considers how accessible they are for observation. In this respect they are not unlike colonial birds. The synchronization of their activity with tidal fluctuation has made them valuable subjects for the study of space use, social communication, and reproductive behavior (e.g., von Hagen, 1962; Salmon, 1965; Linsenmaier, 1967; Horch and Salmon, 1969; Zucker, 1974; Hyatt a, b, 1977; Wolcott, 1978).

We also discuss the behavior of the other (non-ocypodid) land crabs within each major section of this chapter. These non-ocypodid species constitute the majority of crabs adapted to varying degrees for a terrestrial, and even arboreal, existence. In contrast to the relatively uniform life-style of the ocypodids, it is among the coenobitids, gecarcinids, grapsids, xanthids, and related groups that we discover the full range of behavioral responses by crustaceans to a terrestrial life.

Though other crabs also exhibit a wide variety of social behaviors, comprehensive reviews (*sensu* Crane, 1975) have not been compiled for the terrestrial and semiterrestrial non-ocypodid crabs. However, Bliss (1968, 1979) provides an extensive treatment of the respiratory behaviors of selected land crabs. In particular, she focuses on the Gecarcinidae. The behaviors of isolated groups such as the Coenobitidae, though exhibiting a nonsocial population structure, have been considered in monographs (see Hazlett, 1966), perhaps owing to their roles as major food sources or as agricultural pests.

Comparisons of land crabs with intertidal, semiterrestrial counterparts are scarce. Perhaps one of the more detailed accounts of this nature is given by Fimpel (1975). He describes habitats, behavior, and physiology of representatives from three families (Ocypodidae, Grapsidae, and Gecarcinidae). However, recent reviews of the Crustacea (see *Biology of Crustacea* series edited by Bliss, 1982ff.) do not explore the behavior of terrestrial non-ocypodid crabs extensively.

2. Feeding behavior

A. *Feeding stimulation*

In order to understand the role of chemical stimuli in initiating feeding in land crabs, one must consider food type and availability for aquatic crabs before discussing modifications that occur in intertidal or terrestrial species. Aquatic crabs commonly seek out large patches of food (e.g., prey or carrion) that may be relatively unpredictable in time and space. An important limitation is patch location. Therefore, aquatic crabs can locate feeding sites through sensitive antennular chemoreceptors, which detect even minute concentrations of the low-molecular-weight compounds asssociated with animal flesh.

On the other hand, beach foragers such as fiddler crabs, which are primarily deposit feeders, feed on such benthic microorganisms as algae, bacteria, ciliates, and nematodes as well as detritus. An important limiting variable for intertidal crabs is time for foraging, since decisions must be made about when to initiate feeding and when to move on. Contact chemoreception is more likely to be useful than distant antennular chemoreception in this feeding mode. It is typical of deposit-feeding crab species to tap the substrate frequently with the outer surface of the major cheliped as they move across the beach (Trott and Robertson, 1984). Moreover, adult *Uca* have fewer antennular chemosensilla than do aquatic crustaceans (Ghiradella, Case, and Cronshaw, 1968; Kinoshita and Okajima, 1968; Zimmer-Faust, 1987). In experiments to test chemosensory preference, *Uca* responded most strongly to compounds that are reflected in the composition of its diet. For example, L-serine, an especially potent feeding stimulant for *Uca* (Robertson, Fudge, and Vermeer, 1981), is a dominant amino acid in the benthic diatoms eaten by fiddler crabs.

Trott and Robertson (1982, 1984) investigated the feeding response of a ghost crab, *Ocypode quadrata*, to experimental administration of chemical stimulants. Stimuli were applied to the major chela, minor chela, and the dactyls of the first walking legs. The stereotyped major and minor cheliped flexion response, and a correlated movement by the first walking legs, were used as bioassays for various natural and artificial food-related substances.

Cheliped flexion is a normal component of feeding. The substrate is pinched with either the major or the minor chela, and the sample is placed in the mouth. In Trott and Robertson's experiments (1982, 1984), stimulation of a given appendage first caused the dactyl to close, if it was open, which also occurs in pinching the substrate. Next the propodus was raised to the mouth, the third maxillipeds grasped the pollex and the dactyl, and the maxillipeds wiped the surface of the claw as it was drawn away. The drawing away and wiping motions were often repeated several times in succession. If a stimulus was applied to the dactyl of the first walking leg, the dactyl and the nearer chela moved toward each other, rubbed together, and then cheliped flexion occurred.

The results of Trott and Robertson's experiments (1982, 1984) showed that sugars were more effective stimulants than were amino acids. Among the sugars, disaccharides were more effective stimuli than were monosaccharides or pentose sugars. Asparagine was the most stimulatory amino acid followed by betaine, which resembles an amino acid. Mixtures of the least effective amino acids had a synergistic effect.

Feeding responses of *O. quadrata* are of special interest because this land crab has an unusually broad diet, taking living prey on the beach, as well as being a scavenger, deposit feeder, and marine macrophyte consumer. The wide spectrum of chemical stimuli to which *O. quadrata* responds reflects the plasticity of foraging behavior in this ghost crab. Components of both live and decomposing foods in the diet of *Ocypode* were found to be effective in eliciting the cheliped flexion feeding response (carboxylic acids, carbohydrates, amines, and amino acids). Other species of *Ocypode* probably differ in the breadth of their foraging, but the effective stimuli for their feeding behaviors have yet to be identified.

A comparison of the results from *O. quadrata* with those from a fiddler crab, *Uca pugilator*, illustrates the relationship between effective stimuli for feeding and the chemical composition of the food naturally consumed. *U. pugilator* is primarily a deposit feeder. Both species respond more strongly to disaccharides than to monosaccharides. This may be related to the fact that both the food storage compounds in algae and the extracellular sheath of benthic diatoms, on which both species feed, are disaccharides. *O. quadrata* responds more strongly to the monosaccharides fructose and mannose than does *U. pugilator*. These sugars are abundant in fresh (fructose) and decomposing (mannose) fruits on tropical beaches where *O. quadrata* is found but *U.*

pugilator is absent. *O. quadrata* responds to the amino acids taurine and glycine, whereas *U. pugilator* does not. Detection of these amino acids would enable *O. quadrata* to locate coquina bivalves (*Donax variabilis*) and mole crabs (*Emerita talpoida*) on which it preys.

Olfaction has been studied more extensively in aquatic Crustacea than it has in the land crabs (Bardach, 1975; Atema, 1977; Ache, 1982). Their feeding responses are primarily to amino acids (Case and Gwilliam, 1961; Case, 1964; Hazlett, 1968, 1971a, b; Hindley, 1973; Derby and Atema, 1982). *O. quadrata*, in contrast, shows a relatively weak response to amino acids, differing from the intertidal and subtidal crustaceans. On the beach, autolysis and bacterial and other microbial action break down proteins into small peptides, polypeptides, amino acids, and nitrogenous bases. However, continued anaerobic decomposition consumes amino acids and produces amines, such as cadaverine and putrescine, and short-chained fatty acids, such as butanoic and pentanoic acids. These are good indicators of food sources for *O. quadrata*, and are responded to more strongly than comparable components of fresh tissue, like betaine and trimethylamine.

Terrestrial crabs that do not depend on deposit feeding, such as *Coenobita compressus*, also appear to have limited aerial chemoreception of food odors. Kurta (1982) observed that *C. compressus* does not use olfaction to detect foods over long distances (> 5 m). Instead, feeding activities of one crab apparently cause an aggregation of crabs to form around food items, suggesting that vision may be more important than olfaction.

Banana-baited, enclosed traps, however, attract *C. compressus* from at least 2 m away at night (S. Gilchrist, unpub.). *C. compressus* responds differentially to various stimuli. In laboratory tests conducted with crabs housed in long screen cylinders, *C. compressus* was noted to detect odors from dog and bird feces, rotting bananas and fish, fresh and overripe coconut, fresh papaya, and ripe mangoes from a distance of at least 5 m (Gilchrist, unpub.). Antennal flicking increases dramatically in the presence of such stimuli, perhaps enhancing the response of olfactory receptors. Schmitt and Ache (1979) have made similar observations for aquatic crustaceans. Flicking rate also increases if sight and smell are coupled stimuli. Sight alone increases flicking rate for a short time (< 5 min), but without odor stimulus the rate dropped back to normal. *Coenobita clypeatus* can easily locate potential food sources (bananas and apples) in both horizontal and vertical mazes on the basis of airborne odor cues alone (D. Dunham and H. Schoene, unpub.). However, Rittschof, Barlow, and Schmidt (1985) performed field and laboratory experiments on chemoreceptive capabilities of *C. rugosus*, showing this species to be receptive to odors from fruit, feces, and rotting fish although feeding responses were initiated only after contact with nonvolatile compounds (such as sucrose).

B. Modes of feeding

The best-studied feeding mode of the beach- and mudflat-dwelling ocypodids and their relatives is deposit feeding, in which scoops of substrate material are manipulated by the mouthparts, and benthic microorganisms are removed and ingested (also see Chapter 10).

In *Uca*, a portion of the substrate is scraped up with the small cheliped and placed between the inner edges of the third and second maxillipeds (Crane, 1975). Females use their two chelipeds alternately, but at a slower rate than the male uses his one. If his minor cheliped is missing, he uses the major one in feeding. One to more than four scrapes are made before the material is carried to the mouth. Smaller crabs feed faster than larger ones, and higher temperatures also cause faster feeding (Miller, 1961; Crane, 1975).

Deposit feeding is the principal feeding mode for *Mictyris* (the soldier crab), *Scopimera* (the sand-bubbler crab), *Heloecius* (the semaphore crab), *Macrophthalmus* (the sentinel crab), and *Uca* (the fiddler crabs). *Ocypode* spp. use this feeding mode, but not exclusively. A characteristic of deposit feeding is the deposition of sand pellets, which remain after the extraction of organic matter from the substrate (Miller, 1961; Crane, 1975). Crabs move slowly ahead as they feed. Sand pellets formed by the mouthparts are deposited in a line.

Because *Uca* and *Ocypode* differ in their specializations for deposit feeding, there has been some interest in comparing their use of this mode of feeding. Wolcott (1978) showed that, contrary to previous reports, *O. quadrata* does make some limited use of deposit feeding in its foraging repertory. Robertson and Pfeiffer (1982), using videotape analysis to investigate feeding in *O. quadrata* in the field, confirmed the use of deposit feeding, as well as the more widely observed predatory behavior.

Robertson and Pfeiffer (1982) also describe deposit feeding in detail for *O. quadrata*, and compare it with deposit feeding in two other *Ocypode* species and in the sand fiddler crab, *U. pugilator*. The comparison shows, surprisingly, that the deposit-feeding behavior of *O. quadrata* in some respects resembles that of *U. pugilator* more than it does that of the congeneric *O. ceratophthalmus* and *O. gaudichaudii*. *O. quadrata*, unlike *U. pugilator*, uses apposition of the movable dactyl of the cheliped against the pollex when scooping up a portion of substrate. *O. ceratophthalmus* and *O. gaudichaudii* also use the cheliped, but not the articulation of the dactyl with the rest of the chela. After feeding, *O. quadrata* removes feeding pellets from the buccal region with a flick of the closed minor chela. *U. pugilator* also removes feeding pellets from the mouth with the minor chela, actually grasping the feeding pellet and placing it on the substratum. *O. ceratophthalmus* and *O. gaudichaudii* simply let the pellets drop and do not flick them off with the minor chela.

In terms of feeding efficiency, *O. quadrata* takes up more substrate in a single scoop than does *U. pugilator*, making larger feeding pellets with fewer movements (Robertson and Pfeiffer, 1982). Because the feeding rates of the two species are similar, it is estimated that the ghost crabs process the sediment more than 10 times faster than the sand fiddler crabs. Once a pellet is in the buccal region, food must be separated from nondigestible matter, and the mouthparts used in sorting the substrate particles are similar in *Ocypode* and *Uca*. Ingested particles are trapped by the setae of the first maxillipeds. Coarse, inorganic particles, which are swept away by the second maxillipeds, form pellets, which are discarded. Setae on the maxillae pass the remaining edible material back to the mandibles for further ingestion (Miller, 1961; Crane, 1975). *O. quadrata* is less efficient than *U. pugilator* in extracting algae from the substrate, but nonetheless it extracts up to 70% of the available algae (Robertson and Pfeiffer, 1982). Deposit feeding is an important supplement to predatory feeding in *O. quadrata*.

Like many of the ocypodids, the gecarcinid crab *Cardisoma guanhumi* tends to hunt for food in the vicinity of its burrow. These crabs "taste" many objects in the environment by touching the minor chela to the object and then to the mouthparts, suggesting contact chemoreception is important in initiating the feeding response. Herreid (1963) noted that the crabs either ate an acceptable item immediately or moved quickly to the burrow with the object. He also noted that these crabs are cannibalistic, as has been observed for other terrestrial crab genera (D. L. Wolcott and T. G. Wolcott, 1984; see the Appendix for a list of foods consumed by selected crab species). *Cardisoma carnifex*, a relatively large land crab species, appears to ingest mud, although they often pull leaves from bushes and sedges (Grubb, 1971). We speculate that they might be digesting microflora from surfaces of sediment grains and perhaps from leaves as well.

Lee (1985) reported that *C. carnifex* is an important local dispersing agent for *Pandanus tectorius*, a plant with fleshy, fibrous fruit. These crabs emerge from the litter at night to consume the fruits that have fallen to the ground. The crabs did not remove fruits from trees, although it was noted that they would climb trees for short distances. However, Lee (1985) describes the crab population as living under leaf litter rather than in burrows. The dispersal activities of these crabs may be limited in other cases by their habit of remaining near a fixed burrow (Herreid, 1963).

Sesarma reticulatum and *S. cinereum* have complex behavioral patterns related to feeding. The former tears food before moving pieces to the mouthparts, and the latter apparently scoops finer food particles directly to the mouthparts (Alexander and Ewer, 1969; Seiple and Salmon, 1982). Abele (1973) noted that individuals of *S. curacaoense* are sluggish and feed on detrital material within their immediate vicinity, whereas the sympatric species *S. ricordi* are very active, trapping insects as well as feeding on detritus.

Carnivorous shore crabs, such as *Carcinus maenas*, can apparently learn to manipulate prey to decrease the handling time of each item. Elner (1978) and Jubb, Hughes, and Rheinallt (1983) describe a series of motions relating to size selection of prey by these crabs. They suggest two alternative behavioral hypotheses for explaining size selection: the prey-evaluation hypothesis whereby prey rejection is by size, and the relative-stimulus hypothesis, which is dependent on the relative strength of the resistance of the shell of the molluscan prey. According to Cunningham and Hughes (1984), *C. maenas* has at least five different methods of attacking mussels and at least three ways of consuming gastropods. Such crabs are apparently able to transfer learned patterns to similar (size, shape) prey items.

Pagurus traversi, a high intertidal hermit crab that apparently spends a great deal of its feeding time on land, shreds or plucks marine algae with the chelae before ingestion (Schembri, 1982). He describes opportunistic scavenging of gastropods and crustaceans from algae as well as attraction to and feeding on carrion by these crabs.

Coenobita clypeatus feeds on a variety of items in the field, including fruits, vegetation, insects, and carrion. There appears to be a slightly different approach to feeding for large as opposed to smaller (carapace length < 4 mm) crabs as well as a variation in appendages employed, depending on whether the material is liquefiable or solid. Parasitized crabs also appear to have some feeding anomalies. One of us (S. G.) has made extensive observations on the feeding habits of this crab both in the laboratory and in the field. The following account is a summary of notes from 172 hours of field observation of feeding and 427 hours of observations in the laboratory.

When consuming a fruit, a large crab holds the fruit against the substrate with the major chela. The minor chela is used to pinch the flesh of the fruit. If the minor chela is removed from the crab, the first pereiopod opposite the major chela is used to hold the fruit and the flesh is pinched with the major chela. If the major chela is removed, the first pereiopod opposite the minor chela is used to hold the fruit and the minor chela is used to pinch the fruit flesh. Removal of both chelae results in a palpation of the fruit with the pereiopods and a direct transfer of "juice" to the mouth region via the endopodites of the third maxillipeds. In the intact crab, the pinch is made with the palm of the claw upward and nearly horizontal with the fruit surface. The shape of the closed claw forms a "cup" for the fluid obtained from squeezing the fruit. The minor chela is moved toward the mouthparts. The endopods of the third maxillipeds are used to move the fluid to the mouth region. However, smaller crabs may transport the fluid directly from the minor chela to mouth region. In larger crabs, the maxilliped on the same side as the minor chela sweeps the juice toward the tip of the claw. The endopodite of the other third maxilliped is moved toward the claw tip as well until the two endopodites are crossed. The crossed maxillipeds are moved to the mouth region where the maxillae move across the densely packed serrate setae (and to some extent, simple setae) of the third maxilliped endopodites. The third maxillipeds are extended again and the process is continued. From the time the fruit is pinched to the time of the next pinch varies considerably for individuals. It does appear, however, that if several individuals are consuming a fruit, the rate is more than doubled per individual. Rarely will individuals directly interfere with a similarly sized conspecific feeding on the same fruit; however, large crabs may flick a much smaller crab away from a fruit with the first pereiopod.

Land hermit crabs are one of the few terrestrial crustaceans that directly drink water, and water drinking by larger crabs appears to follow the same pattern as fruit consumption.

When eating carrion or an insect, the hermit crabs will hold the food item with the major chela while stripping and plucking the item with the minor chela. The minor chela is used to transfer bits of food to the mandibular region directly. If the item is large relative to the crab or if tearing is difficult, the crab may hold the item with the major chela as well as with the first pereiopods. The other pereiopods are used to anchor the pulling motion. If crabs are feeding while in a tree, the pereiopods are not used to hold the food item. Only on three occasions were crabs observed feeding upside down on a limb. The material being consumed was a saplike residue from the tree. The material was collected with the minor chela and moved directly to the mouth region.

The largest land crab, *Birgus latro*, feeds by crushing and stripping using the powerful, large chela. The smaller chela is used to move particles to the maxillipeds, which reach outward to grab food items. Chelate fourth pereiopods are used in support and in carrying prey items (Grubb, 1971). This crab species appears omnivorous, consuming a wide variety of plant and animal material. It has adapted to the presence of humans on islands by becoming a major agricultural pest, foraging in fields as well as attacking young domestic birds.

In addition to deposit feeding and predation, herbivory is common among land crabs (Mattson, 1982; Giddins et al., 1986; also see Chapters 3 and 10). Hartnoll (1965) and Beever, Simberloff, and King (1979) describe a combination of herbivory and predation (on insects and conspecifics) by the mangrove tree crab *Aratus pisonii*. However, they pointed out that individuals may be very limited in their food selections, even though the species as a whole exploits a wide variety of foods in different habitats. Another mangrove crab, *Chiromanthes onychophorum*, also consumes mangrove leaves. Malley (1978) reported that up to 95% (by volume) of the rectal and proventricular contents of this crab species consisted of leaf material. The chelipeds of this species are similar to those of *A. pisonii* in that they appear to function more efficiently in grasping than in tearing. Thus it is speculated that the maxillipeds and/or mandibles may function in reducing materials to smaller particles for ingestion.

C. *Feeding in time and space*

The ocypodids and their relatives typically inhabit well-constructed burrows, whereas many other terrestrial crabs may seek shelter in leaf litter, crevices, and beneath rocks. Beach foragers are dependent on the changing tides to bring the microorganisms on which most of them feed from the open water up into the intertidal zone. When the tide recedes, these beach foragers emerge from their burrows and engage in deposit feeding. Some feed near the burrow, using it as a foraging center, whereas others move down into the intertidal zone, sometimes in aggregations of many individuals. The most spectacular are those of the forward-marching soldier crabs (*Mictyris*), which, with their brilliant blue, purple, and white coloration, emerge to feed in large numbers (see Healy and Yaldwyn, 1970). Kelemec (1979) has shown that temperature is an important determinant of when *M. longicarpus* emerge from their burrows and begin their long feeding treks. In the laboratory, more individuals emerged at higher temperatures, within the range 14.5–29.0° C. Kelemec (1979) also found that feeding rate covaried positively with temperature, being most efficient (in terms of time invested) at the highest temperature in this range. Kraus and Tautz (1981) established experimentally that *M. platycheles* will usually maintain their individual distance in aggregations by avoidance of contrasting objects with a given height-to-width ratio.

Harada and Kawanabe (1955) describe *Scopimera* foraging in exclusive territories that overlap only under experimentally induced very high densities. The feeding areas are more restricted, in terms of angular sector transversed from the burrow entrance, under natural high densities than under natural low densities. However, even under low densities, feeding forays directly toward other burrow entrances are avoided. Availability of food may influence reproductive cycles, as has been suggested for *Dotilla myctiroides* (Hails and Yaziz, 1982). The two annual peaks in reproduction occur when food abundance, affected by monsoon winds and sea currents, is greatest (also see Chapter 5).

Movements of aggregations from the burrowing area down into the intertidal zone, and back again after feeding when the tide is rising again, are often referred to as "droving" or "herding." In some areas *Ocypode* may move long distances, from burrows hundreds of meters inland and more than 30 m above the high-tide level, down to the sea to forage and back again (Healy and Yaldwyn, 1970). Murai, Goshima, and Nakasone (1982) discuss differences between the sympatric *Uca vocans* and *U. lactea* in droving and feeding. The former burrows in sandy habitat that is poor in microorganisms and organic matter. It therefore typically moves down to the water's edge, feeding in droves. Not only is the food supply richer than near the burrows, but branchial and body water can be replaced *ad libitum* from the seawater at the bottom of the burrow. When *U. vocans* does forage near the burrow, it makes frequent descents into the burrows for water replacement. *Uca lactea* burrows in areas richer in organic matter, and because of this high food density is not dependent on droving to the water's edge. It defends exclusive territories in which it feeds. *U. vocans*, on the other hand, has overlapping home ranges around the burrow entrances, expending more energy in droving than in territorial defense (Nakasone, 1982, and Murai, Goshima, and Nakasone, 1983, present a more detailed analysis of this species). This comparison is an instructive example of how differences in feeding behavior can affect social behavior and the use of space. Similarly, Salmon and Hyatt (1983a) point out the importance of food supply in different intertidal zones in determining territorial feeding versus droving in *U. pugilator*. Feeding requirements take precedence over territoriality in the time–energy budget of these intertidal crabs, and therefore locally abundant food is a prerequisite for a strict territorial social system. In vertebrates (including fish, lizards, birds, and mammals), although breeding territory size may be inversely related to food density when feeding is done on the territory, if food is patchy and therefore not easily predictable in time or space, foraging cost is high in time and energy and defended breeding territories may be quite small (Davies, 1978).

Sympatry can involve adaptations to different microhabitats, including differences in feeding appendages and gut structure, as shown by Icely and Jones (1978) for four species of fiddler crabs in Kenya. *Uca* spp. share their habitat not only with congeneric species but with other crabs as well. Bursey (1982) describes resource partitioning between the mud fiddler crab, *U. pugnax*, and the purple marsh crab, *Sesarma reticulatum*. Both are primarily vegetarians, but the former forage from the surface of the substrate, whereas the latter forage on vascular plants and some animal matter. They also time-share, *Uca* feeding at low tide and *Sesarma* at high tide.

Salmon (1984) presents a valuable detailed discussion of the behavior and ecology of *U. vocans*, one of the "narrow-fronted" *Uca* species, previously thought to be more primitive than the "broad-fronted" *Uca* species. (Narrow and broad refer to the morphology of the carapace and eyestalks.) Female *U. vocans* forage during low tides, bringing detritus back to the burrow, presumably feeding on it during subsequent high tides (Nakasone, 1982; Salmon, 1984). It is also possible that they use this material to line the burrow, and thus decrease the probability of burrow collapse during high tide (N. Zucker, pers. comm.). This resource independence and other factors influence the form of reproductive and agonistic behavior (as discussed in sections 4 and 5).

3. Grooming by non-ocypodid crabs

In addition to feeding by the previously discussed methods, terrestrial crabs may ingest some materials during grooming and cleaning. Land crustaceans experience a different fouling regime than their marine counterparts. Parasitization from flies and mites is not uncommon (see Bright and Hogue, 1972, for a literature review of internal and external parasites of selected crustaceans).

However, Bauer (1981) suggests that terrestrial crustaceans are under fewer epizoic pressures than aquatic crustaceans. This observation deals with juvenile and adult phases only; epizoic pressures on aquatic larval forms of terrestrial crustaceans have not been determined.

Individual *S. reticulatum* groom the anterior region frequently. Felgenhauer and Abele (1983) describe this anterior grooming in detail. Briefly, the merus and carpus are used to remove large debris from the branchiostegite to areas where the respiratory current eliminates the material. The third maxillipeds function in removing particles from the epistomal region while the flagella of the first maxillae may aid in cleaning the gills. It appears that terrestrial species in general do not groom as much as their aquatic counterparts.

Holmquist (in press) has described the grooming structures and functions in selected terrestrial Crustacea. His review of the current literature on morphology of grooming structures and the behaviors involved in grooming for *Coenobita clypeatus* and *Cardisoma guanhumi* is summarized in the following discussion.

As in feeding, the third maxillipeds have major roles in grooming. Frontal grooming is common. In *C. clypeatus*, the eyestalk is moved downward and is groomed by hooking one or both third maxillipeds over it. The serrate setae of the mesial propodus and the dactylus are moved over the surface of the structure. The antennules are groomed in a similar manner, brushing the surface with the serrate setae of the maxillipeds. The chelipeds and pereiopods groom themselves. This is accomplished by rubbing the appendages together in twos or threes.

The fifth pereiopods are chelate and very flexible. Holmquist (in press) describes virtually all movements of these highly setose structures as "scrubs." These appendages groom the basal segments of the chelipeds, the second and third pereiopods (exclusive of the dactyli), all of the fourth pereiopods, the posterior two-thirds of the carapace (including the branchial chamber), and most of the abdomen. In gravid females, the fourth and fifth pereiopods as well as the minor chela are used to groom the eggs (S. Gilchrist, unpub.). This appendage is also used for the very important function of scrubbing the interior of the shell as well as the shell lip region. The fifth pereiopods were groomed by the third maxillipeds.

Holmquist (in press) noted that the grooming actions of *C. clypeatus* are quite similar to its marine counterparts. The major difference appears to be in the increased amount of grooming performed on the eyestalks by the terrestrial crabs. He suggested that an increased importance of visual cues for terrestrial crustaceans, as opposed to aquatic crabs, may result in this behavioral modification.

As in the coenobitids, it is the third maxillipeds of *C. guanhumi* that are featured prominently in the grooming behaviors of this crab. This species uses all five pairs of pereiopods as well as the mouthparts to groom the body. Similar to the coenobitids as well as other terrestrial crabs (Koepcke and Koepcke, 1953; Alexander and Ewer, 1969; Crane, 1975), the eyestalks receive a great deal of attention. The third maxillipeds scrub around the eyestalks and interorbital region. The chelipeds are also used in grooming the eyestalks. Debris from the interorbital region and the eyestalks may be either dropped to the ground or consumed by the crab.

Gill cleaning, as typical of brachyurans in general, is carried out by the epipods of the maxillipeds. These setose structures are moved across the gill surface to remove debris. Other groups of terrestrial crabs (e.g., *Pachygrapsus crassipes*) use the setiferous epipods on the maxillipeds to groom the gills even though general body grooming may not occur (Bauer, 1981).

Though the terrestrial crabs do groom without water present, Holmquist (in press) comments extensively on the role of water as a cleaning agent for these crabs. He noted that grooming activities were much more vigorous following a rain or when the animal was standing in water.

For a more detailed account of grooming behavior, the reader is referred to the work of Holmquist (in press) and Bauer (1981). However, it is clear that this behavior has received little attention in general for land crab species.

4. Reproductive behavior

Social behavior of terrestrial crab species consists primarily of agonistic behavior (fighting, fleeing, and associated displays) and reproductive behavior. In both cases, the immediate functions are related to breeding success. Since the ocypodids have received a great deal of attention in this regard, they are treated separately from other terrestrial crabs.

A. Courtship display in ocypodids

Agonistic behavior among male ocypodids is descriptive of interactions that result in the obtaining of a burrow, and the defense of it from other males. Behavior associated with courtship and territoriality is quite conspicuous to the human observer, and was discussed in the now classical papers on ocypodid behavior (e.g., Crane, 1941a, b, 1943, 1957; Altevogt, 1955, 1957a; von Hagen, 1962; Salmon, 1965). In most *Uca* spp. courtship takes place on the territory, and copulation on the territory surface or in the burrow proper. Territory size is largest in species with complex displays, where the cheliped is extended laterally and the animal moves back and forth near the burrow. It is smallest in species that perform less expansive, vertical movements of the cheliped. These males leave their own territories when courting females. In the latter species, male defense of the burrow and territory may be weak or absent, e.g., in *U. deichmanni* (Zucker, 1983) and *U. vocans* (Salmon, 1984). (The following discussion of ocypodid social behavior is based on Crane's 1975 review, except as otherwise documented.)

Figure 4.1 illustrates the three different stages of courtship in a representative fiddler crab, *Uca panacea*. The most conspicuous part of courtship is stereotyped movement (termed "waving") of the very large, brightly colored major cheliped of males. Species occupying open, flat habitats assume a light body coloration, which enhances their contrast against the background when displaying. Species inhabiting dense vegetation may enhance their conspicuousness by displaying from elevated sites, and sometimes incorporate visible and audible display components into their pursuit of females (Salmon and Hyatt, 1983b). An audible display compensates for loss of visual contact between a courting male and the female in *Uca tangeri* (von Hagen,

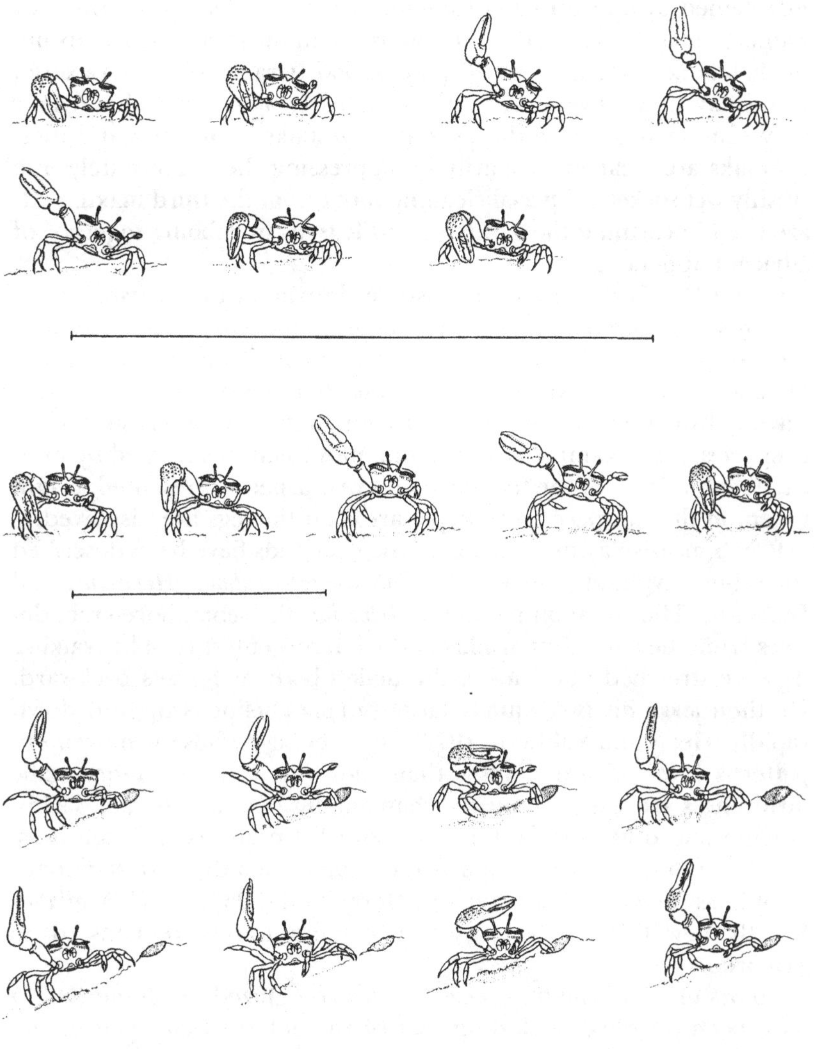

Fig. 4.1. Waving displays of *Uca panacea*. *Top*: low intensity; *middle*: medium intensity; *Bottom*: high intensity. Lines indicate duration of sequences; bottom line spans ca. 400 msec. (Drawing by R. Tuckerman, modified from Salmon et al., 1978.)

1962; Salmon and Atsaides, 1968). A crab of appropriate size without the typical large cheliped of a male is a sufficient stimulus to elicit courtship. Waving occurs in agonistic contexts with other males as well as in courtship, but it reaches its highest intensity only during courtship.

Conspicuousness of colored areas of the carapace and chelae is maintained by appropriate grooming behavior. Cleaning of these by immersion in water in the burrow is common in a soiled crab immediately before starting a display period (Crane, 1975). Crabs not in display phase (wandering phase or early aggressive phase) may only remove mud from the eyes and eyestalks. "The eyes and their eyestalks are cleaned primarily by depressing them alternately into freshly wet sockets." Special cleaning motions of the third maxillipeds are used in cleaning the carapace and legs, as in rubbing together of adjacent appendages.

Crane (1975) discusses the possible derivation of courtship components. Displacement feeding movements, movements that produce acoustic signals, cleansing movements, threat postures, and incomplete locomotory movements are implicated. Several components of waving display are restricted to courtship, but these vary from species to species. One example is a revolving movement performed by male *Uca beebei* in front of the female. As his rear aspect is presented, bright green on the carapace and purple areas on the legs are displayed.

Rhythmic display movements of the chelipeds have been described for other ocypodids, such as *Dotilla, Macrophthalmus, Heloecius,* and *Scopimera.* The common name for *Heloecius,* the semaphore crab, derives from the courtship display at the burrow entrance. The walking legs are stretched out, raising the male's body as he tilts backward. He then jerks his two equally large purple chelipeds up and down rapidly (Healy and Yaldwyn, 1970). In other ocypodids the movement patterns involved are simpler than those in *Uca,* and interspecific differences are less pronounced than among *Uca* spp. (Crane, 1975). Waving and other displays involve both form and color/brightness. Fiddler crabs can discriminate shapes, even when they are stationary stimuli, as shown by Langdon and Herrnkind (1985) for *U. pugilator.* Exactly how this ability is used in communication remains to be explored.

Colors of 59 of the 62 species of *Uca* recognized by Crane (1975) have been decribed, including that of the enlarged cheliped of the male. During periods of display some species do not alter their colors, other species change their colors somewhat, and still others brighten their coloration greatly. (There are also important population or subspecific differences in this regard.) Relevant variables within populations are age, sex, behavioral phase, and season. Color changes can occur on the carapace, appendages, or both. Bright white, yellow, orange, copper, gold, red, rose, crimson, purple, intense blue, green, and pink constitute the palette of *Uca.* In some, but not most, species females also become quite brightly colored in reproductive phase. von Hagen (1970) showed that the shape and color of female models were important in stimulating male courtship. A precise definition of the role that brightness and/or color play in social signaling awaits studies addressing these variables. Promise in this area is offered by Hyatt's (1975) work on *Uca* species. He demonstrated discrimination of blue, red-orange, white, and ultraviolet in *U. pugilator*, and suggested that color perception may function in close-range communication. In *U. vocans*, where both sexes are brightly colored, both are also quite aggressive toward members of the same sex (Salmon, 1984).

Nondisplay colors are more cryptic browns and grays. The endogenous rhythms of lightening and darkening have been the subject of endocrine physiological studies (see Crane, 1975). In all *Uca* there are hormonally mediated darkening–paling rhythms, one diurnal and one tidal, which when superimposed produce a semilunar rhythm of approximately two-week periods.

Component movements and postures used in waving displays, and their temporal organization, are species specific. However, in some cases species differences may be subtle, as demonstrated by von Hagen (1983) in the sympatric sibling species *Uca mordax* and *U. burgersi.* Doherty (1982) carried out a detailed analysis of the spatiotemporal patterning of waving in *U. minax* and *U. pugnax.* He found both stereotyped and variable components, which he suggests could provide information on species identity and motivation, respectively. Temperature affected some temporal components, e.g., generally covarying negatively with wave duration and ascending and descending wave times. However, components that seem to function as graded signals between interactants, such as chela waving directly in front of the body with no lateral extension of the cheliped, were relatively independent of temperature. The latter varied with the sex and distance of the stimulus animal. Crane (1975) describes components of waving display of *Uca* in detail. She recognizes eight different patterns of movement by the major cheliped, five patterns of ambulatory leg movement, waving by the minor cheliped, and four different parameters in the timing of waves. In addition there are movements associated with sound production (see section 6) or derived from such movements, which add to the visual effect of waving.

Recent work has shown that species characteristics such as acoustic display are heritable (Salmon and Hyatt, 1979). Species with extensive geographic ranges may show great consistency in the morphology of their waving display, as in *U. tangeri*, which ranges from Spain to Angola. However, species that are sympatric with other *Uca* species may show local differences in a form that presumably aids in species differentiation, as in *U. rapax*, which is found from Florida through Brazil. The implication is that females discriminate conspecific males from other males (see Zucker and Denny, 1979), although, as Salmon (1983) points out, this is still in need of direct confirmation.

Crane (1975) recognizes two broad categories of waving form, one with principally a vertical component and one with a lateral component predominating. However, intermediate types of waving display also occur. Among the Indo-Pacific *Uca*, species in the subgenera *Deltuca* and *Thalassuca* exemplify the simplest displays, consisting of very low vertical waving movements that are typically given in series but that can also occur as single jerks. The cheliped elevation is not high, and other elaborations are absent (although a nearly vertical tilting of the carapace back toward the female may be seen in these and perhaps all subgenera of *Uca*). In these species the male follows the female across the substrate, patting, plucking at, or stroking her carapace prior to copulation in the open. Waving display does not always occur.

The next stage of elaboration can be seen in other Indo-Pacific species, especially in the subgenera *Australuca, Thalassuca,* and *Amphiuca*. As the low vertical waves are performed, the carapace is raised up by extending the legs, either with each wave or held high during a series of waves. The cheliped may be moved to a semilateral position if the animal addressed by the display is behind or to one side of the actor. Surface copulation, as mentioned earlier, is typical.

In the remaining, principally American subgenera, the waving display movements are chiefly lateral in orientation, and the lateral-straight waving shifts to high circular movements at the higher intensities, becoming lateral-circular. Two Indo-Pacific species of *Celuca* also belong to this group of lateral-circular displayers. Although vertical wave forms occasionally occur in displays of lateral wavers, lateral waves are not given by vertical wavers.

When a female *Uca* approaches a male's burrow, he performs other visual displays involving special movements of the ambulatory legs, of the chelae, and rhythmic bobbing movements of the carapace. A receptive female follows the male down his burrow and copulation takes place underground. Nocturnal courtship, which includes important sound components, commonly ends in copulation above ground in temperate zone species or populations. Yamaguchi, Noguchi, and Ogawara (1979) report both surface copulation and (presumed) burrow copulation in *Scopimera globosa* in Japan.

B. Female response in ocypodids

In the Indo-Pacific subgenera, the male does not attract females to his territory, but rather visits females' burrows. Copulation takes place on the surface near the burrow. The female's active participation in courtship is primarily one of response to male initiation. A male may visit many females before achieving a single copulation (Salmon, 1984). Individuals of a pair may maintain adjacent burrows for up to several days. In other subgenera the receptive female enters a wandering phase during which she visits males on their territories. Wandering females will approach a male when he begins intensive waving display. A receptive female will eventually follow a male down into his burrow after courtship, which may lead to copulation.

"Directing" (*sensu* Zucker, termed "herding" by Crane) is a strategy in which the male dispenses with typical courtship waving and maneuvers the female toward his burrow with quick darting and zigzagging movements, at the conclusion of which he may push her into the burrow and then enter. It occurs sporadically in both lateral and vertical wavers. In *U. deichmanni*, the male may actually scoop her up between the major cheliped and body, and carry her down the burrow (Zucker, 1983). The female makes her final decision about whether to mate with a male after she is in the burrow. If she leaves his burrow, she is competed for and directed by one of the other males waiting nearby. Zucker observed repeated occurrences of this pattern. She suggests that this courtship pattern may occur in low-density populations where there is insufficient stimulation from courting males to evoke following of males by females.

Female *Uca* make individual selection of a male when they respond to courtship (Greenspan, 1975, 1980; Hyatt, 1977a, b; Christy, 1979, and 1980 as cited by Salmon, 1983). In *U. pugilator*, females respond most strongly to larger males, which are found in the higher parts of the intertidal zone (Hyatt, 1977a). However, this preference can vary with tidal amplitude in a most instructive fashion. Christy (1979, 1980) showed that when tides are high, larger males monopolize the relatively few burrows in the upper intertidal zone. Here female preference for the larger males is most pronounced. When tidal levels are lower, and more relatively high burrows exist, they are occupied by a greater size range of males. Here female preference for larger males is less pronounced. Relative burrow height in the intertidal is a good predictor of burrow collapse following tidal inundations. Since females in the burrow require several days for secure egg attachment to their pleopods, burrow collapse at this time causes heavy egg loss. *Uca* in salt marsh habitats, where the more stable substrate protects females' brood chambers from collapse, court and mate within the intertidal zone, and not just above it (Salmon and Hyatt, 1983a). In *U. vocans*, where males and females maintain separate burrows, females show no size preferences in males (Salmon, 1984). Zucker (1984) presents experimental evidence that both competition from larger males and female selection suppress courtship in small male *U. terpsichores*.

A responsive female moves in brief, jerky runs, with the body low and ambulatories held close to the body. A female *Uca* may signal nonreceptivity by a chela display (Powers, 1979), or by two quite different, specialized postures, "high-rise" and "legs-out." They make it very difficult for the male physically to place the female in the copulatory position, or to dislodge her from her burrow, respectively. Zucker and Denny (1979) discuss use of "repetitive high-rise" display in avoiding both conspecific and heterospecific courtship.

Although decalcification of the female's vulvar opercula is necessary for mating and egg deposition in *U. pugnax*, female responsiveness to male courtship begins before this stage (Greenspan, 1982). In *U. vocans*, the opercula do not fully calcify, and females are continuously receptive (Salmon, 1984).

C. Copulation in ocypodids

Copulation proper begins with the male climbing on the female's carapace from the rear. His minor cheliped is used in tapping, stroking, or plucking the anterior part of the female's carapace. The male performs stroking or tapping movements with his ambulatories. After a few seconds or minutes of stroking, the male turns the female upside down and holds her with his ambulatories. The tips of his gonopods are inserted into her gonopores, and the crabs may remain thus for a few minutes to an hour or longer. Copulation may not always be required for female oviposition. Greenspan (1982) presents evidence of sperm retention over several reproductive cycles in *U. pugnax*. In contrast, Salmon (1984) found that *U. vocans* males and females commonly mate more than once a fortnight.

D. Reproductive cycles

The timing of terrestrial activity in general depends on periods of low tide, when *Uca* emerge from their burrows. Their courtship activity does not go on during every low-tide period, however. It is usually restricted to one of the daily low-tide periods, the particular low-tide period varying among species. Zucker (1976) showed experimentally that courtship waving in some tropical species occurs at particular times of the day. Since it continues for only a few days at a time, but recurs twice a month, it shows a semimonthly rhythm. Differing period of low-tide courting has been viewed as facilitating reproductive isolation. The selective disadvantage of hybridization has been documented for *U. pugilator* and *U. panacea* (Salmon, 1978; Salmon and Hyatt, 1979). Cyclical female receptivity also means cyclical periodicity of larval release (also see Chapter 5). Christy (1978) and Zucker (1978) have interpreted this in terms of the optimal tidal conditions for larval dispersal. Greenspan (1982) interprets the coordination between male and female reproductive cycles in similar terms. Bergin (1981) presents evidence for endogenous cyclicity in larval release timing and discusses the importance of both daily and semilunar reproductive cycles in *Uca* relative to tide periods.

Species differ in their apparent resistance to overheating and desiccation, which are most severe later in the day. Less-resistant species ("primitive" *sensu* Crane, 1975) court during early-morning low tides, whereas others court later in the day. Two species of *Ocypode (O. kuhli* and *O. guadichaudii)* are diurnally very active and perform waving displays (Wright, 1968; Vannini, 1976b). As Salmon and Hyatt (1983b) point out, they are also the only ghost crabs that have intertidal burrows, probably extending below the water table. As in *Uca*, availability of water at the display site is important in combating desiccation stress and may be a permissive variable for consistent diurnal display. (See Chapter 7 for further information on desiccation tolerance in land crabs.)

E. Larval release in non-ocypodids

Owing to the variety of habitats exploited by nonocypodid land crabs, reproduction is modified in many ways (see Ducruet, 1976; also Chapter 5). There is one commonality among most terrestrial and semiterrestrial forms: larvae go through an aquatic phase. As usual, there is an exception to this rule.

Koba (1936) noted that among the potamonid species there is direct development. Large, yolky eggs produce young crabs that are carried by the female on the carapace and the ambulatory legs. During heavy rains, young (measuring about 4 mm) leave the female and aggregate at edges of vegetation (Fernando, 1960).

Arboreal crabs such as *Sesarma jarvisi, S. cookei* (Abele and Means, 1977), and *Metapaulias depressus* (Hartnoll, 1964), and "desert" crabs such as *Holthuisiana transversa* (MacMillen and Greenaway, 1978), appear to have abbreviated periods of larval development owing to the ephemeral nature of freshwater habitats in which larvae are released. Females of *A. pisonii* move down to the edge of mangrove areas, cling to mangrove roots with the abdomen extended over water, and vigorously vibrate the abdomen, releasing prezoeal larvae. Apparently, larval hatching stimulates the female to move to the water's edge to release the larvae, with much of the actual hatching occurring before reaching water (Warner, 1967). Female *Gecarcinus lateralis* migrate to the seashore to spawn (described in detail by T. G. Wolcott and D. L. Wolcott, 1982a). Spawning females avoid deep water for release, fanning the abdomen vigorously on initial inundation and moving to dry substrate after larval release.

Other terrestrial crabs such as *Sesarma dehaani, S. haematocheir*, and *S. intermedium* migrate to riverbanks to release their larvae. Saigusa (1981) reported that *S. intermedium* and *S. haematocheir* have a strong semilunar component to the release of larvae (similar to that reported for some *Uca*; Christy, 1978), while the sympatric species *S. dehaani* showed no such pattern. The high tolerance of fresh water by larvae of *S. dehaani* may explain the differences in the timing of larval release for these three species.

F. Copulation in non-ocypodids

Prior to copulation, females of some terrestrial crab species release pheromones. Ducruet (1976) gives the most recent literature review of pheromone work, including examples from both aquatic and terrestrial crab species. The sexual attractant for *Parathelphusa hydrodomus* reportedly is 5-hydroxytryptamine or a similar substance (Sundara, Santhanakrishnan, and Shyamalanath, 1973). Kittredge, Terry, and Takahashi (1971) made observations on the pheromone activities of *Pachygrapsus crassipes*, finding that β-ecdysone (a common arthropod molting hormone) is a sex pheromone for this species and for the aquatic crabs *Cancer antennarius* and *C. anthonyi*. Seifert (1982) examined the precopulatory activity of male *Carcinus maenas* in the presence of β-ecdysone excreted by premolt females. He found that ecdysones do not act as sex pheromones for *C. maenas*. However, his results did indicate that an unknown pheromone plays a role in initiating precopulatory behavior of male crabs, perhaps over a distance of up to 10 cm. Tactile and visual cues appear important for this species. Owing to difficulties in identifying unique pheromones, pheromonal attraction has not been demonstrated to be ubiquitous among terrestrial crabs.

Salmon (1982) suggested the following generalizations concerning copulatory behavior of crustaceans: (1) Aquatic brachyurans usually have prolonged courtship beginning with attraction of mates mediated by pheromones; males usually attend females for several days. (2) In some terrestrial crabs, pair bonding lasts only minutes with no prior or postcopulatory attendance. (3) Chemical signals are probably most often employed by aquatic species whereas visual and acoustical signals may be primary among land crabs.

There are glaring exceptions to these tenets among both the terrestrial and semiterrestrial crab species. Berrill and Arsenault (1982) describe, in detail, the mating behavior of *C. maenas*. The male carries the female in a precopulatory embrace for 2–16 days (mean 5.5 days). Following molt of the female, copulation (lasting on average of 2.7 days) occurs with the female actively participating in positioning and maintaining the copulatory pairing. A brief postcopulatory embrace may occur. This embrace is longer in the presence of another male. Females fan the developing egg mass to oxygenate it (Wheatly, 1981).

In the potamonid crabs (*Parathelphusa* spp.), a male and a female occupy a single burrow prior to mating. If the crabs select an old burrow, a new chamber may be added. Fernando (1960) observed that male crabs in burrows with females are very aggressive during the period just before and after mating. It is not clear from the descriptions whether the female must molt to facilitate copulation.

Helfman (1977a) described the copulatory behavior of *B. latro* from observation of a single event. Both male and female crabs were in intermolt phase during copulation. The male approaches the female slowly, clasps the dorsal meri of the chelipeds, and quickly moves forward to turn the female onto her back. Abdomens are extended, the male deposits the spermatophore, and the pair disengages.

The copulatory behavior of other coenobitids apparently lasts much longer than for *Birgus*. Hazlett (1966) and De Wilde (1973) depicted the migration and reproductive behaviors of *Coenobita clypeatus*, but no copulatory activity was recorded. Observations of mating of *C. clypeatus* in the field, confirmed by the presence of a spermatophore on the female, have been made (S. Gilchrist, unpub.). Initiation begins by the male grasping the aperture of the female's shell and moving her shell from side to side. A series of rocking and tapping motions either stimulates the female to extend from the shell, in which case mating proceeds ventral to ventral, or the female retracts farther into the shell and the male releases the shell. Page and Willason (1982) noted similar copulatory behavior in *C. perlatus*. Mating occurs during migration to the sea preceding larval release. Mating is ventral to ventral with both crabs about three-quarters out of their shells. Males pass the spermatophore to the females using the modified pereiopods. Mating may occur before release of the developed egg mass.

Male *Gecarcinus lateralis* drum the substrate to attract females. Receptive females will vigorously court males immediately prior to ovulation (Klaassen, 1973) and may also use substrate-borne acoustical signals in the process. Copulation usually occurs ventral to ventral, sometimes taking place a short distance up a tree trunk (Bliss, 1979).

Aratus pisonii is one of the few arboreal crab species dexterous enough to carry out elaborate mating rituals on tree trunks. Warner (1967) described the display and copulatory behaviors of the crabs on the trees in detail. Hartnoll (1965) concluded that copulation occurs only when the female is in molt phase.

Male *Sesarma reticulatum* and *S. cinereum* court females by making a precopulatory approach with the pereiopods extended. Seiple and Salmon (1982) observed that the males tap themselves or the receptive females, orient front to front with the female, and then wrap the second and third pairs of walking legs around the female. The tapping behavior during courtship is observed in other species as well (von Hagen 1967a, b, 1975). Males palpate the dorsal carapace and mouthparts of the females, the female is dragged on top of the male, and copulation occurs if the female opercula are soft. *S. cinereum* produce bubbles during mating, although *S. reticulatum* do not. It is uncertain why the bubbles are produced or if they play a role in the copulatory process. Copulation generally lasts for less than one hour for both species. Female *Sesarma ricordi* also mate shortly after molting. As noted by Hartnoll (1965) ovaries are slow to recover from egg laying and probably do not recover fully by the time the larvae hatch. Hartnoll (1965) reports that sperm plugs packed within the females situate sperm in optimal positions for fertilization. He speculates that sperm may be stored within females over several ovulations in that sperm are not lost during female ecdysis.

Grapsids tend to have a hard operculum that covers the genital aperture. Thus, it is unlikely that copulation can occur during intermolt. Those crabs that have a chemical component to the mating process generally copulate immediately following a molt by the female. There are grapsids, however, that do copulate during intermolt. *Pachygrapsus transversus* has a reduced operculum that can be pushed aside, allowing copulation during intermolt (Hartnoll, 1965). Although similar in appearance, female *Pachygrapsus gracilis* can copulate only immediately following a molt. These crabs have well-developed opercula. *Cyclograpsus integer* and *C. punctatus* represent yet another variation. These crabs have fully developed, hardened opercula that are movable (Brockhuysen, 1941; Hartnoll, 1965). Copulation could therefore take place at any time. It seems that those crab species that can copulate during intermolt tend to have females with ovaries capable of quick recovery, whereas those that mate after a molt have ovaries that take a much longer time to refill.

5. Agonistic and resource-related behaviors

A. Use of burrows by ocypodids

The agonistic behavior of ocypodid crabs, which includes fighting, fleeing, and the various forms of ritualized behavior that animals use in settling conflicts (termed "displays"), is most frequently observed in the context of burrow "ownership." The burrow is the focus of most of the life of adult ocypodids, and therefore the outcome of agonistic conflicts can affect feeding behavior, escape from predation, and reproductive success (see Dingle, 1983, for a comprehensive overview of agonistic behavior in crustaceans).

Uca have simple burrows extending diagonally downward. They range from a few centimeters to over a meter in depth, larger animals and species digging the longer ones. Chambers, forks, and dual entrances are not common. Substrate type and distance from water can influence the shape, branching, and diameter of burrows, as Chakrabarti (1981) has shown for *Ocypode ceratophthalmus* and Chakrabarti and Das (1983) have shown for *Macrophthalmus telescopicus*. Digging by *Uca* is done with the legs on the minor side, and the crab carries the excavated dirt away from the burrow. Both sexes dig burrows, and Becker and Hinsch (1982) have shown sex differences in the burrows of male and female *U. pugilator* regarding angle of entry, burrow length, and diameter. These differences might be expected, considering different uses during reproduction and size differences between the sexes. *Uca* commonly plugs the entrance to its burrow when the tide rises or during the midday heat. Very young crabs, often found near the water's edge, do not burrow, but bury themselves in the mud or escape into others' burrows. As Powers and Bliss (1983) indicate, less well known genera of ocypodids, such as *Dotilla* (Tweedie, 1950), *Heloecius* (Griffin, 1968), *Hemiplax* (Beer, 1959; Griffin, 1968), *Ilyoplax* (Ono, 1965), *Macrophthalmus* (Ono, 1965), *Deiratonotus* (Ono, 1965), and *Scopimera* (Harada and Kawanabe, 1955; Ono, 1965; Fielder, 1970), all show burrow or territorial defense to some degree, although definitive studies have yet to be performed.

Kraus (1982) discusses the different functional aspects of burrow utilization in ocypodid crabs, and relates them to semimonthly tidal cycles and ecological and social factors. Although use of burrows for mating and incubation is typical early in the season, use for protection during molting predominates late in the season, as in *Uca minax*. In *Scopimera* and *Dotilla*, a new burrow is usually dug at every low tide, but this burrow is defended only by crabs in reproductive phase, as is typical for *Uca*. During the nonreproductive phase, the burrow is used primarily as shelter from predation, heat, and desiccation. In contrast to that of *Uca*, the burrow of *Dotilla* may not reach the water table. Ocypodids have specialized setae on the fourth abdominal segment that collect interstitial water from the sand (Hartnoll, 1973; see also Chapter 7). If a crab feeds near its burrow during low tide, it normally retains possession of the same burrow for some time. However, it can be dispossessed by another, larger crab, in which case it first searches for an empty burrow. If this fails, it may in turn dispossess a smaller crab, prying it out with the major cheliped or digging it out with the ambulatories, or it may dig a new burrow itself.

B. Shelter from predation in ocypodids

Shelter from predation is important because fiddler crabs are conspicuous not only to humans but to other animals as well (also see Chapter 3). For example, Begg (1981) reports predation on *Uca annulipes* by the Cape wagtail, *Motacilla capensis*. This ground-foraging passerine actively chases and catches fiddler crabs. They are beaten against a stone, inducing leg autotomy, which allows the wagtail to eat the chelipeds, ambulatory legs, and carapace separately. A male, plucked from the mouth of its burrow by its large chela, was reported "too aggressive to overpower," and most predation was on females. Bosche (1982) observed predation on *Uca stenodactyla* by sandpipers, plovers, terns, and egrets. Juvenile crabs were taken selectively. Powers and Bliss (1983) report predation on *Uca* colonies by raccoons, rails, egrets, and gulls. Other predators are fishes and predaceous crabs (Salmon and Hyatt, 1983a, and refs.), and also dogs and humans (see Chapter 3).

Uca's chief defense is entering a burrow – its own or any other nearby. Owners briefly tolerate intruders. Vibrations and moving objects nearby inhibit reemergence. If it is away from any burrows, the crab sinks into the mud with a rotating motion. If the substrate is too hard for this, it quickly digs a burrow with the ambulatories. Alternatively, it may run into the water until it is entirely submerged. *Uca*'s final defense is rearing back and spreading out both chelipeds with chelae open. If a moving object comes within reach, it is seized with the chelae.

Remaining in a burrow is no guarantee of escape from predation. Subramanian (1984) describes an Indian monophagous snake eel, *Pisoodonophis*, that preys on *U. annulipes*. This predatory specialist emerges from its burrow during high tide and captures *Uca* by entering the burrows where fiddler crabs reside until low tide.

C. Agonistic behavior in ocypodids

Crane (1975) reviews agonistic behavior in *Uca* in considerable detail. The behavoral acts in displays include both body posture and movement, and therefore involve most of the body parts. She lists 15 basic agonistic acts and describes each in detail. Table 4.1 gives a very brief description of these acts. Among these is the "lateral-stretch" posture of the major cheliped, a classical threat posture not only of ocypodids but of brachyuran crabs in general. Crane (1975) dissects the anatomy of the fights themselves, looking in detail at the forms of animal interactions. She considers the ways in which physical force is used, as well as ritualized conventions in interacting, and describes 15 of the latter. The probable derivation of agonistic components from precursory behaviors is detailed.

Males fight only when they are aggressive wanderers, or when they are burrow holding. When a wanderer engages a burrow owner, use of physical force in the form of pushing and gripping is frequently a component of the interaction, and one of the animals may even be flipped over onto his carapace. Physical fighting only

Table 4.1. *Agonistic acts in* Uca *(after Crane, 1975)*

Manus rub	Both rub the outer surface of the manus against that of the opponent.
Pollex rub	Both rub the outer surface of the pollex against that of the opponent.
Pollex under-and-over slide	Actor rubs the inner edge of his pollex along the ventral edge of the opponent's pollex, then rubs the ventral edge of his pollex along the inner edge of the opponent's pollex.
Subdactyl and subpollex slide	Actor slides the dorsal edge of his dactyl along the inner edge of the opponent's dactyl while sliding the inner edge of his pollex along the ventral edge of the opponent's pollex.
Pollex-base rub	Inner edge of the actor's pollex rubbed against the base of the opponent's pollex.
Dactyl slide	Inner edge of the actor's dactyl slides back and forth along the inner edge of the opponent's dactyl.
Upper-and-lower-manus rub	Dorsal and ventral edges of opponent's manus rubbed by the inner edges of actor's dactyl and pollex.
Dactyl-submanus slide	Actor slides the inner edge of his dactyl along the ventral edge of the opponent's manus.
Interlace	The open chela of each is inside the gape of the other, with the bases of the gapes in contact and the fingers overlapping the opponent's manus.
Pregape rub	Tips of the actor's dactyl and pollex rub the manus of opponent.
Heel-and-hollow	Tip of the actor's pollex is inserted in the hollow at the base of the opponent's pollex.
Heel-and-ridge	Actor's pollex taps the inner edge of the opponent's manus.
Supraheel rub	Actor's dactyl rubs the upper proximal surface of opponent's manus.
Dactyl-along-pollex-groove	Actor rubs groove on opponent's pollex with his dactyl.
Subdactyl-and-suprapollex saw	Actor rubs inner edge of his dactyl across inner edge of opponent's pollex.

rarely results in injury. Puncture wounds on the manus can be the result of squeezing by the opponent's chela tips, probably most often during the eviction of a burrow owner by an intruder. Powers and Bliss (1983) report that neighboring male *U. pugilator* approach an intruder engaged in displacing a resident from his burrow. The result is that a number of owners interact with the intruder, which often results in his leaving the area. Encounters between neighboring burrow owners, on the other hand, are almost always restricted to ritualized exchanges. In the latter, acts by the two opponents are usually alternated, first one executing an act and then the other. The larger crab usually wins the encounter, but size does not always indicate the winner, as Hyatt (1983) documents in his survey of asymmetries in agonistic outcomes. Although the larger male usually wins, a smaller burrow owner may have a competitive advantage over a larger wanderer. Also, males may displace females from burrows independent of size differences.

Size is also important for agonistic success in *Ocypode*. Brooke (1981) reports that large male *O. ceratophthalmus* defeat smaller males in burrow contests. Smaller males were more successful when they dug their burrows later on the ebbing tide, when most burrow-related fighting was over, and also in less populated (inferior) habitat.

The extent to which males defend burrows varies across species. This is not only related to species differences in the role of the burrow in reproduction, but may also reflect habitat differences in the relative abundance of burrows, and in the energetic cost of digging new burrows (Hyatt and Salmon, 1978). Salmon (1984) reports that male *U. vocans* in Australia rarely defend burrows, and their fights are brief when they occur during courtship among groups of females. In this species, the male's burrow is not the site of copulation or oviposition. Both sexes are brightly colored, and it is the females that may have long fights, when they meet during foraging or contest burrow ownership. Females have as large an agonistic repertory as males, including some sex-specific displays. In contrast to Salmon's findings (1984), Nakasone (1982) reports intense fighting between male *U. vocans* over burrow occupancy in the population he observed.

The highly ritualized nature of typical agonistic exchanges in *Uca* is well illustrated in this excerpt from Crane (1975) on *U. rapax*: "An instigator, whether wanderer or neighbor, approaches a burrow-holder. A rub by one or both crabs, outer manus against outer manus, usually follows. Next the instigator sometimes holds perfectly still while his opponent slowly eases his chela into the actor's slide position; the two crabs may then reverse the role, the shift being accomplished slowly, without fumbling, and with the apparent cooperation of the crabs. In a few moments they may progress to a similar alteration of heel-and-ridging or, in heteroclawed encounters, to an alteration of heel-and-ridging with interlaces." (Heteroclawed refers to two males in which different chelipeds, the right or the left, have become the greatly enlarged major cheliped.)

Low-intensity threat displays by adult feeding males are used toward encroaching females or young crabs (Crane, 1975). Young and females themselves use threat displays similarly. Unreceptive females use threat displays toward males. They are also used interspecifically.

When a male is in courtship phase, a small territory centering on the burrow mouth is defended from other males (Crane, 1975). Visual displays, including waving of the major cheliped, and sounds given on the surface and underground, are thought to function in territorial defense (and in courtship as well). Agonistic encounters between males usually occur near the burrow entrance, although border disputes between neighboring males can take place on the periphery of the territory. Females also defend burrows from other females, especially when the resident is carrying eggs. They use visual displays, stridulation, and sometimes contact fighting with the ambulatories. The actual size of the territory defended varies across species with differences in the type of display performed, and therefore the extent of space required for displaying. Adult females, young, and heterospecific males may be tolerated when passing through territories. A display territory may be defended for only part of one low-tide period or held for a number of weeks.

D. Hood construction by ocypodids

In several species of *Uca* (e.g., *U. terpsichores, U. latimanus, U. beebei*), individuals carry substrate scooped up with ambulatory legs on the minor side to a position near the burrow entrance. They use this material to form a structure, which may consist of a wall surrounding the mouth of the burrow, termed a chimney, or of a small heap beside the entrance, termed a pillar, or of a more elaborate, arched hood that is parabolic on the side facing the entrance (Crane, 1941b; von Hagen, 1968; Zucker, 1974). The latter is reminiscent of the back wall on an outdoor stage. The occurrence of these constructions varies among populations, and even among individuals in the same populations (Zucker, 1974, 1981).

Zucker (1974, 1977) described agonistic situations in *U. terpsichores* that lead to hood construction. Closely neighboring males may fight until one leaves, and the other then fills in the latter's burrow. This is also known in *Ocypode* (e.g., Lighter, 1974). If such dislodging is not successful or neighbor density is high, males will build hoods that may visually shelter their burrow somewhat from that of neighbors. The males concentrate their displaying in the area in front of the hood, which decreases the time and energy investment in agonistic behavior. Hood construction is an integral part of courtship in *U. latimanus* (Zucker, 1981). This would appear to be a functional analog to the classical "reversed movements" described by Tinbergen (e.g., 1951) in gull behavior. The reduction in visual stimulation was said to result in diminished agonistic interaction between members of a pair at the nest in gulls. Here we see a similar reduction in agonistic territorial behavior, and an increase in courtship activity.

The sand pyramids built by male ghost crabs near the burrow entrance (like hoods in *Uca*) have an attracting effect on females and can stimulate avoidance in males (Linsenmaier, 1967).

E. Analysis of ocypodid aggression

Hyatt (1983) reviews crustacean aggression, and although his review is broad and includes the very important studies that have been carried out on stomatopods and hermit crabs, he gives due emphasis to work on *Uca*, the most intensively studied genus in the class. Hyatt's (1983) emphasis is on methods of analysis, and he deals with information-theoretic statistics applied to sequential acts during agonistic encounters. In such ritualized fights the behavior of two interactants is treated as a sequence of actions, in which behavior by one is reacted to by the other. Information statistics are based on the Shannon-Wiener equation, and Losey (1978) provides a comprehensive and creative discussion of their use in analyzing animal behavior. These statistics provide conservative estimates of the immediate effect of the behavior of reactants on each other during the course of interactions. Because measurements are made in neutral units (bits), which are independent of the context or the type of behavior, they permit comparisons across functional contexts such as fighting and courtship, and across species, provided certain conditions are met.

Data from the studies of Hyatt and Salmon (1978, 1979) on *U. pugilator* and *U. pugnax* are examined in special detail to illustrate one method of dealing with the vexing problem of nonstationarity in data taken over prolonged interactions in animal behavior. The problem of nonstationarity refers to the fact that the frequency of occurrence of various behaviors changes during the course of interactions between animals. Whether in fighting, courtship, or other contexts, some behaviors will be more common at the beginning or middle or end of encounters. Because their probability of occurrence is not stationary, the interpretation of mean values of association over the entire length of the interaction, as done in information theory, is highly questionable. Hyatt (1983) illustrates a "natural" method for subdividing interactions into smaller, more homogeneous data sets, by taking advantage of a natural demarcation of a change in the behavior of interactants that occurs when the male *Uca*, defending his burrow from a wandering male, first retreats into the burrow. Prediction of the ultimate winner is very much easier after this point. Information analysis of the structure of interactions before and after this event shows how much the behavior of the animals has changed, illustrating the need for subdivision of the data and the biological relevance of this approach.

F. Ecological effects of burrowing

Fiddler crabs are certainly the most conspicuous burrowing organisms in salt marshes, and can also represent the largest biomass of burrowing animals in this habitat (Katz, 1980). Katz (1980) determined that *U. pugnax* turned over approximately 18% of the upper 15 cm of marsh sediment in a year. Burrows increased the surface area of the marsh by 59%. *Uca* must certainly have significant effects on their immediate environment through the physical and chemical consequences of this level of manipulation of sediment. Although burrowing behaviors among non-ocypodid crabs have been described, they have received comparatively little attention (however, see Bright and Hogue, 1972). Mud crabs, in general, use burrows for shelter but do not typically display around this resource. It appears nonetheless that protection from predation or from desiccation is still the primary function of the burrows for these animals. *Helograpsus haswellianus* occupy burrows on southern Australian mudflats. Those crabs that possess burrows have an advantage over those without burrows in that predators appear to selectively consume crabs exposed on the surface. When crab densities are very high, two or more crabs will occupy the same burrow, reportedly to prevent collapse of the surrounding substrate (McKillup and Butler, 1979). The Australian amphibious *Holthuisana transversa* uses the burrow as protection from severe desiccation from drought by trapping moist air within the chamber. The crab can assume a dormant state within the humid chamber for many months (see MacMillen and Greenaway, 1978, and Chapters 7 and 8 for a full account). It burrows into the banks of streams, swamps, and waterholes penetrating up to 1 m.

The burrow can be effectively closed with a plug of clay, allowing the crabs to survive extended periods of drought (Bishop, 1963). The burrow of *Gecarcinus lateralis* serves many purposes. Bliss (1968, 1979) described the burrow as a refugium from predators as well as from extremes of abiotic conditions. Although the predominantly nocturnal behavior of *G. lateralis* partially circumvents problems of desiccation, the use of a burrow is especially important in water conservation during the molting and premolting periods (Bliss, 1979). During proecdysis, these crabs spend several weeks within the moist burrow, presumably to conserve energy, avoid predation, and decrease desiccation stress. Saigusa (1978) described the architecture and use of burrows by *S. haematocheir, S. intermedium,* and *S. dehaani* in detail. Sesarmids use the burrow in a manner similar to gecarcinids.

In some cases, many individuals were found occupying a common burrow or rock crevice. However, in areas where densities of burrows were high, only one or two crabs (a male or a pair of females) occupied a single burrow. The crabs appeared to hibernate during periods of environmental stress near the bottoms of the burrows, although none of the crabs were noted to hibernate in burrow waters.

Some hermit crabs burrow under leaf litter. *Coenobita spinosa* will burrow deeply under dead vegetation during cold weather and seemingly hibernate within the burrow (George and Jones, 1984). Although hermit crabs do not burrow in the same sense as a mud crab or a fiddler crab, it is not uncommon to find *C. compressus* buried shallowly (between 2 and 5 cm below the surface) prior to molting. *C. compressus* prevented from burying themselves prior to molting either in the field or laboratory suffer high mortality from parasitism (primarily from sarcophagid flies; S. Gilchrist, unpub.). In a sense, those hermit crabs also use the adopted molluscan shell, functionally, as a mobile burrow. The shell provides protection from predation and desiccation as well as serving as a reservoir for nurturing developing eggs. Abrams (1978) added to the descriptions of shell use and associated behaviors made by Ball (1972) for *C. compressus*. He observed that this crab modifies the inside of marine shells to increase internal volume. However, further information on shell acquisition and use is limited for this species. Although many authors point out the importance of the shell in environmental resistance for several species of semiterrestrial and intertidal hermit crabs (McMahon and Burggren, 1979; P. Taylor, 1982; McMahon and Wilkens, 1983), little information is available on shell quality and use for terrestrial hermit crabs.

6. Acoustic display

Crane (1975) discusses various means of sound production in *Uca*, principally of two general kinds: stridulation, in which two body parts such as the legs are rubbed together, and drumming or rapping, in which a body part such as the major chela is struck against the substrate (or the carapace). Some subgenera typically use rapping and others use stridulation; some species employ both methods.

Although the use of sound signals has been discussed more in the context of courtship, *Uca* uses sound in agonistic situations as well. For example, a male *Uca* fleeing from capture will enter an occupied burrow. The resident male gives acoustic signals in burrow defense, but will tolerate the intruder for a brief period of time (Crane, 1975). Opinion differs as to whether acoustic display functions "primarily" in agonistic or sexual contexts (Crane, 1975; cf. Salmon and Hyatt, 1983b), but it is clearly important in both.

In some *Uca* species courting males produce sounds when females are close by. After the female has responded to waving by close approach, the male switches to sound signals before she descends into the burrow (von Hagen, 1961, 1962; Salmon, 1965). Males of temperate species of *Uca* court during nocturnal low tides and signal exclusively with sounds at that time (von Hagen, 1962; Salmon, 1965). Salmon (1965) showed that males stimulate each other in acoustic signaling, increasing their calling rates in response to the increased acoustic courtship intensity of other males. Females of *U. pugilator* can detect a male 50–100 cm away (Salmon and Horch, 1972).

Both fiddler crabs and ghost crabs (*Ocypode* spp.) transduce acoustic stimuli with Barth's myochordontonal organ, which appears as a thin-walled, triangular structure located on the meral segment of each walking leg (Horch, 1972; Salmon, Horch, and Hyatt, 1977). Ghost crabs respond to sound transmitted through the substrate and also through the air, up to 10 m away, whereas fiddler crabs seem to respond primarily to substrate-transmitted acoustic signals, over a range of about 1 m. Moist sand is a better sound conductor than dry sand (Salmon et al., 1977), so moisture content is a relevant ecological variable.

The nature of an acoustic signal transmitted as pressure waves through the substrate changes not only as a function of the properties of the ground through which it travels, but also as a complex function of the structure of the crab that receives the signal (inertia of body parts, muscle tensions, hemolymph pressure). Recent studies of unrestrained *U. pugilator* using a noncontact optical method of measurement (laser Doppler vibrometry) have shown that the substrate signals undergo a major transformation as they travel up the leg (Aicher et al., 1983). When the responding crab assumes the "upright" posture normally elicited by a drumming stimulus, there is a large amplification of low frequencies corresponding to the natural rap repetition rate. Higher frequencies are attenuated by the same posture. Furthermore, responding crabs exhibit subtle postural changes, the "altering response," as rap series continue, fine-tuning their bodies to the input, which enhances low-frequency transmission even more and extends the potential range for signal detection. Since rapping rate is a species-typical character, it would be interesting to know whether species of *Uca* differ in the details of signal amplification relevant to this parameter.

Von Hagen (1984) has documented differences in principal mode of sound production in two sibling species in Trinidad. *Uca mordax* produces rapping sounds by ambulatory percussion, whereas *U. burgesi* makes "howling" sounds through quivering movements of the cheliped. Each can produce signals similar to those of the other. Comparison with acoustic signals from other parts of the species' ranges suggests that there is character displacement where they are in sympatry. Differences in their visual displays are less extreme (von Hagen, 1983).

Crane (1975) discusses various means of sound production in *Uca*, principally of two general kinds: stridulation, in which two body parts such as the legs are rubbed together, and drumming or rapping, where a body part such as the major chela is struck against the substrate (or the carapace). Some subgenera typically use rapping and others use stridulation; at least one species employs both methods. Crane (1975) recognizes 16 different movement patterns in sound production, which she describes in detail. Sound is especially important in nocturnal courtship, where it replaces waving, as well as while signaling from the burrow. Males can be induced to switch from visual to acoustic signaling by covering the displaying males to exclude light, and the converse switch occurs if males calling at night are illuminated by floodlights (Salmon and Atsaides, 1968).

Different species of *Ocypode* produce sound by either rapping or stridulation (Horch, 1975). Recent studies on the acoustic communication of ghost crabs have shown adaptations to counter both physical distortion of the sound signal when traveling from sender to receiver, and enhancement of its detection by the receiver (Horch and Salmon, 1969; Horch, 1971, 1975). Sound emission from the male's burrow is an important part of courtship, although the timing of chorusing and the degree to which acoustic diplays are used vary greatly across species (Salmon and Horch, 1973; Horch, 1975). Frequency of the calls is just above the pitch where the background noise would interfere maximally, and just low enough to minimize frequency-dependent attenuation, e.g., 300–600 Hz in *Uca minax*. Species distinctiveness is encoded in the different temporal patterning of calls (e.g., among *Ocypode ceratophthalmus, O. cordimana,* and *O. laevis*). Hearing sensitivity drops off rapidly just below the fundamental frequencies of the calls (where the sound energy is concentrated), further minimizing detection problems by selectively filtering. Neighboring males alternate calling with both conspecifics and other species, which decreases acoustic interference. This may also facilitate localization of an individual male by an approaching female, which is a further problem in any colonial breeding system. More detailed accounts of bioacoustic studies of ocypodid crabs, chiefly *Uca* and *Ocypode*, can be found in reviews by Dumortier (1963), Horch and Salmon (1969), Guinot-Dumortier and Dumortier (1966), Markl (1968), Burke (1954), and Aicher et al. (1983). *Dotilla, Ilyoplax, and Macrophthalmus* also produce sounds, and sound production is probably widespread in branchyurans, although largely unstudied in most groups.

Stridulation in conjunction with posturing is common in aggressive displays of *Coenobita* (Hazlett, 1966; S. Gilchrist, unpub.). Clicking by rapping appendages together and by tapping the shell are integral parts of aggressive encounters of *C. clypeatus* and *C. compressus*. When alarmed, *Birgus latro* briskly stamps the second pereiopods. At other times, even when not apparently alarmed, this crab produces continuous clicks (Grubb, 1971). This may be a proximity warning to conspecifics. *Gecarcinus quadratus* and *Potamocarcinus richmondi* produce sounds by stridulating if their retreats are disturbed (Abele, Robinson, and Robinson, 1973). Abele et al. (1973) describe the mechanism of sound production by *P. richmondi* as somewhat unusual, in that it involves the second and third maxillipeds. *Gecarcinus lateralis* produces sounds (Klaassen, 1973) in a different manner than does *G. quadratus*. Although both scrape the subhepatic regions, *G. lateralis* uses the palms of the chelipeds and *G. quadratus* uses the merus of the cheliped. Self-tapping, performed after a successful aggressive encounter and during courtship (von Hagen, 1975; Seiple and Salmon, 1982), may be common among the grapsids.

7. Locomotory behavior

There have been a number of interesting studies on the ability of crustaceans to move through their varied environments in an organized and directed fashion. Considerable attention has been paid to the mechanics and energetics of movement, which are treated in detail in Chapter 10.

Work with non-ocypodid crabs reveals a variety of locomotory adaptations to land. Although most crabs are adapted for walking sideways, all are capable of movement in any direction (Manton, 1977; Clarac, 1982; Clarac and Barnes, 1985). Unlike other terrestrial crabs, the hermit crabs typically walk in a forward direction. Such movement may be facilitated by the lateral compression of the body. The energetics of hermit crab forward motion have been extensively investigated (Herreid and Full, 1986a, b, see also Chapter 10). *C. compressus* moves forward using a tripod gait similar to that observed in insects. This type of interleg coordination may be a requirement for support in walking forms (McEvoy and Ayers, 1982).

One of the unique locomotory modifications of the land crab is the development of tree-climbing abilities (only one shrimp genus, *Merguia*, exhibits tree-climbing behavior; Abele, 1970). Such activities extend potential habitat exploitation of five major crab families: Coenobitidae, Gecarcinidae, Grapsidae, Ocypodidae, and Xanthidae.

Arboreal species can be categorized into three major groups by the extent of their use of the habitat. Of the 560 tree-climbing species recorded in the literature (see von Hagen, 1977, for a partial list), about two-thirds use the trunk region for feeding or as a refugium from predators (Yaldwyn and Wodzicki, 1979; Wilson, 1985; see Table 4.2). The second group, consisting of approximately one-fourth of the species, is associated almost exclusively with the root regions, below 1.5 m. These animals feed on epizoans and epiphytes of the roots as well as forage on the substratum when the tide is low.

Table 4.2. *Use of trees as refugia by land crabs*

Crab species	*N*	Tethered position	Average time to consumption (in min)	Comments
Uca pugilator	25	Trunk	—	Not consumed during experiment
	25	Base of mangrove	47 ± 4	2 escaped
	25	Open substrate[a]	28 ± 7	3 crabs were partially buried; these were consumed last
Sesarma reticulatum	25	Trunk	—	Not consumed during experiment; 3 escaped
	25	Base of mangrove	32 ± 6	Once one crab was consumed, other predators gathered
	25	Open substrate[b]	18 ± 5	
Aratus pisonii	25	Trunk	—	Not consumed during experiment
	25	Base of mangrove	36 ± 6	
	25	Open substrate[c]	24 ± 6	

[a] *U. pugilator* on open substrate remained motionless after inundation; 3 crabs were partially buried in the substrate.
[b] *S. reticulatum* strained at the tether when inundated.
[c] *A. pisonii* thrashed on tether after inundation.

The remaining species inhabit the canopy and upper trunk regions. This third group is composed of two distinct subgroups (Thienemann, 1935; von Hagen, 1977). The phytotelmic species reproduce in fresh water trapped by trees and other plants. Adults rarely venture down to the ground. Larvae have accelerated development that takes place entirely in the temporary freshwater habitat of micropools. The second group, called canopy crabs, lives as adults in the canopy, leaving this habitat to return to the sea for reproduction. The larval stages are planktonic, with the crabs returning to the arboreal habitat as juveniles.

Unlike typical terrestrial locomotion in crabs, the primary direction of movement is straight ahead for tree-climbing crabs, with little sideways movement. *Gecarcinus lateralis* (usually smaller individuals) will climb trees for short distances to avoid predation. However, motion is more laterally oriented than in many other crabs (S. Gilchrist, unpub.). Von Hagen (1977) described the tree-climbing behavior of five species of grapsids, noting that *Goniopsis cruentata* ascends and descends in an anterior down position. The sesarmids he observed, however, ascend in an upright position and descend with the fronto-orbital region oriented downward. The ability of tree-using crabs to jump has been related anecdotally (Miller, pers. comm.). Short jumps (< 10 cm) from a root or piling to the substrate or laterally between objects appear common in *Sesarma reticulatum* and *S. cinereum*. However, movement in this manner from a lower point has not been documented.

The arboreal hermit crabs of the Coenobitidae tend to climb with eyes directed forward and to descend with that same orientation. *C. compressus* and *C. clypeatus* tend to move upward with a spiral rotation, typically in the direction of the shell aperture opening (Gilchrist, unpublished data). This seems especially noticeable when the shell is large relative to the crab size. It is not uncommon for shelled hermit crabs to lose their grip and drop to the ground during descent. Watt, Dunham, and Schoene (1985) have made similar observations.

The freshwater crabs of the genus *Parathelphusa* are semiterrestrial, occurring in or near streams from the sea coast to an elevation of about 7,000 feet. These crabs climb only a short distance up the stalks of plants, presumably to topple the plant. Leaves and portions of the plants are then consumed (Fernando, 1960). Other crabs such as *Callinectes sapidus* also climb for short distances to topple plants. These predators, however, do not eat vegetation. They consume gastropods stripped from the stalks (Warren, 1985) or charge into herding *Uca*, returning to the water to consume their prey.

Like other facets of terrestrial crustacean behavior, locomotory rhythms may be strongly influenced by environmental conditions. Limiting dessication may be one of the primary factors affecting locomotory rhythms. For example, the aquatic/land crab *Potamon potamios* exhibits size-related differences in activity correlated with water loss (Warburg, Goldberg, and Rankevich, 1982). Medium-sized crabs tend to be nocturnal in their locomotory patterns, whereas larger crabs are diurnal in such patterns. Taylor and Wheatly (1979) suggest that emigration from water to land in *Carcinus maenas* could function in thermoregulatory cooling, with a subsequent reduction in oxygen demand. Allometric constraints, imposed by surface area to volume ratio of respiratory and absorptive surfaces, in geometrically similar organisms adapted to terrestrial habitats, may require adjustments of behavior, favoring nocturnal activity patterns in smaller organisms (Wernick, 1982; Pelligrino and Stoff, 1983; Pelligrino, 1984).

In addition to correlations with temperature, rhythmic locomotory activities may be linked to tidal fluctuations. This is especially evident in the semiterrestrial, intertidal forms. For example, the shore crabs *Hemigrapsus edwardsi* (Naylor and Williams, 1984), *H. oregonensis* (Batie, 1983), and *Pachygrapsus crassipes* (Stutz, 1978), among others, show a very strong circatidal component to locomotory rhythmicity compared to the much weaker circadian component. Some coenobitid hermit crabs appear to have a strong diurnal rhythm (Rebach, 1983). However, periods of greatest activity for most semiterrestrial crabs are correlated both with the tide and with illumination. Terrestrial crabs associated with freshwater, such as *H. transversa* (Bishop, 1963; MacMillen and Greenaway, 1978) and *H. agassizi*, do not have a tidal rhythm but do increase activity with heavy rainfall.

Several studies have been made of the abilities of land crabs to transduce and use environmental information in locating goals, and in determining the appropriate direction for movement. Among the ocypodid crabs, much interest has been shown in the rhythmicity and direction of movements that bring the animals down into the intertidal zone for feeding at low tide, and return them to higher beach locations to wait out the high-tide periods in their burrows. These specific aspects of locomotory behavior will now be discussed.

8. Orientation

Comprehensive reviews of crustacean orientation are given by Herrnkind (1983) and Rebach (1983). Generally, orientation during locomotion is a function of proprioceptive, statocystic, and visual input (Neil, 1982). Schoene (1971) and Vannini (1975) have suggested that there is a complex intergradation of data from many systems used to detect crab position, perhaps rivaling that of the vertebrates (McEvoy and Ayers, 1982).

The position of the sun and polarized light (useful when the sun is not directly visible) are environmental cues that ghost crabs and fiddler crabs can use in orientation to the beach. In *U. pugilator*, landward–seaward directional choices are guided by a well-developed, time-compensated sun-compass mechanism and the ability to use polarized light (Herrnkind, 1968). This has also been shown in *Uca tangeri* (Altevogt and von Hagen, 1964; Altevogt, 1965), *Ocypode ceratophthalmus* (Daumer, Jander, and Waterman, 1963), and *O. quadrata* (Schoene and Schoene, 1961), and probably occurs in other species of *Uca* and *Ocypode*.

Visual landmarks are used by *Uca* (Herrnkind, 1968, 1972) and by *Ocypode* (*O. ceratophthalmus*; Hughes, 1966). *Ocypode* returns to its burrow after foraging, covering total distances of up to 300 m. *Uca* can cover distances of over 100 m when moving from the burrow zone downslope to foraging areas. In both *Uca* and *Ocypode*, "foraging" movements such as these also commonly involve social interactions with other conspecifics. Ghost crabs are the fastest-running crustaceans, and can move at speeds of up to 3.4 m/sec (Herrnkind, 1983).

They recognize their burrows by surrounding visual landmarks. If these are displaced by the investigator, the crabs search where the landmarks have been transposed (Hughes, 1966). *U. tangeri* has difficulty in returning when it is displaced from its burrow (Altevogt and von Hagen, 1964).

Sand pyramids constructed by males near the burrow are used as landmarks by female *Ocypode saratan* (Linsemaier, 1967). *Uca* can orient to gross visual landmarks such as mangroves, other trees and shrubs, and clumps of beach grass (Creutzberg, 1975). In tests with different-shaped stimuli on the beach, they approach those with vertical components resembling these types of vegetation. They avoid stimulus objects with rounded contours, which crudely resemble predatory shorebirds (Langdon, 1971, as cited by Herrnkind, 1983). Form discrimination in *Uca* is well established (Langdon and Herrnkind, 1985).

The use of kinesthetic memory of past movements, enabling an organism to "retrace its steps," is a process termed idiothetic orientation. The use of visual cues is not required. This is of great value for the ocypodids because many species are active at night, avoiding diurnal predators and higher diurnal water and salt loss. At the approach of a predator, a crab can instantly locate its own burrow, unoccupied and of the correct size. In animals whose population density can exceed $100/m^2$, efficacy of burrow locating is of great value (Herrnkind, 1983). Idiothetic orientation has been demonstrated in *U. tangeri* (Altevogt and von Hagen, 1964) and *U. rapax* (von Hagen, 1967), has been implicated in the movements of *U. pugilator* (Herrnkind, 1972), and is probably present in other *Uca* species. *Ocypode* also uses kinesthetic memory to return to its burrow (von Hagen, 1967a, b).

Uca can use the slope of the substrate in orientation (Kropp and Crozier, 1928; Young and Ambrose, 1978). *Uca* also uses geotaxis in burrow construction, in which the vertical component of movement is influenced by gravity. These crabs burrow at a fixed angle to the direction of the force of gravity (Schoene, 1961). Perception of the effects of the force of gravity on the body is primarily through organs such as statocysts, and through proprioceptive feedback from supportive structures such as the legs, that are differentially affected when the body position changes relative to the vertical force of gravity (see Schoene, 1980, 1984).

There seems to be a paucity of information on orientation of the arboreal species. *Aratus pisonii* with covered eyes appear to orient by geotaxis while in trees, restricting lateral movement. These temporarily blinded crabs easily climb trees if placed on mangrove prop roots. However, location of roots by temporarily blinded crabs is what

Table 4.3. *Orientation by* Aratus pisonii

Condition	*N*	Average time to ascent (min)
Control	30 (19♀, 11♂)	6.2 (4.8) ± 1.3
Temporarily blinded (total)	52 (40♀, 12♂)	79.4 (67.1) ± 19.6
Temporarily blinded (partial)	16 (9♀, 7♂)	7.5 (4.4) ± 2.7

Note: Control crabs were collected and released within 3 m of a mangrove prop root; treatment crabs were similarly handled. Crabs were blinded with duct tape over the eyestalks. Partially blinded crabs had tape over only one eyestalk. The average time (in minutes) to contact a root and begin ascent was recorded. There was no significant difference between males and females, thus these data were combined. Times in parentheses are times for crabs to locate roots. Standard deviation is recorded for overall time to ascent.

would be expected by chance alone, suggesting that visual cues are critical for these animals in orientation (Table 4.3). However, data on shore crabs and selected coenobitids are common (see Rebach, 1983, for an overview). In general, slope is used by a number of terrestrial and semiterrestrial crabs for orienting to the sea (van Tets, 1956). For example, nocturnal activities carried on near the shore (e.g., feeding, larval release; see De Wilde, 1973) may rely on the accurate detection of slope. In *Coenobita clypeatus*, the behavior threshold for slope detection lies between 0° and 2.5° (Dunham and Schoene, 1984). Chemotaxis is also important in orienting hermit crabs on land (Kurta, 1982), as well as other terrestrial and semiterrestrial crabs. As in many other adaptations to land observed in hermit crabs, the aesthetascs of these animals are more similar structurally to those of insects than they are to those of close marine congeners (Ghiardella et al., 1968). Related to the chemotactic orientation is the ability of some hermit crabs to use the wind direction (anemotaxis) to orient to the sea. Vannini and Chelazzi (1981) describe the flight behavior of *C. rugosus* on Aldabra. They suggest that the compass reaction for returning to the home beach is a learned response based on celestial cues and landmark recognition. They suggest that such cues are not available at night, however. Doujak (1985) demonstrated that the shore crab *Leptograpsus variegatus* could distinguish stars of 0.5 magnitude or brighter. While there are only a few stars of this magnitude, orientation by these cues is possible at night. The work by Doujak (1985) points out the need for evaluating nocturnal vision of other semiterrestrial and terrestrial crabs, especially those that search for food at night. This is the first demonstration that an invertebrate with a compound eye has the physical capability of using celestial navigation.

9. Summary and conclusions

Although there is a great deal of information concerning the behavior of the highly visible Ocypodidae and related groups, comprehensive syntheses of most other terrestrial and semiterrestrial crab families are lacking. We have reviewed here the most representative studies of the behavior of land crabs. It is, first of all, apparent that the Ocypodidae have received a great deal of attention, probably due to their diurnal conspicuousness and abundance. However, even in this group, interactions with the environment (such as in feeding), and with each other (as in communication), are only now beginning to be understood. We really know almost nothing about the behavior of most land crabs, including the most common species. When these neglected groups have been studied, their behaviors can be compared with those of their marine relatives. That will provide a rich source of information about the constraints and permissions encountered when diverse crab groups emerged from the sea to take up a terrestrial existence.

What generalizations can we make about the behavior of those terrestrial crabs that have been studied?

1. How, when, and where feeding takes place is determined by celestial cycles, the microhabitat, physiological rhythms, and social behavior. Contact chemoreception is an important stimulus in initiating feeding. Deposit feeders and detritivores tend to burrow, but carnivores and herbivores typically do not burrow into the ground.

2. Grooming and cleaning behaviors seem to be minimal in terrestrial species.

3. Reproduction, even in the most terrestrial crabs, is dependent on returning to the sea, or at least use of standing fresh water, for larval or postlarval development.

4. The range in apparent social structures among land crabs is great, from the complex systems of *Uca* to the solitary existence of *Birgus*.

5. Males are the more active sex in courtship, but females exercise selection of males, or their defended resources, such as burrows.

6. Communication can involve visual, chemical, and acoustic signals. Signaling systems are sometimes elaborate and may involve construction of edifices near a burrow.

7. Burrowing appears to be a function of distance from water, with those species farther from streams or the sea showing less burrowing activity.

8. Lateral movement in locomotion is common for most terrestrial species. However, land hermit crabs tend to move forward, as do their

marine counterparts. Species that exploit arboreal habitats have added a novel dimension to crab locomotion as a consequence.

The lack of information on many groups makes the preceding conclusions very tentative. However, they should be viewed as stimuli for generating studies of the effects of terrestriality on the behavior of land crabs.

Acknowledgments

We express sincere appreciation to T. Greely and P. Bryant (both of New College) for their tireless assistance in compiling the literature review for the non-ocypodid crabs. We thank J. Holmquist of the National Audubon Society at Tavernier, Florida, for sharing valuable information on grooming structures and functions of terrestrial crustaceans. S. Ye (Toronto), D. Matthews (New College), and G. Dubbert (New College) assisted with typing. We are grateful to R. Tuckerman (Toronto) for the care with which he prepared Figure 4.1, and to N. Zucker (New Mexico) for critically reviewing an earlier draft of the chapter. Some of the information on the Coenobitidae was gathered during tenure as a postdoctoral fellow by one of us (S. G.) at the Smithsonian Tropical Research Institute (Panama, S.A.). Funding from the Natural Sciences and Engineering Research Council of Canada to D. D. is gratefully acknowledged.

5: Reproduction and development

RITA G. ADIYODI

CONTENTS

1. Introduction

Anomuran and brachyuran crabs show varying degrees of terrestrial adaptation, ranging from species that venture out of water only briefly to those that actually inhabit arid or semiarid lands and depend on bodies of water for reproductive purposes only. Among the Anomura, considerable terrestrial adaptation is seen in the family Coenobitidae (land hermit crabs). Adults of the genus *Coenobita* are permanently confined to molluscan shells in which provision is made to trap water, whereas *Birgus* exhibits the shell-dwelling habit only during the early part of its life. The coconut crab *Birgus latro* returns to the ocean only to release the brood. Among the Brachyura there is a wide range of variations in habitat. For example, the shore crab, *Carcinus maenas*, wades through shallow water and wanders over the beach, hiding in burrows made in wet sand. The Ocypodidae are amphibious crabs inhabiting littoral waters and estuaries. Among grapsids, some are inhabitants of rocky shore and some are terrestrial. Some frequent mangroves and yet others live in fresh water. The Gecarcinidae are best adapted for terrestrial life, but they turn to the sea to breed. The freshwater crabs *Parathelphusa hydrodromus* and *Gecarcinucus steniops* depend on fresh water of the rice fields primarily for brood release and for the development of the young. Not surprisingly, the tremendous diversity observed in habitats of crabs is reflected in their reproductive and developmental biology.

2. Breeding seasons of land crabs

The breeding cycles of crabs are programmed to a great extent in accordance with cues received from diverse environmental factors, such as availability of food and water, temperature, salinity, lunar phases, tidal cycles, and competition between congeneric species (also see Chapter 3). Generally, species confined to the tropics and having access to perennial sources of water have prolonged breeding seasons that may extend for several months. Monsoon rains, however, affect the breeding pattern of estuarine and coastal decapods in peninsular India (cf. Pillai and Nair, 1971) because changes in salinity affect the survival of their larvae. The breeding season of the crabs inhabiting the shores of Cochin in southwestern India (e.g., *Uca annulipes, U. marionis, Sesarma quadrata,* and *Ilyoplax gangetica*) extends from August to April (Pillai and Nair, 1968). *Ucides cordatus*, living in mangrove swamps of Brazil, breeds from January to May (Alvesmota, 1975). Some species breed throughout the year: for example, *Pachycheles tomentosus* and *P. natalensis*, inhabiting the rock crevices in the low-water-mark areas of Karachi (Ahmed and Mustaquim, 1974); *Emerita portoricensis* (Goodbody, 1965) and *E. holthuisi* (Subramoniam, 1979), living on sandy beaches of Jamaica and the southwest coast of India, respectively; and *Aratus pisonii* (Warner, 1967) of the mangroves of Jamaica.

Land crabs inhabiting colder regions are generally seasonal breeders, however, and avoid breeding and development during the winter months. Thus, ovigerous females of the spider crab *Libinia emarginata* from Massachusetts occur only from late May/early June, and breeding activity ceases before the beginning of winter (Hinsch, 1972).

Again, a comparable pattern of breeding activity has been reported in the stone crab, *Menippe mercenaria*, in which the peak spawning season in Florida is from May/June to October (Cheung, 1969). In the land crab *Gecarcinus lateralis*, inhabiting the southwest coast of Florida, August and September consititute the chief spawning season (Bliss et al., 1978).

Crabs that depend on seasonal rainfall for breeding purposes program their reproductive and developmental biology in such a way that the brood is released only during the rainy season. In the state of Kerala along the Malabar coast of southwestern India, two geographically separated populations (Trivandrum and Calicut) of the freshwater crab *P. hydrodromus* have been found. Although both inhabit rice fields, these two populations exhibit different breeding patterns, apparently in response to variations in the pattern of the monsoon. For example, Trivandrum (southern Kerala) receives two moderately heavy monsoons, one from the southwest during June-August and the second from the northeast during October-November, plus some scattered rains in between. The Trivandrum population of *P. hydrodromus* has been thus able to accommodate two reproductive cycles in a year, the brood release in each case taking place at the onset of the monsoon. In contrast, the Calicut population of *P. hydrodromus* breeds only once a year, releasing the young during June. The southwest monsoon is very strong in Calicut and the whole of North Kerala, while the northeast monsoon is weak. As a result, there is insufficient water in the rice fields in Calicut during October-November to support a second breeding cycle.

The Madaras population of *P. hydrodromus* breeds only once a year like the Calicut population, but in Madaras, unlike Calicut, the young are released at the onset of the northeast monsoon (October-November) (Pillai and Subramoniam, 1984). The possibility that two populations of *P. hydrodromus* in different geographical areas, distinguished by breeding either once a year or twice a year, may differ genetically has not been determined. This can be verified only by transferring natural populations of one locality to the other and studying their breeding patterns under field conditions, over a period of say, two or three years, but to our knowledge this has not been attempted. As Pillai and Subramoniam (1984) maintain, it may be that vitellogenesis and spawning in *P. hydrodromus* take place when temperatures are high and availability of water low; brood release takes place only when water is abundant and the environmental temperature low. Monsoon-dependent breeding has been also reported in other land crabs from peninsular India such as *Barytelphusa cunicularis* (Diwan and Nagabhushanam, 1974; Mathad, 1983), *G. steniops* (Samuel, 1984; Santhamma, 1985), and *Barytelphusa [Parathelphusa] guerini* (Ali, 1955).

Whereas land crabs preferring fresh water (e.g., *P. hydrodromus)* wait for the monsoon to arrive to release their brood into fresh water, saltwater-preferring land crabs such as *G. lateralis* and *Cardisoma guanhumi* (Gifford, 1962a) migrate to salt water to release their brood.

3. Programming of reproduction and molt

Unlike natantians, in brachyurans reproduction and molting are programmed as "antagonistic" or mutually exclusive processes (K. G. Adiyodi and R. G. Adiyodi, 1970; R. G. Adiyodi, 1978, 1985; Adiyodi and Subramoniam, 1983). In the Calicut population of *P. hydrodromus*, for example, June/July is the molting season and March/April the spawning season (Anilkumar, 1980). In *Menippe mercenaria*, the peak spawning season falls in August and September, and molting takes place primarily in November and January (Cheung, 1969). A similar temporal separation of breeding and molting occurs in *C. maenas* (Bauchau, 1961; Demeusy, 1963; Cheung, 1966) and *G. lateralis* (Weitzman, 1964).

Detailed studies of the biology of the Calicut population of the freshwater crab *P. hydrodromus* have brought to light some fascinating aspects of the nature and extent of this temporal separation (K. G. Adiyodi and R. G. Adiyodi, 1981; Kurup and Adiyodi, 1981, 1984; Kurup, 1983; Mathad, 1983). In many instances these findings seem applicable to several other brachyuran species. In Calicut, the young of *P. hydrodromus* are liberated from the brood pouch of the mother, as mentioned earlier, in June with the beginning of the monsoon. The physiological emphasis being largely on somatic growth, the juvenile crabs go through a rapid succession of ecdyses and finally attain sexual maturity by about September of the same year. In young females, the spacing between molting and reproduction becomes apparent following the first mating in postmolt during September/October. By December these young females, which are now in intermolt (stage C_4), enter the first vitellogenic cycle and spawn in March. They remain in the berried state until the beginning of the monsoon in June, when the brood is released. These females enter premolt soon afterward and undergo ecdysis by late June/July. In these crabs, now over a year old, mating is scheduled during postmolt. They enter intermolt by late July/August and remain in that stage until the next molting season in the following June/July.

A general analysis of the molting cycle reveals that in the Brachyura intermolt is chiefly devoted to reproduction, whereas the premolt/postmolt is for somatic growth. In crabs such as the stone crab, *Menippe mercenaria* (Cheung, 1969), the Trivandrum population of *P. hydrodromus* (R. G. Adiyodi, 1968a; G. Anilkumar and K. G. Adiyodi, un-

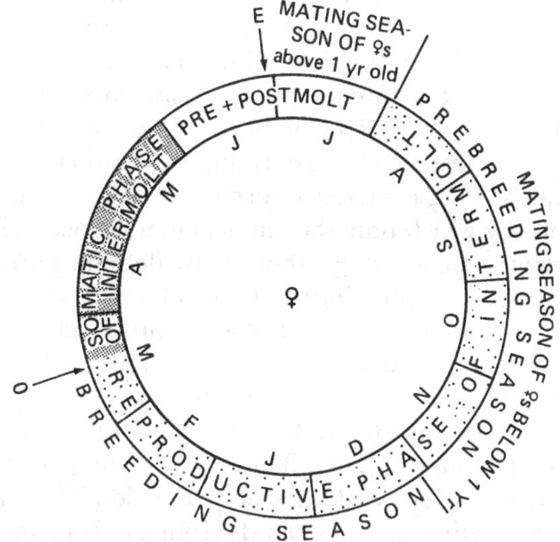

Fig. 5.1. Schematic diagram of the physiological calendar of adult female *Parathelphusa hydrodromus*. *E*, annual ecdysis; *J–D*, January to December; *O*, oviposition. (Modified from Anilkumar, 1980.)

pub.), and the mud crab *Rhithropanopeus harrisii* (Morgan, Goy, and Costlow, 1983), intermolt accommodates more than one reproductive (ovarian) cycle.

In sexually mature, mated female *P. hydrodromus*, the intermolt (stage C_4) itself is physiologically divisible into two distinct phases: a reproductive phase and a somatic phase (Kurup and Adiyodi, 1981) (Fig. 5.1). During the reproductive phase, which usually encompasses the first half of the intermolt, many aspects of the physiology of the crab appear supportive of reproduction. The breeding season per se (characterized by vitellogenic activity in the ovary) begins only in December in *P. hydrodromus*. However, even from August (i.e., early intermolt stage), the hepatopancreas, spermatheca, and hemolymph, tissues that are either directly or indirectly involved in reproduction, show high concentrations of metabolites. Furthermore, bilateral eyestalk excision precipitates precocious vitellogenesis in females during this season. Thus the period extending from August to November, i.e., the prebreeding season, is clearly a preparatory period for vitellogenesis in *P. hydrodromus*. The reproductive phase is remarkably unfavorable for somatic growth in this species; limbs autotomized during this period show only minimal levels of regenerative growth.

The reproductive phase, which begins in August, extends all through the breeding season, December to March. Following spawning, sometime during March/April, the phase of somatic growth sets in. Walking legs autotomized during this phase undergo regenerative growth with considerable vigor. Bilateral eyestalk ablation results in accelerated somatic growth culminating in ecdysis. During the somatic phase of the intermolt, the effect of bilateral eyestalk ablation on ovarian growth is far less spectacular compared to that on somatic growth. The somatic phase of the intermolt extends from March/ April to May/June, by which time the animal enters proecdysis. Thus in female *P. hydrodromus*, although there exist definite periods earmarked for somatic growth (molting) and reproduction (vitellogenesis and spawning), an adjustment in physiological processes in favor of the ensuing physiological event (reproduction and molting) takes place long before the actual occurrence of the event in question.

In *P. hydrodromus*, mating occurs between hard (intermolt) males and soft (postmolt) females. To facilitate this, molting is normally programmed in males a few weeks ahead of females. Thus, in crabs over a year old, the mating period extends from late June to August. By September, young virgin females (about six to seven months old) become available in the population, and their mating season extends from September to October/November. Obviously, June to November is the period when adult males of the species are sexually active.

In adult male *P. hydrodromus*, the vas deferens factor, defined as

$$\frac{\text{weight of the vas deferens in mg}}{(\text{carapace width})^3 \text{ in cm}} \times 100,$$

reaches its highest value by August/September and drops to its lowest by December. The glandular epithelium lining the genital duct, which is responsible for the production of the seminal plasma, is very active from June/July to October (K. G. Adiyodi and R. G. Adiyodi, unpub.). Protein, the chief organic component of semen, attains its maximum value during August/September (Mathad, 1983). Histological and gross-morphological analyses reveal that the androgenic gland becomes hypertrophied and remains in an apparently active phase throughout the period extending from June/July to September/October. Signs of large-scale degeneration, marked by nuclear pyknosis and glandular hypothrophy, have been found from December to May (R. G. Adiyodi and K. G. Adiyodi, 1976; Adiyodi, 1984). In the testis, there is an appreciable decline in gametogenic activity from January to May. Nuclear pyknosis and cellular degeneration are common among the germinal cells during this period. Testicular activity is revived by June, but gametogenesis per se reaches peak levels only by September/October.

Male *P. hydrodromus* older than one year remain in intermolt from July to May. The active as well as the inactive phases of semen pro-

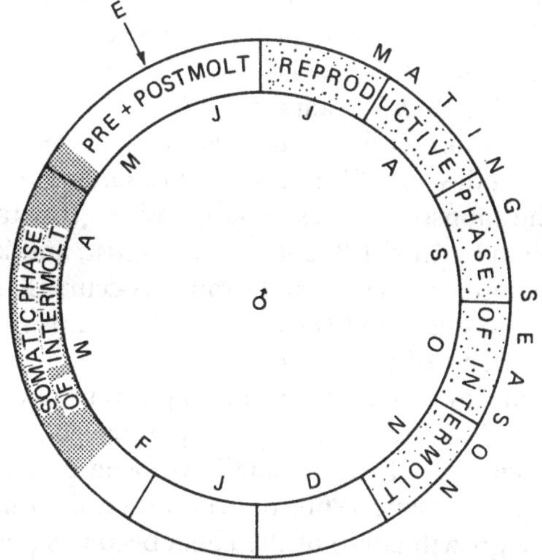

Fig. 5.2. Schematic diagram of the physiological calendar of adult male *Parathelphusa hydrodromus*. *E*, annual ecdysis. (From Mathad, 1983.)

duction and androgenic gland activity are accommodated within the same intermolt stage (C_4). From the point of view of reproductive physiology, the first half of intermolt extending from July to November/December represents the active phase and the second half extending from December to May the relatively inactive phase in males.

Incidentally, in these males, the emphasis on somatic growth (as evidenced by the ability to regenerate lost limbs) is rather low from July/August to February and is high from late February to May. Thus, male crabs are evidently in the reproductive phase of intermolt from July to November/December and in the somatic phase of intermolt from February to May (Kurup and Adiyodi, 1981, 1984; Kurup, 1983; Mathad, 1983) (Fig. 5.2). This physiological pattern of male crabs is comparable to that of female crabs belonging to the same age group.

4. Sexual differentiation, sexual maturation, and mating

A. Sexual differentiation and maturation

An individual is regarded as sexually mature only if it can accomplish successful mating. The age and size at which sexual maturity is attained vary both between and within species, and also with geographic distribution of the populations. Females of *Cancer magister*, for example, become sexually mature on attaining a carapace width (cw) of 10 cm in British Columbia, but only 8 cm in Washington

(Mackay and Weymouth, 1935; Clever, 1949). In India, *Barytelphusa cunicularis* and *B. [Parathelphusa] guerini* reach sexual maturity at a cw of 4.4–4.5 cm (Diwan, 1973), the latter occurs at age one to two years (Gangotri, Vasantha, and Venkatachari, 1971). The Calicut population of the freshwater crab *Parathelphusa hydrodromus* oviposits in March/April; juveniles about 2 mm in cw are liberated in June at the beginning of the monsoon. These young crabs grow fast and attain a cw of 1.0–1.5 cm by July, 1.8–2.4 cm by August, and 2.8–3.1 cm by September when puberty (sexual maturation) occurs (Kurup and Adiyodi, 1981). The spider crab *Chionoecetes opilio* attains sexual maturity only on reaching a cw of about 5 cm.

Sexual dimorphism generally becomes apparent in crabs only with sexual maturity. The differential growth patterns of the chela of the male and the abdomen of the female have been extensively studied in Brachyura (see Teissier, 1960, for references). In males, the discontinuity of the growth curve of the chela becomes perceptible with sexual maturation. In *P. hydrodromus*, one of the chelae shows a disproportionate increase in size with the approach of sexual maturity. This dimorphism, which first becomes obvious at a cw of about 2.6 cm, is steadily accentuated by molts in sexually mature males (S. Gupta and K. G. Adiyodi, unpub.). In *C. maenas*, the characteristic form of the female abdomen is acquired following puberty molt to sexual maturity, which is one or several molting cycles preceding the one in which the first ovarian cycle commences (Veillet, 1945; Démeusy, 1958). In some groups, this discontinuous nature of the growth curve is realized with a single molt. This is quite unlike most decapods in that secondary sex characters are generally not fully developed with a single molt, but only consecutively in the course of several molts.

Information on postembryonic development, sexual differentiation, and sexual maturation is limited only to a few species of crabs. Payen (1974) has studied these aspects in some detail in four species of brachyuran crabs: *Callinectes sapidus, Carcinus maenas, Rhithropanopeus harrisii*, and *Menippe mercenaria*. She found that the reproductive system remains completely undifferentiated in the course of embryonic, larval, and the first two postlarval (juvenile) stages. However, the sex could be differentiated externally in both *C. sapidus* and *R. harrisii* in the second postlarval stage in that the first pair of pleopods appear in the male; males also lack the ventral ectodermal invagination in the sternum of the sixth thoracic segment (which is the rudiment of the female genital opening). The sexually undifferentiated postlarva possesses only four pairs of pleopods (pleopods 2–5). In *M. mercenaria* also, although the first pair of pleopods is differentiated in the male in the second postlarval stage, in the female the rudiment of the genital opening appears only by the fourth postlarval stage

(Payen, 1974). In *P. hydrodromus*, the first pair of pleopods becomes distinct in the male on attaining a cw of 4 mm (S. Gupta and K. G. Adiyodi, unpub.). In *C. sapidus, C. maenas, R. harrissi,* and *M. mercenaria*, differentiation of gonads and their ducts becomes obvious from the third postlarval stage (about 3 mm carapace length). During this period the female's gonad develops a posterior extension on either side. The gonad, at this stage, is composed of protogonia and mesoderm cells, and the gonoducts are represented by a pair of ductal anlages. In the ovary of *R. harrisii*, in the third postlarval stage (about 3 mm in cephalothoracic length), oogenesis seems to have advanced, and it was possible to detect gonia in mitosis, oocytes in early prophase of meiosis and also somatic cells. A comparable stage in ovarian development is achieved in *C. sapidus* and *C. maenas* only on attaining a carapace length of 1.5 cm and 1.0 cm, respectively (Payen, 1974). In *C. sapidus* the oviduct is formed during the third postlarval stage by the proliferation of mesoderm cells of the gonad. They penetrate the muscles of the sternum of the sixth thoracic segment as a very thin duct with a tiny lumen. Their distal region is formed by the invagination of the sternal epidermis and is provided with a cuticular intima. The female genital opening is formed by the third postlarval stage in *C. sapidus*, by the fourth in *R. harrisii*, and on attaining a carapace length of 4.2 mm in *C. maenas* (Payen, 1974).

Compared to the oviduct, the vas deferens of crabs differentiates toward the sternum of the eighth thoracic segment more slowly. In the third postlarval stage, it has a tiny lumen and terminates blindly in a zone showing intense mitotic activity.

After initial sex differentiation, the abdominal appendages of both sexes undergo further changes. In males, for example, the exopodites of the second to fifth pleopods disappear and a tiny groove appears toward the posterior region of the first pleopod, by the fourth postlarval stage in *R. harrisii*, and by the sixth in *C. sapidus*. In *R. harrisii*, the gonopods become calcified by the sixth postlarval stage, and the second pleopod accommodates itself into the groove of the first pleopod. The gonopod thus acquires its final and definite morphology in *R. harrisii* only by the sixth postlarval stage and in *C. sapidus* by the eighth at a carapace length of 1.9 cm. In *C. maenas*, a comparable degree of gonopod development is detained on attaining a carapace length of 12.5 mm (Payen, 1974).

Female *C. sapidus* show four pairs of small biramous pleopods (second to fifth) with nonpedunculate endopodites until the eighth postlarval stage. In *R. harrisii* and *C. maenas*, the segmentation of the endopodites and exopodites is apparent by the fifth postlarval stage, and the bristles adorning them appear only by the sixth (Payen, 1974).

In the fourth postlarval stage (3.1–3.6 mm in carapace length) of

R. harrisii, the androgenic gland can be distinguished in the subterminal region of the vas deferens. The gland develops further by the fifth postlarval stage (4–4.6 mm carapace length) as the male genital duct forms coils in the anterior region and spermatogenesis begins in the testis. By the sixth postlarval stage (4.75–5.6 mm in carapace length), regional differences become obvious in the vas deferens and the testis develops lobes and lobules, connected by collecting ducts. The androgenic gland develops further at a carapace length of 6.0–7.5 mm, and spermatogenesis begins at 12.5 mm. In *C. sapidus*, the gonads are in an undifferentiated state during the first five postlarval stages. In males, the androgenic gland (the source of the male sex hormone) becomes distinct as a slender cord in the sixth postlarval stage (4.75–5.6 mm in carapace length). Further significant changes occur in the male genital system only by the eighth postlarval stage, on attaining a carapace length of 1.9 cm. Spermatogenesis is complete as crabs reach a carapace length of 3 cm; spermatophore formation is evident in individuals with a carapace length of 4.3 cm (Payen, 1974). In *P. hydrodromus*, spermatogenesis begins in crabs having a cw of 1.2 cm, and spermatophore formation is evident only from 1.6 cm upward (S. Gupta and K. G. Adiyodi, unpub.).

In female *R. harrisii*, further differentiation of the oviduct, spermatheca, and vagina occurs rapidly. In the ovary, oogenesis progresses actively from the fifth postlarval stage. During the sixth postlarval stage there is an increase in the number of oocytes in the early phase of vitellogenesis ("previtellogenesis," as reported by Payen, 1974), and the ovary acquires a swollen appearance. In *C. sapidus*, the process of sex differentiation and sexual maturation appears to proceed relatively slowly compared to *R. harrisii*. In fact, little change was noted in *C. sapidus* during the first eight postlarval stages (8.0 mm carapace length). Oogenesis begins in this species only on attaining a carapace length of 1.5 cm. Meanwhile, the oviduct becomes distinct from the spermatheca and vagina. Primary growth of the oocytes begins in *C. sapidus* with a carapace length of 2.5 cm and greater. In *C. maenas*, on the other hand, oogenesis starts at a carapace length of 9–10 mm (Hartnoll, 1968) and in *P. hydrodromus* from a cw of 1.2 cm upward (S. Gupta and K. G. Adiyodi, unpub.).

In young female *P. hydrodromus* having a cw of 1.5 cm, the spermatheca has a saccular appearance; the columnar epithelial cells, which line the spermathecal lumen, are only moderately secretory and produce an intrinsic secretion (S_1). On attaining a cw of 2.0 cm, the spermatheca assumes the shape of a fluid-filled sac; the cells in the secretory epithelium are now much larger and are actively secreting. In the spermathecal lumen of crabs having a cw of 2.5 cm, apart from S_1, certain basophilic secretory droplets (S_2) have been found in the lumen during the intermolt stage (Krishnakumar, 1985).

An appreciable increase in total body mass has been observed during larval development of the spider crab, *Hyas araneus* (Anger et al., 1983). Loss in organic body weight was noticed during the last period preceding metamorphosis to the juvenile crab stage. After metamorphosis, biomass accumulation occurs in the juvenile crab at a rate far exceeding that in all larval stages. Possibly, this is due to the absence of further metamorphic changes. Protein constitutes 50–68% of the ash-free dry weight, lipid 17–30%, and chitin 1.6–3.7%. The percentage of protein increased during zoeal development but declined subsequently. Lipid, on the other hand, decreased during larval development.

Studies on the hemocyanin of megalops and young juvenile stages of *C. magister* revealed the close resemblance in molecular weight and copper content between the hemocyanin of young individuals and that of adults (Terwilliger and Terwilliger, 1982).

Sexual maturation is apparently a long process, often extending through two or more ecdysial cycles. In males, the major changes associated with sexual maturation generally coincide with the development of the androgenic gland. This suggests that a hormone or hormones produced by this gland may play a role in male sexual maturation. The control mechanisms involved in larval and juvenile ecdyses may be largely comparable to those in the adults. For instance, 20-hydroxyecdysone stimulates the initiation of premolt (apolysis) in the third and fourth instar larvae of *R. harrisii.* In vitro experiments have further revealed the presence of molt-inhibiting hormone (MIH) in extracts of the eyestalks of third and fourth instar zoeae and megalops (Freeman and Costlow, 1984). Eyestalk extirpation accelerates molting in megalops and juveniles (first crab instars) (Freeman, West, and Costlow, 1983).

B. Mating

Mating is programmed as a seasonal event in the vast majority of terrestrial and aquatic crabs. Generally, crabs copulate only after attaining sexual maturity. Morphological and physiological factors often prevent premature copulation. For example, in the Majidae the vulvae are known to acquire the right size only after the pubertal molt (Hartnoll, 1969).

The physiological phase of the crab at the time of mating varies between species and within the species between sexes. For example, in a large number of species (e.g., *Carcinus, Cancer, Parathelphusa*), pairing takes place between hard (intermolt) males and soft (postmolt) females. However, in crabs such as *Pachygrapsus marmoratus, Aratus pisonii,* and *Sesarma ricordi,* in which partners are uniformly in intermolt stage, decalcification of the opercula that normally close the vulvae occurs only during a brief period in intermolt when mating is to take place (Hartnoll, 1969). In grapsids, ocypodids, and xanthids, mating is generally scheduled during intermolt in either sex, although certain exceptions have been reported. Thus in the xanthid *M. mercenaria* and the grapsid *Pachygrapsus crassipes,* the mating females are uniformly in postmolt.

In several crabs, mating follows a period of courtship (also see Chapter 4). This is especially apparent in those groups in which mating females are in postmolt stage. Elaborate courtship behavior is not very common in those groups in which the mating partners both are in intermolt. Exceptions do occur: In *Corystes,* the male shows a tendency to carry the hard female during the brief period in which she undergoes opercular decalcification. In *Uca pugnax,* decalcification of the opercula is a prerequisite for mating and oviposition, but this is not always linked to behavioral responsiveness to courting males, which often precedes decalcification (Greenspan, 1982).

Aspects of sexual recognition and courtship have been extensively investigated in a number of terrestrial crabs (also see Chapter 4). In *M. mercenaria,* for example, the male patiently waits at the burrow occupied by the premolt female and copulates soon after she has completed the molting process. In crabs such as *Eriocheir sinensis* and *Grapsus grapsus,* courtship is associated with a complex dancing pattern. The males of *Sesarma rectum* produce drumming sounds to attract the opposite sex. Males of *Dotilla* attract the females by waving the chelae (Altevogt, 1957b). In *Ocypode saratan,* it is the pyramid made by the male that attracts the female (Hartnoll, 1969). The most complex courtship pattern, however, has been observed in *Uca.* The males display themselves prominently by waving and beckoning movements of their large chela along with dancing and drumming (Crane, 1957). Thus visual, tactile, and auditory stimuli are made use of by male *Uca* in attracting the attention of females.

The period of courtship can be either very brief, as reported in *P. crassipes,* or rather prolonged, as in *Corystes,* where it extends for several days. The females are active partners in mating in those species in which there exists an elaborate courtship behavior. Thus, in *Corystes* and *P. crassipes* the female cooperates with the male. In the broad-fronted group of *Uca,* it is the female that takes the initiative to mate.

Courtship behavior generally is observed only in those species in which there is an appreciable reduction in duration of the period of receptivity of the female. This significantly reduces the chance of mating, unless special adaptive behavior patterns are evolved to advertise the presence of the receptive female. Some evidence indicates that sex attractants could play a useful role. A sexual pheromone inducing searching behavior in males is said to be released into the water by molting females of *Portunus sanguinolentus* (Ryan, 1966) and by grapsids (see Hartnoll, 1969). Female sex pheromones possibly function as a stimulus for inducing complex behavioral patterns in the male, thereby ensuring successful mating. However, there is some doubt regarding the copulatory readiness of males throughout the intermolt (C_4). In the Calicut population of *P. hydrodromus*, for instance, the male is sexually active only during the first half of intermolt (reproductive phase), which extends from June/July to October/November. During the second half of intermolt (i.e., December to May), the male is sexually inactive (R. G. Adiyodi and K. G. Adiyodi, 1976; Kurup, 1983; Mathad, 1983; Adiyodi, 1984; S. Gupta and K. G. Adiyodi, unpub.). It is not clear if such a pattern exists in other land crabs, but it is not unlikely in species with a long intermolt.

5. Reproduction in the female

Although considerable information exists on reproduction in female land crabs, our understanding is not complete, and the following accounts must also draw on information available for terrestrial crabs.

A. Oogenesis

Oogenesis encompasses the train of events beginning with germinal activity in the ovary, preparatory processes preceding yolk deposition, and vitellogenesis leading to the production of eggs that are ready for fertilization.

The basic ovarian structure varies with groups. A regional demarcation into germarium and vitellarium is not obvious in the ovary of most crabs (see Adiyodi and Subramoniam, 1983). Among terrestrial crabs, the germarium is scattered as germinal nests throughout the ovary in *G. lateralis* (Weitzman, 1966), *P. hydrodromus* (Anilkumar, 1980), and *G. steniops* (Santhamma, 1985). In *P. marmoratus* (Rouquette, 1970) and *L. emarginata* (Hinsch, 1970), however, the germarium is confined to the central region of the ovary.

Germinal activity appears to be a seasonal phenomenon in a number of species. Nests of germ cells in various phases of division and also cords of young oocytes occur in the ovary of *G. lateralis* during the postovulatory period (September/October) (Weitzman, 1966). This also appears to be the case in *P. hydrodromus* during March/April, which is the postspawning season of the species (G. Anilkumar and K. G. Adiyodi, pers. comm.) and during May/June in *Potamon koolooense* (Joshi and Khanna, 1982a). Germinal nest activity is rather prolonged in *G. steniops* (Santhamma, 1985). The nests, for instance, are in a state of high activity during the postspawning period (March/April), decrease from May to August, and are resumed by September. Germinal activity is low in *G. steniops* throughout the active vitellogenic period, which extends from October to December/January, but a revival of activity has been noticed during the final phase of vitellogenesis in February (Santhamma, 1985).

Cytophysiological events that succeed initial germinal activity and precede vitellogenesis proper may be either accelerated to take place within a few days or may progress leisurely over a span of several months. Vitellogenesis often interrupts the completion of meiosis. The oocytes produced after oogonial mitosis enter the meiotic prophase. In *P. marmoratus*, these oocytes seem to proceed to the zygotene stage, whereupon their subsequent progress becomes arrested and they move toward the periphery of the ovary. By autumn, they enter the early phase of vitellogenesis. The ovary acquires a milky or creamy color and, with further progress of vitellogenesis, turns yellow. This relatively slow early phase of vitellogenesis extends all through the winter and early spring. By late April/early May, spring molt interrupts the course of ovarian growth. The final stage of vitellogenesis ensues soon afterward and progresses for a month or so; oviposition finally takes place by June/July. It is only within about 48 hours to oviposition that the oocytes enter the metaphase of the first meiotic division. In *P. marmoratus*, the second meiotic division is suspected to occur only following oviposition (Pradeille-Rouquette, 1976).

In the Calicut population of *P. hydrodromus*, germinal activity occurs during the postspawning season (March/April), but vitellogenesis per se commences only by the following December. During the relatively long period extending from March/April to December the ovary remains in the avitellogenic phase. Adult females undergo annual ecdysis in June, when their ovaries are in an avitellogenic phase and the bulk of the oocytes are small, ranging in diameter from 0.1 to 0.2 mm. Ultrastructural details of changes that occur in oocytes during this period are not available; there is an unmistakable systematic gain in size of the oocytes as they move away from the germinal nests. The growth of oocytes is not synchronized within the ovary. As a result, oocytes in various stages of development are found in a graded pattern: Those produced earlier from the germinal nest are in a more advanced stage of development, and those formed subsequently are smaller and less developed. The gradation of oocytes into various stages of growth is discernible even as oocytes enter the vitellogenic phase in December and is seemingly retained throughout the early vitellogenic phase (0.6–1.3 mm in diameter) (February). In *P. hydrodromus*, the growth becomes synchronized in oocytes only as they enter the final phase of vitellogenesis (> 1.3 mm in diameter) by late February-March (S. Gupta and K. G. Adiyodi, unpub.).

An almost comparable pattern of germinal activity is observed in *G. steniops*, a species sympatric with *P. hydrodromus* in distribution in at least the northern parts of Kerala. Oviposition and annual ecdysis of this species coincide with those of *P. hydrodromus*. The avitellogenic phase, which is characterized by slow oocyte growth and immediately follows initial germinal activity in March/April, extends up to August in *G. steniops*, whereas the vitellogenic phase extends from September to March. Monthly analysis of the ovary has revealed that all oocytes are not in the same stage of development throughout the avitellogenic phase or through the major part of the vitellogenic phase (until January/February). In fact, oocyte growth becomes synchronized in *G. steniops* only during the final phase of vitellogenesis (Santhamma, 1985). The cytoarchitectural changes that occur during the avitellogenic phase of the oocyte of land crabs have yet to be investigated in detail.

Yolk (vitellus), the nutritive reserve produced to meet the energy requirements related to embryonic development, accumulates in growing oocytes of crabs only over a prolonged period. The rate of vitellogenesis is not uniform throughout this period: In many instances, there occurs an initial phase in which the vitellus accumulates only at a slow pace and a second phase during which the yolk amasses rapidly. In *P. hydrodromus*, we have called the first phase Vitellogenesis I (V_1) and the second Vitellogenesis II (V_2). V_1 immediately succeeds the avitellogenic phase and extends from December to January/February. We have subdivided V_1 into three stages based on size, cytology, and color of the oocytes. The ovary is whitish in stage 1 (V_1S_1), and the majority of the oocytes acquire a diameter ranging from 0.5 to 0.6 mm. The ovary turns pale yellow in stage 2 (V_1S_2) (oocytes 0.6–0.8 mm in diameter). In stage 3 (V_1S_3), the ovary is orange, and oocyte diameter ranges from 0.81 to 1.33 mm. V_2 extends from late January/February to March, and it is during this period that the oocytes rapidly increase in size (1.31–2.0 mm in diameter) by a massive accumulation of yolk. The early vitellogenic phase of *P. marmoratus*, which extends from autumn to spring molt (Pradeille-Rouquette, 1976), seems to correspond to V_1 of *P. hydrodromus*.

B. Composition of yolk

The composition of yolk in land crabs varies with species. Water is the chief component; in *P. hydrodromus*, 43.5% of the wet weight of the mature egg is water (Anilkumar, 1980). Among solids, the yolk contains all organic and inorganic components needed for normal embryonic development. Proteins and lipids comprise the bulk of the

Table 5.1. *Major chemical constituents in the ovary of* Parathelphusa hydrodromus *during the vitellogenic cycle* ⋆ ± *SD (expressed as percentage of ovarian wet weight)*

	Vitellogenesis I			
Compound	Stage 1	Stage 2	Stage 3	Vitellogenesis II
Water	76.84 ± 9.9	70.30 ± 4.24	53.36 ± 4.69	43.50 ± 4.35
Lipids	8.45 ± 2.42	11.35 ± 2.24	19.25 ± 2.89	22.55 ± 2.08
Proteins	12.04 ± 1.73	12.96 ± 3.32	24.81 ± 4.69	24.02 ± 8.19
Free amino acids	0.72 ± 0.39	2.06 ± 0.58	0.73 ± 0.17	0.69 ± 0.10
Oligosaccharide fraction	2.34 ± 0.84	1.39 ± 0.29	0.86 ± 0.12	0.88 ± 0.17
Polysaccharide fraction	0.50 ± 0.23	1.01 ± 0.27	0.29 ± 0.02	0.40 ± 0.14

Source: Anilkumar (1980).

organic constituents of yolk, as indicated in Table 5.1 for *P. hydrodromus* during the final phase of vitellogenesis (Anilkumar, 1980).

Lipovitellin (LV), the major high-density lipoprotein/glycolipoprotein fraction present in crustacean yolk, has an average molecular weight of 350,000 daltons (Wallace, Walker, and Hauschka, 1967). LV is often conjugated to carotenoid pigments. In *C. sapidus*, the LV molecule contains only a single carotenoid component, but a mixture of all egg carotenoids is found in *Cancer pagurus* (Zagalsky, Cheesman, Ceccaldi, 1967). The principal ovarian carotenoid of *Potamon dehaani* is astaxanthin (Matsuno, Yoshiko, Mahahiro, 1982). The amino acid composition of the LV molecule of *P. crassipes* is shown in Table 5.2.

Lipovitellin appears to be a composite protein fraction made up of several subunits, which often show a tendency to dissociate the reaggregate. Purified LV of *Pachygrapsus* is composed of three principal subunits, LV_1, LV_2, and LV_3 with molecular weights of 118,000, 105,000, and 83,000, respectively.

Apart from LV, the yolk may contain other protein fractions. For instance, three glycoproteins, comparable to the ovomucoids of vertebrates, and three simple proteins have been found in the ovary of *P. hydrodromus* toward the final phase of vitellogenesis (V_2) (Adiyodi, 1968c).

In oocytes, LV is confined within membrane-bound bodies called yolk platelets, while the glycoproteinaceous component of the yolk remains outside the yolk platelets. Lipid spheres are freely suspended in the ooplasm (Eurenius, 1973; Dhainaut and De Leersnyder, 1976).

C. Vitellogenesis

Vitellogenesis is the process by which yolk, complete in all its organic and inorganic constituents, is accumulated in the egg. Information

Table 5.2. *Amino acid composition of lipovitellin of* Pachygrapsus crassipes

Amino acid	mole (%)
ASP	9.9
THR	6.3
SER	10.6
PRO	6.3
GLU	12.1
GLY	6.2
ALA	8.0
VAL	8.0
CYS	0.0
MET	0.0
ILE	5.7
LEU	10.9
TYR	2.7
PHE	4.3
LYS	4.2
HIS	1.4
ARG	3.3

Source: Lui and O'Connor (1977).

on vitellogenesis in land crabs is chiefly centered around the pattern of buildup of proteinaceous yolk.

Ultrastructural, histochemical, and autoradiographic evidence leaves no doubt regarding the ability of crab oocytes to produce yolk proteins autosynthetically. In a number of species, the oocytes show the presence of the requisite infrastructure for large-scale protein synthesis, at least during some period in the course of their development. The oocytes become endowed with the necessary cytoarchitecture for intra-oocytic yolk protein synthesis during the early phase of vitellogenesis in *Libinia emarginata* (Hinsch and Cone, 1969), *P. hydrodromus* (Adiyodi, 1969a), *G. steniops* (Santhamma, 1985), *Cancer pagurus* (Eurenius, 1973), *Emerita analoga, P. crassipes* and *Cancer* sp. (Kessel, 1968). In the spider crab *L. emarginata*, there is cytological evidence for transport of nuclear material into the cytoplasm during the initial phase of oocyte growth. At first the nucleolus appears as a mass of granules. Subsequently, there is a progressive compartmentalization of the nucleolus into granular and fibrous areas. Later, the nucleolus appears ring-shaped, the central area enclosing a homogeneous substance. Granules seem to be emitted from the nucleolus, and such granules tend to aggregate beneath the nuclear membrane as membrane-bound vesicles. Subsequently, these vesicles release their contents into the cytoplasm by blebbing of the outer nuclear membrane. Both granules and fibers have been frequently observed to migrate through the nucleopores (Hinsch, 1970).

Morphological evidence for the elaboration of protein droplets in developing oocytes during the early phase of vitellogenesis (V_1S_1) has been obtained in *P. hydrodromus* (Adiyodi, 1969a). Oocytes become multinucleolated, and the nucleoli show a tendency to move toward the periphery of the germinal vesicle, in preparation for the production and expulsion of tiny RNA-positive granules into the ooplasm. There is also a progressive increase in vacuolation of the nucleoli. Large, pyroninophilic, protein yolk precursors appear in the ooplasm during the late stage 1 of V_1. Yolk proteins are then formed as isolated droplets among the large precursor bodies. At the ultrastructural level, a vesicular type of granulated endoplasmic reticulum containing disk-shaped intracisternal granules has been observed by Kessel (1968) in oocytes of *E. analoga, P. crassipes,* and *Cancer* sp. In these species the yolk is reportedly formed in these intracisternal granules. The rough endoplasmic reticulum is elaborately developed in growing oocytes of *C. pagurus* during the early phase of vitellogenesis (Eurenius, 1973). Disk-shaped intracisternal granules are present in distended regions of the rough endoplasmic reticulum throughout the cytoplasm. These granules later transform into type 1 yolk platelets.

Lui and O'Connor (1977) have made an autoradiographic analysis of the ovary of *P. crassipes* containing eggs ranging from 0.2 to 0.3 mm in diameter, cultured in vitro. ^{3}H-Leucine from the medium is incorporated into two principal subunits of the lipovitellin (LV_2 and LV_3), confirming the ability of oocytes to autosynthesize lipovitellin (Lui and O'Connor, 1977). More or less comparable results have been recently obtained in the fiddler crab *U. pugilator* by Eastman-Reks and Fingerman (1985).

Crab oocytes are endowed with appreciable heterosynthetic ability (Fig. 5.3) Ultrastructural studies of the vitellogenic oocytes clearly point out that selective sequestration of presynthesized protein molecules from the hemolymph can occur by micropinocytosis. Autosynthesis precedes heterosynthesis in developing oocytes of *P. hydrodromus* (Adiyodi, 1969a; Krishnakumar, Vijayalakshmi, and Adiyodi, 1979), *L. emarginata* (Hinsch and Cone, 1969), *C. pagurus* (Eurenius, 1973), and *E. sinensis* (Dhainaut and De Leersnyder, 1976). Autosynthetic ability is reduced or lost toward the later phase of vitellogenesis because oocytes cannot accommodate the requisite subcellular machinery involved in protein synthesis, the ooplasm becoming increasingly filled with yolk.

Biochemical and immunological evidence has further substantiated heterosynthesis as a mode of protein yolk deposition in crab oocytes. The presence of a protein (vitellogenin) in hemolymph with electrophoretic mobility comparable to that of the chief yolk protein (lipovitellin) in oocytes was first observed in *P. hydrodromus* (Adiyodi, 1968b)

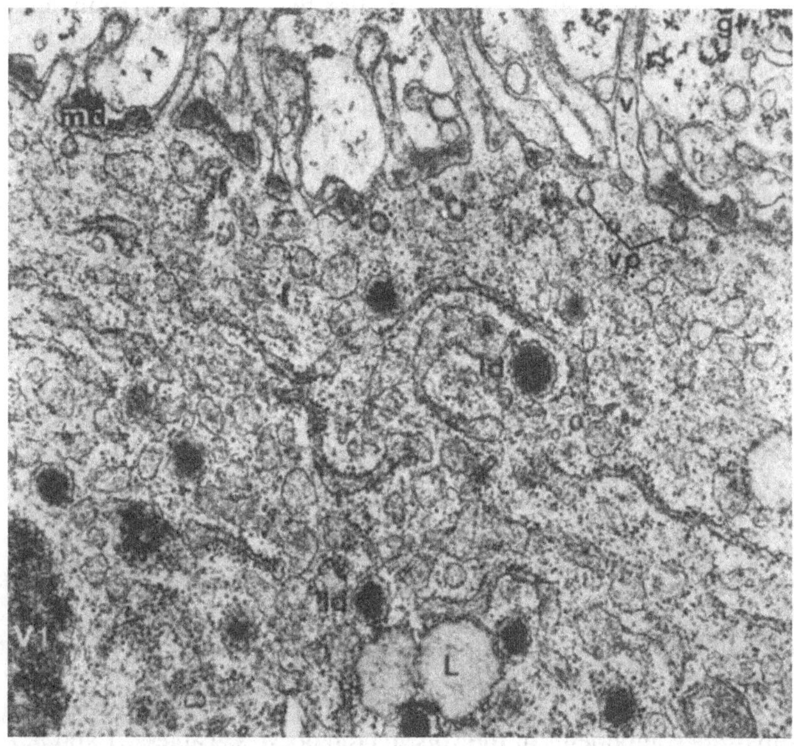

Fig. 5.3. Electron micrograph showing the beginning of heterosynthetic yolk synthesis in oocyte of eyestalkless *Eriocheir sinensis*. *gl*, glycogen of follicle cells; *L*, lipid; *ld*, vesicles with dense core; *md*, accumulation of dense material along the oolemma; *v*, microvilli; *Vl*, autosynthetically produced yolk; *vp*, micropinocytotic vesicles × 28,000. (From Dhainaut and De Leersnyder, 1976.)

and *C. sapidus* (Kerr, 1968). Vitellogenin is serologically identical to lipovitellin in *C. sapidus* (Kerr, 1969), *U. pugilator*, and *L. emarginata* (Wolin, Laufer, and Albertini, 1973). Vitellogenin (fraction 4/5) appears in hemolymph of *P. hydrodromus* only toward sexual maturity and shows cyclic quantitative fluctuations in relation to vitellogenesis (Adiyodi, 1968b). In *L. emarginata* (Wolin et al., 1973) and *U. pugilator* (Fielder, Rao, and Fingermann, 1971; Wolin et al., 1973), vitellogenin has been detected only in females having ovaries in vitellogenesis. Ovarian fluorescence has been experimentally induced in *U. pugilator* by incubating the vitellogenic oocytes in fluorescein-conjugated female serum; similarly treated male serum was, however, not appreciably absorbed (Wolin et al., 1973). Vitellogenin is seemingly a heterogeneous fraction highly sensitive to degradation. The remarkable similarity in physicochemical characteristics between vitellogenin and lipovitellin suggests that once absorbed, vitellogenin possibly undergoes only minimal macromolecular changes within the ooplasm.

The site of synthesis of crab vitellogenin remains enigmatic. In *C. sapidus*, vitellogenin is reportedly synthesized in the hemocytes (Kerr, 1968). A protein immunologically identical to vitellogenin has been detected in hepatopancreas of vitellogenic females of *L. emarginata* (Wolin et al., 1973). In *P. hydrodromus*, fluctuations occur in TCA-precipitable total protein content of the hepatopancreas in relation to ovarian growth, but we have not succeeded in locating vitellogenin per se in the hepatopancreas of this species (R. G. Adiyodi and K. G. Adiyodi, 1972). Paulus and Laufer (1984), however, have more recently reported identification by immunolocalization of the yolk-producing cells, vitellogenocytes, in the intertubular spongy connective tissue in hepatopancreas of *C. maenas* and *L. emarginata*. These cells are PAS-positive, "resemble reserve-inclusion cells, lipoprotein cells, or cyanocytes." Double immunoprecipitation with lipovitellin-specific antibody suggests that these cells "are most actively producing yolk during intermediate stages of ovarian maturation." They have also demonstrated a temporal relationship between the progress of vitellogenesis in the oocytes and vitellogenin synthesis by the hepatopancreas (Paulus and Laufer, 1987).

Much less is known about the mode of synthesis of lipid and carbohydrate components of the yolk in crab oocytes. Tables 5.1 and 5.3 present a comparison of the profiles of the total lipid contents of the hepatopancreas and ovary in the course of vitellogenesis. Clearly, there is an accumulation of lipid in the hepatopancreas during the beginning of the ovarian cycle and a sharp decline toward its final phase as lipid stores build up in the ovary (Anilkumar, 1980). The patterns of phospholipids in the ovary and hepatopancreas of this species are comparable: phosphatidylcholine, phosphatidylethanolamine, and lysophosphatidylethanolamine of the hepatopancreas underwent cyclic fluctuations related to vitellogenesis (R. G. Adiyodi and K. G. Adiyodi, 1970a).

In *P. hydrodromus*, vitellogenesis-related cyclic fluctuations have also been found in the free sugar content of the hepatopancreas, which appears to be the predominant form in which carbohydrate is stored in this organ (R. G. Adiyodi and K. G. Adiyodi, 1970b). The total oligosaccharide content of the hepatopancreas declined from 186 ± 28 to 49 ± 6 per 100 g of animal wet weight toward the final phase of vitellogenesis (Anilkumar, 1980).

Concerning egg coverings, two types of cortical vesicles are observed in the eggs of *C. maenas*. On contact of the spermatozoon with the egg plasma membrane, these cortical vesicles release by exocytosis two

Table 5.3. *Total lipids in the hepatopancreas of* Parathelphusa hydrodromus *during the vitellogenic cycle (mg/100 g animal wet weight;* $\bar{x} \pm SD$*)*

Vitellogenesis I				
Stage 1	Stage 2	Stage 3	Vitellogenesis II	Spent
3653.39 ± 118.35	3604.58 ± 103.40	2846.36 ± 185.13	1981.14 ± 472.62	731.51 ± 73.20

Source: Modified from Anilkumar (1980).

different exudates, namely, a mass of ring-shaped granules synthesized intracisternally and a fine granular material. Initially, two superimposed vitelline envelopes surround the egg: an outer thin envelop (1a) and an inner thick envelope (1b). The third layer (envelope 2) is secreted by the newly laid egg, possibly after fertilization. These three superimposed envelopes, which comprise the embryonic capsule of *C. maenas*, surround the embryo throughout embryonic development (Goudeau and Lachaise, 1980a, b; Goudeau and Becker, 1982). The funiculus, which facilitates the attachment of the newly laid eggs to the ovigerous setae, is formed from vitelline envelopes 1a and 1b. It is the strong adhesive property of vitelline envelope 1a that permits gluing of eggs to the setae (Goudeau and Lachaise, 1983).

D. Oosorption

In natural habitats, ova left in the ovary following spawning are usually resorbed. Experimental conditions such as inanition, unscheduled hastening of somatic growth, and hormone withdrawal also lead to resorption of yolky eggs (Adiyodi, 1969a; Kurup, 1983; G. Anilkumar and K. G. Adiyodi, unpub.). Hydrolytic enzymes of follicular or hemocytic origin appear to be involved in this lytic process (Carayon, 1941; Hort-Legrand, Berreur-Bonnenfant, and Ginsburger-Vogel, 1974). When subjected to continuous inanition for a total period of two months, the ovaries of 25% vitellogenic females of *P. hydrodromus* responded by resorption of the yolk. The ovaries lost their bright orange color and acquired a loose, saclike appearance. Discelectrophoretic analysis of the homogenates of the ovaries revealed loss of prosthetic groups from conjugated protein fractions 1–5, including the lipovitellin (Adiyodi, 1969).

E. Endocrine control of vitellogenesis

The pioneering work of Panouse (1943) brought to light the role of eyestalks in inhibiting ovarian growth in decapod Crustacea. In males, bilateral eyestalk ablation leads to precocious spermatogenesis and enlargement of the vasa deferentia (Démeusy, 1953; Gomez, 1965). Because eyestalks from males and females are reciprocally effective in restoring the normal state of the gonads in eyestalkless individuals, the ovarian/testicular growth-inhibiting substance was renamed gonad-inhibiting hormone (GIH) by K. G. Adiyodi and R. G. Adiyodi (1970). The presence of another hormone having a stimulating effect on gonadal growth was initially proposed by Otsu (1963) in *Potamon dehaani*, a finding subsequently confirmed by Gomez (1965) in *P. hydrodromus*, by Hinsch and Bennett (1979) in *L. emarginata*, and by Eastman-Reks and Fingerman (1984) in *U. pugilator*. This substance, which seems to be secreted by neurosecretory cells of the brain (Gomez, 1965) and/or thoracic ganglia of crabs, has been called gonad-stimulating hormone (GSH). In crabs, reproduction is hormonally modulated by the antagonistic actions of GIH and GSH.

Land crabs, like most crustaceans, continue to grow and molt even after the attainment of sexual maturity (also see Chapter 6). Brachyurans, including land crabs, generally program ovarian growth (reproduction) and somatic growth (ecdysis) as mutually exclusive events, the process of ovarian growth being largely confined to the intermolt stage (i.e., C_4) and somatic growth to premolt/postmolt (see section 3, this chapter). Molting in decapod crustaceans is also essentially under a dual hormonal control. Molt-inhibiting hormone (MIH), released in all probability from the X-organ neurosecretory cells of the eyestalk, inhibits molting by suppressing secretory activity of the Y-organ (see Soumoff and O'Connor, 1982). Ecdysone (molting hormone, MH), released from the Y-organ, promotes molting. Thus in the Brachyura, ovarian growth takes place when MIH and GSH are in high concentrations and MH and GIH are at low levels; somatic growth takes place when MIH and GSH are in low concentrations and MH and GIH are at high levels (K. G. Adiyodi and R. G. Adiyodi, 1970; see also R. G. Adiyodi, 1985).

In sexually mature *P. hydrodromus*, the physiological emphasis during the first half of the intermolt is on reproduction, and during the second half on somatic growth. The physiological condition that prevails at any given time, in fact, determines how the individual responds to eyestalk ablation. Bilateral eyestalk ablation during the reproductive phase of the intermolt results in accelerated ovarian growth, and during the somatic phase of the intermolt in accelerated somatic growth. The factors responsible for bringing about this shift in physiological condition in aquatic or terrestrial crabs are not known, but it is suspected that changes in the hormonal milieu during specific phases alter responsiveness of target tissues. Support for this assumption has come from experiments using exogenous β-ecdysone.

Administration of β-ecdysone at premolt concentrations significantly accelerates somatic growth in eyestalkless female *P. hydrodromus* during the somatic phase of the intermolt, but similar treatment fails to evoke any response in females in the reproductive phase of the intermolt (Kurup and Adiyodi, 1981; Kurup, 1983).

The role, if any, of ecdysteroids in vitellogenin synthesis in aquatic or terrestrial crabs is not clearly understood. The occurrence of ecdysone-dependent vitellogenesis in some dipteran insects, and the fact that in some groups of crustaceans vitellogenesis proceeds in a milieu rich in ecdysteroids, have tempted some investigators to the general conclusion that ecdysteroids may be essential for the synthesis of yolk in all crustaceans. It may be recalled that vitellogenesis in crabs is generally programmed during a period when ecdysteroid concentrations in the hemolymph are low. In *G. lateralis*, the level of β-ecdysone in the hemolymph was estimated to be only 10 $ng.ml^{-1}$ in intermolt (McCarthy and Skinner, 1977). Again, the Y-organs are not indispensable for the successful completion of vitellogenesis in *C. maenas* (Démeusy, 1962). Accumulation of surprisingly high concentrations of ecdysteroids has been reported in the ovary of this species as vitellogenesis progresses (Lachaise and Hoffman, 1977). The ecdysteroids in mature ovaries include ecdysone, 20-hydroxyecdysone, and ponasterone (Lachaise et al., 1981). Ovarian accumulation of ecdysteriod during vitellogenesis has also been observed in the blue crab, *Callinectes sapidus* (Soumoff and Skinner, 1983). An increase in concentration of ecdysone and 20-hydroxyecdysone, both at the beginning of vitellogenesis and also as the oocytes reach their maturity, has been reported in the ovary of the spider crab, *Acanthonyx lumulatus*. It may be that there exists in this species some relationship between ecdysone and 20-hydroxyecdysone on the one hand and the onset of vitellogenesis on the other. Interestingly, ecdysone and 20-hydroxyecdysone are also present in spawned eggs of *A. lumulatus*. In the embryo, levels of 20-hydroxyecdysone are high at the beginning of naupliar development and also at metanauplius stage III when the metamerization molt occurs. The hormone, again detected in graying eggs when the first (embryonic) exoskeleton is secreted, reaches peak concentrations at the end of the development of the zoea (Chaix and De Reggi, 1982).

In *C. maenas*, in addition to the embryonic capsule comprised of vitelline envelopes 1a, 1b, and 2 (see section 5.3, this chapter), four additional layers are successively secreted beneath it by the embryonic ectoderm cells. The presence of these additional layers gives us some clue regarding the successive molting cycles of the embryo. It is reported that the secretion of each of these envelopes occurs in the presence of high concentrations of the ecdysteriod, ponasterone A

(Goudeau and Lachaise, 1983). The egg of *C. maenas* contains high concentrations of ponasterone A together with small quantities of 20-hydroxyecdysone and ecdysone at various stages of embryonic development (Lachaise and Hoffmann, 1982).

Eyestalk extirpation, but not removal of the Y-organs, results in a marked rise in hemolymph ecdysone levels of *Eriocheir sinensis*. In females with intact eyestalks, removal of the Y-organs had little influence on ovarian growth. However, when Y-organ removal was coupled with removal of the eyestalks, the early phase of ovarian growth was adversely affected. Normal oogenesis could be restored in such instances by administration of exogenous ecdysone. Thus, it is apparent that (1) high concentrations of ecdysone are required in destalked females to maintain the early phase of oogenesis, and (2) in females with intact eyestalks small quantities of ecdysone present in the hemolymph after removal of the Y-organs are sufficient to permit normal oogenesis (De Leersnyder, Dhainaut, and Porcheron, 1981).

Recent studies have shown that GIH cannot be considered simply as a substance that inhibits ovarian growth in female crabs. The presence of eyestalks is most essential in females to ensure normal progress of vitellogenesis. This becomes apparent right from the beginning of the ovarian cycle during the postspawning season (somatic phase of the intermolt) in *P. hydrodromus*. Though emphasis during this period is very largely on somatic growth, deprival of eyestalks leads to some acceleration in vitellogenesis, but this is less spectacular compared to the induced ovarian growth during the prebreeding season. The pattern of oocyte growth throughout the ovary during the postspawning season is asynchronized, resulting in the production of yolky eggs in various stages of vitellogenesis, randomly dispersed throughout the ovary (Kurup, 1983). By 15–20 days postoperation, only a few oocytes could reach V_1S_3; the bulk of them were in the avitellogenic phase, and some were in the early phase of vitellogenesis (V_1S_1 and V_1S_2). In a similar experiment conducted five to six months later (i.e., during the prebreeding season), the ovary of *P. hydrodromus* was found to reach V_1S_3 or V_2 by 15 days postoperation. The asynchronized pattern of oocyte growth persisted but was less spectacular compared to the postspawning group. The yolk, precociously produced following bilateral eyestalk ablation during the prebreeding season, was biochemically impoverished. No spawning was recorded even 45 days after eyestalk excision (Anilkumar and Adiyodi, 1980). Interestingly, when bilateral eyestalk excision was conducted at a still later phase in the ovarian cycle (i.e., during the first half of the breeding season), notwithstanding the acceleration in ovarian growth, the ovary of *P. hydrodromus* retained its apparently normal nature in both histological and biochemical terms until 15 days postoperation. The oocytes, distributed in the ovary, showed a synchronized growth pattern roughly comparable to that of normal crabs. Abnormality set in only afterward, and it became conspicuous by 30 days postoperation. None of these females spawned during this period (Anilkumar and Adiyodi, 1985).

In contrast, if eyestalks were extirpated during the final phase of vitellogensis, although accelerated vitellogenesis was not perceived, the majority of females without eyestalks spawned normally (Anilkumar and Adiyodi, 1985). It thus becomes obvious that the ovary of *P. hydrodromus* is dependent on eyestalk factors (possibly GIH) at all stages of the ovarian cycle, except perhaps at the final phase of vitellogenesis. The asynchrony that becomes conspicuous in the growth pattern of oocytes in the ovary of crabs without eyestalks is quite understandable when we consider that oocytes in various stages of growth occur in a graded manner throughout the ovary of normal crabs all through the avitellogenic phase and the early vitellogenic phase. The presence of the eyestalk is seemingly vital for bringing about the final synchronized growth of the oocytes in the ovary.

F. Spawning

Spawning in land crabs is generally programmed with great precision so that brood release occurs at a time most appropriate for the growth of the young (also see Chapter 3). Species inhabiting the temperate regions reproduce in the warmer season. Thus, in *L. emarginata*, the first oviposition is scheduled in late May or early June and the subsequent ones follow at 25-day intervals until early September (Hinsch, 1968, 1972). The peak spawning season of *M. mercenaria* falls in August and September (Cheung, 1969). *G. lateralis*, inhabiting the southeastern coast of Florida, also spawns more or less during the same period (Bliss et al., 1978).

Spawning migrations are not uncommon among those land crabs that move down to the sea for brood release. The occurrence of such mass migrations of ovigerous females has been reported in *Gecarcinus lalandii* in the Indo-West Pacific (Johnson, 1965), *C. guanhumi* of southern Florida (Gifford, 1962a), and *G. planatus* of Clipperton Island (cf. Klaassen, 1975).

The process of brood release has been studied in some detail in *G. lateralis* (Klaassen, 1975; Bliss et al., 1978). On approaching the incoming breakers, the ovigerous female stands high on her claws. As soon as the wave bathes her she raises and retracts her abdomen and moves her body up and down, permitting the larvae to be washed off by two or three consecutive waves. Once brood release has been achieved, the female quickly returns to the shore.

Bergin (1981), who studied the hatching rhythms of *U. pugilator*, observed that gravid females maintained in 14:10 (L:D) light regime show a significant positive correlation between the time of hatch and the time of nighttime high tide. As far as gravid females are concerned, this appears to be an adaptive character developed as a predator-avoidance mechanism.

Ovigerous females show behavioral modifications associated with brood care. Thus, in shallow hypoxic seawater, ovigerous female *C. maenas* reverse the direction of normal ventilation, taking in air at the normally exhalant opening and passing it via the base of the last walking legs and the egg mass (Wheatly, 1981; also see Chapter 8, this volume). This raises the oxygen partial pressure of the water surrounding the eggs and promotes oxygen uptake by the developing embryos. If the female is forced to remain in hypoxic seawater, larval release is suspended until the well-aerated condition has been restored.

Ovigerous females of *R. harrisii* show a circadian rhythm of larval release. There possibly exists an interaction between the mother and her eggs in the brood that is responsible for synchronizing the development of the embryos in the brood. At the time of hatching, an active substance is apparently released by the embryo that induces abdomen pumping by the female (a behavior observed at the time of brood release) and synchronizes larval release (Forward and Lohmann, 1983).

In at least *P. hydrodromus*, oviposition takes place only if the eggs to be laid are normal (viable?). Although bilateral eyestalk ablation induces precocious out-of-season vitellogenesis in this species, as in many other crustaceans, the biochemical composition of the yolk thus produced is not normal, and invariably in all such instances the females refrain from spawning (Anilkumar and Adiyodi, 1980, 1985). This appears to be an adaptive character developed by females of *P. hydrodromus* to avoid an otherwise wasteful process.

Sexually mature female *P. hydrodromus*, which continue to remain in the reproductive phase of intermolt until the final phase of vitellogenesis, switch over by March/April to the somatic phase of intermolt, nearly a week before the spawning season. Though the crabs have entered the somatic phase by spawning season, the presence of a brood in the brood pouch exerts an inhibitory effect (indirectly) on somatic growth during this period. That forcible removal of the brood significantly accelerates somatic growth is obvious from the increased growth rate of the regenerating limb buds following brood removal. This effect on somatic growth by the brood, called the "brood effect," could be negated by bilateral eyestalk excision. Conceivably, the brood effect is mediated through the eyestalks, possibly by MIH (Kurup and Adiyodi, 1980; Kurup, 1983).

6. Reproduction in the male

A. Spermatozoa

Crab spermatozoa are immotile and are frequently bizarre in shape. They are characterized by the absence of a flagellum, but they may possess varying numbers of immobile processes or arms extending radially from a central zone (Dhillon, 1968). Eight radial arms, for instance, have been detected in the spermatozoa of *C. sapidus* (Brown, 1966), six in *G. steniops* (Varghese, 1984), and three lateral arms and a posterior median process in *L. emarginata* (Hinsch, 1969). These arms, which may or may not contain chromatin, are arranged around a centrally placed nuclear region (Fig. 5.4A, B). The latter, in turn, surrounds a refringent acrosomal vesicle. This vesicle readily undergoes eversion at the time of fertilization and also when exposed to certain chemical solutions, sometimes even saline (Brown, 1966; Krishnakumar, 1985).

The acrosomal vesicle of crab spermatozoa arises by coalescence mainly from cisternae of rough endoplasmic reticulum; some smooth endoplasmic reticulum and nuclear membrane are also added to it (Fig. 5.5). An osmiophilic granule, which comes to lie in the apex of the vesicle, gives rise to the opercular sphincter. The central duct (acrosomal tubule) is formed by invagination of the basal membrane. The central duct houses a percutor organ comparable to the perforatoria of flagellate spermatozoa. Two centrioles could be detected at the base of the percutor organ. The vesicle is basally surrounded by a nucleo-chondrio-polymicrotubular (NCT) complex, which is an ensemble of membranous sheaths, mitochondria, and microtubules (Pochon-Masson, 1962, 1968a, b, 1983).

A prominent membrane system, continuous with the nuclear envelope and located close to the base of the acrosome tubule, has been described in the spermatozoa of *C. maenas* (Goudeau, 1982). At the time of acrosome reaction, when the initial contact is made between the spermatozoon and the egg, the outer layer of the everted acrosomal vesicle interdigitates deeply with the microvilli present along the oolemma. At the point of contact, a very tiny fertilization cone is formed in the oolemma, caused by the intrusion of the acrosome tubule.

Following the acrosome reaction, the arms are withdrawn and the nuclear material assumes a spherical configuration. The acrosomal tubule, which is now in direct contact with the plasma membrane, seems to facilitate fusion of gametes (Brown, 1966).

B. Semen

Information on composition of semen of land crabs is limited to a few species. Spermatozoa are generally passed to the female in tiny

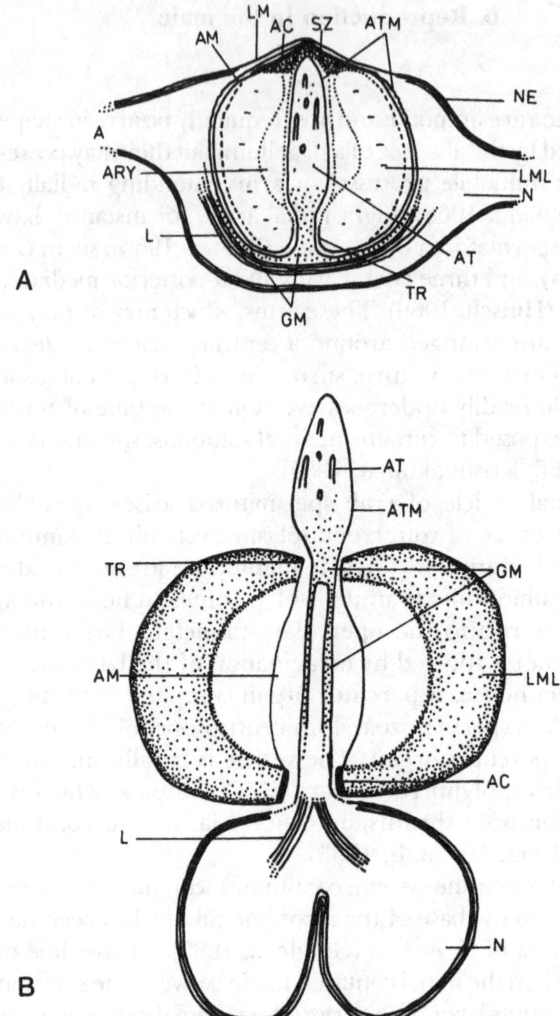

Fig. 5.4. A. Diagram of the sperm of *Callinectes sapidus*. B. Eversion of *Callinectes sapidus* sperm. *A*, radial arms; *AC*, apical cap; *AM*, acrosomal membrane; *ARY*, acrosomal rays; *AT*, acrosomal tubule; *ATM*, acrosomal tubule membrane; *GM*, granular material of acrosomal tubule; *L*, lamellae; *LM*, limiting membrane; *LML*, large microtubular layer of acrosomal vesicle; *N*, nucleus; *NE*, nuclear envelope; *SZ*, subcap zone; *TR*, thickened ring. (Redrawn from Brown, 1966.)

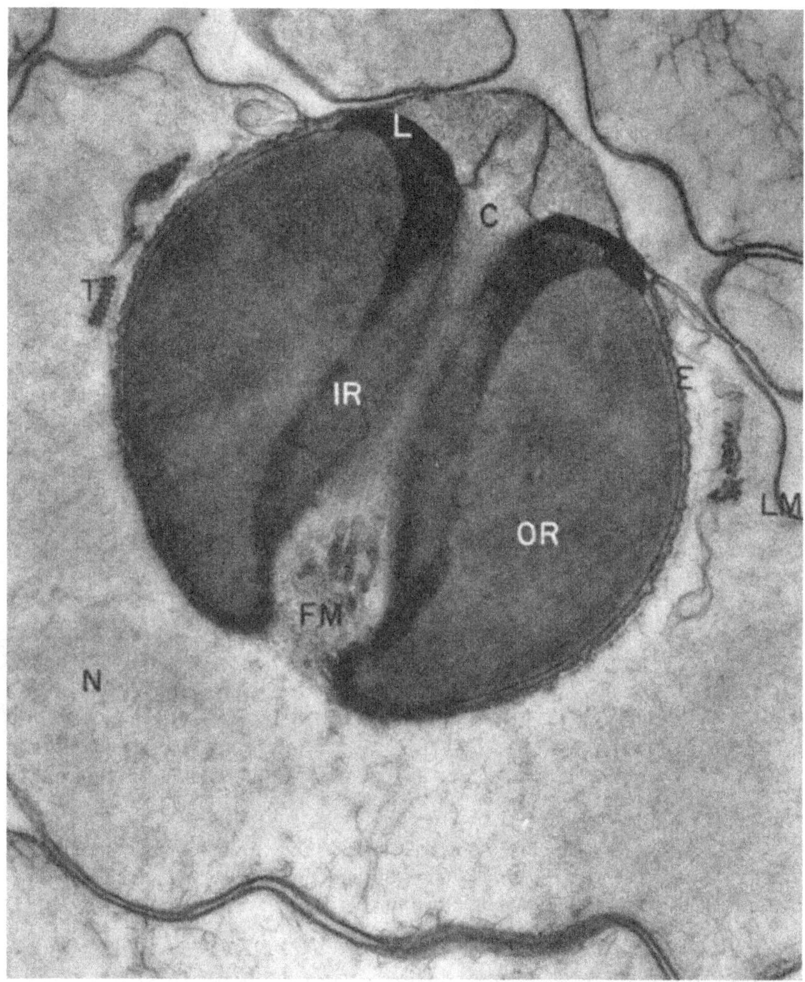

Fig. 5.5. Electron micrograph of the mature sperm of *Cancer borealis*. *C*, acrosomal core; *E*, nuclear envelope; *FM*, fibrous material; *L*, acrosomal lip; *LM*, limiting membrane; *N*, nucleus; *OR*, *IR*, outer and inner regions of the acrosome; *T*, tubules in the nucleus. × 33,000. (Reproduced from *The Journal of Cell Biology*, 1969, 43, pp. 575–603.)

packets known as spermatophores, which are often suspended in a rich fluidy medium, the seminal plasma.

The mode of synthesis of semen has been studied in some detail in *P. hydrodromus* (Mathad, 1983) and in the ghost crab, *Ocypode platytarsis* (Sukumaran, 1985). From the testicular acini, the sperm mass reaches the anterior region of the vas deferens, coated in a mucous

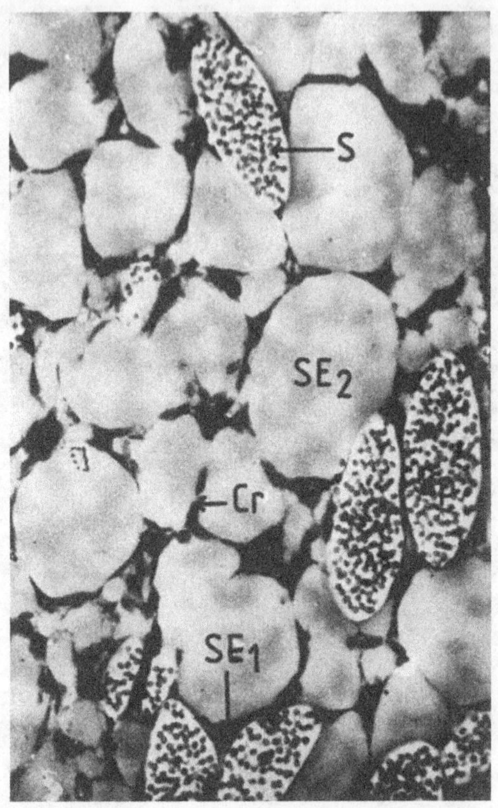

Fig. 5.6. Cross section of the mid vas deferens of the ghost crab, *Ocypode platytarsis*. *Cr*, SE_1, and SE_2 represent various components of the seminal plasma. *S*, spermatophore. Photomicrograph. × 220. (From Sukumaran, 1985.)

product. Slightly posteriorly, in *P. hydrodromus*, the wall of the sperm duct secretes a fluidy medium (SS_1), rich in basic proteins, which individually coats each spermatozoon. Farther down, a predominantly proteinaceous secretion (SS_2) with tyrosyl residues and some carbohydrates is produced by the sperm duct epithelium. A comparable secretory product (SE_1) has also been described from the vas deferens of *O. platytarsis* (Sukumaran, 1985) (Fig. 5.6). In *P. hydrodromus*, as more and more SS_2 is produced and added, the sperm mass breaks into smaller groups, the first step in formation of spermatophores. Farther posteriorly, a highly basophilic proteinaceous component (SS_3), with few tyrosyl residues, carbohydrates, alcianophilic substances and lipids, makes its appearance as distinct droplets, randomly distributed in SS_2. Membranous walls enclosing the spermatophores are evident from the posterior part of the anterior vas deferens. In *O. platyarsis* also, it is in the anterior vas deferens that the spermatophore walls become distinct for the first time. The spermatophore wall of *P. hydrodromus* is soluble in dilute solutions of potassium hydroxide, sulfuric acid, hydrochloric acid, and 1% saline, but insoluble in 0.7% saline and water. Semen shows hardly any change in consistency posteriorly along the sperm duct (Mathad, 1983).

The spermatophores of anomurans are generally pedunculate. For instance, in *Emerita asiatica*, the spermatophores are dimorphic and pedunculate and arranged to form a spermatophoric ribbon embedded in a jellylike matrix. However, spermatophores of *Albunea symnista* are nonpedunculate (Subramoniam, 1984).

Besides SE_1, a few more components are added to the seminal plasma of *O. platytarsis* along the sperm duct. A little behind the site of formation of the spermatophores, a β-metachromatic hyaline substance, poor in protein, is secreted by the secretory epithelium of the anterior vas deferens. As more and more hyaline substance is produced and added, SE_1 is reduced to thin strands delimiting spherical or oval units of the hyaline substance. The chief component of the seminal plasma of the ejaculated semen of *O. platytarsis*, however, is a fibrous protein. This substance is secreted by paired coral-shaped accessory glands that open into the vas deferens at the point of origin of the ejaculatory duct. The mixing of secretory products of the coral-shaped accessory gland with the semen is limited in vas deferens of *O. platytarsis*, as the vast bulk of the secretory products are stored in the glandular ducts themselves (Fig. 5.7) (Sukumaran, 1985).

Remarkable regional differences have been observed in physicochemical properties of the seminal plasma stored in different regions of the vas deferens of the freshwater crab *G. steniops*, suggesting that different components of the seminal plasma are secreted by epithelial cells lining different regions of the vas deferens. Spermatophores of this species exhibit much flexibility in shape and size (Jacob, 1985).

The semen of *P. hydrodromus* is a white or cream-colored fluid having a density of 1.23 and a pH of about 7.0 (Mathad, 1983). It has a tendency to solidify on exposure to air. Protein, the chief organic component of the semen, ranges from 12.1 to 18.6 mg per 100 mg. The total free amino acid content of *P. hydrodromus* semen is estimated to be 0.30–0.76 mg per 100 mg. Total oligosaccharides constituted 0.20–0.23 mg per 100 mg. The semen of *P. hydrodromus* also contains 0.60–1.68 mg per 100 mg of chloroform-extractable lipids. Qualitative analysis has further revealed the presence of fairly large quantities of ascorbic acid and inositol in semen. Major components of the seminal lipid are triglycerides, cholesterol, and cholesterol esters (Mathad, 1983).

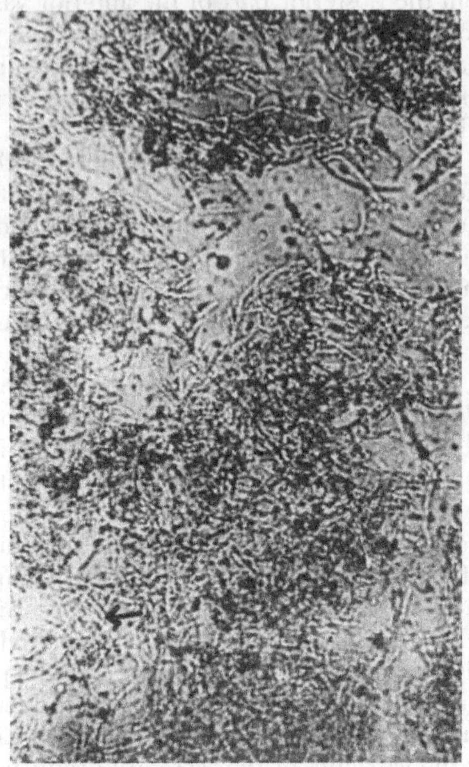

Fig. 5.7. Fibrous secretion of the coral-shaped accessory sex gland of *Ocypode platytarsis*. Photomicrograph. × 300. (From Sukumaran, 1985.)

C. Cyclicity in spermatogenesis and semen production

Among crustaceans, seasonal cyclicity in reproductive activity has been best documented in females (see section 2), but some male crustaceans also exhibit such cyclicity (Carpenter and De Roos, 1970; Payen and Amato, 1978). Periods of intensive spermatogenic activity and semen production alternating with "resting" phases are clearly more marked in representatives from colder regions, where temperature fluctuations are more pronounced. Such fluctuations do exist in species inhabiting the warmer regions as well, though their sperm ducts may appear filled with semen year-round. This is particularly true of land crabs, which are seasonal breeders and depend on temporary bodies of water for reproductive purposes.

In males of the Calicut population of *P. hydrodromus*, the mating season extends from June/July to October/November, which actually encompasses two virtually continuous periods: (1) June to August (with females older than one year), and (2) September to November (with young females, five-eight months old). However, male genital ducts appear filled with semen throughout the year. Sperm duct epithelium, the site of production of various seminal components, is in peak secretory activity from June/July to October (K. G. Adiyodi and R. G. Adiyodi, unpub.). The vas deferens factor reaches the peak value by August/September and drops to its lowest by December (Mathad, 1983).

The testes and vasa deferentia of *P. hydrodromus* function synchronously, all major organic components in both tissues reaching their highest levels throughout the male reproductive system by August/September. However, levels of protein, the chief organic component of the testis, do not fluctuate drastically in the male genital system of *P. hydrodromus*. A steady rise in testicular protein content has been detected, with March values of 5.0 mg per individual rising to 6.4 mg per individual through August/September. The value declined to 4.3 mg by December, the postmating period, possibly due to large-scale release of spermatozoa from the testicular acini.

Protein content of the semen of *P. hydrodromus* remains more or less steady from December to June/July (December, 19.4 mg; June/July, 19.2 mg). The value reaches 35.3 mg by August/September. Testicular and seminal free amino acid content doubled, compared to the levels in March, by August/September (1.3 mg and 0.6 mg, in testis and semen, respectively). A more or less comparable trend has been found in the content of carbohydrate (oligosaccharides + polysaccharides) in the testis and semen. Oligosaccharide contents of testis and semen during August/September are 0.6 mg and 0.5 mg, and polysaccharide contents are 0.7 and 0.6 mg, respectively. Generally, fluctuations in lipid content of the semen through different seasons are not significant. In the testis, however, the highest total lipid value recorded is in August/September (1.6 mg per individual), and the lowest is in December (0.8 mg per individual) (Mathad, 1983).

Information on seasonal fluctuations in testicular activity of land crabs is limited. An annual cyclicity in spermatogenic activity has been recorded in *U. pugnax* from Chappaquoit Marsh (United States), with active spermatogenesis occurring during the summer months (Young, 1974). Short photoperiod or excess illumination prevents spermatogenesis in this species. Spermatogenic activity is also seasonal in the freshwater crab, *Potamon koolooense* (Joshi and Khanna, 1982b). Spermatogenesis begins in this crab in January/February, progresses slowly all through March, and reaches its peak by April/May. Spermiation takes place during May and June. Spermatogenesis ceases gradually by December and the testes enter a brief period of rest.

In *P. hydrodromus* the annual spermatogenic activity is revived during July/August (i.e., during the first half of the intermolt) but reaches its peak only by September/October. A perceptible decline in spermatogenic activity is apparent by November/December. The number of acini containing dividing spermatocytes is significantly reduced; spermatocytes and spermatids show condensation; and there is a clear fall in the number of spermatogonia. Signs of spermatozoan degeneration have appeared in some acini, though in others spermiogenesis progresses normally. By February/March, male *P. hydrodromus* enter the second half of the intermolt and there is an almost total suspension of all mitotic and meiotic activities. From February/March through May, the testis contains relatively few spermatogonia. Although several acini contain spermatocytes and spermatids during this period, their nuclei show chromatin condensation; furthermore, spermatolysis is common in all testicular acini containing the sperm masses. By June, there is a slight rise in the number of spermatogonia, but spermatolysis continues to be intense throughout June (termination of intermolt) declining only by July/August (i.e., at the commencement of the intermolt of the following molt cycle) (S. Gupta and K. G. Adiyodi, unpub.).

In *P. hydrodromus* seasonal changes occur in the levels of activity of the androgenic gland, the source of the male sex hormone, correlated with changes in semen production. The androgenic gland remains in a regressed state from January through May, which is the sexually inactive phase of the animal. Nuclear pyknosis and cytoplasmic vacuolation are predominant especially from January to March. The androgenic gland hypertrophies significantly by June/July and attains the maximum size by August/September. In the hypertrophied state, the androgenic gland shows regional differences: The bulk of the gland is occupied by the cells with large nuclei and with cytoplasm rich in basophilic granules; some regions show nuclear pyknosis and cellular degeneration. The androgenic gland undergoes considerable atrophy by November/December; the cytoplasmic granules are now hardly distinct and the cells show signs of degeneration (R. G. Adiyodi and K. G. Adiyodi, 1976; R. G. Adiyodi, 1984).

In the Calicut population of *P. hydrodromus*, the molting season begins with the arrival of the monsoon in June. Both the active and the inactive phases of semen production of this species are accommodated within the same intermolt. The first half of the intermolt, extending from July to November/December, represents the reproductively active phase of the male (reproductive phase) (Mathad, 1983). During this period the relative emphasis on somatic growth (as tested by the ability of the crab to regenerate lost limbs) is very low (Kurup and Adiyodi, 1981). The second half of the intermolt, extending from December to May, includes the somatic phase of the male. During this period, the emphasis is largely on somatic growth (Kurup and Adiyodi, 1981) and much less on reproduction (Mathad, 1983) (Fig. 5.2). This leads us to suspect that temporal separation of somatic growth and reproduction in females also exists in males of *P. hydrodromus*.

D. Semen storage

The considerable time lag between production of spermatozoa in the testes and fertilization necessitates the storage of semen in the genital system of either sexes for varying lengths of time. Although spermatogenesis and mating are seasonal events in *P. hydrodromus* (see sections 5.C and 6.C), the semen is stored in bulk quantities in the genital duct year-round (Mathad, 1983; S. Gupta and K. G. Adiyodi, unpub.).

Specially developed chambers for semen storage (seminal vesicles) associated with the sperm ducts are not generally found in land crabs. In *P. hydrodromus*, *O. platytarsis*, *Ocypode ceratophthalmus*, *G. steniops*, and *Barytelphusa cunicularis*, all investigated in our laboratory, it is the vas deferens that serves for storage of semen. In *P. hydrodromus* and *B. cunicularis* there is no difference in consistency between semen stored in the vas deferens and ejaculated semen (Mathad, 1983). However, in *O. platytarsis* (Sukumaran, 1985), *O. ceratophthalmus* (Sreenarayanan, 1980), and *G. steniops*, final mixing of the various components of the semen occurs only at ejaculation (Varghese, 1984). It may be that the components stored separately in accessory sex glands of *O. ceratophthalmus* and *O. platytarsis* and in the posterior region of the mid vas deferens of *G. steniops* have important roles to play in the female system after ejaculation.

The metabolic requirements of spermatozoa of land crabs during their prolonged sojourn in male and female sex ducts seem to be met from both intracellular and extracellular resources. Because they are immotile and compactly stacked within the spermatophores (and, in the female, stored in the protected environment of the spermathecal lumen), crab spermatozoa have limited energy demands. Intracellular storage of glycogen occurs in spermatozoa of *P. hydrodromus*, *O. platytarsis*, and *O. ceratophthalmus*. That the spermatozoa retrieved from the spermathecal lumen of *P. hydrodromus* during the prespawning season contain nearly as much glycogen reserve as those collected from the vasa deferentia suggests that, at least in this species, the spermatozoa depend little on endogenous glycogen reserve during their prolonged period of storage in male and female sex ducts. Their metabolic requirements appear to be met largely with extracellular resources. The spermatophore wall of the marine crab *Scylla serrata* is permeable to substances of low molecular weight (Uma and Subramoniam, 1979), suggesting that nutrients from the seminal plasma can pass through the spermatophore wall. Biochemical analyses of semen taken from male sex ducts of crabs indicate that there is no dearth of energy-rich compounds for stored spermatozoa in the male system. However, there is, to our knowledge, no conclusive evidence that spermatozoa of land crabs make use of these compounds.

In the Brachyura, the ejaculated semen is received in a special chamber, the spermatheca, which is associated with the female reproductive tract. Semen received during coitus appears to undergo a series of changes within the spermatheca. In *P. hydrodromus*, for example, the dissolution of the spermatophore wall occurs within a few hours after mating. In *O. platytarsis* the spermatophore wall is apparently retained for a longer period. Within the spermatheca, the spermatozoa crowd themselves near the exit of the spermathecal duct. They form a thin layer lining the luminal surface, while they are distributed in groups of 2–100 in the middle of the lumen.

Histological and histochemical evidence indicates production of an intrinsic secretory product by the spermatheca of several species of crabs. The secretion may provide nutrients for spermatozoa stored in the spermatheca (K. G. Adiyodi and R. G. Adiyodi, 1975). The spermatheca is devoid of any accessory glandular organs, but the luminal surface is either completely or partly lined by a secretory epithelium that is multilayered and holocrine in *O. platytarsis* (Krishnakumar, 1985), and columnar and one layer thick in both *P. hydrodromus* (Krishnakumar, 1985) and *G. steniops* (Santhamma, 1985).

In *P. hydrodromus* the spermatozoa are stored in the spermatheca for a minimum of eight to nine months as mating is scheduled during June/July and spawning occurs only in March/April of the next year (Fig. 5.8). Studies of female crabs cultured in the laboratory from early stages, with one group in isolation and the other with access to males, have clearly established that transmolt retention of spermatozoa occurs in *P. hydrodromus*, as in *Libinia* (Cheung, 1968). Female *P. hydrodromus* that fail to mate at the appropriate time make use of the reserve spermatozoa to fertilize the eggs during the following spawning season. In such instances, spermatozoa remain viable for a total period of 20–21 months within the spermatheca (Krishnakumar, 1985). In the tanner crab, *Chionoecetes bairdi*, 97% of the females produced viable egg clutches using sperm stored in the spermatheca for one year, and 71% were similarly successful using 2-year-old stock of spermatozoa (Paul, 1984). Females of *U. pugnax*, separated from males during the period when opercular decalcification occurred, often produced fertile clutches, suggesting that stored sperm from previous mating could fertilize eggs in a later cycle (Greenspan, 1982). In *P. hydrodromus*, the energy requirements during this prolonged period of storage in the female system appear to be at least partly met by

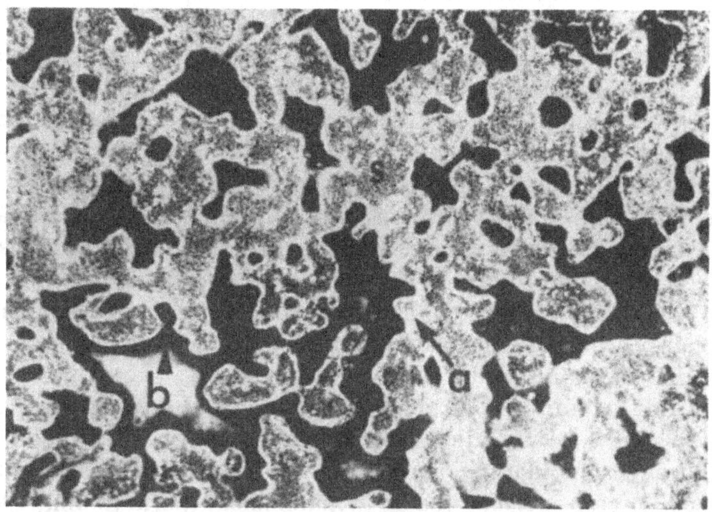

Fig. 5.8. Photomicrograph showing the nature of the contents of the spermathecal lumen of adult female *Parathelphusa hydrodromus*. *a*, acidophilic secretion; *b*, basophilic secretion; *S*, spermatozoa. × 170. (From Krishnakumar, 1985.)

intrinsic secretions of the spermatheca. Two peaks of secretory activity have been observed in spermathecal epithelium of *P. hydrodromus* (Krishnakumar, 1985) and *G. steniops* (Santhamma, 1985). One peak is related to annual ecdysis and mating in both species. The other peak is related to the prebreeding season (August/September) in *P. hydrodromus* and the early vitellogenic period (September) in *G. steniops*. Studies of seasonal fluctuations in chief organic components of the spermatheca of *P. hydrodromus* have shown that levels of all components rise by late premolt above the minimum value recorded during the postspawning season.

The intrinsic secretory product of the spermatheca of *P. hydrodromus* is composed of an acid mucopolysaccharide component (S_1) and a basophilic, proteinaceous component (S_2). One peak of secretory activity by spermathecal epithelial cells occurs in time for the arrival of a fresh batch of spermatozoa during the annual mating in early postmolt. Besides organic reserves, intrinsic secretory product of the spermatheca may contain enzymes and/or substrates required to produce substances from freshly ejaculated semen that are needed for the dissolution of the spermatophore wall. Spermathecal products are replenished with seminal contents, especially proteins, following mating during the postmolt. The change in consistency of the spermatheca during this period may be due to biochemical changes arising from mixing of semen with spermathecal luminal contents.

The second peak of secretory activity of the spermathecal epithelium occurs in *P. hydrodromus* by August/September (i.e., during the prebreeding season) and leads to a spectacular rise in chiefly proteinaceous organic reserves of the spermatheca (Krishnakumar, 1985). This is concomitant with a rise in protein content of the hemolymph and hepatopancreas (Anilkumar, 1980). The bulk of spermathecal proteins are mobilized by the beginning of the breeding season in December. This is followed by a fall in protein content of the hemolymph as well as hepatopancreas (Anilkumar, 1980). The significance of these fluctuations in protein content of the spermatheca is unclear at present. It may be that proteins from spermatheca are used to meet the energy requirements related to spermatozoan maintenance or are mobilized to the ovary for vitellogenic purposes.

A steady fall in the free amino acid content of the spermatheca of *P. hydrodromus* occurs during the breeding season (Krishnakumar, 1985), suggesting that free amino acids may be utilized for respiratory function by crab spermatozoa, as in mammalian sperm (Mann, 1964). Interestingly, beginning with premolt, the spermathecal carbohydrates also showed a steady decline that lasted until the final phase of the breeding season. The hard appearance of the spermatheca, apparent in female *P. hydrodromus* soon after mating, is progressively lost, becoming soft during the early phase (January) and almost degenerating to a fluid-filled sac toward the final phase of vitellogenesis (March). The fluid nature of the spermatheca toward the end of vitellogenesis likely serves to increase the internal hydraulic pressure and thereby facilitates ejaculation of spermatozoa by a mild contraction of the spermathecal musculature at the time of ovulation (Anilkumar, 1980; Krishnakumar, 1985).

E. Endocrine control of reproduction in the male

Historically, crustaceans occupy an enviable position in the study of invertebrates, in that the presence of an invertebrate sex hormone was conclusively demonstrated for the first time in the amphipod *Orchestia gammarella* (Charniaux-Cotton, 1952, 1953). This hormone, responsible for the differentiation and maintenance of the primary and secondary sex characters of the male, is produced by the androgenic gland, superficially attached (except in certain oniscoid isopods) to the subterminal ejaculatory part of the vas deferens (Charniaux-Cotton, 1954; Charniaux-Cotton and Payen, 1985).

In crabs, the androgenic gland is composed of linearly arranged cords of cells organized as simple, parallel, somewhat sinuous strands, as in *P. hydrodromus* (Adiyodi, 1984); as sinuous strands forming random lumps, as in *O. platytarsis* (Thampy and John, 1970) and *Callinectes sapidus* (Payen, Costlow, and Charniaux-Cotton, 1971); or as anastomosing cellular cords of irregular diameter with the cells arranged in rows or masses, as in *R. harrisii* (Payen et al., 1971).

Under light microscopy, the androgenic gland of crabs is composed of a uniform type of cell, rather loosely arranged, showing morphological changes linked to their state of activity. The gland appears hypertrophied during the active phase in *P. hydrodromus*. The cells have large nuclei and many basophilic cytoplasmic granules. The inactive phase of the gland is characterized by atrophy, cytoplasmic vacuolation, total disappearance of basophilic granules, nuclear pyknosis, and cellular degeneration (Adiyodi, 1984). Degenerating cells often coexist with normal ones in *O. quadrata* (Payen, 1972) and *P. hydrodromus* (Adiyodi, 1984).

The androgenic glands of *P. crassipes* (King, 1964), *C. maenas* (Meusy, 1965a, b), *R. harrisii*, *C. sapidus*, *M. mercenaria*, and *O. quadrata* (Payen et al., 1971; Payen, 1972) have several ultrastructural features in common. The gland cells interdigitate and their surfaces are covered by fibrous basal lamina. Intercellular spaces are dilated and no secretory material is discernible. A vesicular or lamellar type of rough endoplasmic reticulum, devoid of electron-dense intracisternal material, is present. The mitochondria are spherical or elongated and have relatively short cristae. A discrete Golgi complex with flattened saccules is also present. Spherical or ovoid multivesicular bodies (MVB) have been observed in varying numbers in practically all cells. MVBs show acid phosphatase activity in *P. crassipes*, suggesting that they may play a role in cellular degeneration.

The rough endoplasmic reticulum is replaced by the smooth type in degenerating cells of the androgenic gland of *O. quadrata*. Other ultrastructural features of degenerating cells include vacuolation of mitochondria, formation of vacuolar bodies, and the presence of autolytic vacuoles, residual bodies, and nuclear pyknosis (Payen, 1972).

The chemical nature of the androgenic hormone of crustaceans has not been settled. The active constituents, extracted from the androgenic gland (dissected with vas deferens) and hemolymph of *C. maenas*, are said to contain two lipid substances (C_{18} isoprenoid ketones), a hexahydrofarnesyl acetone and a farnesyl acetone. These lipoids inhibit vitellogenesis in sexually active female *Orchestia gammarella* and induced carotenoid pigment deposition on the second antennae (a male character) of the female isopod *Talitrus saltator* (Berreur-Bonnenfant et al., 1973; Ferezou et al., 1977). A second substance extracted from the male reproductive system, including androgenic glands, of *Armadillidium vulgare* is a protein having a molecular weight of about 16,000 (Katakura, Fujimaki, and Unno, 1975; Katakura and Hasegawa, 1983). Administration of this substance caused not only external masculinization but also development of the testis, seminal vesicles, and vasa deferentia in sexually inactive females. A water-soluble, thermostable compound having a molecular weight ranging between 1,200 and 8,000, and capable of inducing external masculinization of females, has been extracted from androgenic glands of intersexes of *A. vulgare* (Juchault, Maissiat, and Legrand, 1978).

The androgenic gland is indistinguishable during the beginning of male sex differentiation in Brachyura and Anomura (Payen, 1974; Le Roux, 1976). In genetic male crabs, an "androgenic inductor" released from gonadal mesoderm cells induces the differentiation of gonia into spermatogonia. Initiation of spermatogenesis in young crabs, as evidenced by the emergence of gonia from the germinative zone and production of further gametogenic stages, can progress only if the androgenic gland begins to secrete its hormone (Payen, 1974; Payen and Amato, 1978).

That the androgenic gland is responsible for testicular differentiation in crustaceans is borne out by experiments involving inversion into testes of the ovaries of *O. gammarella* implanted into normal or castrated males (Charniaux-Cotton, 1953, 1954). Implantation of androgenic glands has been reported to induce external masculinization in *C. maenas* (Charniaux-Cotton, 1958) and *R. harrisii* (Payen, 1969, 1975). In *C. maenas*, however, spermatogenesis progresses unhindered and the germinative zone remains unaltered in adult testis transplanted into females (no androgenic hormone). This suggests that androgenic hormone is not necessary for the maintenance of spermatogenic activity in this species (Payen, 1974).

The possible presence in *P. hydrodromus* of a gonad-stimulating hormone in brain and thoracic ganglion, capable of promoting testicular and ductal activity, was proposed by Gomez (1965). More recently, Joshi and Khanna (1984) have reported that administration of an extract of thoracic ganglia during the inactive phase of testicular activity results in a rise in gonadal index, hypertrophy and hyperactivity of the androgenic glands, and promotion of spermatogenesis in the crab, *Potamon koolooense*.

The exact role of eyestalk principles on male reproduction is not understood. In sexually mature males of *U. pugnax*, spermatogenesis is seasonal and normally occurs during summer. Eyestalk removal in summer causes a hasty entry into spermatogenesis much ahead of this event in normal crabs during this season (Young, 1974). In young crabs in which androgenic hormone is necessary for the initial burst of spermatogenic activity, the eyestalks may restrain androgenic gland activity, even in larvae (Payen et al., 1971; Payen and Amato, 1978). Bilateral eyestalk ablation in zoeae of *R. harrisii* and megalops of *C. sapidus* leads to a hypertrophy of the androgenic gland due to hyperplasia and increase in cell size (Payen et al., 1971). Ultrastructural studies reveal that such gland cells have a well-organized rough endoplasmic reticulum, mitochondria crowded in the perinuclear area, and electron-dense material accumulated in intercellular spaces. In *R. harrisii*, lipid droplets also are present in cytoplasm, and a thinning of the basal lamina is evident. That glandular hypertrophy leads to hypersecretion is obvious from the precocious acquisition of male physiology by such individuals lacking eyestalks (Payen et al., 1971).

In *R. harrisii*, the restraining influence of the eyestalk on the androgenic gland appears to decline with sexual maturity. Thus, in pubertal male *R. harrisii*, eyestalk removal leads to only a mild hypertrophy of the androgenic gland by 42 days after the operation; fine structural images of the gland show that the cells are nearly normal. Spermatogenesis is little affected (Payen et al., 1971). However, in *C. maenas*, glandular hypertrophy induced by eyestalk removal in adult males leads to degranulation of the extremely well developed endoplasmic reticulum, to hyposecretion, and to cytolysis (Meusy, 1965a). Thus, eyestalks appear to be necessary in adult males for the normal functioning of the androgenic gland.

Adult males of the Calicut population of *P. hydrodromus* are sexually inactive during the somatic phase of intermolt, February through May. Bilateral eyestalk removal in February revives germinal activity and results in a substantial increase in the number of spermatogonia (S. Gupta and K. G. Adiyodi, unpub.). At 10 days after the operation, there are clear signs of an increase in several organic components (e.g., proteins, amino acids, oligosaccharide fraction, polysaccharides) of the testis and semen (homogenate of the vas deferens). Secretory cells lining the lumen of the sperm duct become hypertrophied, and the testis and sperm duct gain weight (Mathad, 1983). Obviously, eyestalk substances play a key role in the synthesis and release of various seminal components in *P. hydrodromus*. However, it is not clear if the eyestalk factor(s) exerts its influence directly or if the action is only mediated through the androgenic gland.

Similar experiments conducted by Mathad (1983) on male *P. hydrodromus* during August/September, when spermatogenic activity is already high, gave no evidence of a further acceleration in activity. However, a significant increase in semen production was observed 10 days postoperation. There was a gain in weight of testis and sperm duct, a hypertrophy of the glandular epithelium lining the lumen of the genital duct, and a rise in total TCA-precipitable proteins, chloroform-extractable lipids, free amino acids, and polysaccharides and oligosaccharides of the testis as well as the sperm duct. Testicular and ductal weights, however, returned to near-normal values by 20 days after the operation. The size of the ductal epithelium was restored, although the epithelial cells appeared exhausted, possibly as a result of hyperactivity. Copulation was ruled out as the cause for loss of seminal contents, since experimental crabs were not allowed to mate. By 30 days postoperation, seminal and testicular oligosaccharide fraction and lipids, seminal free amino acids, and testicular polysaccharides declined to values less than those of normal crabs (Mathad, 1983).

In *P. hydrodromus*, bilateral eyestalk removal leads to hyperphagia for 7–10 days postoperation; feeding becomes normal again for a brief period, but the animals soon become hypophagic. This is coupled with an increased mobilization of resources (possibly to cope with the increased rate of utilization) from various sources including hepatopancreas and hemolymph (Adiyodi, 1969b; Anilkumar, 1980). The hepatopancreas becomes pale in color due to excessive mobilization of metabolites and cellular degeneration. Organic materials are mobilized from various sources, which in *P. hydrodromus* seem to include the seminal plasma.

In *P. hydrodromus*, the role of eyestalks in semen production is not purely inhibitory. The spurt in spermatogenic activity observed in February and the initial hypertrophy of the ductal cells that occurs in February as well as in August/September, all following eyestalk ablation, argue for an inhibitory role for eyestalk secretions in spermatogenesis and semen production, but the subsequent changes (as already reviewed) hardly substantiate such a view. It is more logical to consider the eyestalk of *P. hydrodromus* the source of a "restraining" substance whose presence is essential in females for maintaining normal vitellogenesis and in males for maintaining normal semen production.

It is difficult to specify the role of the androgenic gland in semen production in *P. hydrodromus*. Although the hypertrophy of the gland during the sexually active season and subsequent hypotrophy during the inactive season can be cited as indirect evidence for a role for the androgenic gland in male reproduction, there is nothing in the cytology of the gland suggestive of either an increased or decreased level of activity following eyestalk ablation (Mathad, 1983).

7. Development

A. Embryonic development

Crustaceans show great diversity in embryonic development, especially owing to the remarkable variations in egg size, duration of development, and cleavage patterns (see Anderson, 1982, for review). Information on embryonic development of crabs is limited to only a few species, and is particularly lacking for land crabs. Centrolecithal cleavage and direct blastoderm formation have been reported in some decapods (Zehnder, 1934a, b). There is remarkable similarity in the fate maps of crustaceans and variations in presumptive areas being

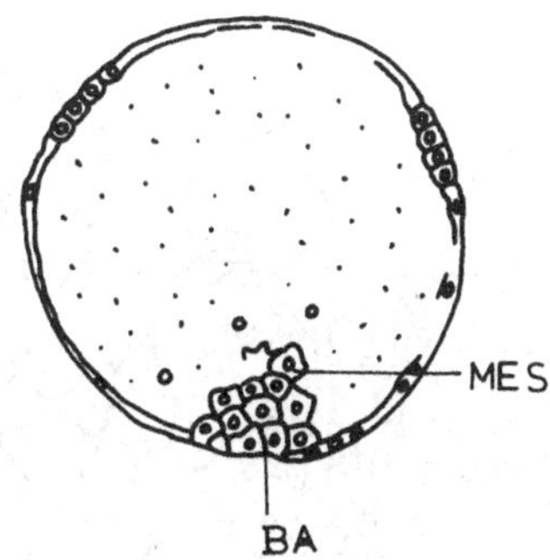

Fig. 5.9. Diagrammatic representation of the embryo of the anomura *Eupagurus prideauxi* in the process of gastrulation. *BA*, blastoporal area; *MES*, mesendoderm. (Redrawn from Scheidegger, 1976.)

limited to presumptive midgut and presumptive extraembryonic ectoderm.

In many decapods, presumptive midgut is part of the blastoderm; it migrates inward by cellular proliferation (Krainska, 1936; Lang, 1973; Scheidegger, 1976). Generally in crustaceans, cells of the presumptive stomodaeum are located in front of the presumptive midgut, and the presumptive proctodaeal cells are located posteroventrally in the middle of the presumptive ectoderm. Upon gastrulation in the anomuran *Eupagurus prideauxi*, the presumptive mesoderm and midgut become distinct as a composite mesendoderm (Scheidegger, 1976). In several malacostracans, including *Maja squinado* (Lang, 1973), and in *E. prideauxi* (Scheidegger, 1976), the presumptive extraembryonic ectoderm surrounds the yolk mass (Fig. 5.9).

It is during gastrulation that the presumptive areas occupy their definitive organ-forming positions. In species in which blastoderm formation is direct, the cells of the presumptive midgut proliferate and enter the yolk mass (Krainska, 1936; Lang and Fioroni, 1971; Lang, 1973). During gastrulation, the presumptive mesoderm cells move toward the interior anteriorly under the ventral ectoderm (Lang, 1973).

As development proceeds, the stomodaeum and proctodaeum shape out as cuticle-lined foregut and hindgut epithelium, respectively, and become linked to the midgut. The anterior midgut rudiment, which enters the yolk mass, forms the vitellophage epithelium. A posterior midgut rudiment proliferates and becomes linked to the vitellophage epithelium. The rudiment grows to form the posterior midgut tube connected to the proctodaeum. At the junctions of the posterior midgut tube and stomodaeum with the vitellophage epithe-

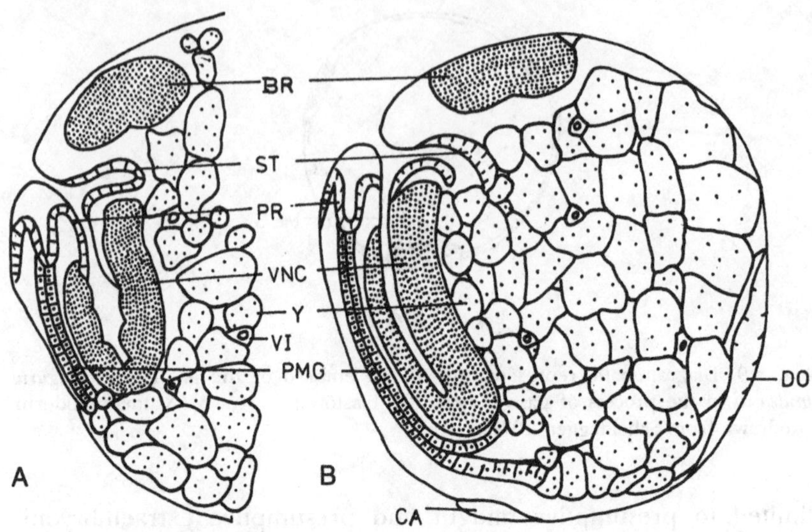

Fig. 5.10. A, B. Diagrammatic sagittal sections of two developmental stages of the embryo of *Eupagurus prideauxi*. *BR*, brain; *CA*, carapace; *DO*, dorsal organ; *PMG*, posterior midgut; *PR*, proctodaeum; *ST*, stomodaeum; *VI*, vitellophage; *VNC*, ventral nerve cord; *Y*, yolk. (Redrawn from Scheidegger, 1976.)

lium, the latter forms the anterior and posterior rudiments of the digestive gland. Finally, almost the entire vitellophage epithelium gets incorporated into the digestive glands (Krainska, 1936; Fioroni, 1969, 1970a, b; Lang and Fioroni, 1971; Lang, 1973; Scheidegger, 1976) (Figs. 5.10A, B and 5.11).

Vitellophages also appear from the germ band during the initial formation of the nauplius rudiment and also from mesoderm of the head lobes at a later stage. These vitellophages collectively absorb the yolk (Fioroni, 1970a, b; Fioroni and Bandaret, 1971; Lang and Fioroni, 1971; Lang, 1973).

After gastrulation, presumptive mesoderm is seen as a group of cells underlying the postnaupliar ectoderm. Naupliar mesoderm arises from a portion of cells of this group and forms paired antennulary, antennal, and mandibular somites. In Malacostraca, eight of the remaining mesoderm cells enlarge to form mesoteloblasts arranged in pairs, and they give rise to the postnaupliar somites. Coelomic cavities, often vestigial, develop in these somites. Cardiac, perivisceral, and pericardial hemocoels become distinct. The walls of the heart, as well as the muscles of the alimentary canal, develop from the pericardial septum, which lies beneath the pericardial coelom. Limb and ventral longitudinal muscles are developed from major part of each somite.

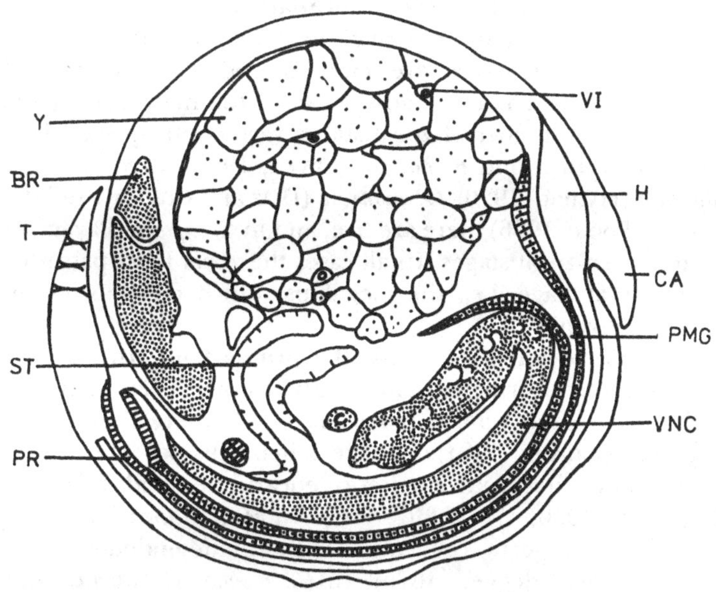

Fig. 5.11. Diagrammatic saggital section of the embryo of *Eupagurus prideauxi* at an advanced stage of development. *BR*, brain; *CA*, carapace; *H*, heart; *PMG*, posterior midgut, *PR*, proctodaeum; *ST*, stomodaeum; *T*, telson; *VI*, vitellophage; *VNC*, ventral nerve cord; *Y*, yolk. (Redrawn from Scheidegger, 1976.)

Somatic musculature of the antennulary, antennal, and mandibular segments originate from the somites of naupliar segments. Although preantennulary mesoderm has been detected in several crustaceans, it is not reported to exist in *E. prideauxi* (Scheidegger, 1976).

B. Larval development

Starting with the nauplius stage, crustaceans pass through a number of larval stages in the course of their development. In species with direct development, all the larval stages are embryonized and the individual hatches out as a juvenile that greatly resembles the parents except in aspects related to sexual maturity. This developmental pattern is followed by thalassinids such as *Upogebia savigni* and *Callianasa kraussi* (Gurney, 1937; Forbes, 1973) and brachyurans such as *Pilumnus novae-zealandiae, Sesarma perraceae, P. hydrodromus, Gecarcinicus steniops*, and potamid crabs (Wear, 1967; Adiyodi, 1968a; Soh, 1969; Pace, Harris, and Jaccarini, 1976; Samuel, 1984). In species with indirect development, the individual emerges either as the naupliar larva or in a more advanced larval stage by embryonizing the initial one or more larval stages. Thus in some thalassinids, pagurids, and brachyurans, both nauplius and the subsequent protozoeal stages are embryonized and the individual emerges as a zoeal larva, a modification of the mysis stage (Anderson, 1982). The number of zoeal stages varies with species. For example, in *C. maenas* and *R. harrisii* (Payen, 1974) there are four, in *U. pugilator* (Hyman, 1920), *O. quadrata* (Diaz and Costlow, 1972), and *C. magister* (Poole, 1966) there are five, and in *C. sapidus* (Payen, 1974) there are seven zoeal stages. In all cases the zoeal larva passes on to the next larval stage, i.e., the megalops, which metamorphoses to become the juvenile.

In most species of crustaceans, the primordial germ cells make their appearance during or after gastrulation. After their migration, they reach the ventral pericardial wall where they form the early gonad (Fioroni, 1981). The germinal cells become distinct in the maxillary or first trunk somites of eucarid decapods (Nair, 1941, 1949; Shiino, 1950; Weygoldt, 1961). In *P. prideauxi* and *M. squinado*, primordial germ cells occur in the mesendodermal mass (Lang, 1973; Scheidegger, 1976). In *C. maenas*, *C. sapidus*, and *R. harrisii*, the genital rudiment can be observed as early as 10 days before eclosion of the zoea as two cell groups. Each group is composed of three to four germinal cells, located in the second thoracic segment beneath the pericardial septum and anterior to the heart. Some mesoderm cells with ovoid nuclei can also be detected among the protogonia (Payen, 1974).

In the third zoeal stage (Z_3) of *C. sapidus*, Z_2 of *R. harrisii*, and Z_1 of anomurans, two cell groups, each composed of four to six germinal cells, can be identified ventral to the heart. The gonadal rudiment of *C. sapidus* becomes more extensive by the seventh zoeal stage. Posteriorly, the two germ cell masses unite, and a collection of somatic cells, possibly representing the genital duct rudiment, is seen in association with the protogonia. A comparable pattern has been also observed in Z_4 of *R. harrisii*.

The genital rudiment holds its association with the pericardial septum, even as *C. sapidus* reaches the megalops stage. The germinal cell groups have extended anteriorward; posteriorly, the genital ducts develop to form two mesodermal cords. The gonadal rudiment is now composed of protogonia, gonia, and mesoderm cells, their numbers far exceeding those in the zoea. The rudiment of the androgenic gland can be distinguished as a tiny swelling composed of 6–10 somatic cells with ovoid nuclei.

Only five abdominal segments (and a telson without appendages) are present in the first and second zoeal stages of *C. sapidus*. The sixth segment becomes apparent by Z_3 and the pleopod (second to sixth) rudiments by Z_6. It is in the megalops stage that the pleopods (second to fifth) become biramous.

8. Conclusions

Among all crustaceans, only limited information is available on reproduction and development, and the data available are largely morphological rather than physiological or biochemical. This is a serious handicap in drawing comparisons and conclusions on reproductive and developmental physiology of decapod crustaceans, especially the land crabs.

Though there have been past references to the physiology of crabs favoring somatic growth or ovarian growth during different seasons of the year, recent studies on the freshwater crab *P. hydrodromus* have brought to light the clear-cut demarcation of the intermolt itself into two distinct phases, reproductive and somatic. That bilateral eyestalk ablation induces accelerated ovarian growth during certain seasons and accelerated ecdysis during others in other crabs also suggests that this type of phasing of the intermolt may be more universal. The change in hormonal milieu responsible for this shift in physiology is not understood, and is a challenging theme for future investigators.

Biochemical and cytological aspects of vitellogenesis of crabs have been far less studied than those of insects. Information on cytoarchitectural changes that occur in crab oocytes preceding vitellogenesis is woefully inadequate. Such knowledge is vital in understanding the prolonged initial slow-growing phase of oocytes spanning the postspawning period and the prebreeding season of seasonal breeders, such as *P. hydrodromus* and *G. steniops*. Experiments have shown that the effect of total deprivation of eyestalk secretions on ovarian growth depends on the stage of development the oocytes have reached at the time of eyestalk ablation. However, cytophysiological studies on oocytes at various levels of development are needed to understand the role of eyestalk hormones in controlling oogenesis.

Recent studies have revealed progressive accumulation of ecdysteroids in oocytes. This suggests a possible role for these substances in embryonic development. The role of ecdysteroids in crab vitellogenesis is yet another problem deserving critical study. Seminal chemistry of land crabs is still rudimentary and the data available on embryonic and postembryonic development are few and mostly out of date.

Clearly, virtually any study of reproduction and development of decapod crustaceans, and terrestrial species in particular, will yield valuable new information.

6: Growth and molting

RICHARD G. HARTNOLL

CONTENTS

1. Introduction

Land crabs are no different from other crustaceans in that growth can take place only by means of intermittent shedding of the more or less inextensible integument – the process of molting or ecdysis. For aquatic crabs, molting is a time of stress and mortality, resulting both from the dangers inherent in the molt process itself, and from the high risk of predation while the newly molted crabs are soft and relatively immobile. For land crabs there are added complications: The risk of desiccation is greater at this time, and there is the problem of obtaining the quantity of water needed for the postmolt increase in size, which must occur rapidly before the new integument hardens.

Although molting appears superficially as a short and intermittent interlude, it has a pervasive effect on the whole of the life cycle, and the period between molts is one of continuous morphological and physiological change (also see Chapter 5). These changes enable the crab to prepare for molting, and to recover from it. Some basic elements relating to molting and growth of crabs will now be outlined;

a more detailed account for crabs in general is provided by Hartnoll (1982, 1983).

The molting cycle can be divided into the following stages.

Premolt (proecdysis). The new integument is being laid down beneath the old, and stored energy reserves are being mobilized to enable the new structures to be formed. Calcium is being resorbed from the old integument and stored in the tissues.

Molt (ecdysis). The old integument is shed, and the crab rapidly increases in size by the absorption of water. The process takes at most only a matter of hours.

Postmolt (metecydysis). The new integument hardens using the stored calcium, plus additional supplies obtained from the water and food. Extra tissue is formed to build up the muscles and other structures in the new and larger body.

Intermolt. Formation of new tissue is completed, and energy reserves are accumulated and stored pending the next molt. Final hardening of the integument is accomplished. During times of rapid growth the intermolt is relatively short and is known as a diecdysis. During slow growth, such as over the winter, a long intermolt termed an ancecdysis occurs. (Further discussion of the division of the intermolt stages is found in Chapter 5).

This cycle of events results in a series of molts that allow for the continued growth of the crab. The rate of growth will depend upon two variables. One is the amount by which the size increases at a molt – the molt increment. This can vary from imperceptible amounts to as much as 80% of the carapace length, and generally the percentage increment falls with increasing size. The second variable is the time between molts – the intermolt period. This can vary from a few days to two or more years, and usually becomes longer with greater size. This combination of reducing increments and lengthening intermolt periods results in growth slowing as size increases (Hartnoll, 1978a).

The succession of molts that makes up the life cycle of a crab is punctuated by particular molts that mark distinctive stages in its life. The transitions from the zoeal larva to the megalops, and from the megalops to the juvenile crab, are metamorphoses, i.e., molts at which substantial morphological changes occur. After a number of juvenile crab instars a particular molt, usually marked by morphological changes, occurs following which the crab is sexually mature; this is called the puberty molt. In some crabs molting eventually ceases: The final molt is the terminal ecdysis, and the crab is afterward in a state of terminal anecdysis.

A special feature of crustacean growth is the ability to shed certain limbs and then to regenerate them. In crabs the chelae and walking legs, if restrained or damaged, can be shed at a breaking plane near the base with minimal damage and where a specialized muscular sphincter stops hemolymph loss; this is referred to as autotomy. The limb is then regenerated. It grows as a nonfunctional limb bud during the intermolt, and at the next molt appears as a reduced but complete and functional appendage. However, several molts are often needed to restore the limb to its original size.

A final aspect of growth is that as well as increasing in size during its life, change is seen in bodily proportions; this is termed relative or "allometric" growth. It particularly affects the secondary sexual characters such as the chelae and abdomen, in which changes can be especially obvious at the puberty molt.

2. Molting in land crabs

A. The molt process

Most of the basic phenomena of ecdysis and the molt cycle are very similar in land crabs and aquatic species. A comprehensive review of molting in Crustacea is given by Skinner (1985). This chapter will discuss only those special aspects and problems of molting related to a terrestrial environment, i.e., the risk of desiccation and the acquisition of the water needed for size increases, as well as the calcium to harden the new integument.

Many land crabs minimize the risk of desiccation during molting by remaining in a burrow, which can provide a cool and moist microhabitat, as well as protection against predation. For example, there are several reports that the robber (coconut) crab, *Birgus*, molts within sealed burrows, both in the wild (Reyne, 1939; Johnson, 1965; Alexander, 1976) and in captivity (Held, 1963). Held (1963) observed a specimen of *Birgus* that molted four times in a terrarium, remaining sealed in the burrow for between 30 and 39 days each time. Before unplugging and leaving the burrow the crab completely consumed its cast exuvium, which may explain why these casts are not found in the wild. When *Birgus* enters the burrow prior to molting, the abdomen is highly distended, often causing it to extend behind the carapace (Reyne, 1939; Held, 1963; Alexander, 1976). This distension is partly due to stored fat (the accumulation of energy reserves prior to molting is a general phenomenon). *Ocypode ceratophthalmus* and *O. cordimana* both molt within plugged burrows (Fellows, 1973). Grubb (1971) found a molting specimen of *Geograpsus grayi* inside the burrow of a *Birgus*; *Gecarcinus lateralis* molts within its own burrows (Weitzman, 1963).

Some land crabs have water readily available during ecdysis. Thus, *Cardisoma guanhumi* molts within its burrow, but in this case the base of the burrow reaches down to the water table and provides not only a refuge from predation but also a source of water (Henning, 1975b; see also Chapter 3). *Coenobita clypeatus* remains within its shell for most of the molt, a habit that similarly serves both to protect against desiccation and to provide a reserve of water (De Wilde, 1973; see also Chapters 7 and 8). *Madagopotamon humberti* molts only during the wet season, and does so in small temporary pools collecting in irregularities of the limestone rock of its habitat (Vuillemin, 1970).

However, by no means do all land crabs have supplies of standing water available during the molt; *Birgus* (Alexander, 1976) and *Gecarcinus* (Weitzman, 1963), for example, clearly do not. In such cases the necessary water to enable the postmolt increase in size must in some way be accumulated in the tissues during the premolt period. In *Birgus* it must be assumed that the greatly swollen abdomen accommodates water stored as hemolymph, but there are no detailed studies. However, extensive research has determined the mode of water storage in several Brachyura.

The key structures involved in water storage are the pericardial sacs, present in all brachyurans but notably developed in land crabs such as *Cardisoma*, *Gecarcinus*, and *Ocypode* (Bliss, 1963, 1968; Bliss and Boyer, 1964; Bliss, Wang, and Martinez, 1966; Bliss and Mantel, 1968). The pericardial sacs are pouchlike organs lying on the thoracic epimera at the rear of the branchial chambers; they are covered by cuticle continuous with the chitinous wall of the branchial chamber and have a cavity communicating with the pericardial sinus. In marine species the pericardial sacs are narrow and elongate (Fig. 6.1), but in terrestrial species they are enlarged (Table 6.1). In *Ocypode quadrata* they are expanded laterally; in *Cardisoma guanhumi* there is also a posterior setae-fringed extension reaching the ventral surface of the abdomen; in *Gecarcinus lateralis* the lateral extension has the form of an expanded lobe, and there is a posterior extension similar to that of *Cardisoma* (Fig. 6.1). The surface area of the pericardial sacs of *Gecarcinus* and *Cardisoma* is significantly greater than that of marine species when both carapace width and body weight are compared (Table 6.1). The largest are found in *Gecarcinus*, which molts in a burrow without access to standing water. During premolt this crab accumulates water, which it may take up from damp sand, using setae at the base of the abdomen and the extensions of the pericardial sacs (Bliss, 1963; Bliss and Boyer, 1964; T. G. Wolcott, 1976, 1984; see also Chapters 3 and 7). This water is possibly absorbed by the pericardial sacs. These structures swell so that they may extend beyond the posterior edge of the carapace, accommodated by the very extensible arthrodial membrane between the carapace and abdomen. This absorption of water results in an increase in the weight of the crab

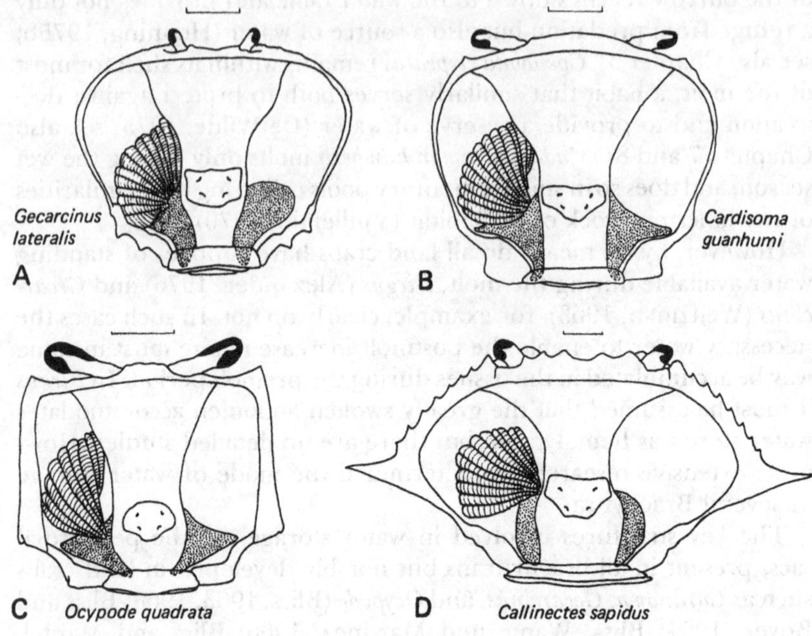

Fig. 6.1. The pericardial sacs (*stippled*) of three land crabs (A–C) and one marine crab (D). Gills shown on left side only. Scale bar = 10 mm. (After Bliss, 1963, 1968.)

Table 6.1. *Size- and mass-specific areas of pericardial sacs from crabs in different environments*

Species	Habitat	Area of pericardial sacs: $mm^2 \cdot cm^{-1}$ carapace width	Area of pericardial sacs: $mm^2 \cdot g^{-1}$ body mass	
Gecarcinus lateralis	Land	73	8.6	Bliss (1963)
Cardisoma guanhumi	Land	58	4.7	
Ocypode quadrata	Land	31	10.0	
Callinectes sapidus	Marine	29	2.6	
Cancer spp.	Marine	32	2.3	
Ocypode cordimana	Land	36		Rao (1968)
Ocypode macrocera	Intertidal	25		
Ocypode platytarsis	Intertidal	22		

Note: The methods of calculating pericardial sac area are not fully explained in the original papers, so the results of the two studies may not be directly comparable.

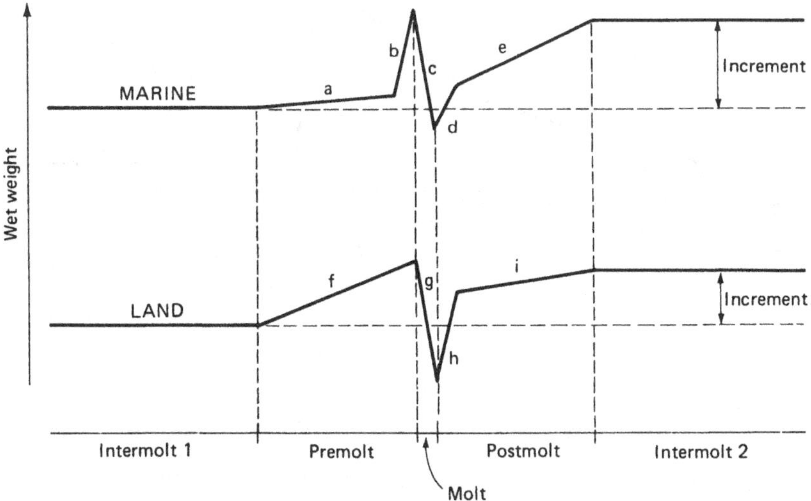

Fig. 6.2. Diagrammatic representation of changes in wet weight of a marine and a land crab during the molt cycle. The main cause of weight change is indicated for each section of the line by (a) formation of new integument; (b) uptake of water; (c) casting of integument; (d) further water uptake; (e) recalcification of integument using outside sources; (f) formation of new integument and uptake of water; (g) casting of integument; (h) consumption of old integument; (i) recalcification using outside sources. (Partly after Travis, 1954, and Skinner, 1966a.)

throughout the premolt period, in contrast to marine species where the increase is confined to the very late premolt (Fig. 6.2). Experimental studies in the laboratory with *G. lateralis* have shown that this increase in weight can exceed 25%, and that the uptake of water required for ecdysis can be obtained entirely from damp substrates (Bliss et al., 1966). It is not altogether clear how this water taken from the substrates is absorbed. Certainly, the pericardial sacs are important in conducting it into the rear of the branchial chambers, but their role in uptake is doubtful, the posterior gills probably being more significant (Bliss and Mantel, 1968; Mason, 1970). Mason's (1970) experimental work indicates that the pericardial sacs become permeable to any degree only when the old cuticle starts splitting in the last few hours before ecdysis. If surface water is available, it could also be taken up by drinking and absorbed through the foregut (Bliss and Mantel, 1968).

This store of water in the pericardial sacs of land crabs is used to bring about the rapid increase in size that must take place quickly after ecdysis. There is a movement of water from the hemolymph into the gut at ecdysis, under the control of neurosecretions from the

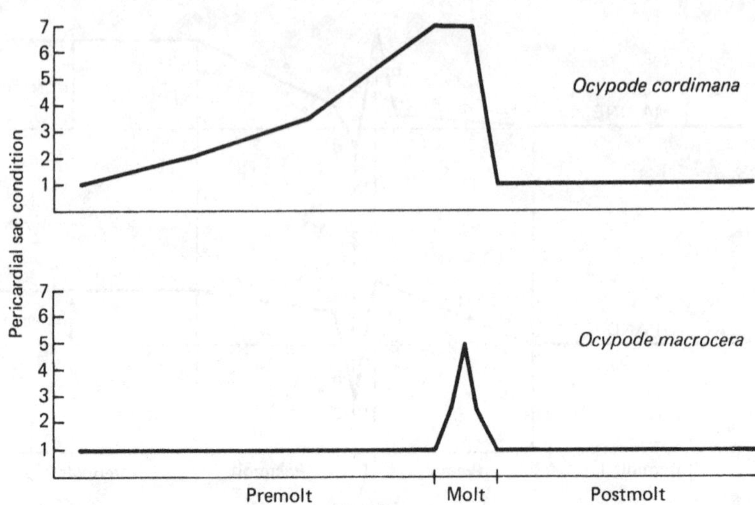

Fig. 6.3. The relationship between the swelling of the pericardial sacs (1 = unswollen; 7 = fully swollen) and the stages of the molt cycle in the terrestrial *Ocypode cordimana* and the intertidal *Ocypode macrocera*. (After Rao, 1968.)

thoracic ganglionic mass (Bliss, 1968). However, this relocation of water is not the only means by which size increase is achieved. In *G. lateralis* the anterior chamber of the proventriculus is distended by gas immediately after ecdysis, though the nature and origin of this gas are not known. Both the gas and the fluid in the posterior proventriculus facilitate expansion of the new carapace to produce the large epibranchial space characteristic of land crabs (Bliss, 1968).

The importance of the pericardial sacs in the molting of terrestrial crabs is further illustrated by the work of Rao (1968), who compared intertidal and supratidal species of *Ocypode*. The supratidal *O. cordimana* has larger pericardial sacs than the intertidal *O. macrocera* and *O. platytarsis* (Table 6.1). In *O. cordimana* the pericardial sacs become swollen during the premolt period (Fig. 6.3), whereas in the other species they show swelling only during ecdysis itself, as is the general case in marine crabs (Drach, 1939). If the swelling of the pericardial sacs in *O. cordimana* is reduced prior to ecdysis by desiccating the crabs, then the size increase at ecdysis is reduced, or the crabs even decrease in size.

The final problem facing land crabs at the time of molting is the provision of adequate supplies of calcium salts for the hardening of the new integument. Land crabs tend to be heavily calcified for their size – presumably connected with the rigors and stresses of the terrestrial environment. Yet they do not have the chance to absorb the necessary supplies from the surrounding water as has been shown to occur in marine crabs (Bliss, 1968). The problem is countered by two mechanisms.

The first is to reabsorb as much calcium salts as possible from the old integument prior to ecdysis. Thus, whereas the marine *Carcinus maenas* can lose over 90% of its body calcium at ecdysis (Graf, 1978), this would be unacceptable for terrestrial forms. The reabsorbed calcium is initially stored in the hemolymph; in *Sesarma haematocheir* and *Holthuisana transversa*, for example, it reaches 50–150 times the intermolt value during the premolt (Numanoi, 1940; Greenaway, 1985). The calcium may subsequently be stored in structures that are not shed at ecdysis, the precise location varying with species. In *Parathelphusa hydrodromus* (Adiyodi, cited in Bliss, 1968) and in *Sesarma dehaani* (Numanoi, 1940), it is stored in the hepatopancreas, but in other species special storage structures (termed gastroliths) are formed. These are developed between the epidermis and the cuticle of the foregut, where they appear as whitish concretions. At ecdysis the cuticle lining the foregut is shed, and the gastroliths are freed into the gut lumen where they are dissolved and absorbed. As a result there is a second rise in hemolymph calcium level following ecdysis (Fig. 6.4) when this stored calcium is mobilized and incorporated into the hardening exoskeleton. Gastrolith formation occurs in *Sesarma haematocheir* (Numanoi, 1940) and in *Gecarcinus* spp. (Skinner, 1962; Weitzman, 1963). In *G. lateralis* the formation of gastroliths commences about 30 days before ecdysis, and they have disappeared in the three days following the molt (Skinner, 1962). On the other hand in *C. guanhumi* large gastroliths are not formed, and Henning (1975b) considers that calcium is stored in the hemolymph and midgut glands.

The second method for conserving calcium is for the newly molted crab to consume the cast integument, thereby recovering the remaining calcium that was not resorbed. This occurs in a number of marine crabs, but it is particularly prevalent in terrestrial forms. This phenomenon has already been mentioned for *B. latro*, and is recorded for *C. guanhumi* (Henning, 1975b) and *G. lateralis* (Bliss, 1968). However, *Gecarcinus* does not necessarily consume its shell when living on calcareous sand, which presumably can be ingested as an adequate alternative calcium source.

B. The molt increment

A single specimen of *B. latro* observed through four molt cycles in captivity increased in size from 20.8 to 30.0 mm cephalothorax length, with molt increments of between 8.4% and 11.5% in length (Held, 1963). In *C. guanhumi* the length increment decreased with size, from about 10% at a carapace width of 10 mm to less than 5% when the

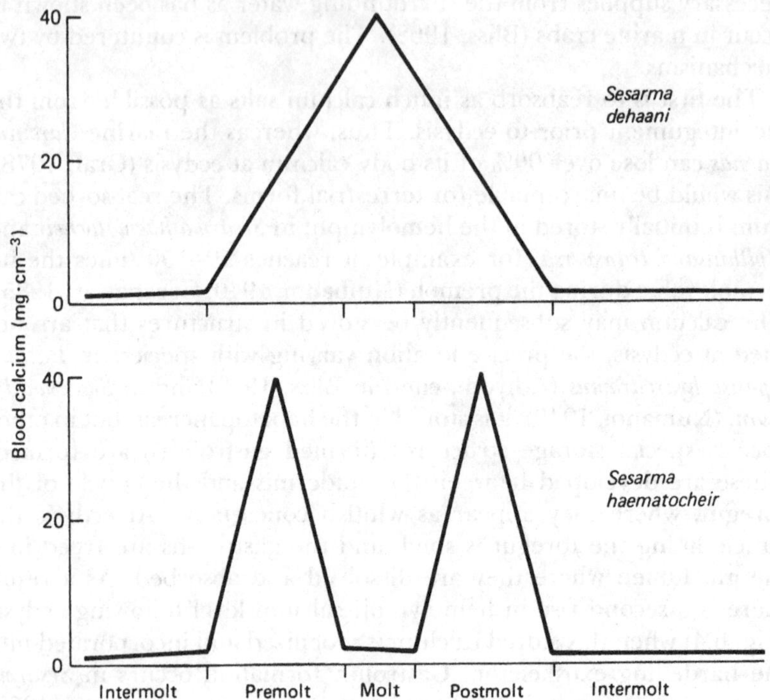

Fig. 6.4 Hemolymph calcium levels during the molt cycle in *Sesarma dehani* (which does not form gastroliths) and *Sesarma haematocheir* (which does). (After Numanoi, 1940.)

carapace width exceeded 50 mm (Henning, 1975b). In the arboreal grapsid *Aratus pisonii* the increment ranged from 10% at 12 mm carapace width to 3% at 18 mm carapace width (Warner, 1967). However, these are values for mature females, and in all crabs the increment is generally low in females after maturity (see Hartnoll, 1982). In the upper-shore grapsid *Pachygrapsus crassipes* the increment ranges from about 15% in smaller specimens to 5% in specimens of 40 mm carapace width (Olmstead and Baumberger, 1923; Hiatt, 1948). These are minimal data on which to base any conclusions, but it would appear that in comparison with marine crabs of the same size range the molt increments are low – in marine species size increments of 20% or more tend to be general (Hartnoll, 1978a, 1982, 1983). Additional observations are clearly needed to confirm this.

There are several reasons why the molt increment might be smaller in terrestrial crabs than in fully aquatic species. One possibility is the restriction in the supply of water to produce the rapid postmolt size increase. Crabs that molt out of contact with water must depend upon water stored in the pericardial sacs and in the hemolymph generally, water stored in the pericardial sacs and in the hemolymph generally, and this may limit the size increase. Another factor is that the constraints of the terrestrial environment such as the risk of desiccation, lack of the support provided by the bouying effect of water, and pressure of predation may all favor a rapid hardening and return to full activity, which may be facilitated by a smaller molt increment. This is a difficult hypothesis to test, however, since there are limited data on molt increments in land crabs, and enormous variation in the recorded increments in aquatic species.

C. *The intermolt period*

There is as little information on the intermolt period in land crabs as on the molt increment. In a captive specimen of *B. latro* the period ranged from about 90 to 120 days (Held, 1963). In *C. guanhumi* the typical pattern of lengthening intermolt period with size was observed, ranging from around 35 days at 10 mm carapace width to over 100 days in specimens with carapace widths of 50 mm and greater (Henning, 1975b). Warner (1967) calculated that the average intermolt period for mature females of *A. pisonii* was 53 days. For *P. crassipes*, Hiatt (1948) estimated an intermolt period increasing with size from about 20 to 45 days. In *Hemigrapsus sanguineus* the period increases with size from 20 to 80 days (Kurata, 1962). The data for fully terrestrial crabs are thus minimal, and since intermolt period varies so much with factors such as temperature, no general conclusions are possible.

In many crustaceans the frequency of molting is influenced by seasonal factors. In aquatic forms temperature variation is the main factor, with rapid molting often occurring during the summer period, and a reduction or total suppression of molting during the winter (see Hartnoll, 1982, for examples). In land crabs the seasonal availability of water is more significant than temperature, and molting may be restricted to the rainy season, as in *B. latro* (Andrews, 1900; Seurat, 1904; Reyne, 1939; Alexander, 1976). In *Madagapotamon humberti* molting is restricted to the wet season when small, temporary pools of water become available (Vuillemin, 1970). *Holthuisana transversa* remains inactive in its burrows during droughts lasting a year or more, and becomes active to forage, breed, and molt only during the short and irregular wet periods (Bishop, 1963; Greenaway and MacMillen, 1978). On the other hand, crabs whose burrows normally reach the water table are not under the same constraints; thus *C. guanhumi*, for example, molts throughout the year (Henning, 1975b).

D. *The control of molting*

Molting is directly controlled by the endocrine system, and so far as is known this is the same in land crabs as in other brachyurans. The control mechanism for crabs in general is discussed in detail by Skinner (1985), and only the outlines will be presented here. Normally the crab is held in the intermolt condition, through the action of a hormone produced by the X-organ–sinus gland complex in the eyestalks (see Passano, 1960a). The X-organ consists of specialized neurosecretory cells in the ganglia of the eyestalk, and the axons of these cells terminate in the sinus gland, which abuts hemolymph sinuses of the eyestalk. The complex produces molt-inhibiting hormone (MIH), though this compound has yet to be isolated or characterized. The other structures involved are the Y-organs or molt glands, small bodies a few millimeters in size lying laterally in the carapace near the external adductor muscle of the mandible. MIH acts on the molt glands by suppressing their activity. The Y-organs secrete the hormone α-ecdysone, and this is converted in other tissues to the active molting hormone, 20-OH-ecdysone. Thus, when the production of MIH falls, the production of ecdysone rises and premolt is initiated.

This regulatory mechanism is activated or suppressed by a combination of intrinsic and extrinsic factors. Intrinsic factors include sexual maturity and the loss and regeneration of limbs (as discussed in section 5). Extrinsic factors include various environmental variables such as light, temperature, and rainfall. Their importance can be deduced from seasonal effects, but has also been investigated by laboratory experiments. Bliss and Boyer (1964) have studied molt initiation in *G. lateralis* in some detail. Factors that stimulate the onset of proecdysis are darkness, moderate temperature, and solitude, conditions typical in the burrows where molting occurs and that reduce the risks during ecdysis. On the other hand, light, high temperature, and the presence of other crabs inhibit proecdysis, which will have the adaptive value of delaying the molt until a suitable refuge is available. Perhaps surprisingly, moisture does not seem to exert an overriding influence. A dry substrate delays proecdysis but does not prevent it. However, it does stop the uptake of water in premolt so that the usual size increase associated with a normal molt fails to occur. In *O. macrocera* premolt is similarly delayed by low temperature, and in larger crabs by lack of solitude, but it is stimulated by darkness. The effect of light depends on the color of the background, producing inhibition only on a dark as opposed to a light background (Rao, 1966). In *Uca pugnax* molting is inhibited by low temperatures (Passano, 1960b). Molting is suppressed in *U. thayeri* maintained in groups, and it is delayed in *U. rapax*, though no such effects are apparent in *U. pugnax* and *U. pugilator* except in groups of adult males (Weis, 1976, 1977). Weis (1976, 1977) relates this to the environment, suggesting that temperate species can less afford to suppress molting because of the short season available for growth.

These responses are much as expected, in that molting is stimulated by conditions conducive to successful molting and reduced risk of predation.

3. Patterns of growth

Patterns of growth are dictated by many factors such as the puberty molt, the occurrence of a terminal ecdysis, growth rates, and population structures. It is difficult to obtain information on age and growth rates for crustaceans (see Hartnoll, 1982). The primary factor is the loss of all hard parts at each ecdysis, so that no persistent annular structures occur, and tags or marks tend to be shed with the integument. It is particularly difficult to obtain unbiased information on land crab populations due to their burrowing and cryptic habits because the capture of specimens depends on their activity, which often varies with size, sex, and reproductive state. As a consequence of this shortage of data this section must be regarded as a very provisional analysis.

A. The puberty molt

The puberty molt is a useful landmark in the growth of crabs in that it indicates the size at which sexual maturity is achieved. This is not always readily deduced from more direct indications such as the ripening of gonads, the occurrence of copulation, or the carrying of eggs by females. The puberty molt is detectable in the males of various species of crabs, but sometimes only after fairly intensive anatomical study. In females, on the other hand, the puberty molt is usually readily recognized due to enlargement and broadening of the abdomen, and increased setation of the pleopods (see Hartnoll, in press). The distinctiveness of the puberty molt varies both within and between the various families of land crabs.

Morphological changes associated with maturity have not been studied in detail in land hermit crabs. In *Coenobita perlatus* and *C. rugosus*, the condition of the pleopods was used as an index of maturity in the females (Page and Willason, 1982), and the "mature" pleopods appeared at about the same size as the females first became ovigerous. However, it is not clear whether the "mature" pleopods appear abruptly at a distinctive puberty molt.

In the grapsid crabs a defined puberty molt occurs in most terrestrial species examined, typically with changes such as those seen in *Metopaulias depressus* (Fig. 6.5). Further examples are *Sesarma bidentatum* and *S. verleyi* (Hartnoll, 1964), and *Aratus pisonii*, *S. ricordi*, *Pachygrapsus transversus*, and *P. gracilis* (Hartnoll, 1965). In some of these species, puberty molt is marked only by small changes; in *A. pisonii*,

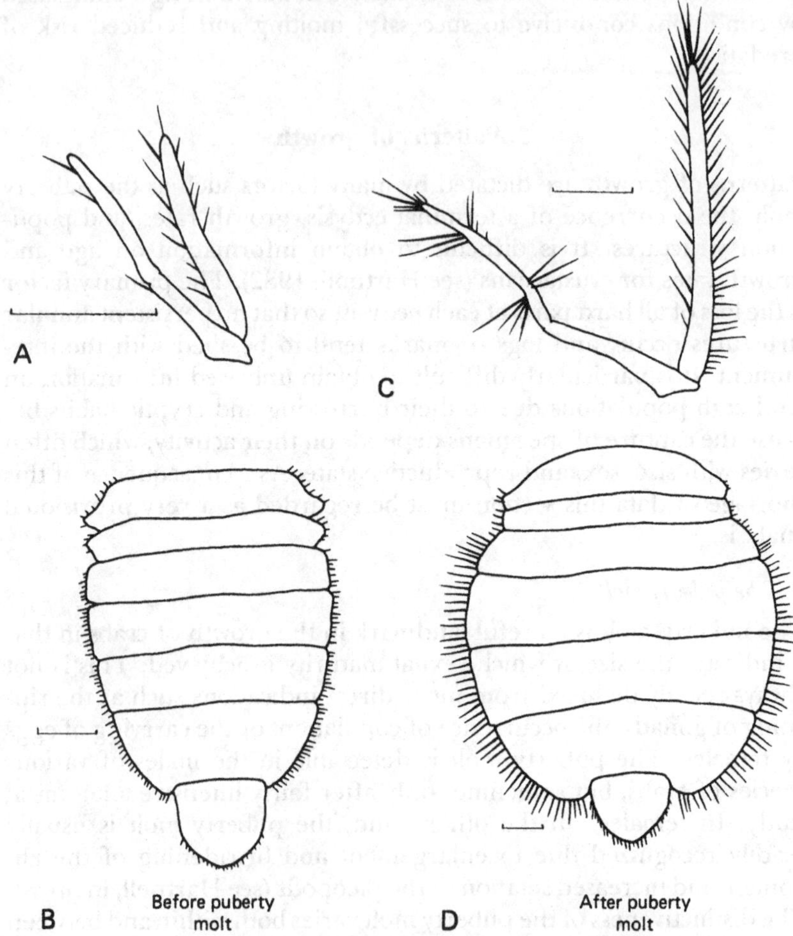

Fig. 6.5. *Metopaulias depressus*. A, B. Second pleopod and abdomen of a female before the puberty molt, C, D. The same organs in a postpuberty female. Scale bar = 1 mm. (After Hartnoll, 1964.)

for example, there are only differences in the form of the pleopods. In contrast, others such as *P. gracilis* exhibit the full range of changes shown in Figure 6.5, including broadening of the abdomen and lengthening of the marginal setation, increase in curvature and setation of pleopods, and enlargement of the genital openings. However, the puberty molt is not obvious in all grapsids; it could not be distinguished in *Cyclograpsus integer*, for example, (Hartnoll, 1965).

In the Ocypodidae a puberty molt can similarly be recognized in females. Haley (1969) noted that in *Ocypode quadrata* the width of the abdomen increased by some 25% at puberty, and to a somewhat lesser extent in *O. ceratophthalmus* (Haley, 1973). In male *Uca*, despite the considerable interest that the exceptional growth of the major chela has stimulated (see section 4, and Crane, 1975), there is no clear demonstration of the presence or not of a puberty molt. In females of *U. pugnax* there is a broadening of the abdomen around the size of sexual maturity (Morgan, 1923b; Huxley, 1924, 1932), but it is not clear whether this occurs at a single molt.

In the Gecarcinidae characteristic changes in the abdomen and pleopods of the females at the size of sexual maturity have been described both for *G. lateralis* (Bliss et al., 1978) and for *C. guanhumi* (Henning, 1975b). In neither case, however, it is made clear whether a distinct puberty molt occurs, though the changes in the abdominal width are substantial.

The evidence relating to a puberty molt in land crabs is thus rather equivocal. In males there is little sign of marked change, but this is characteristic of crabs in general; only in certain groups such as the Majidae are the changes in the male chelae sufficiently clear-cut and decisive to enable a single molt to be readily designated the puberty molt. In females the changes in the abdomen and associated structures are more abrupt, and in some species there is an unambiguous puberty molt. It is likely that studies of further species of land crabs will show this to be of fairly general occurrence.

B. The terminal ecdysis

In some crustaceans growth eventually ceases because molting stops, the last molt being the terminal ecdysis and the crustacean being subsequently in terminal anecdysis. The terminal ecdysis may coincide with the puberty molt, as in the Majidae, for example, or it may occur later, as in *Carcinus*. The occurrence of the terminal ecdysis is discussed in detail by Hartnoll (1982, 1983), from which it is clear that an understanding of the growth and life cycle of a species requires a knowledge of whether a terminal molt takes place. Unfortunately, the terminal molt is often difficult to determine. However, molting after sexual maturity is usually a reliable indication that there is no terminal molt.

There is no evidence for a terminal ecdysis in land crabs, and the fact that a number of species are known to continue molting well beyond puberty indicates that a terminal ecdysis is generally absent. In *Coenobita perlatus*, females mature at 4 mm carapace width but grow to over 24 mm (Page and Willason, 1982). In various grapsids there are observations that mature specimens have molted and appear to continue doing so indefinitely (Hiatt, 1948; Vernet-Cornubert, 1958; Hartnoll, 1964, 1965). Of the gecarcinids, *G. lateralis* becomes mature at around 30 mm carapace width but reaches nearly twice that size (Bliss et al., 1978). In *C. guanhumi* females mature at around 35 mm carapace width and grow to some 70 mm (Henning, 1975b). *O. quadrata* matures at 25 mm carapace width, and continues to grow to over 50 mm (Haley, 1969), as does *O. ceratophthalmus* (Haley, 1973). *Dotilla fenestrata* matures at 5 mm carapace width and grows to 10 mm (Hartnoll, 1973). Crane (1975) provides data for a number of species of *Uca* to show that females reach a substantially greater size than that at which they can become mature.

Of all the crabs described as having any degree of terrestrial adaptation, only the intertidal shore crab *C. maenas* is known to undergo a terminal ecdysis (Carlisle, 1957). Yet its terrestrial habits are so limited that its relevance to this point is marginal.

C. *Age and growth rate*

Observations or estimates of age and rate of growth are available only for a few species of land crabs.

Some deductions have been made regarding the growth rate of *Birgus*. Based on data in Harms (1932, 1938), Gibson-Hill (1947) concludes that the earliest free-living stages found on land are approximately a year old and that maturity is reached in the fourth year, after the eighth molt. Held (1963) observed that the cephalothorax of one captive specimen grew from 20 mm to 30 mm in just over 400 days. From this admittedly limited evidence he concluded that the commonly found specimens with cephalothorax lengths exceeding 100 mm must be some five years old, and it follows that the larger specimens approaching 200 mm could be considerably older. It seems that this largest of land crabs clearly has a relatively long life span, though perhaps not one that would be disproportionately long in relation to marine crabs and other decapods of comparable size. Alexander (1976) notes that on Aldabra, *Birgus* reaches a substantially greater mean weight in areas with coconut palms (654 g) than in areas without (390 g). He suggests that this is due to faster growth rates in areas with improved food supply, and in the absence of obvious differences in mortality, this seems a reasonable supposition. Certainly laboratory studies on many Crustacea have demonstrated faster growth under higher feeding rates (see Hartnoll, 1982). I have found no data on age or growth rates in *Coenobita*. Page and Willason (1982) reported marked differences in the size distribution of nearshore and inland populations for both *C. perlatus* and *C. rugosus*. These, however, would seem to be the result of migration, rather than of different growth rates in the two environments.

Information on growth in land brachyurans is similarly fragmentary. In *Gecarcoidea natalis* a carapace width of about 20 mm is reached after 10 months. The smallest mature females, at 33 mm carapace width are thus probably two years old (Gibson-Hill, 1947). Henning (1975b) has used data on molt frequency and molt increment in *C. guanhumi* from Colombia to estimate growth rate. He calculated that the largest female found, 98 mm carapace width, would have an age of 13 years. Females become mature at 35–40 mm carapace width, and observations on captive specimens would indicate an age for this size of two to three years. It is unwise to extrapolate laboratory observations of growth to the field, but the indications are that gecarcinids are relatively long lived and slow growing.

4. Relative growth

In Crustacea the presence of the rigid outer skeleton, while making the study of absolute growth difficult because of the loss of all hard parts at every molt, does facilitate the accurate measurements needed for the analysis of relative growth. Consequently, relative growth – the differential growth of various parts of the body – has been widely studied, particularly in the Brachyura. General reviews of relative growth in crabs are provided in Hartnoll (1974, 1978b, 1982), and the basic concepts and methods of analysis are explained there in detail, so only a few major points will be made here. Relative growth is most marked in those structures that are directly or indirectly associated with reproduction, particularly the chelae and the abdomen. Differences within a species occur between the sexes, and within each sex between immature and mature specimens. Growth can usually be described by the allometric growth equation, $y = ax^b$, where y is the size of the organ under analysis, and x is the body size. This relationship results in a linear plot after logarithmic transformation, and b, the slope of the resulting line, is a useful measure of the level of allometric growth. The attainment of sexual maturity is often marked by a change in slope, or a discontinuity, in the plot of relative growth.

A. Relative growth of the chelae

The relative growth of chelae in crabs is generally characterized by the chelae of males growing relatively faster than in females, so that with progressively increasing body size the male chelae become increasingly larger than those of the females. Additionally, the rate of allometric growth in males displays a marked increase at the attainment of sexual maturity. This pattern is widespread in marine crabs (Hartnoll, 1974) and occurs in those terrestrial crabs that have been studied. Examples of the latter include *C. guanhumi* (Gifford, 1962a; Herreid, 1967), *A. pisonii* (Hartnoll, 1965), *M. depressus* (Hartnoll, 1964), *S. bidentatum* (Hartnoll, 1964), *S. dehaani* (Hamai and Hirai,

1940), *S. ricordi* (Hartnoll, 1965), *O. quadrata* (Haley, 1969), *O. ceratophthalmus* (Haley, 1973), and various species of *Uca* (see Crane, 1975). Figure 6.6 illustrates the relative growth of the chelae in *A. pisonii*. The marked difference in chela size between large specimens of the two sexes is obvious, as is the clear increase in the value of *b* for chela length from 1.15 to 1.75, which occurs in males when they become sexually mature at a carapace width of about 13 mm. It is generally accepted that the sexual difference in chela size is related to the use of the chelae as secondary sexual organs, being important in territory defense, display, combat, courtship, and copulation (see Chapter 4). This explains their greater size in males, particularly sexually mature ones. Since the sense of sight is more generally effective in the aerial than in the aquatic medium, and visual display is important in some of the activities mentioned, it is natural that chelar sexual dimorphism is well developed in terrestrial crabs.

A further feature of the relative growth of the chelae is that in some crabs one claw is larger than the other, a phenomenon known as heterochely. This occurs in a wide range of decapods (see Przibram, 1905; Schaefer, 1954), but is particularly well developed in some of the land crabs, notably *Cardisoma* and *Uca*. Extreme heterochely is limited to the male and is associated with signaling and combat (see Chapter 4), and so for the reason mentioned earlier it is likely to be developed in the aerial environment. Chelar growth in *C. guanhumi* is described by Gifford (1962a) and Herreid (1967). The major chela is not only larger than the minor chela, but it has a higher rate of relative growth so that the difference becomes exaggerated with greater body size. In females the heterochely is far less marked (Fig. 6.7). The major chela occurs on the right and left sides with equal frequency. There is a similar pattern of heterochely in *C. carnifex* (Alexander, 1976). The major chela is white and thus very conspicuous, though it was not observed being used for obvious display. It was, however, often used in pushing contests and in more serious conflicts between males. It was also used to block the burrow entrance, the crab entering the burrow with the major chela last. In *Uca* heterochely is limited to the male. Females possess two small chelae of similar size that are used for feeding, spooning up the substrate to be sorted by the mouthparts (see Chapters 3 and 10). Males have one minor chela similar to those of the female, and a major chela that is enormously enlarged. This is used not for feeding but for the very distinctive signaling patterns of the male, and for combat between males. Crane (1975) provides a detailed account of the structure of the chelae in species of *Uca*, and of the signaling displays. Species of *Uca* generally do not display handedness (Crane, 1975), but populations of predominantly left-handed *U. burgersi* (Gibbs, 1974), and

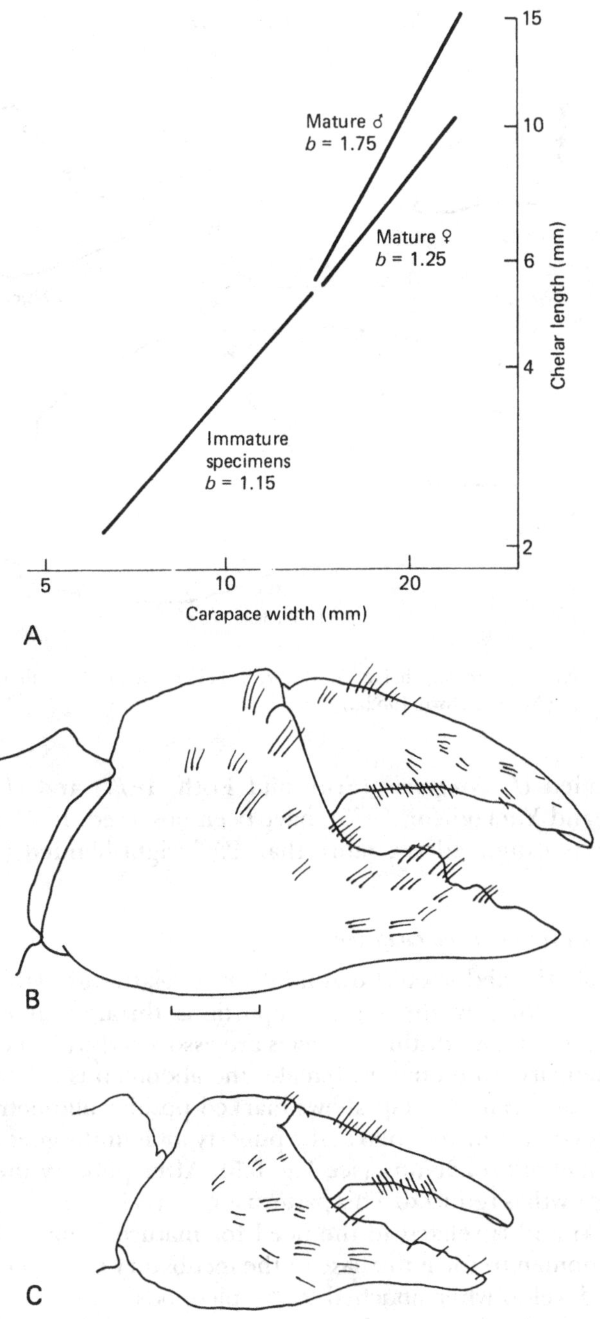

Fig. 6.6. *Aratus pisonii*. A. Log chelar length plotted against log carapace width. B. Chela of mature male of 20.5 mm carapace width. C. Chela of mature female of 21 mm carapace width. Scale bar = 2 mm. (After Hartnoll 1965.)

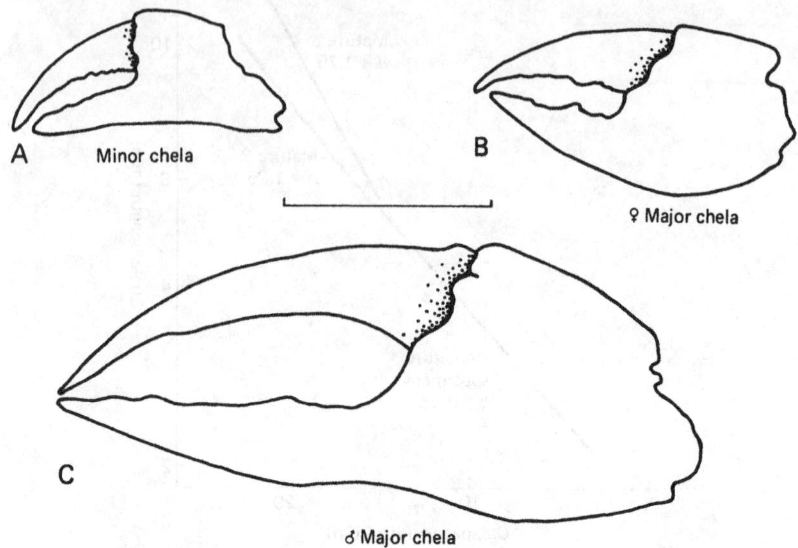

Fig. 6.7. *Cardisoma guanhumi*, chelae. A. Minor. B. Female major. C. Male major. Scale bar = 50 mm. (After Gifford 1962a.)

right-handed *U. tetragonon* (Frith and Frith, 1977) and *U. marionis* (Takeda and Yamaguchi, 1973), have been observed. In *U. vocans* all populations examined are more than 95% right-handed (Williams, 1981).

B. Relative growth of the abdomen

In the male the abdomen shows no distinct relative growth patterns, maintaining roughly the same proportions throughout ontogeny. However, in females distinct changes are associated with the onset of sexual maturity. In immature females the abdomen is relatively narrow in small specimens, but shows marked positive allometry so that it increases in width, and then at the puberty molt undergoes a sudden and substantial broadening (see Fig. 6.5). After puberty the level of relative growth is reduced. This pattern is general in Brachyura (Hartnoll, 1974) and is related to the need for mature females to have a broad abdomen to form the base of the incubatory chamber in which the eggs develop while attached to the pleopods.

Among land crabs the relative growth of the abdomen has been investigated in *A. pisonii* (Hartnoll, 1965), *S. dehaani* (Hamai and Hirai, 1940), *S. ricordi* (Hartnoll, 1965), *O. quadrata* (Haley, 1969), *O. ceratophthalmus* (Haley, 1973), *G. lateralis* (Bliss et al., 1978), and *C. guanhumi* (Henning, 1975b). Generally these crabs demonstrate the pattern

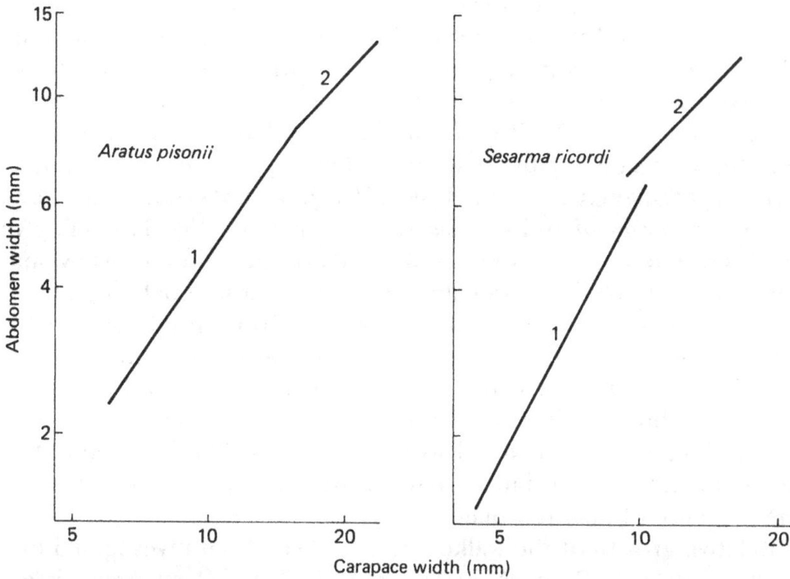

Fig. 6.8. Log abdomen width plotted against log carapace width (1 = prepuberty; 2 = postpuberty) for females of *Aratus pisonii* and *Sesarma ricordi*. (Partly after Hartnoll, 1965.)

of a broader abdomen in mature females. A marked increase in relative size at the puberty molt is apparent in *S. ricordi*, *O. quadrata*, and *G. lateralis*. In *A. pisonii* and the two species of *Sesarma* a reduction in the level of positive allometry follows puberty, although the other species listed show an increase (Hartnoll, 1974). The relative growth of the abdomen in females of *A. pisonii* and *S. ricordi* is shown in Figure 6.8.

C. Relative growth of other structures

Various workers have investigated how relationships between different carapace dimensions vary with body size. Gifford (1962a) studied *C. guanhumi* and found little change in the relation of carapace length to carapace width, as did the rather more detailed analysis of Herreid (1967). However, Herreid (1967) found that the orbital width, relative to the carapace width, decreased with size. In *O. ceratophthalmus* carapace length and width grow isometrically (Haley, 1973). In *Gecarcinus ruricola* and *G. lateralis*, the orbital width becomes relatively smaller with size compared to carapace width (Britton, Kroh, and Golightly, 1982). There is no obvious adaptive advantage in changes in carapace proportions, and it is not surprising that such changes are relatively trivial.

The importance of acute vision for land crabs is obvious, and there have been several studies on the relative growth of the eyes. Henning (1975a) studied this in *C. guanhumi* and found that the diameter of the eye grew faster than the carapace width, so that visual acuity increased with size. In *Ocypode*, on the other hand, there is evidence that the cornea is relatively larger in small specimens, for example, *O. ceratophthalmus* (Cott, 1929; Haley, 1973), *O. saratan* (Sandon, 1937), and *O. gaudichaudii* and *O. occidentalis* (Crane, 1941a). This will give small specimens better vision than if the eyes grew isometrically, and such precocity could be valuable in an active crab inhabiting an exposed environment with minimal shelter from predators. Haley (1973) also studied the relative length of the eyestalk in *O. ceratophthalmus*. This is approximately isometric in females and immature males, but shows distinct positive allometry in mature males as a result of the elongation of the distal styliform process. The function of this process is not known, but perhaps it indicates to other conspecifics the presence of an adult male.

Relative growth of the walking legs has also been investigated in a few land crabs. In *C. gaunhumi* (Herreid, 1967) and *O. quadrata* (Haley, 1969) it is isometric. However, in several other species of *Ocypode* – *O. ceratophthalmus* (Cott, 1929; Haley, 1973), *O. saratan* (Sandon, 1937), and *O. gaudichaudii* and *O. occidentalis* (Crane, 1941a) – the walking legs show negative allometry. The same adaptive advantage applies here as to the precocious development of good vision, in that it will facilitate the escape of the juveniles from predation. An interesting growth pattern occurs in the fourth pereiopod of *Birgus*. In hermit crabs this appendage is much reduced and has lost its locomotory functions. This limb is similarly reduced in the young shell-inhabiting *Birgus*, but once the free-living habit is assumed this leg elongates considerably and is used in walking (Wolff, 1961).

5. Autotomy and regeneration

These two processes are closely linked, in that autotomy is the process by which an appendage can be lost with minimal damage to the crab as a whole, and regeneration is the process by which the appendage can be regrown. These special aspects of growth can be important in facilitating escape from predators and in the repair of severely injured limbs.

A. Autotomy

Autotomy occurs generally in decapods, including the Anomura and Brachyura, and is therefore a property of all land crabs. They are capable of shedding the chelae and walking legs, which become separated at a special breaking plane across the basi-ischiopodite. The mechanism and control of autotomy in Crustacea are fully described by McVean (1982), and details will not be repeated here. It will suffice to point out that the process is under nervous control and does not necessarily require the limb to be damaged or held. The only detailed study of the autotomy mechanism in a land crab is on *C. guanhumi* (Moffett, 1975). Moffett suggests that the mechanism differs in some respects from that in *Carcinus*, notably in the precise role of the posterior levator muscle. In *Carcinus* the contraction of this muscle initiates autotomy, whereas Moffett (1975) suggests that in *Cardisoma* its contraction protects against autotomy, but this is by no means certain (see discussion in McVean, 1982).

The terrestrial habit exposes land crabs to different predation risks, and this could influence the incidence of autotomy, but there are few studies available. On Aldabra, Grubb (1971) recorded that only 1 specimen of *Birgus*, out of 50 examined, showed evidence of having lost a limb. In a more extensive study of 1,500 specimens, Alexander (1976) found that only 4 had lost or were regenerating chelae, but that 85 had lost walking legs; thus about 5% had missing or regenerating limbs. This low figure is not surprising, since *Birgus* has few enemies on Aldabra. The small loss of chelae is also explicable, since *Birgus* uses the first walking legs, rather than the chelae, for threat and stamping displays. In contrast, *Cardisoma carnifex* showed signs of limb loss in 25% of specimens on Aldabra, with the chelae most commonly affected (Alexander, 1976). In the intertidal *C. maenas* the chelae are also most often lost (McVean, 1982). Limb loss was commonest among *Cardisoma* in densely populated areas, suggesting that intraspecific conflict was the main cause. Weis (1977) noted a difference between species of *Uca* in the readiness to autotomize limbs. The relatively inactive *U. thayeri* was reluctant to do so, in distinct contrast to the very social and active *U. rapax*.

An interesting pattern of behavior, termed "attack autotomy," has been described in land crabs by Robinson, Abele, and Robinson (1970). This was initially observed when *Potomocarcinus richmondi* was attacked by an otter, which was repelled with an autotomized chela firmly gripping its skin. Laboratory experiments showed that this was a consistent response to a simulated predator, and that the chela was deliberately shed, rather than being discarded as the "predator" was pulled away. The same response was found in *G. lateralis*, and it is suggested that this behavior has evolved in some land crabs as a deterrent to mammalian predators. It did not occur, however, in the more cryptic *Cardisoma crassum*, which perhaps is not exposed to the same risks.

The loss of limbs by autotomy can affect the timing of the next

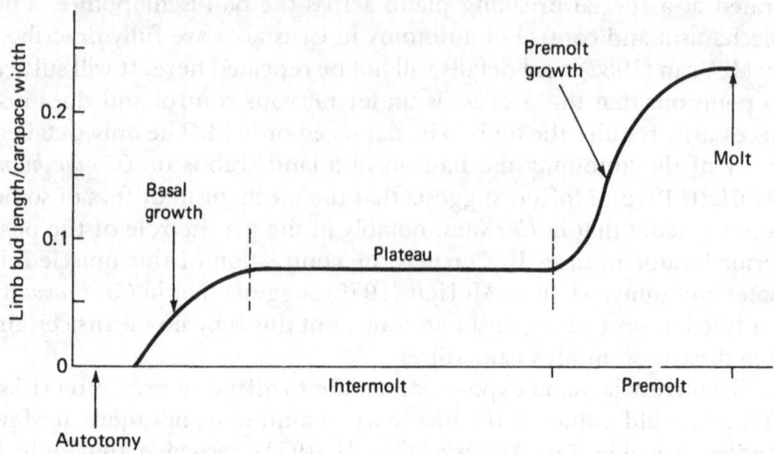

Fig. 6.9. Limb bud growth (expressed as limb bud length/carapace width) in *Gecarcinus lateralis* in relation to the molt cycle. (Based on Bliss and Boyer, 1964.)

molt. This feature is not peculiar to land crabs and it merits some discussion. However, regeneration, rather than autotomy, is the aspect of most interest in this context.

B. Regeneration

The process of regeneration following autotomy is described in detail for brachyurans by Skinner (1985), the account being based partly on studies of *G. lateralis* (Hodge, 1958). In the course of regeneration a limb bud is developed from the breaking plane. It consists of a flexible chitinous sac, which contains a small regenerated limb, folded twice upon itself. At ecdysis the chitinous sac is shed with the cast integument, and the regenerated limb straightens and becomes functional, though still smaller than before its loss. The full size is regained at succeeding molts. Following autotomy the limb bud develops to a small size – basal growth – and then growth ceases until proecdysis is initiated. A period of rapid proecdysial growth then follows, leading up to the molt. The stages in the growth of the limb bud in *G. lateralis* are shown in Figure 6.9. Basal growth is thought to occur independently of the presence of ecdysone, and in *Sesarma reticulatum*, at least, it can take place in Y-organectomized crabs (Passano and Jyssum, 1963). Proecdysial growth is dependent on the increased titer of ecdysone produced during proecdysis.

After multiple loss of limbs, many species show an acceleration of the following molt. This has been recorded in a number of land crab – *Ocypode macrocera* (Rao, 1966), *U. pugnax* (Skinner and Graham,

1972), *U. pugilator* (Hopkins, 1982), and *G. lateralis* (Skinner and Graham, 1972), to name a few. The loss of several pereiopods is obviously a severe handicap, and the advantage of regenerating as quickly as possible is clear. In land crabs, where movement is more difficult without the buoyant support of water and there is often the need to avoid active and keen-sighted predators, the need to regain full mobility will be particularly pressing. Thus, in *G. lateralis*, the most extensively studied species in this respect, the loss of five or more pereiopods reduced the duration of the molt cycle from 150–300 days to 55 days (Skinner and Graham, 1972). Interestingly, Y-organectomized *Gecarcinus* will molt under the stimulus of multiple limb loss (Soumoff and Skinner, cited in Skinner, 1985). This suggests that the regenerates, or some other tissue, can act as a secondary source of molting hormone.

The question of regeneration has raised some interesting points in species where there is distinct heterochely, particularly in *Uca*. This concerns the mechanism by which "handedness" is determined as well as the nature of regeneration following the loss of a major chela. In some crustaceans, such as *Alpheus* and *Homarus*, when the large chela is lost the remaining small chela develops into a large one, and a new small chela replaces the lost one (see examples in Hartnoll, 1982). However, this is not so in *Uca* once heterochely is established – the autotomized claw is always replaced by one of the same type (Morgan, 1923a; Vernberg and Costlow, 1966; Yamaguchi, 1977). There is less agreement concerning the early stages of this genus, however. In those cases listed earlier, where there is very marked handedness, it must be assumed that this is determined genetically. In species where there is no clear bias, the first suggestion was that handedness was fixed by the early loss of one chela, the one that remained developing into the major chela (Morgan, 1923a). Later studies on *U. rapax* and *U. pugilator* (Vernberg and Costlow, 1966), and *U. lactea annulipes* (Feest, 1969), did not show that early chela loss determined handedness, and suggested a genetic basis. Subsequently, an extensive study of *U. latea* (Yamaguchi, 1977) has shown that in this species, at least, the loss of one cheliped in the stage prior to differentiation does fix handedness. This variation may result from differences between the mechanism of species, or from differences in the size at which autotomy was experimentally induced in young crabs.

6. Conclusions

The most obvious difference in the way the growth of land crabs varies from that of aquatic forms relates to the problems and consequences of molting in air, rather than in water. The need for water to produce the rapid postmolt increase in size is dealt with by behavioral means in some species, which utilize supplies of water in their burrow or some other casual source. However, the main adaptation is the ability to accumulate and store water in the pericardial sacs during premolt, so that no additional water is needed at molting. The second problem is retaining calcium salts for the hardening of the new integument. These are conserved by reabsorption before molting, storage in particular sites (such as the gastroliths), and incorporation into the new integument.

Thus, the problems of molting in air have been overcome, but only, it would seem, at some cost. The molt increment tends to be smaller than in aquatic crabs. The frequency of molting is limited not only by seasonal temperature constraints, but also by seasonal lack of water. Although the data are very limited, the impression in land crabs, at least in more highly adapted ones, is of relatively modest growth rates. Nevertheless, many species reach a large size, with *Birgus* the most outstanding example.

The relative growth of land crabs has several adaptive features. In the case of the chelae, sexual dimorphism is well marked, and male heterochely can be strongly developed. These changes are important in signaling and display, territoriality, and intraspecific combat: The terrestrial habitat, with its emphasis on the visual sense, involves an increase in the significance of such activities. In some land crabs there is a precocious development of visual and locomotory ability. This is particularly important in air, providing the young with an improved chance to locate predators, and to escape from them and find shelter.

Autotomy and regeneration occur readily in land crabs, though there is no actual evidence that overall they are more effective than in aquatic species. Nevertheless, the need to escape active predators, and rapidly to regain effective powers of locomotion, can be seen as more pressing in air. Certainly the facility to autotomize in land crabs bears some correlation with the level of risks. Attack autotomy is known only from land crabs, and seems an adaptation to mammalian predators.

7: Ion and Water Balance

PETER GREENAWAY

CONTENTS

1. Introduction

Species of aquatic crabs live in waters ranging in concentration from freshwater (0–15 $mOsm \cdot kg^{-1}$ H_2O) to hypersaline media (>1,000 $mOsm \cdot kg^{-1}$ H_2O). These waters may be of relatively constant composition (offshore seawater and large lakes) or of fluctuating composition due to tidal or climatic factors. Crabs (and other aquatic organisms) have developed mechanisms to maintain the preferred concentrations of their tissue fluids in the particular medium in which they live. Several distinct patterns of osmotic and ionic regulation are evident (see reviews by Greenaway, 1979; Mantel and Farmer, 1983) and will be briefly summarized.

Species characteristic of sublittoral marine habitats have body fluids similar in concentration and composition to seawater. They are relatively permeable to water and ions and produce urine that is isosmotic with the hemolymph. Limited regulation of the ionic composition of the hemolymph is generally evident with magnesium and sulfate concentrations kept lower than in seawater by preferential elimination in the urine. Concentrations of the other major ions are often slightly higher in the hemolymph than in seawater. The animals may withstand limited dilution of their medium but show no ability to regulate the concentration of their extracellular fluids, which remain isosmotic with the medium. Crabs showing this pattern of osmoregulation are referred to as osmoconformers.

Many euryhaline crabs regulate the concentration of their hemolymph in dilute media. In seawater and concentrations higher than seawater, their extracellular fluids are isosmotic with the medium, but in media of lower concentration they attempt to maintain the osmolality of the hemolymph above that of the medium. This results both in loss of ions across the body surface down electrochemical gradients and in osmotic influx of water. Generally, permeability to ions is lower than in osmoconforming species, thus reducing the size of these fluxes. Lost salts are replaced by absorbing ions from the water by means of ion-transporting enzymes (ATPases) situated in the gill epithelium. Excess water entering the body is removed by increasing the rate of output of urine, which, however, remains isosmotic with the hemolymph. This pattern of osmoregulation is known as hyperosmotic regulation.

Freshwater crabs also regulate hyperosmotically, maintaining hemolymph concentrations well above those of the medium. They have much lower permeability to ions and water than euryhaline species and also have reduced the osmotic and ionic gradients across the body surface by lowering the concentration of their hemolymph. Salts lost are replaced from the medium, and the ATPases in the gills have a high affinity for ions to enable net uptake from the very low levels found in fresh water. Water influx is reduced to a very low level, and excess is eliminated via the isosmotic urine or by an extrarenal mechanism.

Hyposmotic regulators attempt to minimize fluctuations in the concentration of the hemolymph and generally inhabit brackish waters. Usually the hemolymph concentration is maintained below that of seawater, and the permeability to salts and water is lower than in osmoconforming crabs. In dilute media they utilize the same mechanisms to maintain salt and water balance as described earlier for hyperosmotic regulators. In hyperosmotic media, their hemolymph concentration is lower than that of the medium, resulting in outflux of water and influx of ions. To compensate, the crabs drink the medium and absorb its salts and water across the gut. Excess ions are then eliminated across the gills or via the gut.

Crabs that have colonized land probably originated from all three of these osmoregulatory categories: osmoconformers, hyperosmotic regulators, and hyposmotic regulators representing marine, brackish, and freshwater habitats. Thus the osmoregulatory abilities of different species of crabs that have colonized land are far from uniform.

Life out of water poses severe problems of water balance, since air is a desiccating medium and species with low permeability to water (osmoregulators, particularly freshwater species) will be best able to cope with terrestrial conditions. There are also problems in maintaining salt balance. In aquatic species adjustments to ion content are made principally by salt transport across the gills, and the renal organ makes only a minor contribution. In terrestrial animals the gut and renal organs become the main sites of ionic regulation. The ionic regulatory mechnisms of aquatic species must be altered to cope with this changed situation.

The life-style of crabs covers the spectrum from fully aquatic through various degrees of dependence on water to fully terrestrial (see Chapters 1–3). This chapter first considers the difficulties of maintaining water balance on land and then the mechanisms crabs have developed to cope with these problems. The second section is devoted to osmotic and ionic regulation in intermolt crabs and a brief consideration of the problems entailed in molting out of water. Emphasis will be placed on the more terrestrial species. Here the problems of maintaining salt and water balance are greatest because the full rigors of the terrestrial environment must be faced continually. Amphibious species can readily avoid adverse conditions by retreating to water and usually show more affinity with aquatic osmoregulators than with the terrestrial ones. Nevertheless, amphibious species may show some physiological adaptations to semiterrestrial life and thus provide information on the routes by which the more terrestrial patterns of osmoregulation have been achieved.

2. Water Balance

A. *Evaporative water loss*

There is a steep gradient of water potential* across the body surface of crabs in air, except in very moist habitats (Table 7.1). This results in evaporative loss of water, the rate of which is given by the equation

*Water potential is defined as the potential energy per unit mass of water with reference to pure water at zero potential. Units are $J \cdot kg^{-1}$.

Table 7.1. *The water potentials and their equivalent osmolalities and relative humidities of the hemolymph of crabs (hydrated and dehydrated), air and water 25° C*

Species	Osmolality ($mOsm \cdot kg^{-1}$, H_2O)	Relative humidity (%)	Water potential ($kJ \cdot kg^{-1}$)	Reference
Holthuisana transversa (FW/T)	524–1018	99.09–98.05	1.3–2.53	Greenaway & MacMillen (1978)
Gecarcinus lateralis (T)	610–1060	98.9–98.1	1.51–2.63	Skinner et al. (1965)
Coenobita clypeatus (T)	941–1005	98.3–98.2	2.34–2.49	Wheatly et al. (1984)
Carcinus maenas (M)	1027	98.1	2.55	Zanders (1980)
Air	0	100	0	Calculated
Air	2.8×10^3	95	7.03	Calculated
Air	15.9×10^3	75	39.41	Calculated
Air	38.3×10^3	50	94.96	Calculated
Air	76.5×10^3	25	189.92	Calculated
Seawater	954	98.3	2.37	Calculated
Fresh water	0–15	100 – 99.97	0–0.037	Calculated

Notes: M, marine; T, terrestrial; FW, fresh water.
The osmolalities and water potentials of various media were calculated using the following equations:

$$\psi = \frac{RT}{M} \ln \frac{P}{P_o} \quad \text{and} \quad \text{Osmolality} = \frac{\psi}{RT}$$

where R is the gas constant, T the absolute temperature, P/P_o the relative humidity and ψ the water potential (Campbell, 1977).

$$E = \frac{(p_{vs} - p_{va})}{r_v} \ldots 1,$$

where E is the water vapor flux density, r_v is the resistance of the cuticle to vapor diffusion, and p_{vs} and p_{va} are the vapor pressures (densities) of the evaporating surface (cuticle water) and the air, respectively (Campbell, 1977). Any factor that affects either p_{vs} or p_{va} alters the vapor pressure difference and consequently the rate of evaporative loss. Thus, changes either in the temperature of the body surface or in the vapor pressure of the air will be important in determining the rate of evaporative loss. Additionally, convection of air across the body surface will increase loss by restricting the formation of moist, unstirred layers next to the body surface, thus maximizing the vapor pressure difference (Edney, 1977).

Limited changes in permeability have been reported for certain

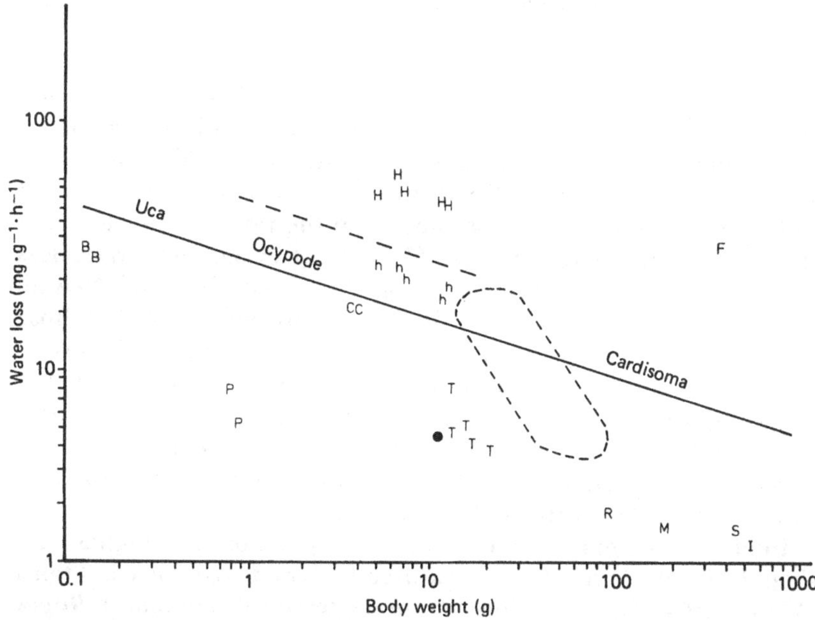

Fig. 7.1. Rates of evaporative water loss from terrestrial crabs compared with other animals. All data were collected under similar conditions. The *solid line* (y = 37 m–0.31) represents loss from semiterrestrial crabs, and the *broken line* (y = 73.8 m–0.27) represents loss from intertidal crabs (*Sesarma cinereum* and *Panopeus herbstii*). The area enclosed by a *broken line* represents data for the terrestrial crab *Gecarcinus lateralis* and the *solid circle* (●) represents water loss from the freshwater/land crab *Holthuisana transversa*. *B*, the isopod *Porcellio scaber*; *C*, the cockroach *Blaberus cranifera*; *F*, the bullfrog *Rana catesbeiana*; *H*, the hermit crab *Clibanarius vittatus* without shell and *h*, *Clibanarius vittatus* with shell; *I*, the lizard *Iguana iguana*; *M*, the musk turtle *Stenotherus odoratus*; *P*, the cockroach *Periplaneta americana*; *R*, the kangaroo rat *Dipodomys spectabilis*; *S*, the snake *Lampropeltis getulus*; *T*, the tarantula, *Eurypelma* spp. (Data from Herreid, 1969a, and MacMillen and Greenaway, 1978.)

euryhaline crabs transferred between media of different osmolality (Smith, 1976). However, in terrestrial crabs, it is not known whether the value of r_v (= 1/permeability) can be altered on a short-term basis to cope with changes in water availability in the environment. Marked differences are apparent between species (Fig. 7.1).

In addition to these physical factors, mass-specific evaporative water loss decreases with increasing body size (Herreid, 1969a; Calder, 1984). Because mass-specific metabolic rate follows a similar relationship, mass-specifiic evaporative water loss from respiratory surfaces is also a function of body size.

All of these factors affect the rate of evaporative water loss. If effective comparisons of loss rates are to be made between species, it is essential that measurements are made under similar conditions and that body size is considered.

i. Rate of water loss. Most published data, in fact, have been collected under differing experimental conditions, and few measurements of evaporative water loss from crabs are strictly comparable. Some comparable data are presented in Figure 7.1. Each published data set indicates an inverse relationship between the rate of water loss and the terrestriality of the species. The rate of evaporative water loss from aquatic crabs is very high (Fig. 7.1; Ahsanullah and Newell, 1977), but in semiterrestrial species the rates are only 20–50% of those of aquatic crabs. In intertidal fiddler crabs (*Uca* spp.), water loss is inversely related to the terrestriality of the habitat, with those living higher up the littoral zone having lower rates of water loss (Edney, 1961). In the most terrestrial species such as *Gecarcinus lateralis* and *Holthuisana transversa,* the rate of water loss is lower still at 10–50% of the rate for semiterrestrial species (Fig. 7.1).

Intertidal anomurans lose water by evaporation at a higher rate than do aquatic brachyurans measured under comparable conditions (Young, 1978; Fig. 7.1). However, in the terrestrial anomuran *Birgus latro*, the rate of water loss is similar to that of terrestrial brachyurans (Harris and Kormanik, 1981). The mollusk shells inhabited by hermit crabs are effective in restricting evaporative loss of water, and animals lose water three times more rapidly if removed from their shells (De Wilde, 1973).

Mass-specific water loss is inversely related to body size in aquatic and semiterrestrial crabs (Ahsanullah and Newell, 1977; MacMillen and Greenaway, 1978; Warburg and Goldenburg, 1984; Fig. 7.1). The rate of water loss is affected both by the temperature of measurement and the vapor pressure of the air surrounding the crab (Dandy and Ewer, 1961; Wilkens and Fingerman, 1965; Herreid, 1969a; Warburg and Goldenburg, 1984), since these factors affect p_{vs} and p_{va}, respectively. This is most clearly seen in the data for the aquatic crabs *Portunus marmoreus* and *Carcinus maenas*, where the two factors were assessed independently (Ahsanullah and Newell, 1977). Unfortunately, in most studies, the effects of temperature and vapor pressure are not separated.

Comparative data on the effects of air movement on evaporative water loss from crabs are largely absent in the literature. In the amphibious freshwater crab, *H. transvera*, water loss is enhanced by increasing the speed of air movement over crabs tethered in a wind tunnel (P. Ikin, pers. comm.). Nocturnal activity of land crabs on Christmas Island (10°30′ S, 105°35′ E) is noticeably reduced under windy conditions (P. Greenaway, unpub.), and the degree of air movement may thus be important ecologically. Since crabs are recent colonists of land (see Chapter 2), it is interesting to see how effective they are at restricting evaporative water loss compared with some of the longer-established terrestrial species, both vertebrates and invertebrates. Where data are comparable (Herreid, 1969a; Fig. 7.1), it is apparent that semiterrestrial crabs lose water much more rapidly than do mesic insects, isopods, or tarantulas. However, rates of water loss in the most terrestrial crabs are similar to, or lower than, the latter groups of arthropods. Even the most terrestrial crabs for which data are available (*G. lateralis* and *H. transversa*) lose water three to five times more rapidly than reptiles and mammals of similar size, and rates of loss are higher than in most insects (Herreid, 1969a). A high permeability to water may be inferred from the preceding data, but quantitative data for permeability to water in air are available only for the fiddler crabs *Uca annulipes* ($235 \times 10^{-4} cm\, s^{-1}$), and *U. marionis* ($586 \times 10^{-4} cm\, s^{-1}$) by Edney (1961, 1977) and *U. pugilator* ($5{,}555 \times 10^{-4} cm\, s^{-1}$) by Smith and Miller (1973). These values are similar to those for some terrestrial isopods, myriapods, and hygric insects in particular, but are much higher than those for most insects and arachnids (Edney, 1977). Measurements of the rates of water loss from portions of isolated carapaces of crabs indicate that the permeability of terrestrial crabs is lower in the more terrestrial species (Herreid, 1969b).

ii. Site of water loss. Evaporative water loss occurs from all exposed parts of the body surface, and the rate of loss from any particular portion will depend on its permeability. The most permeable parts of the body are probably the respiratory surfaces (gills and lungs) due to the requirements for a thin permeable surface for gas exchange. Water loss from respiratory surfaces will be enhanced by their large surface area and active ventilation with air. External surfaces are generally modified to protect underlying tissues and are likely to be less permeable. Several attempts to partition loss between the external surfaces (cuticular loss) and the respiratory surfaces have been made either by coating the external surfaces with a water-impermeable substance or by comparing water losses from living crabs with those from dead animals, where the respiratory component is assumed to be eliminated (e.g., Dandy and Ewer, 1961; Herreid, 1969b). The results obtained are highly variable, often conflicting, and give no clear picture of relative losses of water. Both methods have their drawbacks. Evaporative water loss is not completely eliminated with the coverings used (petroleum jelly and enamel paint), so respiratory loss will be overestimated by this technique. Use of dead animals may also result in serious errors. There is evidence that cuticle water is actively regulated in insects and some terrestrial Crustacea (Isopoda).

Transpiration may be hormonally controlled in insects and dead insects, and arachnids lose water more rapidly than their living counterparts (Lindqvist, Salminen, and Winston, 1972; Edney, 1977, Hadley, 1981). Clearly, some more reliable method of separating respiratory and cuticular evaporative losses is needed in crabs (e.g., use of a mask to separate respiratory and cuticular air streams during measurements of water-loss rate). Experiments using cuticular sealants do provide information on the relative contributions to water loss made by different portions of the cuticle. Thus in *Cardisoma guanhumi*, legs contributed 36%, chelae 26%, and the rest of the exoskeleton 38% of total exoskeletal water loss (not including losses from the branchial chambers), whereas water loss from the arthrodial membranes was negligible (Herreid, 1969b). The abdomen of female *Potamonautes depressus* contributes 17% of total evaporative water loss (Dandy and Ewer, 1961). Both the respiratory surfaces and integument probably contribute significantly to water loss of terrestrial crabs, but the available data are inadequate either to decide proportions or to discern ecological trends.

In insects, low permeability to water is largely due to lipids present in the epicuticle, although the structure of the lower layers of the skeleton and the activity of the epidermal cells may also be important (Edney, 1977; Hadley, 1981). Although lipids are present in the epicuticle of crabs, the well-developed surface layer of lipids typical of insects has not been observed, perhaps explaining their high permeability to water. Thickness and density of the exoskeleton could affect its permeability. However, heavy calcification clearly is ineffective in reducing water permeability in the Crustacea, perhaps because the calcified layers of the exoskeleton are permeated by huge numbers of microvilli (pore canals) from the hypodermal cells (ca. $9.5 \times 10^5 \cdot mm^{-2}$) (Roer, 1980). The thickness of the exoskeleton is variable in different parts of the body. In *H. transversa*, for example, the cuticle of the lung is 0.2–0.25 μm thick, the gill cuticle is 2–3 μm thick and that of the carapace 50–100 μm (Taylor and Greenaway, 1979; also see Chapter 8). Although there is no direct evidence that this affects water permeability, it seems probable. The epicuticle is reduced in thickness in the gills and lungs (Mary and Krishnan, 1974; Taylor and Greenaway, 1979), but again there are no data for an effect on water permeability.

Fluid loss

i. Urinary loss. The paired antennary glands, which serve as the excretory organs of crabs, lie in the antennal segment of the head. They open to the exterior via nephropores, which are generally closed by a flap of cuticle. Each gland consists of several morphological parts,

Table 7.2. *Rates of urine production in amphibious and terrestrial crabs*

Species	Habitat	Urine flow (ml·100 g^{-1} body mass·day^{-1})	Conditions	Reference
Carcinus maenas	M/Aq	4.4	In SW	Binns (1969)
Hemigrapsus nudus	M/Aq	4.15 32.0	In SW 320 mOsm SW	Ramamurthi (1977)
Goniopsis cruentata	M/Am	7.0 16.6	In SW 265 mOsm SW	Zanders (1978)
Cardisoma carnifex	Am	3.2	800 mOsm SW moist sand	Kormanik & Harris (1981)
Cardisoma guanhumi	Am	1.89	SW moist sand	Harris (1977)
Gecarcoidea lalandii	T	3.4	80% SW moist sand	Kormanik & Harris (1981)
Gecarcinus lateralis	T	10.5	SW moist sand	Harris (1977)
Gecarcinus lateralis		ca. 2.0[a]	Dry sand	Harris (1977)
Potamon edulis	Am	0.56	In FW	Harris (1975)
Holthuisana transversa	Am	0.47	In FW	Greenaway (1980)
Birgus latro	T	6.6[a]	Fresh drinking water	Kormanik & Harris (1981)

Note: M, marine; Aq, aquatic; Am, amphibious; T, terrestrial.
[a]Inulin clearance rate.

the coelomosac, labyrinth, nephridial tubule, and bladder. The coelomosac receives hemolymph from the antennal arteries, and this is filtered through a layer of podocytes into the lumen of the sac to form the primary urine. This fluid moves from the coelomosac through the labyrinth and nephridial tubule to the bladder. The primary urine is modified by resorption and secretion of ions, water, and organic materials during its passage, particularly during its temporary storage in the bladder. For further details see Riegel (1972) and Mantel and Farmer (1983).

Urine is a major route of loss of free water in most crabs. Rates of urine flow are similar in marine and amphibious species, but in terrestrial crabs on moist sand rates are lower, with the exception of *G. lateralis* (Table 7.2). Freshwater species have extremely low rates of urine flow (Shaw, 1959; Table 7.2). Rates of production of urine have not been measured in terrestrial crabs under desiccating conditions because of the difficulty of collecting urine under such circumstances, suggesting that the urine is unlikely to be a significant component of water loss during dehydration.

ii. Nonurinary loss. Loss of free water by routes other than the antennal organs is largely confined to the feces, but terrestrial coenobitids have glands that produce a fluid thought to moisten the respiratory surfaces (Borradaile, 1903a; Harms, 1932; Storch and Welsch, 1984). Fecal water losses have not been quantified, and it is not known if fecal water content is regulated. De Wilde (1973), however, reports that the fecal pellets of *Coenobita clypeatus* are very dry during water deprivation and suggests that resorption of water may occur in the hindgut.

In freshwater potamoid crabs, there is considerable indirect evidence for excretion of water at sites other than the antennal organs, but these sites have not been identified (Shaw, 1959; Thompson, 1970; Greenaway, 1980). Two possible sites for nonurinary water excretion are the anterior and posterior ceca of the midgut and the gills. The gill axes of *H. transversa* contain nephrocytes, similar in structure to podocytes (Taylor and Greenaway, 1979). The role of the nephrocytes is obscure, and it is difficult to see how they could be responsible for water elimination, given their location. The structure of the midgut ceca of certain marine crabs is suggestive of a transport function (Mykles, 1977; Smith, 1978; see also Chapter 10), and their involvement in osmoregulation has been suggested (Heeg and Cannone, 1966; Dall, 1967). However, no studies have been carried out on potamoid crabs.

C. Water content and tolerable water loss

For a complete understanding of water loss in terrestrial crabs it is necessary to consider how much water they contain and the degree of water loss they can withstand. Clearly, if desiccation is likely in a particular habitat, both high body-water content and tolerance to loss of body water are desirable. In their normal habitat, many semiterrestrial and terrestrial crabs have ready access to free water. Thus it is not surprising to find that their water contents are similar to those of aquatic species (Table 7.3). The Australian arid-zone crab, *H. transversa*, has a rather higher water content. This may be an adaptation to the greater probability of desiccation in its arid habitat. Values for maximum tolerated water loss (Table 7.3) also reflect conditions within the habitat. Thus, amphibious, maritime, and rain forest species show a low tolerance to desiccation whereas the coenobitids, which often occupy quite dry areas (De Wilde, 1973), are more tolerant (Table 7.3). *H. transversa*, which may spend several years in an unsealed burrow without access to free water (MacMillen and Greenaway, 1978), is the most tolerant brachyuran studied, surviving loss of almost half its body water (Table 7.3).

Table 7.3. *The water content of the body and maximum tolerable water loss or lethal water loss[a] in land crabs*

Species	Habitat	Body water ($ml \cdot 100g^{-1}$ body mass)	Maximum water loss (% body water)	Reference
Cardisoma carnifex	Am	65.5	15–18	Harris & Kormanik (1981) Burggren & McMahon (1981a)
Cardisoma guanhumi	Am	60–70	10–15[a]	Gifford (1962b)
Gecarcoidea lalandii	T	60.0	18	Harris & Kormanik (1981)
Gecarcinus lateralis	T	66.2	31.7[a]	Bliss (1968), Flemister (1958)
Ocypode quadrata	T	69.9	20[a]	Bliss (1968), Flemister (1958)
Potamonautes depressus	FW/Am	68.4	22[a]	Dandy & Ewer (1961)
Sudanonautes africanus	FW/Am	58.3	34[a]	Lutz (1969)
Holthuisana transversa	FW/T	73–77	43	Greenaway & MacMillen (1978), Greenaway (1980)
Petrolisthes elongatus	M/IT	60.3	20.8[a]	Jones & Greenwood (1982)
Pagurus longicarpus	M	68.0	46[a]	Young (1978)
Clibanarius vittatus	M/IT	66.0	50[a]	Young (1978)
Coenobita brevimanus	T	—	28[a]	Burggren & McMahon (1981)
Coenobita clypeatus	T	—	30[a]	De Wilde (1973)
Birgus latro	T	65.6	21–22	Burggren & McMahon (1981a) Kormanik & Harris (1981)

Note: IT, intertidal. Other abbreviations as in Table 7.2.
[a]These values may be overestimates as loss may include some shell water in coenobitids and branchial chamber water in *C. carnifex*.

D. Water gain

Water may be obtained by a variety of different mechanisms. These will be discussed here, and their importance in various groups of crabs is summarized in Section 2.E.

i. Immersion in water. Immersion is a common method of water uptake by intertidal crabs and species that live in the splash zone or on ocean beaches. *Ocypode quadrata* and *O. ceratophthalmus* visit the sea at night and immerse themselves for short periods (Borradaile, 1903a; Flemister, 1958; Bliss, 1968), but this is neither essential nor the rule in the former species (T.G. Wolcott and D.L. Wolcott, 1985a). Water may be absorbed across the gut or gills in these species. The coenobitids *Coenobita perlatus* and *C. scaveola* also may immerse themselves in the sea at night and store water in their shells (Seurat, 1904; Gross, 1964a; Volker, 1965) or absorb it across the diaphanous membranes of the abdomen (Harms, 1932). Many intertidal species carry water in the branchial chambers while active out of water, e.g., some species of *Uca* (Teal, 1958; Gross, 1964a; Bliss, 1968), *Heloecius cordiformis* (D. Maitland, pers. comm.) *Cyclograpsus punctatus* (Alexander and Ewer, 1969), and *Cardisoma carnifex*, (Wood and Randall, 1981a). This habit is also discussed in Chapter 8. The amphibious, freshwater crabs also spend considerable time in water if it is available (Dandy and Ewer, 1961; Greenaway, 1980).

Other crabs have burrows that descend to the water table, *C. guanhumi* and *C. carnifex* regularly immerse themselves in the water contained in the burrow, e.g., two hours per day in *C. guanhumi* (Gifford, 1962a; Herreid and Gifford, 1963; Cameron, 1981a). These species are restricted to areas where the crab can burrow down to the water table. This is also true of *C. hirtipes*, which is restricted to freshwater seepage areas on Christmas Island (Hicks, Rumpff, and Yorkston, 1984).

ii. Drinking. The more terrestrial crabs, such as *Gecarcoidea natalis*, do not immerse themselves on any regular basis and may instead drink water. To drink, *G. natalis* dips its chelae in water and transfers this to the mouth with a spooning motion (Gibson-Hill, 1947). *Geograpsus grayi*, *C. guanhumi*, *C. carnifex*, *B. latro*, and the terrestrial coenobitids also drink using the chelae when offered shallow water (Lister, 1888; Gross, 1955; Gifford, 1962a; Gross et al., 1966; Greenaway, unpub.). There is no evidence that the water transferred by the chelae actually enters the gut, and it could equally well be sucked into the branchial chambers and absorbed by the gills. *B. latro* can transfer water from the maxillipeds to the branchial chambers using the last pair of pereiopods. The coenobitids that carry mollusk shells can transfer drinking water to the shell. Water picked up on the chelae is sucked into the branchial chambers and then voided into the shell by raising the carapace and making pulsating movements of the abdomen (De Wilde, 1973).

Not all water is acceptable for drinking, and many species show distinct salinity preferences (Gross, 1955, 1957; Gross and Holland, 1960; De Wilde, 1973). This is discussed further for individual species in Section 3.

iii. Uptake of soil water. Movements of water between air, free-water bodies, soil, and living organisms are more readily understood if the units used are the same for all media. To this end the concept of water potential (Ψ), with units of $J \cdot kg^{-1}$, is commonly used. Water potential is defined as the potential energy per unit mass of water with reference to pure water at zero potential. Water will move down its potential gradient (Campbell, 1977).

Soils contain water that may be free, in the interstices between soil particles (if the soil is saturated), or bound more or less tightly to the particles. Binding forces constitute the matric water potential of the soil, defined as ψ_m (Campbell, 1977). Saturated soils have maximal water potentials (zero), and drier soils have lower (negative) potentials. Soils with larger inert particles and little organic material (sand) exert high ψ_m, whereas clay soils with very fine particles and high organic content have much lower (more negative) potentials.

Many crabs have setae that are strongly hydrophilic and exert low ψ_m (T. G. Wolcott, 1984). If this ψ_m is lower than that of the soil water, water will move from soil to setae. To extract water from the setae into the branchial chambers requires an even lower water potential. This is generated as a pressure potential by the scaphognathites, which can reduce the air pressure in the branchial chambers to below ambient pressure (see Chapter 8). Most intertidal ocypodid and mictyrid crabs absorb water from the substrate in this way (Hartnoll, 1973; Powers, 1975; Quinn, 1980). Tufts of setae between walking legs, or on the abdomen, are lowered into the sand or shallow water and the fluid is sucked from them into the branchial chambers. Since free water is generally available in the substrate, little force is required to extract fluid from the setae.

In the supralittoral crab, *O. quadrata*, tufts of hydrophilic setae are present between pereiopods 2 and 3 and surround the openings into the branchial chambers (T. G. Wolcott, 1976, 1984). This crab, which can develop suction pressures as low as 76 mmHg ($= -10\ J \cdot kg^{-1}$) below atmospheric pressure, can absorb water from sand containing less than 5% water. Extracted water is moved from the branchial chambers to the mouth and ingested (Wolcott, 1984). Uptake from sand containing more than 5% water is rapid ($> 1\%$ body $mass \cdot h^{-1}$), but at lower water contents uptake is reduced. Clearly, there will be a threshold where ψ_m soil $= \psi_m$ setae, and below this net gain will not occur. It is important, then, that burrows be constructed in sand of the appropriate water content (see Chapter 3). *Ocypode cursor* selects sand of high water content in which to burrow, detecting differences in water content of 1% (Warburg and Shuchman, 1979). The ultimate factor limiting the uptake of soil water may be either the ψ_m of the setae or the subambient pressure that can be developed in the branchial chambers. Dehydrated crabs lose the ability to absorb water from sand, due perhaps to loss of the water film that normally seals the carapace against the bases of the walking legs. This would prevent generation of negative pressures in the branchial chambers (Wolcott, 1984). The importance of a "water seal" for the development of subambient pressures has also been reported for *C. guanhumi* (Burggren, Pinder et al., 1985). Uptake of water from the substrate is also reported in *O. guadichaudii* and *O. cordimanus* (Koepcke and Koepcke, 1953; Rao, 1968) and is probably characteristic of the genus *Ocypode*.

The ability to absorb water from the substrate is also well developed in the gecarcinid crabs *G. lateralis*, *C. guanhumi*, and *C. carnifex* (Bliss, 1963; Gross et al., 1966; Wolcott, 1984). In fact, *G. lateralis* can live indefinitely on damp sand without access to drinking water (Gross, 1963). In *G. lateralis*, a channel lined with setae connects tufts of setae at the base of the fourth pair of pereiopods with the posterior margins of the pericardial sacs within the branchial chambers (Bliss, 1963; Mason, 1970). *G. natalis* has similar tufts of setae (Fig. 7.2). In *G. lateralis*, water moves from the substrate into the setal tufts, onto the surface of the pericardial sacs and then forward to the gills, the posterior pairs of which lie on top of the pericardial sacs. Water is absorbed by the gills, possibly by simple osmosis rather than solute-linked transport since dilute water is absorbed readily but seawater is not absorbed at all (Bliss, 1968; Mason, 1970; T. G. Wolcott, 1984). *G. lateralis* can also generate subambient pressures (20–30 mmHg) in the branchial chambers, and bulk uptake of fresh water and seawater is reported under laboratory conditions (T. G. Wolcott, 1984). However, the matric water potential of burrow soil is normally so low that it is unlikely that this mechanism is useful in the field (D. L. Wolcott and T. G. Wolcott, 1987). The setae may be utilized to absorb droplets of dew that are available at night (Bliss, 1979). *C. carnifex* and *C. guanhumi* are capable of high rates of uptake from damp substrates (Gross et al., 1966; T. G. Wolcott, 1984), and it is probable that they too utilize a setal suction method to extract water.

The ability of grapsid crabs to absorb water from the substrate is largely unknown. Many species of *Sesarma* have tufts of plumose setae at the bases of their pereiopods, particularly legs 2–4, and these are used to absorb water (Felgenhauer and Abele, 1983). *Geograpsus crinipes* has very large setal tufts between periopods 2 and 3, but in *G. grayi* these tufts are very small (Fig. 7.2). The former species lives on rocky cliffs, and its long setae may be designed to utilize seepages or films of free water. The latter species is widely distributed on the forest floor, where free water is less available by setal suction.

The freshwater/land crabs are devoid of setae on the ventral surface. Species such as *H. transversa*, which spend long periods out of water, generally burrow in heavy clay soils (Greenaway and MacMillen, 1978). Although clay holds large amounts of water, it is tightly bound,

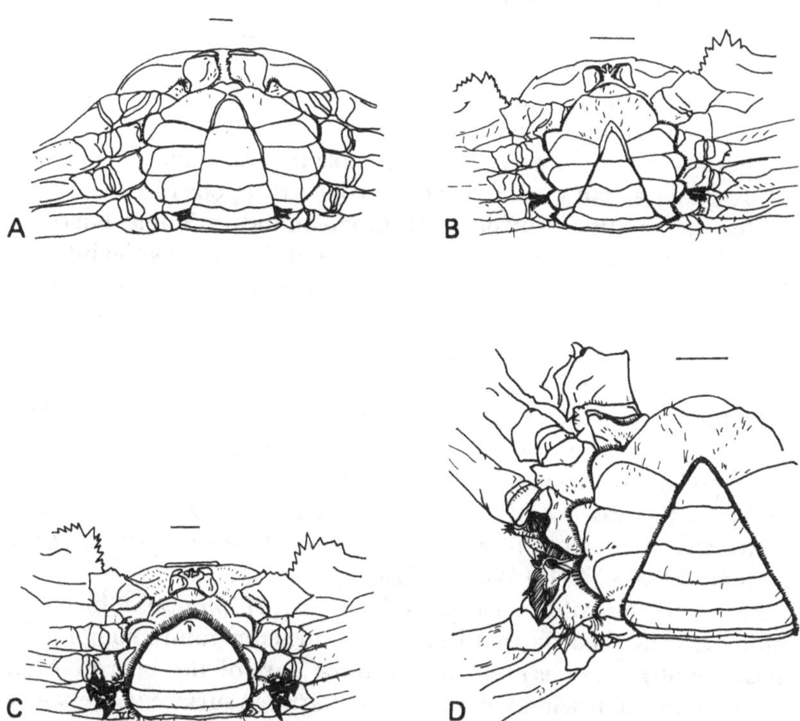

Fig. 7.2. Setal tufts used for water absorption in terrestrial crabs. A. *Gecarcoidea natalis*, adult carapace width 100 mm. B. *Geograpsus grayi*, adult carapace width 45 mm. C. *Geograpsus crinipes*, adult carapace width 60 mm. D. *Geograpsus crinipes* with pereiopods parted to show opening into the branchial chamber. Scale bar = 10 mm.

and the force required to extract water would be beyond the capability of setal suction systems (Greenaway and MacMillen, 1978; T. G. Wolcott, 1984). The absence of ventral setae reflects this situation, and soil water is not directly available to these crabs. *Coenobita rugosus* burrows into damp sand and extracts water using setae on the chelae and third maxillipeds (Vannini, 1975).

iv. Other water sources. Various other sources of water that might be available include metabolic water, uptake of water vapor, preformed water in the food, and condensation. Water produced by the metabolism of food materials is likely to be a very small component of the water requirement since metabolic rates of crabs are often low (see Chapters 8 and 10) and water loss rates high. However, water losses from crabs living in humid burrows will be lower, and metabolic water could contribute substantially toward water balance under these conditions. Several land crabs, notably *B. latro* (Lawrence, 1970), *C. clypeatus* (De Wilde, 1973), and *H. transversa* (Greenaway, 1984b) store substantial amounts of fat, which could be a source of metabolic water during nonfeeding periods.

There is no evidence that crabs are able to absorb water from subsaturated air as can some other terrestrial arthropods (Edney, 1977). Under the conditions of near saturation found in humid burrows, water will move across permeable portions of the exoskeleton into the tissues if the water-potential gradient is favorable. Such movement is unlikely to contribute substantially to short-term water balance, but has not been examined experimentally.

Water contained in food may contribute substantially to water balance, but no data are available for dietary water intake in crabs. The diet of a carnivorous species such as *G. grayi* (largely other crabs and arthropods) would contain at least 60% water (Table 7.3; Edney, 1977). In the absence of drinking water, *C. clypeatus* is reported to derive water by eating fragments of porous limestone with a water content of 20–30% (De Wilde, 1973).

In the burrows of the arid-zone land crab, *H. transversa,* high humidity and marked diurnal fluctuations in temperature combine to cause nightly condensation in the lower part of the burrow even though saturation seldom occurs at the burrow's surface (Greenaway and MacMillen, 1978). It is unlikely that the crabs can utilize water condensing on the burrow walls as this will immediately be bound by soil particles. However, condensation of water within the branchial chambers is utilized by dehydrated crabs (Greenaway and MacMillen, 1978). Nocturnal vertical movements of the crab between the cool upper regions of the burrow and the warm, moist burrow depths could multiply the number of condensation cycles and maximize uptake of water. This source of water is probably utilized to some extent by all burrowing crabs, since water loss within the burrow is likely to be low. Condensation may be of considerable importance to species such as *H. transversa* that frequently have no other source of water.

E. Regulation of water balance

To maintain water balance, a crab must match water uptake to water loss. Amphibious crabs are unlikely to experience serious difficulty in maintaining water balance as water is always close at hand. Similarly, intertidal species are regularly inundated with seawater. Any water lost during terrestrial activity by these species is replaced promptly so that selective pressure to reduce water permeability has probably been slight. Indeed, certain fiddler crabs utilize their high rate of evaporation to lower body temperature in hot environments (Edney, 1961; Wilkens and Fingerman, 1965; Smith and Miller, 1973; Powers and Cole, 1976).

The maintenance of water balance in terrestrial crabs is more difficult and has involved development of both behavioral and physiological adaptations to restrict loss of water and facilitate water uptake. Burrows are of major importance since they provide a shelter from low humidity, high temperature, and moving air and reduce or eliminate evaporative loss of water during resting periods. Evaporative loss during surface activity is limited behaviorally, by selection of moist habitats (e.g., rain forest) and by gearing activity to the most favorable conditions. Thus activity of *C. clypeatus* from dry habitats is largely nocturnal (De Wilde, 1973), whereas *G. natalis* from the rain forests of Christmas Island is active diurnally (Hicks et al., 1984).

Physiological adaptations for the maintenance of water balance are often closely linked with behavior. In *O. quadrata*, a relatively high permeability to water and considerable surface activity lead to high rates of evaporative water loss. By digging a burrow down to damp sand and utilizing a setal suction mechanism, lost water is replaced from the substrate. The high turnover of water characteristic of crabs of this genus imposes ecological limitations. Dependence on the uptake of soil water restricts *Ocypode* to sandy soils from which their setal suction system can extract water. Effectively, the genus is limited to sandy beaches and their backing dunes (see the Appendix and Chapters 2 and 3).

Crabs belonging to the genera *Gecarcinus* and *Gecarcoidea*, as well as *Geograpsus grayi*, penetrate much farther inland and are apparently not restricted by soil type. Evaporative water loss has been reduced, and behavioral mechanisms restrict surface activity to favorable conditions or habitats. During resting periods, burrows or retreats provide a secure, moist environment. The reduced rate of water loss is balanced by uptake of soil water and, if conditions permit, utilization of dew and drinking. *B. latro* also utilizes moist habitats, constructs burrows, and drinks free water. This group has evolved a pattern of water balance involving a moderate turnover of water.

Freshwater/land crabs from arid or monsoonal regions are faced with long periods of very hot and dry conditions when surface activity is not feasible (McCann, 1938; Fernando, 1960; Greenaway and MacMillen, 1978). Dry periods are spent entirely in the burrows, which maintain a humid environment minimizing daytime loss of water by evaporation. At night, condensation provides a water source. The rate of water loss is at the lower end of the measured range for land crabs (MacMillen and Greenaway, 1978; Fig. 7.1). Given these low rates of water loss, metabolic water may make a significant contribution to water balance. A high water content may act as a buffer against brief periods of desiccation. This group has evolved a low-water-turnover approach to maintenance of water balance.

The shelled coenobitids are highly permeable to water but reduce their evaporative losses of water behaviorally by means of the mollusk shell they inhabit. They may also restrict their activity to humid conditions. Moreover, a store of water in the shell enables them to survive short dry periods and allows them to forage farther from a water supply. Water balance in the coenobitids is largely behavioral, and they are generally dependent on a supply of free water for drinking or immersion, although some species do extract water from moist sand.

The relative importance of the various routes of water loss is difficult to gauge accurately, since complete and comparable data are lacking. When evaporative water loss is low (humid conditions) and water is freely available, the urine is probably the major component of loss. Under desiccating conditions urinary loss is reduced and evaporative loss increases. Failure to reduce urine production under dry conditions would increase total water loss by 76% in *Gecarcoidea lalandii* and by 56% in *C. carnifex* (Harris and Kormanik, 1981). With regard to gain of water there is a similar lack of data. However, where water turnover is high, soil water and drinking water are the main sources of supply, whereas under conditions of low water turnover, metabolic water, preformed water in the food, and condensation assume major importance.

3. Osmotic and Ionic Regulation

A. Patterns of regulation

i. Intertidal and splash-zone species. Species inhabiting intertidal or splash zones are either inundated for long periods each day or return frequently to tidal pools or to the sea itself. They are primarily adapted for aquatic osmoregulation and exhibit a wide range of osmoregulatory ability. Thus, when immersed in seawater, many brachyurans are feeble osmoregulators with body fluids generally isosmotic with the medium. They have a limited ability to hyperregulate in dilute media and hyporegulate in media more concentrated than seawater (e.g., *Mictyris longicarpus* and *Macrophthalmus setosus*; (Barnes, 1967). Others osmoregulate powerfully over a wide range of external concentrations maintaining the concentration of their hemolymph more or less constant at a level usually below that of seawater, e.g., *Uca* spp. (Baldwin and Kirschner, 1976a, b; Fig. 7.3; Table 7.4) and *Goniopsis cruentata* (Zanders, 1978; Table 7.4); species such as *Pachygrapsus crassipes* are intermediate in osmoregulatory ability (Gross, 1964b).

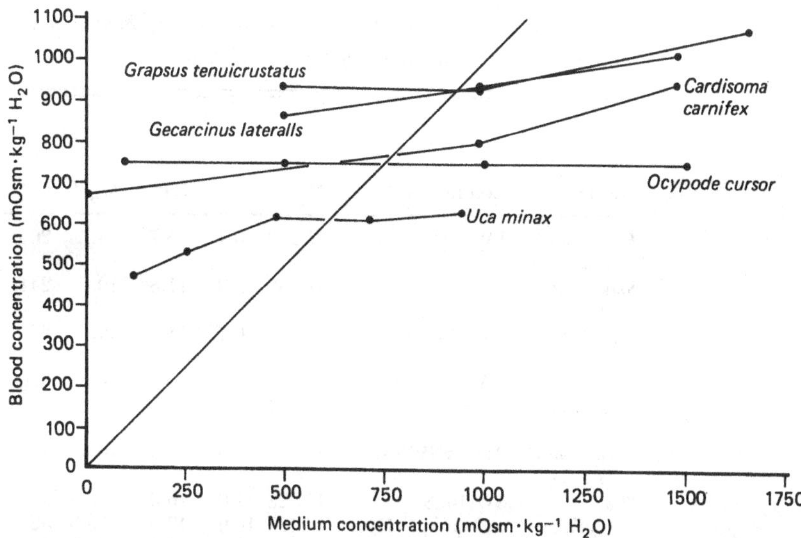

Fig. 7.3. Osmoregulation in amphibious and terrestrial crabs. (Data from Gross, 1964b; Gross et al., 1966; Borut and Neumann, 1966; and Wanson et al., 1984.)

The mechanisms of hyperosmotic regulation in dilute media are well known (Mantel and Farmer, 1983) and have already been summarized (see section 1). The freshwater crabs appear to be the only group of hyperosmoregulating brachyurans to have colonized terrestrial habitats. The osmoregulatory characteristics the freshwater crabs have brought to terrestrial life are a relatively low hemolymph osmolality (Table 7.4), isosmotic urine, a very low permeability to water (Greenaway, 1980), and an ability to absorb ions across the gill epithelium (Greenaway, 1981; Sparkes and Greenaway, 1984).

The principal mechanisms of hyporegulation are also understood (Mantel and Farmer, 1983) and are summarized in section 1. However, certain details remain unclear. Water lost to hyperosmotic media by osmosis is replaced by drinking the medium and absorbing the ingested ions and water from the gut (Dall, 1967; Hannan and Evans, 1973; Baldwin and Kirschner, 1976a; Evans, Cooper, and Bogan, 1976). The absorbed ions, and ions that enter the body across permeable surfaces, must be eliminated. Although excess magnesium, calcium, and sulfate ions are excreted in the urine, the major ions (Na^+ and Cl^-) are not (Gross, 1955; Green et al., 1959; Gross and Marshall, 1960). In *Uca*, both excess Na^+ and Cl^- are removed from the body by active transport (Baldwin and Kirschner, 1976a). Elimination of Na^+ is in exchange for K^+ from the medium (Evans et al., 1976). There is some evidence favoring both the gut and gills as the site of

Table 7.4. *Composition and concentration of hemolymph and urine of terrestrial crabs* [a]

Species	Conditions	Na^+	K^+	Ca^{2+}	Mg^{2+}	Cl^-
Holthuisana transversa	FW	H 270	6.4	15.7	4.7	266
Sudanonautes africanus	FW	H 280	5.9	11.8	10.6	241
Cardisoma hirtipes	Rain forest	H 259	6.1	11.3	5.1	263
Cardisoma carnifex	SW moist sand	H 393	—	—	—	370
		U 393	—	—	—	400
Cardisoma carnifex	Access SW/FW	H 317	8.0	12.0	10.0	408
Uca pugilator	SW(1028 $mOsm \cdot kg^{-1}$ H_20)	H 328	11.0	16.0	46.0	537
		U 276	16.0	17.0	108.0	622
Goniopsis cruentata	SW(1066 $mOsm \cdot kg^{-1}$ H_20)	H 475	8.0	13.0	12.0	428
		U 314	8.0	13.0	154.0	525
Ocypode quadrata	Ocean beach	H 449	7.0	15.7	28.6	475
		U 379	9.8	12.5	32.3	447
Gecarcinus lateralis	FW + SW + Food	H 459	9.4	20.9	13.5	—
		U 470	9.8	17.3	28.4	—
Gecaroidea lalandii	SW moist sand	H 421	—	—	—	401
		U 417	—	—	—	429
Gecarcoidea natalis	Rain forest	H 315	6.7	14.7	9.3	327
Geograpsus grayi	Rain forest	H 367	6.6	16.1	5.9	367
Coenobita perlatus	FW + SW + Food	H 465	10.5	14.7	30.9	—
		U 517	14.9	12.9	40.7	—
Coenobita clypeatus	Access (SW 1000 $mOsm \cdot kg^{-1} H_20$)	H 376	8.0	23.0	59.0	337
Coenobita brevimanus	Access SW + FW	H 444	15.0	22.0	40.0	374
Birgus latro	Rain forest	H 324	8.3	18.1	19.6	346

Note: H, hemolymph; U, urine.
[a] All ions as $mmol \cdot l^{-1}$.

HCO_3^-	SO_4^{2-}	Osmotic pressure (mOsm·$kg^{-1}H_2O$	Protein ($g \cdot l^{-1}$)	Free amino-acids ($mmol \cdot l^{-1}$)	References
9–13	—	524	—	—	Greenaway & MacMillen (1978)
—	—	480	—	—	Lutz (1969)
—	—	565	—	—	Greenaway (unpub.)
15	—	761	83.0	1.72	Kormanik & Harris
—	—	822	—	—	(1981), Henry & Cameron (1981)
—	—	649	—	—	Burggren & McMahon (1981a, unpub.)
—	42.0	497	—	—	Green et al. (1959)
—	47.0	583		—	
—	15.0	—	—	—	Zanders (1978)
—	71.0	—	—	—	
—	24.1	—	66–183	—	Gifford (1962b),
—	19.1	—	—	—	Pequeux et al. (1979)
—	—	—	—	—	Gross (1963)
—	—	—	—	—	
—	—	838	72.9	1.99	Kormanik & Harris
—	—	947	—	—	(1981), Henry & Cameron (1981)
—	—	700	—	—	Greenaway (unpub.)
—	—	786	—	—	Greenaway (unpub.)
—	—	1040–1225	114	2.52	Gross & Holland (1960),
—	—	—	—		Henry & Cameron (1981)
11–19	—	969	114–126	—	Wheatly et al. (1984), McMahon & Burggren (1979)
—	—	807	137	2.82	Burggren & McMahon (1981a, unpub.), Henry & Cameron (1981)
—	—	752	96	1.78	Greenaway (unpub.), Henry & Cameron (1981)

excretion of ions during hyporegulation (Green et al., 1959; Dall, 1967). Thus, the level of the ion-transporting enzyme $Na^+ + K^+$ – ATPase in the gills is increased during hyporegulation in *U. tangeri* (Drews, 1983) and *Eriocheir japonicus* (Watanabe and Yamada, 1980) (suggesting an enhanced ability to excrete ions). Yet, there is no difference in these enzyme levels in the gills of *Metopograpsus thukuhar* acclimated to any medium (Spencer, Fielding, and Kamemoto, 1979). Several other species of *Uca* and *Cyclograpsus henshawi* increase $Na^+ + K^+$ – ATPase in dilute media but not during hyposmotic regulation (Spencer et al., 1979; Hake and Teller, 1983; Wanson, Pequeux, and Roer, 1984; Holliday, 1985). Direct measurements of salt efflux across gut and gills are needed to clarify this matter.

Terrestrial and semiterrestrial brachyurans of marine origin are derived from ancestors with hyposmotic regulatory ability. The terrestrial gecarcinids, grapsids, and ocypodids still display these mechanisms when experimentally immersed in water. The osmoregulatory characteristics with which these crabs began terrestrial life may then have included a hemolymph osmolality slightly below that of seawater, a lowered permeability to water and ions, isosmotic urine, and an ability both to absorb ions across the gills and to excrete monovalent ions either across the gills or through the gut.

The ability of intertidal anomurans to regulate the concentration of their extracellular fluids is relatively limited when compared to the brachyurans. The coenobitid *Clibanarius vittatus* is a weak hyperosmotic regulator but can modulate the activity of $Na^+ + K^+$ – ATPase in its gills in response to salinity changes (Sabourin and Saintsing, 1980). However, the intertidal pagurids studied are osmoconformers whose body fluids remain more or less isosmotic with the medium over a wide range of salinity (Robertson, 1953; Young, 1979), and they must be capable of making large osmotic adjustments in order to maintain intracellular concentration isosmotic with the hemolymph.

Activity out of water, especially in tropical areas, may result in desiccation. On reimmersion, this water must be replaced, and in dilute media this can be achieved either by drinking or by osmotic uptake. In isosmotic and hyperosmotic media, however, crabs that regulate hyposmotically (e.g., *P. crassipes*; Gross, 1955) must be able to gain water without accompanying net gain of salt.

The ability to maintain the concentration of the hemolymph within a narrow range is a common feature of semiterrestrial crabs (Gross, 1964a; Barnes, 1967; Baldwin and Kirschner, 1976a; Mantel and Farmer, 1983). Most species live in areas of fluctuating salinity (e.g., estuaries, lagoons, and mangrove swamps), and it is probable that the strong osmoregulatory powers of these animals represent an evolutionary response to these conditions (Gross, 1964a). Intertidal anomurans are weak hyperosmotic regulators or osmoconformers. Thus, tolerance and intracellular isosmotic regulation are emphasized in this group, representing a different approach to fluctuating conditions. This approach is perhaps aided by the carriage of water in the shells they inhabit and by their ability to restrict osmotic and ionic exchange with the medium by sealing the aperture of their shell with the chelae.

ii. Terrestrial crabs with water-filled burrows. Species of *Cardisoma* require regular contact with water and construct burrows that reach the water table and contain free water. Different species occupy different habitats. *C. carnifex* burrows close to the high-tide mark, and its burrow reaches the saline water table (Cameron, 1981a). *C. guanhumi* shows no salinity preference and burrow water may be either fresh water or seawater (Gifford, 1962a; Herreid and Gifford, 1963), whereas *C. hirtipes* on Christmas Island is restricted to freshwater seepages (Gibson-Hill, 1947).

Cardisoma spp. are strong osmoregulators in water, maintaining the hemolymph hyposmotic in seawater but regulating hyperosmotically in dilute media (De Leersnyder and Hoestlandt, 1963; Gross et al., 1966; Fig. 7.3). The urine is more or less isosmotic with the hemolymph (Table 7.4) and plays little part in osmoregulation of body fluids. Urine is involved in ionic regulation, however, and if *C. guanhumi* is kept in water of high salinity, then Mg^{2+}, SO_4^{2-} and K^+ are preferentially excreted while Na^+ and Cl^- are reabsorbed (Table 7.4). By contrast, there is no resorption of Na^+ or Cl^- by *C. carnifex* kept in seawater of 800 $mOsm \cdot kg^{-1}$ H_2O (Kormanik and Harris, 1981). Although the urine itself is isosmotic with the hemolymph, under terrestrial conditions *C. guanhumi* releases it into the branchial chambers where ions may be resorbed. In water-loaded crabs, this allows the final excretory product to be hyposmotic to the hemolymph (D. L. Wolcott and T. G. Wolcott, 1984).

Urinary loss of sodium in *C. carnifex* is high (51 $\mu mol\ Na^+ \cdot 100\ g^{-1} \cdot h^{-1}$ in seawater and 34 $\mu mol \cdot 100\ g^{-1} \cdot h^{-1}$ in distilled water). Losses of Cl^- are similar (Kormanik and Harris, 1981). These high rates of ion loss might be expected to cause difficulty in the maintenance of ion balance during immersion in fresh water. In practice, the volume of water in the burrow is limited and lost body salts may build up the salt concentration of burrow water to a level at which uptake of ions by the crab balances loss. Na^+ balance is achieved at low external concentrations: 0.5–0.8 $mmol \cdot l^{-1}$ in *C. guanhumi* (Herreid and Gifford, 1963) and 0.1–0.5 $mmol \cdot l^{-1}$ in *C. hirtipes* (Greenaway, unpub.). Clearly, the maximum rate of uptake of ions and the affinity of the ion-transport mechanism are high. It would be interesting to compare the dynamics of ion transport of *Cardisoma* with the freshwater/land crabs, which have a longer evolutionary history in fresh water.

The activity of ion-transporting ATPases (Na^+ + K^+ – ATPase and Cl^-/HCO_3^--activited ATPase) in the gills, intestine, and antennal organs of *C. carnifex* is high relative to levels in other tissues examined (Towle, 1981). Gill ATPases are probably utilized in ion uptake from the medium during hyperosmotic regulation and in absorption of ions from urine released into the branchial chambers. The gills have an ultrastructure suited for ion transport (see also Fig. 7.7). Intestinal ATPases may be concerned with transport of ions and water from gut to hemolymph during hyposmotic regulation. ATPases identified in the antennal organ are presumably concerned with adjustments of the composition of the urine (Towle, 1981).

Clearly, *Cardisoma* spp. can cope physiologically with euryhaline media. They can also osmoregulate behaviorally since they need not remain immersed in suitable media and, while foraging, may have the opportunity to select water of a particular salinity (Gross et al., 1966). *C. hirtipes* maintains a relatively constant composition of the hemolymph when fresh water or fresh water and seawater, is offered as drinking/immersion water. If seawater alone is provided, the concentrations of ions in the hemolymph rise significantly but remain well below those in the water (Greenaway, unpub.). The burrow water may act as a reservoir for salts lost from the body that may then be resorbed as required. Loss of water and ions during foraging can rapidly be made good from this source.

During dehydration, the volume of the extracellular space is maintained at the expense of cell water. The osmolality of the hemolymph rises, indicating movement of solutes into the extracellular space from the cells. Filtration rapidly decreases and soon ceases during desiccation, and there is no indication of ability to produce urine that is substantially hyperosmotic to the hemolymph (Harris and Kormanik, 1981).

Cardisoma spp. have retained and improved the osmoregulatory mechanisms described earlier for intertidal and splash-zone crabs. The construction of a burrow down to ground water and the ability to cope with a very wide range of salinity in this water have allowed the genus to expand its distribution inland away from the sea (see the Appendix and Chapters 2 and 3). Distribution is restricted by the depth of the water table and the maximum depth of the burrow, which may be 1. 5 m in the case of *C. guanhumi* (Herreid and Gifford, 1963). All species of the genus can spend long periods out of water and, as far as osmoregulatory mechanisms are concerned, seem to be potentially capable either of colonizing freshwater habitats or developing a more terrestrial life-style.

iii. Freshwater/land crabs. Many species of the 11 families of "freshwater" crabs recognized by Bowman and Abele (1982) are amphibious (McCann, 1938; Fernando, 1960; Bishop, 1963). Some of these species from moist rain-forest habitats may even be regarded as terrestrial (Holthuis, 1974). Species from monsoonal and arid regions face long periods of dry conditions without access to free water, at which time they lead an inactive fossorial existence (McCann, 1938; Greenaway and MacMillen, 1978; Warburg and Goldenburg, 1984).

The freshwater/land crabs are well adapted for hyperosmotic regulation in dilute waters. The concentration of the hemolymph is lower than in land crabs of marine origin. The urine, although isosmotic, is not a major avenue of salt loss due to the extremely low rate of flow (Shaw, 1959; Thompson, 1970; Harris and Micallef, 1971; Greenaway, 1980; Table 7.4). The pattern of sodium balance indicates a high degree of adaptation to fresh water. The rate of sodium loss is low, the affinity for sodium ions is high, and, coupled with a high maximum rate of ion transport, this enables sodium balance to be maintained at very low external concentrations (Greenaway, 1981).

Most freshwater crabs are tolerant of salt concentrations up to 800–1,000 mOsm.kg^{-1} H_2O (Shaw, 1959; Harris and Micallef, 1971; Greenaway, 1981; Subramanyam and Krishnamoorthy, 1983; Warburg and Goldenburg, 1984), and regulation is hyperosmotic. No information is available for the concentration of burrow water, but it will be essentially fresh water in wet periods, briefly becoming more concentrated as groundwater dries up.

During aquatic and amphibious life, osmotic and ionic regulation present few problems, and any losses of ions and water resulting from surface activity are rapidly rectified by the mechanisms involved in hyperosmotic regulation. The main problems arise in the long dry periods when the crabs may become dehydrated. *H. transversa* is highly tolerant to loss of body water (Table 7.3) and does not regulate the concentration of its hemolymph during dehydration. Thus 43% loss in body water causes a twofold increase in osmolality (Greenaway and MacMillen, 1978). Similarly, the concentration of the hemolymph of *Potamon potamios* from the Negev and Sinai deserts is not regulated during dehydration, and the species is less tolerant of water loss (Warburg and Goldenburg, 1984). The rain forest species *Sudanonautes africanus* loses water preferentially from the hemolymph during mild desiccation with a resultant rise in the concentrations of all ions except Na^+. With further desiccation, water loss occurs preferentially from the tissue compartment, and the Na^+ concentration of the hemolymph rises (Lutz, 1969).

Holthuisana transversa, and probably many other species from arid areas, do not feed during dry periods and, in the absence of burrow water, have no source of ions to replace losses. Continued urine flow under these circumstances would rapidly cause depletion of ions. Urine production has not been measured under terrestrial conditions in any species. In water, *H. transversa* releases urine into the branchial chambers (Greenaway, 1980), and if this were to occur on land it would offer the opportunity for resorption of salts across the gills. Freshwater crabs also excrete water extrarenally (Greenaway, 1980). The route by which this occurs has not been identified, and its relevance in terrestrial osmoregulation is unknown.

iv. Terrestrial crabs. Although many terrestrial crabs can regulate osmotically when immersed (e.g., *Ocypode ceratophthalmus*, *Uca crenulata*, *Grapsus grapsus*, and *Gecarcinus lateralis*; Gross, 1964b), they normally do not enter water and indeed drown if forcibly submerged (Gross and Holland, 1960). This regulation, then, is seen only under experimental conditions, and data providing insight into mechanisms of osmoregulation under natural conditions are preferred here.

There are several species of shell-carrying hermit crabs (*Coenobita*) that live in the supralittoral beach zone or farther inland, often several kilometers or more from the sea (Gross, 1964a; De Wilde, 1973; Vannini, 1975, 1976 a, b; Hicks et al., 1984; see the Appendix and Chapters 2 and 3). The terrestrial hermit crabs carry water in their borrowed molluscan shells. This water can be replaced or replenished by immersion or drinking. Spillage is prevented as the soft abdomen forms a seal against the shell, but water may be lost if the crab has to retract rapidly into its shell (De Wilde, 1973). The concentration of shell water is regulated at a level peculiar to species: e.g., *C. perlatus*, 1,013–1,316 $mOsm \cdot kg^{-1}$ H_2O; *C. clypeatus*, 938–970 $mOsm \cdot kg^{-1}$ H_2O; *C. brevimanus*, 835 $mOsm \cdot kg^{-1}$ H_2O (Gross, 1964a; De Wilde, 1973). A detailed study has been carried out on *C. clypeatus* by De Wilde (1973). The concentration of shell water in this crab is kept constant when the available drinking water is in the osmolality range 105–1,320 $mOsm \cdot kg^{-1}$ H_2O (Fig. 7.4). When supplied only with dilute water, the crabs try to maintain the concentration of the shell solution by enhancing evaporation. To facilitate this, *C. clypeatus* adopts a posture that enhances evaporative loss of water. When the available water is in the range 1,054–1,320 $mOsm \cdot kg^{-1}$ H_2O very large amounts are consumed and used to flush the shell and branchial chambers, thus preventing further concentration of shell water by evaporation. When ion-depleted *C. clypeatus* are offered a range of water salinities, they initially take both high- and low-salinity water but steadily decrease the salinity and amount of water transferred into the shell until after 20 days only fresh water is taken. This behavior first adjusts the concentration of the shell water into the optimal range and allows re-

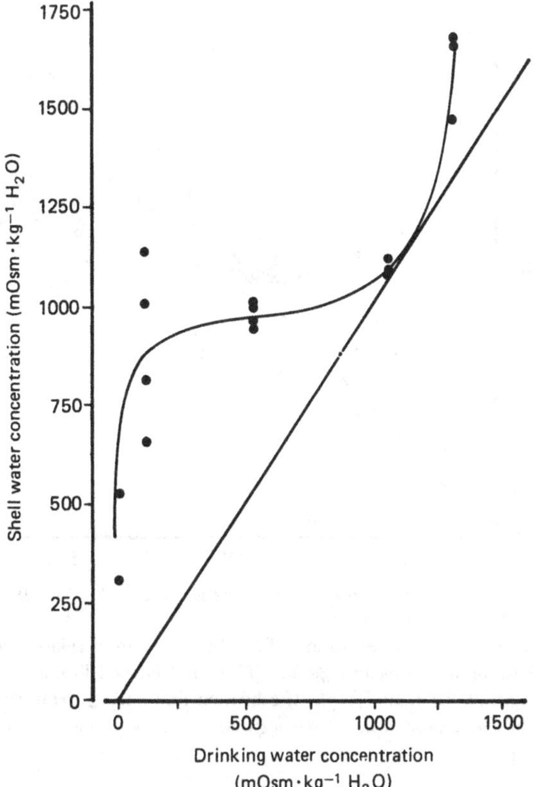

Fig. 7.4. Regulation of the concentration of shell water by the hermit crab *Coenobita clypeatus* when supplied with drinking water of various concentrations. (After De Wilde, 1973.)

covery of lost ions. This level is then maintained by replacing evaporative losses with dilute water.

Coenobitid crabs have a highly developed ability to sense the salinity of both shell and drinking water and apparently distinguish differences in concentration of only 5.5–10.5 mOsm·kg^{-1} H_2O. *C. perlatus* and *C. scaveola*, which live close the the sea, prefer seawater to fresh water for filling the shell, whereas *C. brevimanus*, *C. clypeatus*, and *C. cavipes*, which often penetrate farther inland, prefer fresh water (Gross and Holland, 1960; Gross 1964a; Volker, 1965; Gross et al., 1966; De Wilde, 1973). These preferences reflect the favored concentrations of shell water. It is not known if the ionic composition of shell water of coenobitids is regulated. *C. clypeatus* can also sense and orientate to humidity gradients (De Wilde, 1973), but the receptors involved are unknown. The osmoreceptors enabling salinity choice

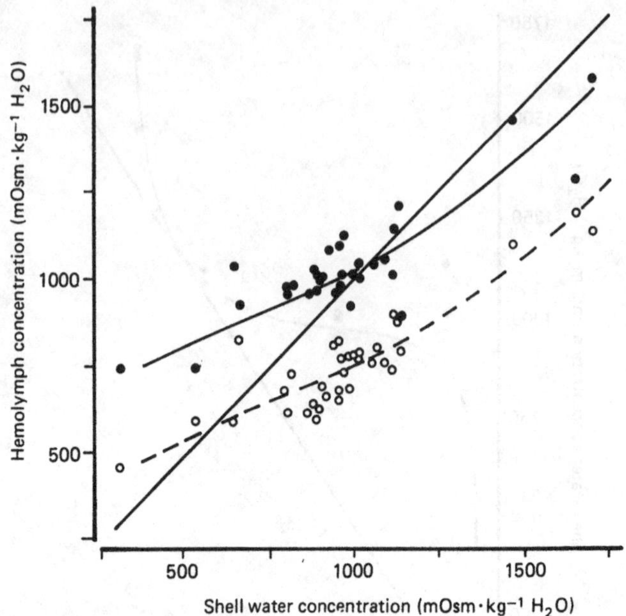

Fig. 7.5. Regulation of the concentration of the hemolymph in relation to the concentration of shell water in *Coenobita clypeatus*. The solid curved line and (●) represent total concentration of the hemolymph; the broken line and (○) represent electrolyte concentrations of the hemolymph. The *straight diagonal line* is the isosmotic line. (After De Wilde, 1973.)

also remain to be identified in coenobitids, but they may well be on the chelae or mouthparts, which are both used in drinking.

In tests of long-term survival, *C. clypeatus* survives well on freshwater (248 days) and seawater of 105 mOsm·kg^{-1} H_2O (253 days) and 527 mOsm·kg^{-1} H_2O (326 days), but relatively poorly on full-strength seawater (1,054 mOsm·kg^{-1} H_2O) (174 days). More concentrated media cause rapid dehydration and death (De Wilde, 1973).

The concentration of hemolymph is regulated against shell water in *C. clypeatus* (Fig. 7.5) and probably in the other species too. *Coenobita* spp. can tolerate large fluctuations in the concentration of their hemolymph, but normally regulate the concentration quite precisely. Thus *C. brevimanus* maintains a concentration of 730–835 mOsm·kg^{-1} H_2O, *C. cavipes* 865–975 mOsm·kg^{-1} H_2O, *C. clypeatus* close to seawater at 970 mOsm·kg^{-1} H_2O, and *C. perlatus* greater than seawater at 1,107–1,336 mOsm·kg^{-1} H_2O (Table 7.4; Gross, 1964a; Gross et al., 1966). Up to 30% of the osmotic pressure of the hemolymph is due to non-electrolytes in *C. clypeatus* (De Wilde, 1973).

The urine is essentially isosmotic with the hemolymph, and minor departures are not significant to osmoregulation (Gross, 1964a; De Wilde, 1973). Data on the ionic composition of urine are limited (Table 7.4) but suggest that the antennal organ may regulate potassium. Urine released could be passed to the branchial chambers or to the shell water from which ions could be reabsorbed. The physiology of urine production in *Coenobita* spp. deserves further detailed examination.

During desiccation, the measured increase in the osmolality of the hemolymph of *C. clypeatus* suggests equal loss of water from cellular and extracellular compartments, with no regulation of the concentration of the hemolymph (De Wilde, 1973). Temperature changes during dehydration reduce changes in osmolality (Wheatly, Burggren, and McMahon, 1984). Considerable ionic adjustments between intracellular and extracellular compartments occur during dehydration in *C. clypeatus* but not in *C. brevimanus*, where adjustments seem to be by changes in levels of organic materials (Burggren and McMahon, 1981a; Wheatly et al., 1984). Further studies, incorporating measurements of ions, organic materials, and extracellular space, are needed to clarify responses to dehydration.

Only the early developmental stages of *B. latro* carry shells, and adults thus lack the reservoir of shell water carried by the adult hermit crabs. Under favorable conditions, an osmolality of 700–800 $mOsm \cdot kg^{-1} H_2O$ is maintained in the hemolymph (Table 7.4). In the field, the crabs prefer fresh water for drinking and do not drink seawater under normal circumstances (Gross, 1964a). Indeed, on Christmas Island many individuals live several kilometers from the sea and probably have access to seawater only on spawning migrations (Hicks et al., 1984). When seawater alone is available for drinking, the concentration of the hemolymph rises rapidly and body mass falls, but normal mass and osmolality are maintained with a supply of fresh water (Gross, 1964a; Harris and Kormanik, 1981). If a choice between seawater and fresh water is offered, fresh water is preferred but some seawater is consumed and the hemolymph is maintained at a relatively high concentration (Gross, 1955). Fresh water probably is adequate to maintain the ion concentration of the hemolymph under normal circumstances, especially as evaporative losses would allow the salts it contains to be concentrated.

The urine in *B. latro* is slightly hyposmotic to the hemolymph (urine to hemolymph ratio of 0.82–0.95) (Gross, 1964a), but composition is unknown. The rate of filtration by hydrated crabs is quite high, but inability to collect urine experimentally has prevented the calculation of flow rates (Kormanik and Harris, 1981). Filtration is not reduced during dehydration. Loss of salt in the urine could potentially lead to salt depletion if the dietary intake of ions is low, but *B. latro* may pass its urine into the branchial chambers for reprocessing as has been reported in other land crabs (Wolcott and Wolcott, 1982b; 1985a). Further work is required to clarify the role of the urine in osmoregulation.

During dehydration to the maximum tolerable level, water is lost almost entirely from the hemolymph. However, the resultant rise in osmolality is only 28% of that expected, and evidently solutes are removed from the hemolymph during dehydration (Burggren and McMahon, 1981a; Harris and Kormanik, 1981). Thus *B. latro* conserves tissue volume at the expense of hemolymph during dehydration. Nonetheless, considerable osmotic readjustments occur to spread the increase in concentration between the hemolymph and the cells.

$Na^+ + K^+$ – ATPase and Cl^-/HCO_3^- – ATPase are present in large amounts in the gills and antennal organs of *B. latro* and in lesser amounts in the intestine (Towle, 1981). ATPases in the gills may be concerned with resorption of ions from the urine, whereas those in the gut and antennal organs could be used in absorption of salts from the diet and resorption of Na^+ from the urine. Certainly, on a seawater regime, levels of ATPases in the gut and antennal organ decline (Towle, 1981). The Cl^-/HCO_3^- – ATPase may be utilized in regulation of HCO_3^- concentration (i.e., acid–base balance) rather than uptake of Cl^-, but this is not clear at present (Towle, 1981).

In summary, coenobitids are tolerant of desiccation and of changes in the concentration of their body fluids. They maintain a water store in their shell, the concentration of which is regulated behaviorally by selecting water of appropriate salinity or by concentrating dilute water by evaporation. The body fluids are regulated against shell water and also behaviorally by selecting the salinity of drinking water. The reservoir of shell water may be important in surviving desiccation. Although the urine is isosmotic with the hemolymph, it is possible that loss of salts and water may be regulated by passing the urine into the branchial chambers or shell water for reprocessing.

Several highly terrestrial crabs belong to the genera *Gecarcinus* and *Gecarcoidea*. The osmolality of the hemolymph of gecarcinids is maintained below that of seawater but varies with the holding conditions. Thus, depending on the salinity of the water available for uptake or drinking, the Na^+ concentration of the hemolymph may range from 281–797 $mmol \cdot l^{-1}$ in *Gecarcinus lateralis* (Gross, 1963). When only seawater is supplied, the Na^+ concentration of the hemolymph of *G. lateralis* rises steadily and reaches a level well above that of seawater. The crabs then seek to immerse themselves in the seawater (which is now hyposmotic to the body fluids) to restrict further concentration of their hemolymph (Gross, 1963). When kept on sand dampened with seawater, *G. lateralis* dehydrates and dies rapidly, although it survives indefinitely on sand dampened with fresh water (Gross, 1963). *Gecarcoidea lalandii* loses body mass slowly on sand dampened with seawater (Kormanik and Harris, 1981). Under natural conditions, *G. lateralis* utilizes relatively dilute media, often interstitial water extracted from the substrate (Wolcott, 1984), and it may have become dependent on this to the extent that it is no longer able to cope with seawater. If offered both fresh water and seawater, *G. lateralis* maintains a constant hemolymph concentration by intake of suitable quantities of both media (Gross, 1963). Bliss, Wang, and Martinez (1966) indicated that *G. lateralis* could survive indefinitely when supplied with water of salinities from 0–878 mOsm·kg^{-1} H_2O.

The urine of both *Gecarcinus lateralis* and *Gecarcoidea lalandii* is virtually isosmotic with the hemolymph although ionic concentrations may differ slightly (Table 7.4). High levels of Na^+ + K^+ –ATPase and Cl^-/HCO_3^- –ATPases in the antennal organs (Towle, 1981) indicate an ability to transport ions between the hemolymph and urine. Urine production is quite high when water is available (Table 7.2), but on dry sand filtration and urine flow rapidly decrease (Harris, 1977; Harris and Kormanik, 1981). Salt loss in the urine is high in hydrated crabs, but there is evidence that urine leaving the antennal gland is passed into the branchial chambers where the gills resorb ions as required (T. G. Wolcott and D. L. Wolcott, 1982b; T. G. Wolcott and D. L. Wolcott, pers. comm.). The epithelial cells of the posterior gills have a structure suited for ion transport (Copeland, 1968) and contain large amounts of the ion-transporting enzymes Na^+ + K^+ –ATPase and Cl^-/HCO_3^- –ATPase (Mantel and Olson, 1976; Towle, 1981). These epithelial cells are probably the site of absorption of ions from groundwater or urine entering the branchial chambers. Water and ions can move between the hemolymph and the lumen of the foregut in *G. lateralis*, and there is evidence that the rate of transport is controlled hormonally (Mantel, 1968). Thus, ions and water from the food and drinking water could potentially be transferred into the hemolymph, or ions and water for excretion could be passed into the lumen. Further studies on the role of the gills and the gut in salt and water balance are needed.

The volume of the hemolymph of *G. lalandii* remains constant during dehydration despite loss of up to 18% of total body water. Osmolality rises by 28%, and this is accompanied by a similar rise in Cl^- concentration and a smaller rise in Na^+ concentration (18%). Thus, water loss is largely from the cellular compartment, and the rise in the concentration of the hemolymph is presumably due to movement of ions and other osmotically active materials out of the cells (Harris and Kormanik, 1981). Equivalent data are not available for *G. lateralis* during desiccation, but there is evidence for some regulation of the concentration of the hemolymph (Gross, 1963).

In summary, the terrestrial gecarcinids do not regulate the concentration of the hemolymph precisely but instead tolerate wide fluctuations. The urine is always isosmotic and can play little part in osmoregulation. However, reprocessing of the urine and the action of the gut bestow some regulatory capability. In dry conditions water and salts are conserved by reducing the rate of urine formation. Hemolymph volume is protected at the expense of tissue water during desiccation, and redistribution of cellular ions occurs to maintain isosmoticity between the two compartments. The animals are dependent on water of low salinity and have lost the ability to survive on concentrated media.

The only highly terrestrial species in the family Ocypodidae belong to the genus *Ocypode*, which is restricted to supralittoral areas of sandy ocean beaches and their backing dunes. Many species only rarely visit the sea and do not have free water in their burrows.

The concentration of the hemolymph is kept below that of standard seawater (Table 7.4) with the exception of *Ocypode saratan* (Spaargaren, 1977). *Ocypode* spp. are strong hyper/hyposmotic regulators when immersed and regulate all major ions and the concentration of protein in the hemolymph (Gifford, 1962b; Gross, 1964b; Pequeux, Vallotta, and Gilles, 1979; Mantel and Farmer, 1983).

The urine of *O. quadrata* in the field is slightly hyposmotic to the hemolymph. The antennal organs conserve Ca^{2+}, Na^{+}, and SO_4^{2-} while excreting K^{+} and Mg^{2+} (Table 7.4; Gifford, 1962b). Urine flow is high (10–40 $ml \cdot 100\ g^{-1} \cdot day^{-1}$) when water is available, but on dry sand, filtration and urine flow rapidly decrease and stop altogether within three hours (Flemister, 1958; Gifford, 1962b). As water generally is available from the substrate (Wolcott, 1984), it is probable that urine flow is normally high except during extended periods of surface activity. The groundwater absorbed is dilute (Wolcott, 1984; T. G. Wolcott and D. L. Wolcott, 1985a) and the potential for salt depletion must be high. However, *O. quadrata* can reprocess its urine and, when water loaded experimentally, reduces the concentration of the final excretory product (reprocessed urine) from the normal level of 900 $mOsm \cdot kg^{-1}\ H_2O$ down to 100–200 $mOsm \cdot kg^{-1}\ H_2O$ (T. G. Wolcott and D. L. Wolcott, 1985a; Fig. 7.6). This is achieved by passing urine into the branchial chambers, where reabsorption of salts and some water occurs across the gills, which have epithelial cells suited for ion transport (Storch and Welsch, 1975; Fig. 7.7).

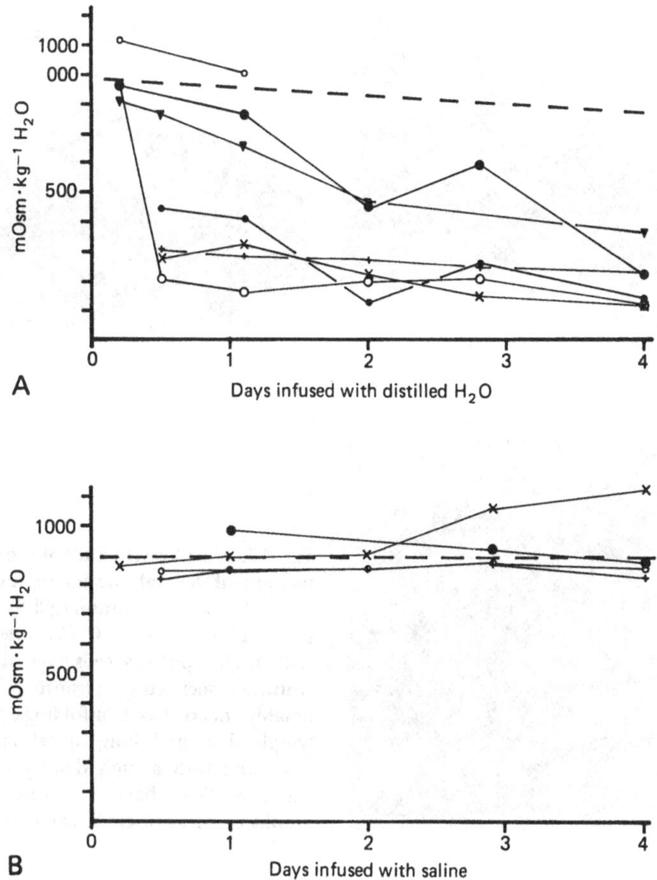

Fig. 7.6. The osmolality of the hemolymph (*broken line*) and the final excretory product (*solid line*) in *Ocypode quadrata* infused with various solutions at 0.3 ml·h^{-1} by means of an infusion pump. A. Infused with distilled water. B. Infused with isosmotic saline. (After Wolcott and Wolcott, 1985a).

Ocypode quadrata could also concentrate water obtained from the soil by using behavioral mechanisms to increase evaporative loss of water (e.g., by spending more time on the surface or selecting zones of low humidity). Regulatory problems due to intake of excess salts from the food could be overcome by increasing uptake of water from the soil and using this to flush out excess salts in the urine.

Ocypode spp. are relatively intolerant of dehydration (Table 7.2), but there is no information on internal osmoregulation during desiccation.

In summary, *Ocypode* spp. are strong osmoregulators if immersed

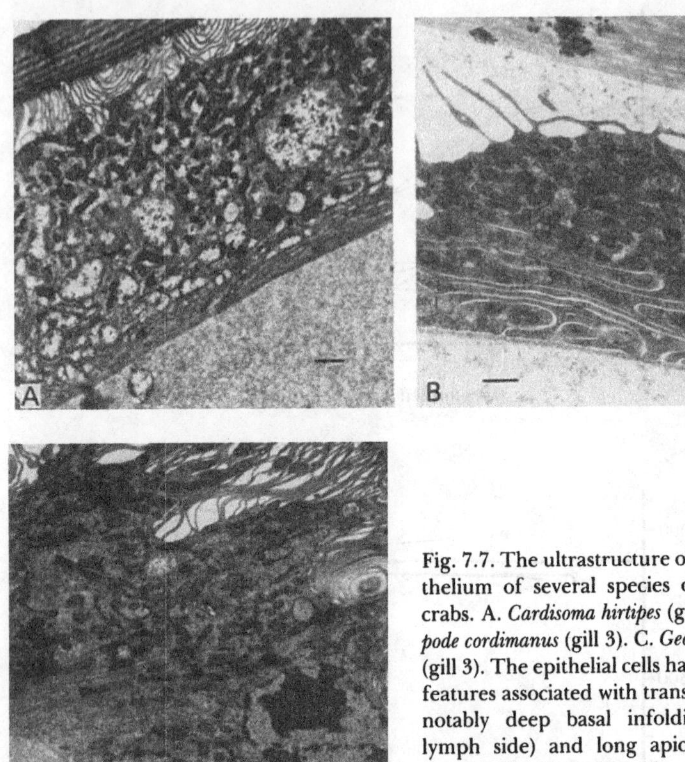

Fig. 7.7. The ultrastructure of the gill epithelium of several species of terrestrial crabs. A. *Cardisoma hirtipes* (gill 3). B. *Ocypode cordimanus* (gill 3). C. *Geograpsus grayi* (gill 3). The epithelial cells have structural features associated with transport of ions, notably deep basal infoldings (hemolymph side) and long apical microvilli associated with a high density of mitochondria. Scale bars = 1 μm. (Micrographs courtesy of C. A. Farrelly.)

and can tolerate a wide range of salinity. On land, they have a high water turnover fueled by the ready availability of dilute groundwater. The antennal organs are involved in ionic regulation, and osmotic regulation is achieved by both physiological and behavioral means, namely, preprocessing final urine and regulating the evaporative loss and uptake of water. The dependence on dilute groundwater restricts the genus to sandy areas where water can be extracted readily from the substrate.

There are a number of highly terrestrial species of grapsid crabs, notably in the genera *Geograpsus* and *Epigrapsus* (Von Raben, 1934) and the montane species of *Sesarma, S. jarvisi* and *S. cookei* (Abele and Means, 1977). However, apart from data on the osmolality and ionic composition of the hemolymph of *G. grayi* (Table 7.4), no information on osmotic and ionic regulation is available for any of these species.

Once again, the ultrastructure of the epithelium of the gills is suited to ion transport (Fig. 7.7).

B. Ion and water balance at molting

At the molt, crabs increase their body volume by absorption of fluid. To maintain osmolality of the hemolymph, absorption of ions must also occur. The molting process is also associated with loss of body calcium. Part of this loss is soluble calcium that has been resorbed from the old exoskeleton to facilitate withdrawal of the soft crab. The bulk is $CaCO_3$ remaining in the rejected exuviae. In aquatic crabs the medium provides a ready source of ions and water as well as the calcium needed to harden the new skeleton (Greenaway, 1985). Crabs with access to large bodies of water (intertidal and splash-zone species, *Cardisoma* spp., and the freshwater/land crabs) generally molt in water and have little difficulty extracting ions from this source. Terrestrial crabs face greater difficulties. Because adequate supplies of water and salts are not available in the short period immediately before and after ecdysis, these materials must be accumulated in advance. Consequently, hemolymph volume increases in premolt, and the extra fluid is accommodated in the pericardial sacs that bulge out at the rear of the carapace (Rao, 1968; Greenaway, pers. observ.). In late premolt, fluid is stored in the lumen of the gut in *G. lateralis* and is reabsorbed into the hemolymph after the molt (Bliss, 1968, 1979; Mantel et al., 1975).

Marine crabs lose most of their body calcium during ecdysis, generally storing less than 10% of their intermolt calcium content. This loss is made good by absorption of calcium from the water in postmolt, a rapid process generally complete within a few days (Greenaway, 1985). The amphibious crab *H. transversa* molts in water but reabsorbs almost 65% of its intermolt calcium content and stores this as tiny spherules in the hemolymph. The rest of its requirements are obtained from the water and from food (Sparkes and Greenaway, 1984). The high degree of storage enables considerable recalcification even if the temporary waters the species inhabits dry up shortly after ecdysis.

The small amounts of water available to terrestrial crabs cannot yield significant amounts of calcium. Increased resorption and storage of body calcium might thus be expected in terrestrial species and have been found in terrestrial amphipods and isopods (Table 7.5). In practice, the terrestrial crabs have not significantly increased calcium storage compared with aquatic species (Table 7.5). Instead, they molt in secure places (burrows, under logs, etc.) and eat most of their cast exoskeleton (Greenaway, 1985). The calcium contained in the exuviae is used to calcify the new skeleton, and hardening is sufficient to allow foraging (additional calcium required presumably comes from the

Table 7.5. *Storage of calcium in terrestrial crabs, amphipods , and isopods expressed as percentage of the total intermolt content*

Species	Habitat	Ca stored (%)	Storage site	Reference
Carcinus maenas	M	8.0	Midgut gland	Lafon (1948), Robertson (1937)
Orchestia gammarella	T	66.7	Midgut caeca	Graf (1968)
Porcellio scaber	T	48.0	General	Auzou (1953)
Holthuisana transversa	FW/T	63.6	Hemolymph	Sparkes & Greenaway (1984)
Cardisoma hirtipes	FW/T	4.5 (7)	NI[a]	Greenaway (unpub.)
Gecarcoidea natalis	T	10.2 (3)	Gastroliths	Greenaway (unpub.)
Ocypode cordimanus	T	15.4 (7)	NI	Greenaway (unpub.)
Geograpsus grayi	T	4.3 (3)	NI	Greenaway (unpub.)

Note: Abbreviations as in Table 7.2. Figures in parentheses indicate number of observations.
[a]NI, site not identified.

diet). Storage of calcium in *G. lateralis*, *C. guanhumi*, and *G. natalis* is in the form of gastroliths (Bliss, 1953; Gifford and Johnson, 1962; Greenaway, unpub.), but these are absent in *C. hirtipes*, *G. grayi*, and *O. cordimanus* (Greenaway, unpub.).

4. General discussion and evolutionary implications

For many years the apparent inability of land crabs to modify urine concentrations was regarded as a major stumbling block in the development of an effective osmoregulatory system for terrestrial life. Inability to regulate salt loss via the urine would necessitate such a high intake of salt in the diet or drinking water that the crabs would be dependent on the sea. The recent work of Wolcott and Wolcott (1985a) has radically changed these beliefs. It is clear that at least some land crabs have considerable ability to alter the concentration and composition of their urine, but only after it is voided from the antennal glands. Thus in *O. quadrata*, urine leaving the nephropores is passed into the branchial chamber where salts and water may be reabsorbed by the gills (T. G. Wolcott and D. L. Wolcott, 1982b; 1985a). In this way the final excretory product can be made markedly hyposmotic to the hemolymph. The gecarcinids have a similar capacity (T. G. Wolcott, pers. comm.). It is probable that all the terrestrial crabs may have this facility although this remains to be demonstrated in the Grapsidae, Coenobitidae, and freshwater/land crabs.

While reprocessing of urine obviously is effective, it is perhaps surprising that the terrestrial crabs have not evolved the capacity to form hyposmotic urine. However, urine released during emersion probably flows naturally into the branchial chambers, and the presence there of ion-transporting cells has presumably favored selection of the gills as the principal site for modification of the urine.

The ability to deal with a high salt intake under terrestrial conditions is poorly known, but neither *G. lateralis* nor *G. lalandii* seem able to cope with concentrated drinking water. This is a little surprising as *G. lateralis* and other terrestrial and amphibious crabs are strong hyposmotic regulators when immersed in concentrated media (Fig. 7.3) and must have the ability to produce hyperosmotic secretions. This aspect of osmoregulation needs further attention.

In practice, water available to terrestrial crabs is usually dilute, even in habitats bordering the sea (Wolcott, 1984). Thus, the ability to conserve ions, rather than to eliminate them, may well have been favored selectively. Excess salt ingested in the food can readily be eliminated as an isosmotic excretory product if dilute drinking water is available. Land crabs are similar to many other highly successful terrestrial animals in lacking the ability to produce hyperosmotic urine and thus to cope with concentrated drinking water.

An interesting, but seldom considered, aspect of urine formation concerns the osmolality of the hemolymph, about 30% of which is due to nonelectrolytes (De Wilde, 1973; Spaargaren, 1975). These molecules may or may not be filtered into the urine, depending on their size. Even if filtered, they are likely to be resorbed rapidly from the primary urine. Initially, the urine will be considerably hyposmotic to the hemolymph, and because the final urine is isosmotic, resorption of water or addition of ions must occur. Available evidence does not favor either of these alternatives, and the matter should be examined in more detail.

In the osmoregulatory responses to dehydration, further physiological adaptations to life on land are apparent. These responses are rather variable between species, but in each case there are adjustments involving temperature-dependent movement of osmotically active materials between compartments. These adjustments involve both ions and organic materials, but changes in the latter have not been investigated. Several species attempt to protect either intracellular or extracellular volume. The responses to dehydration are poorly understood at present and offer considerable scope for further research.

The land crabs of marine origin mostly retain marine larval stages (see Chapter 5). Little is known of the osmoregulatory ability of these larvae, but clearly they must have the capacity to maintain salt and water balance under the salinity conditions that prevail at the spawning and emergence sites, be they marine or estuarine. The larvae of *C. guanhumi*, for example, develop successfully in seawater of 440–1,320 $mOsm \cdot kg^{-1}$ H_2O (Costlow and Bookhout, 1968) and show some powers of hyposmotic regulation (Kalber and Costlow, 1968). The transition from aquatic larva (megalops) to terrestrial crab probably involves minimal changes to the osmoregulatory system. Thus the ion-transporting tissue of the gills, active in ionic exchange with the water in the larvae, is utilized in the reprocessing of the urine in the juvenile crab.

Available evidence from the fossil record indicates that the families of brachyurans with terrestrial representatives are of recent origin (5–60 million years; Warner, 1977), and the terrestrial species are likely to have evolved even more recently (also see Chapter 2). It is thus interesting to consider how well they have adapted to terrestrial conditions, from an osmoregulatory point of view. One way of assessing their success is to determine the range of habitats in which they live, although this is affected by factors other than osmoregulatory ability. Most species occur in damp habitats, but freshwater/land crabs are widely distributed in mesic and xeric environments, and hermit crabs may occupy arid coastal areas (De Wilde, 1973; Vannini, 1975, 1976a, b; see the Appendix and Chapters 2 and 3). Clearly, osmoregulatory strategies have been developed that permit salt and water balance under these conditions. An alternative way of viewing their success is to determine whether osmoregulation is achieved by physiological means or by avoiding the problems behaviorally. Water balance is largely maintained by behavioral mechanisms as discussed earlier, but physiological adaptations are also present, e.g., reduction of permeability (to quite low levels in a few species) and reduction of filtration rate under dry conditions. Ionic regulation is achieved largely by physiological means, and here the main adaptation is reprocessing of the urine, allowing the crab to conserve salts when required. If drinking water of a range of salinities is available, a behavioral component of ionic regulation may also be present. Land crabs are tolerant of wide fluctuations in the concentration of their body fluids. This may be regarded as a useful facet of their osmoregulatory approach, but it could result from an inability precisely to control osmolality. Land crabs have developed a suite of behavioral and physiological mechanisms for the maintenance of water balance that allows them to live in most types of terrestrial habitat; in severe habitats, however, activity is limited to periods of favorable conditions.

8: Respiration

BRIAN R. McMAHON AND
WARREN W. BURGGREN

CONTENTS

1. Introduction

For a majority of animals, obtaining sufficient oxygen (O_2) from the atmosphere and delivering it to the tissues, and removing carbon dioxide (CO_2) produced as a result of oxidative metabolism, are some of life's paramount tasks. These gases, consumed or produced in the tissues, are carried in the hemolymph and exchanged by diffusion at the body surface, often across areas specialized for this purpose. In the majority of aquatic crustaceans this exchange is an important function of the gills. In the more highly evolved decapod crustaceans, including both the brachyuran and anomuran crabs, these have evolved into complex structures beautifully designed for gas exchange with the aquatic environment. Since air contains approximately 40–60 times more oxygen than water at equivalent temperature and oxygen partial pressure (PO_2), one might assume that the gills would continue to function in aerial gas exchange. This, however, is rarely the case. In aquatic crustaceans, and indeed in aquatic animals generally, the gills are designed for aquatic exchange, and they collapse when the buoying effect of water is removed. This reduces the functional surface area available for gas exchange and additionally reduces both perfusion and ventilation of the gills. For most purely aquatic crustaceans, gas exchange in air is thus insufficient and they quickly succumb.

The colonization of land by the decapod crustaceans, as in the vertebrates, thus required the evolution of new structures to allow efficient oxygen uptake from the air. In large part this evolution has involved expansion of the branchial cavity and modification of the branchial epithelium, both of which increase the surface area available for gas exchange. Oxygen uptake, however, is only one of a complex of gill functions, which include CO_2 removal, osmotic and ionic regulation, acid–base regulation, and nitrogenous excretion in aquatic crustaceans (see also Chaper 7). The majority of these functions require water and thus are disrupted when the gills lose contact with their aquatic environment. Often these functions cannot be taken over by the accessory air-breathing organs and thus must be assimilated by other body systems. This assimilation may have involved some relatively complex evolutionary steps for the land crabs, as can be demonstrated by studying a range of semiterrestrial and terrestrial crab species. While the size and surface area of the gills clearly decrease with increasing ability to take up oxygen from the air, gills are nonetheless retained, even in the most terrestrial forms. Thus even the most terrestrial of crabs are, to some extent, dependent on the gills and thereby on at least intermittent contact with water. Land crabs thus apparently show a series of stages, not only in the evolution of air-breathing structures, but also in the devolution of the gills. For this reason this chapter will briefly review the respiratory and circulatory mechanisms typical of aquatic forms before attempting to show the changes that accompany the transition to air breathing.

2. Structure of gas exchange organs

In most terrestrial crabs the architecture of the branchial chamber is extensively modified from the pattern typical in aquatic species. These may be separated into modifications of the gills and of the lining of the branchial chamber.

A. Gills

In adult aquatic crabs gas exchange occurs largely across the surface of the gills since the external body wall is mostly thick and heavily chitinized and calcified. The gills, typically 9 pairs in aquatic brachyuran and 13–14 pairs in aquatic anomuran crabs, arise from the limb bases. The gills on each side lie protected inside a branchial cavity formed by outgrowths of the body wall, the branchiostegites (Milne-Edwards, 1839). Each gill bears a very large number of respiratory lamallae, i.e., flattened plates in which the perfusing hemolymph* is brought into close contact with the ventilatory water stream and over whose surface gas exchange occurs. The diffusion distances reported are variable, ranging from less than 1 μm in *Callinectes sapidus* (Aldrich and Cameron, 1979) to 6 μm in *Carcinus maenas* (Taylor and Butler, 1978). Gill area is relatively large (e.g., 1,367 $mm^2 \cdot g^{-1}$ in *C. sapidus;* Gray, 1957). These values are similar to the range of values reported for fish (Hughes and Morgan, 1973). The efficiency of gas exchange across crab gills is further promoted by the countercurrent flow of water and hemolymph that occurs at the level of the lamellae (Hughes, Knights, and Scammel, 1969). Rates of oxygen uptake across crustacean gills approximate those that would occur across the gills of fish exhibiting similar levels and patterns of activity (McMahon, 1981). Details of crab gill histology have been reported for several aquatic crabs including *Cancer* (Pearson, 1908) and *Callinectes* (Copeland and Fitzjarrel, 1968).

In both anomuran and brachyuran crabs the evolutionary transition from the aquatic to the terrestrial environment involves progressive reduction in the number of gills present (Fig. 8.1) and in the number of gill lamellae, and hence in total gill surface area (Table 8.1; Pearse, 1929; Gray, 1957; Crane, 1975; Hawkins and Jones, 1982; Greenaway, 1984a). Although Greenaway (1984a; Innes and Taylor, 1986e) cautions against too rigorous a comparison of these data, which are rarely mass-specific, these data, in combination with those of Table 8.1, show a general decrease in gill filament number and in gill surface area associated with increase in the capacity for O_2 uptakes from air. In addition to reduction of their number, the respiratory lamellae are also strengthened (sclerotized) (Taylor and Butler, 1978) and often have surface protuberances to prevent collapse of the gill structure

*"Hemolymph" is used here to describe the circulating body fluid of crustaceans to distinguish this fluid from the blood of vertebrates. In the present account, the distinction is made largely on functional grounds, since in vertebrate terms, the circulating fluids of the crab incorporate both hemolymph and interstitial fluid.

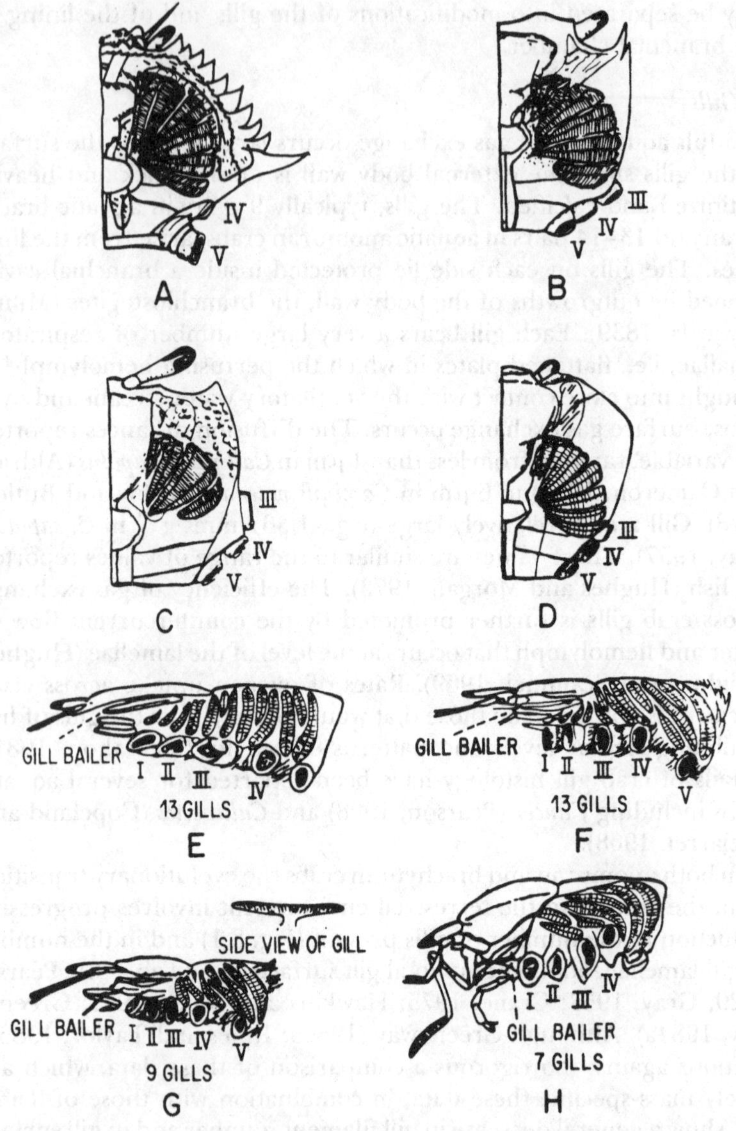

Fig. 8.1. Diagrams of gill and branchial chamber morphology of brachyuran (A-D) and anomuran (E-H) crabs from a series of littoral habitats. Sublittoral: A. *Callinectes marginatus*; E. *Paguristes puncticeps*; F. *Calcinus sulcatus*. Littoral: B. *Pachygrapsus transversus*; G. *Clibanarius tricolor*. Supralittoral to terrestrial: C. *Ocypode albicans*; D. *Gecarcinus lateralis*; H. *Coenobita diogenes*. *Roman numerals* indicate position of limb bases. (Redrawn from Pearse, 1929.)

Table 8.1. *Variation in number of gills and gill lamellae, and in gill surface area in a range of crab species from sublittoral, littoral, and supralittoral habitats*

Species	Habitat	Gill surface area	Filament (no.)	Gill (no.)	Reference
Callinectes sapidus	Sublittoral	1367			Gray (1957)
Ocypode albicans	Supralittoral	325			Gray (1957)
Cardisoma carnifex	Supralittoral	80			Cameron (1981a)
Paguristes oculatus	Littoral		602		Harms (1932)
Paguristes puncticeps	Sublittoral			26	Achituv & Ziskind (1985)
Calcinus sulcatus	Sublittoral			26	Achituv & Ziskind (1985)
Clibanarius tricolor	Littoral			18	Achituv & Ziskind (1985)
Coenobita rugosus	Terrestrial		402		Harms (1932)
Coenobita scaevola	Terrestrial	123		14	Achituv & Ziskind (1985)
Birgus latro	Terrestrial	12			Cameron (1981a)

Note: Gill surface area in $mm^2 \cdot g^{-1}$.

during air exposure (Taylor and Greenaway, 1979; Cameron, 1981a; Taylor and Davies, 1982a). In the case of *Gecarcoidea* spp. the gill spacers have a complex, and almost floral, structure (Cameron, 1981a; C. Farrelly and P. Greenaway, pers. comm.). At least in *G. natalis*, these outgrowths are perfused with hemolymph and may enhance respiratory surface area. In *Ocypode saratan* the gill surface area is clearly increased by many extensions of the surface (A. C. Taylor, pers. comm.), which resemble the respiratory tree adaptations of the gills of some air-breathing fishes (Carter, 1931). Although strengthening of the gills involves addition of extra chitin, the diffusion distance for gas exchange across the gills in the land crab *Holthuisana (Austrothelphusa) transversa* (Taylor and Greenaway, 1979) is similar to that in the shore crab *C. maenas* (Taylor and Butler, 1978). Details of the histology of gills in air-breathing crabs have been reported for *C. maenas* (Taylor and Butler, 1978), *Birgus latro* (Harms, 1932), and *Gecarcinus lateralis* (Copeland, 1968). Fine structure of the gills of land crabs has been studied by Copeland (1968) and Taylor and Greenaway (1979). In many species, including both Anomura and Brachyura, the posterior gills, which often show extensive ion-pumping tissue, show proportionately less reduction than the anterior gills. This suggests that the gills continue to play an important role in ionic uptake and regulation in many land crabs.

B. Branchial chamber lining

Coupled with the loss of gill surface area, the branchial chamber becomes large (Diaz and Rodriguez, 1977), and the lining of the branchial chamber plays an increasingly important role in respiration (Carter, 1931; Edney, 1960; Bliss, 1968, 1979). The development of branchiostegal gas exchangers apparently stems from similar but less well developed structures present in largely aquatic forms such as *Dromia* and *Pagurus* (Wolvekamp and Waterman, 1960). In land crabs the surface topography of the branchiostegal lining may be relatively smooth, especially in those animals with greatly expanded cavities, i.e., *Cardisoma guanhumi, Gecarcinus lateralis, Dilocarcinus dentatus* (Diaz and Rodriguez, 1977). In many other land crabs, however, the branchial cavity is less well expanded, but the surface area lining the branchial cavity has areas that are either extensively perforated (*Eudianella iturbi, Pseudothelphusa garmani garmani*, some other terrestrial pseudothelphusids) or heavily folded (in the very active ocypodid species, e.g., *Ocypode quadrata*) to allow a greatly expanded surface area (Diaz and Rodriguez, 1977; Innes and Taylor, 1968a,b,e). These authors' conclusion (namely, that the wall of the branchial chamber has evolved into an accessory air-breathing organ in a variety of land crab types) is well supported by the physiological evidence presented later in the chapter. C. Farrelly and P. Greenaway (pers. comm.) have carefully studied areas of specialization of the branchial wall in several different land crab species (also see Chapter 9). In gecarcinid and sundathelphusid land crabs the accessory air-breathing organ is formed mostly from the laterally expanded branchiostegal walls. In grapsid land crabs the expansion of the branchiostegites occurs more posteriorly and anterolaterally. A different pattern has been described for the Ocypodidae by Greenaway and Farrelly (1984). In *O. cordimanus*, which seems typical of the family, accessory air-breathing area is further enhanced by additional vascularization of the branchiostegal shelf and thoracic wall as well as by corrugation of the "lung" lining. (Though obviously not homologous to the true vertebrate lung, the use of "lung" to describe the aerial-exchange organ of the expanded branchial cavities has become common.) In Mictyrididae the structure of the air-breathing organ is more spongy, perhaps similar to that described later for *P. garmani*. Details of the circulatory supply to these various air-breathing organs are described in Chapter 9, section

2. The fine structure of the branchiostegal wall has been examined for several air-breathing forms (Diaz and Rodriguez, 1977; Storch and Welsch, 1975, 1984; Taylor and Greenaway, 1979; Innes and Taylor, 1986e). The epidermal and chitinous covering of the epithelium shows marked thinning, providing a very short diffusion path for gases (<1 μm). In the Pseudothelphusidae and some Ocypodidae, development of special "respiratory" areas with many perforations of complex profile is seen in the branchial cavity lining. In the Trinidadian mountain crab *P. garmani* (Innes and Taylor, 1986a,b,e), larger specimens of another pseudothelphusid crab *Eudianella iturbi* (Diaz and Rodriguez, 1977), and *Mictyris longicarpus* (Farrelly and Greenaway, in press), these perforations apparently ramify extensively beneath the surface to form an elaborate spongelike structure that is thought to greatly increase the surface area available for gas exchange (Innes, Taylor, and El Haj, 1987). Modifications to the circulatory system that allow increased perfusion of these branchiostegal wall gas-exchange organs have recently been reported for several terrestrial crabs and are described in detail in Chapter 9.

Similar trends are seen in the anomuran land crabs. Pearse (1929) and Harms (1932) describe progressive reduction of gill area with increase in terrestriality in hermit crabs (Fig. 8.1). The most extensive complex accessory air-breathing structure described for land crabs (Harms, 1932) is reported for the coconut or robber crab, *Birgus latro*. This anomuran land crab differs slightly from the brachyuran pattern described earlier in that it combines a considerably expanded branchial chamber (termed a "lung" by Harms, 1932) with an exceptionally complex branchial lining. The surface of this is very intricately folded to form a respiratory treelike structure almost filling the central cavity (Fig. 8.2). This species also shows the greatest reduction in gill surface area of any land crab (Cameron, 1981a).

Electron micrography of the branchiostegal membranes of several land crabs including *H. transversa* (Taylor and Greenaway, 1979), *B. latro* (Storch and Welsch, 1979, 1984), and *P. garmani* (El Haj, Innes, and Taylor, 1986; Innes, Taylor, and El Haj, 1987; Innes and Taylor, 1986e) confirms that considerable reduction has occurred in both chitin and cytoplasmic layers. This reduces the diffusion barrier between respiratory gas and hemolymph to less than 1 μm, a distance similar to that found in mammalian lung.

The degree of proliferation of lung and reduction of gill area differs between land crab groups and perhaps also species. Crabs of more amphibious habits, such as *Cardisoma* and *Holthusiana*, have a lung surface area equal to only 10% of gill surface area. Measurements of both gill and lung areas are not available for other species, but the

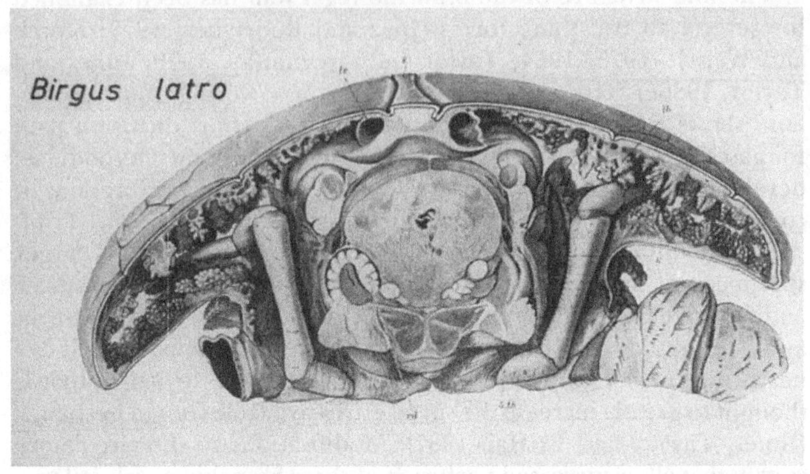

Fig. 8.2. Sagittal section through adult *Birgus latro* showing details of internal anatomy of accessory air-breathing organ ("lung"). *k*, gills in gill pouch; *lb*, lobulated respiratory tissue (treelets); *s*, dorsal (hourglass) sulcus; *lv*, vessels returning hemolymph to pericardium from "lung"; *vd*, vas deferens. *5.th*, fifth pereiopod carried in "lung" chamber. (After Harms, 1932.)

substantial gill reduction and evident "lung" proliferation seen in such animals as *P. garmani* and *B. latro* argue for a reversal of this ratio in more terrestrial species.

C. Other gas exchange surfaces

Although the majority of land crabs rely on modification of the branchiostegal membrane for aerial O_2 uptake, other structures may be involved. For example, a unique modification is reported by Maitland (1986) for the Scopimerinae, a group of ocypodid air-breathing crabs. The gill cavity is not expanded in these air-breathing crabs, but this group has instead evolved complex gas exchange organs on the meral segments of the walking legs (*Scopimera* = "thighs with windows"). In *Scopimera inflata*, an Australian sand-bubbler crab, the cuticle and epidermal layers at the window site are greatly reduced and transparent. Beneath the thinned cuticle lies a complex series of hemolymph spaces that bring deoxygenated hemolymph into very close association with the air. Maitland (1986) reports that the gas diffusion distance across the "windows" is less than 1 μm, and the area of the combined "gas windows" is greater than that of the conventional branchiostegal gas exchanger of a similar-sized *H. transversa*. These crabs are intertidal but use the gas exchange "windows" for air breathing both from air trapped at the burrow at high tide and when active in air at low tide.

These gas exchange organs apparently contribute substantially to gas exchange in air since O_2 uptake decreases up to 60% when the windows are covered with gas-impermeable paint (Maitland, 1986). Other members of this family (e.g., *Dotilla*) have similar modifications – but to the ventral thoracic sternal plates rather than the limbs.

3. Ventilatory mechanisms

A. The scaphognathite pump

In all adult crabs the gills lie protected inside the paired branchial chambers. This internal location has dictated the evolution of efficient pumps to move the respiratory medium past the gills (ventilation) and thus enhance gas exchange. In crabs the ventilatory pumps are a pair of flattened bladelike structures, the scaphognathites, formed from the expodites of the second maxilla (Milne-Edwards, 1839). The scaphognathites oscillate in a narrow channel located at the anterior of each branchial chamber, normally pumping water out anteriorly and thus causing a compensatory flow of water to enter the branchial chamber and flow across the gills. In most brachyuran crabs, this current of water normally enters the branchial chamber via the limb bases, particularly via the large apertures (Milne-Edwards openings) associated with the chelae and passes into the hypobranchial space behind the gills before passing between the gill lamellae where gas exchange occurs (Hughes et al., 1969). An interesting feature of the scaphognathite pumps is that they can be reversed, within a single beat, by a slight change in the output of the ventral ventilatory pattern generators (Young, 1975; McMahon and Wilkens, 1983). This change allows reversal of the direction of water flow across the gills (McDonald et al., 1977). Similar structures are involved in ventilation in the anomuran crabs, and although detailed measurements have not been made, they are thought to operate in an essentially similar manner.

Since most aerial gas exchange in land crabs apparently also occurs in the branchial cavities (see section 2B), it is important to ascertain whether the scaphognathite pumps can adequately ventilate the branchial chambers with air. In fact, the edges of the scaphognathite blades are extremely flexible and in terrestrial, and even some intertidal crabs, provide a sufficiently tight seal that aerial ventilation occurs readily (Cameron and Mecklenburg, 1973; Cameron, 1975; Taylor and Butler, 1978; McMahon and Burggren, 1979; Burggren, Pinder et al., 1985). Although the high O_2 capacity of air usually allows a lower ventilation rate (Table 8.2), the maximum ventilatory performance in air appears equivalent to that in water. For example, the scaphognathites of *B. latro* can pump nearly a liter of gas per kilogram body mass per minute (Cameron and Mecklenburg, 1975), which is

Table 8.2. *Respiratory and circulatory performance in an aquatic crab* Callinectes sapidus *and in several littoral and supralittoral crabs showing increasing dependence on aerial respiration*

Species	Air/water	Mass	Temp.	$\dot{M}O_2$	$\dot{M}CO_2$	F_w	V_w or V_a	% Ex	F_h	TO_2	V_b	Reference
ANOMURA												
Pagurus hirsutiusculus	Water	2.4 ± 0.9	5	2.6								Burggren & McMahon (1981a)
Pagurus hirsutiusculus	Air	2.4 ± 0.9	5	3.8–7.5								Burggren & McMahon (1981a)
Coenobita clypeatus	Air	23 ± 6	20	1.24								Burggren & McMahon (1981a)
Coenobita rugosus	Air	1.01 ± 0.8	20	1.35								Burggren & McMahon (1981a)
Coenobita scaevola	Water	5–12	25	6.57								Achituv & Ziskind (1985)
Coenobita scaevola	Air	5–12	25	4.4								Achituv & Ziskind (1985)
Birgus latro	Air	1285	27–30	1.46		110	57	5.2		0.28		Cameron & Mecklenburg (1973)
BRACHYURA												
Callinectes sapidus	Air	150	15–17	1.32			180	1.8				O'Mahoney (1977)
Callinectes sapidus	Water	150	15–17	3.70			633	39				O'Mahoney (1977)
Carcinus maenas	Water	49 ± 4	15	1.2					92		118	Taylor & Butler (1978)
Carcinus maenas	Air	49 ± 4	15	2.58					86		203	Taylor & Butler (1978)
Eurytium albidigitum	Water	22–33	25	1.14								Burnett & McMahon (1986)
Eurytium albidigitum	Air	22–33	25	0.54								Burnett & McMahon (1986)
Hemigrapsus nudus	Water	5.1	17.5	5.7	3.7							Barnhart & McMahon (unpub.)
Hemigrapsus nudus	Air	5.1	17.5	4.4	3.5							Barnhart & McMahon (unpub.)
Pachygrapsus crassipes	Water	16–26	25	1.86								Burnett & McMahon (1987)
Pachygrapsus crassipes	Air	16–26	25	0.84								Burnett & McMahon (1987)
Ocypode quadrata	Air	44	25	2.25							70	Burnett (1979)
Cardisoma guanhumi	Air	140	24–26	2.6		45	140	5.2				Cameron (1975)
Cardisoma guanhumi	Water	128	24–26	1.98			369	30				O'Mahoney (1977)
Cardisoma guanhumi	Air	128	24–26	2.28			144	3.9				O'Mahoney (1977)
Cardisoma carnifex	Air/rest		25	2.29	1.42	28	122	.11	75	0.4		Wood & Randall (1981a, b)
Cardisoma carnifex	Air/activ.	250–500	25	6.21	7.36	130	230		90	0.81		Wood & Randall (1981a, b)
Holthuisana transversa	Water	20–30	25	1.65			260	46.				Greenaway, Bonaventura et al. (1983)
Holthuisana transversa	Air	20–30	25	0.80			12.5					Greenaway, Taylor et al. (1983)
Pseudothelphusa garmani	Air/rest	10	25	1.51								Innes et al. (1986)
Pseudothelphusa garmani	Air/dist.	10	25	3.76								Innes et al. (1986)
Pseudothelphusa garmani	Estivat.	10	25	0.86								Innes et al. (1986)
Gecarcinus lateralis	Air	30	25	2.06		161	200	2.0	97	.33	137	Taylor & Davies (1981)
Gecarcinus lateralis	Water	30	25	1.08		264	417	85.	54	.143	70	Taylor & Davies (1982)
Gecarcinus lateralis	Air	32	24–26	4.20		302	780	2.3				Cameron (1975)

Note: Mass in g; temperature in °C; $\dot{M}O_2$ and $\dot{M}CO_2$ in $\mu mol \cdot g^{-1} \cdot h^{-1}$; V_w and V_a in $ml \cdot kg^{-1} \cdot min^{-1}$; F_w and F_h in $bts \cdot min^{-1}$; TO_2 in $\mu mol \cdot kg^{-1} \cdot mmHg^{-1}$ ΔP_G: % Ex, percent extraction of 0_2 from water/air.

a volume similar to the maximum water flow generated by a l-kg *Cancer magister* (McMahon, McDonald and Wood, 1979).

Air flow through the branchial cavities has been recorded for a few species of air-breathing crab (Table 8.2). Crabs in air, without access to water, show largely forward-directed pumping of the scaphognathites, drawing air in through the Milne-Edwards openings and pumping it out anteriorly (Taylor and Davies, 1982a; Wood and Randall, 1981a; Burggren, Pinder, et al., 1985). Scaphognathite pumping in air is often intermittent, especially in undisturbed animals. As in aquatic animals (McDonald, 1977), periods of unilateral ventilation, where only one scaphognathite is active, are seen in some air-breathing crabs. Unilateral ventilation appears to be the predominant pattern in resting *Cardisoma carnifex* (Wood and Randall, 1981a) and *C. guanhumi* (A. Smits, pers. comm.), but reverts to predominantly bilateral pumping (both scaphognathites active) during activity or stress (Burggren, Pinder, et al., 1985). Reversed scaphognathite beating (and thus reversed flow of air through the branchial cavities) occurs less frequently in crabs breathing only air but has been noted in the intertidal *C. maenas* (Taylor and Butler, 1978) as well as in land crabs *G. lateralis* (Taylor and Davies, 1982a) and *C. guanhumi* (Burggren, Pinder et al., 1985).

In many land crabs the scaphognathites can effectively pump water when the crab is submerged (Taylor and Davies, 1982b; O'Mahoney and Full, 1984; Table 8.2). In situations where both water and air are available special patterns of scaphognathite pumping are seen that allow ventilation of both media, often simultaneously. Such a pattern is shown in Figure 8.3 for *C. guanhumi* resting in a few centimeters of water, which is sufficient to allow the animal to submerge the Milne-Edwards openings. While the animal has access to water, pressure recordings from the branchial cavities reveal an alternating pattern of forward and reversed scaphognathite pumping (Fig. 8.3; Burggren, Pinder, et al., 1985). However, in the record shown in Figure 8.3, the water has been removed and the animal switches to the pattern of primarily forward scaphognathite pumping described earlier for purely aerial ventilation. Patterns similar to those of Figure 8.3 were recorded by B. McMahon and W. Burggren (unpub.) for the related *C. carnifex* similarly resting in shallow water. In the former study the authors also measured air flow and noted that the alternation of periods of normal and reversed scaphognathite pumping were associated not only with the exhalation of gas from either the exhalant openings or from the Milne-Edwards openings, but were also associated with movement of water into and out of the ventral area of the branchial cavity, which includes the gills. Apparently, periods of forward pumping, occurring when the Milne-Edwards openings are sub-

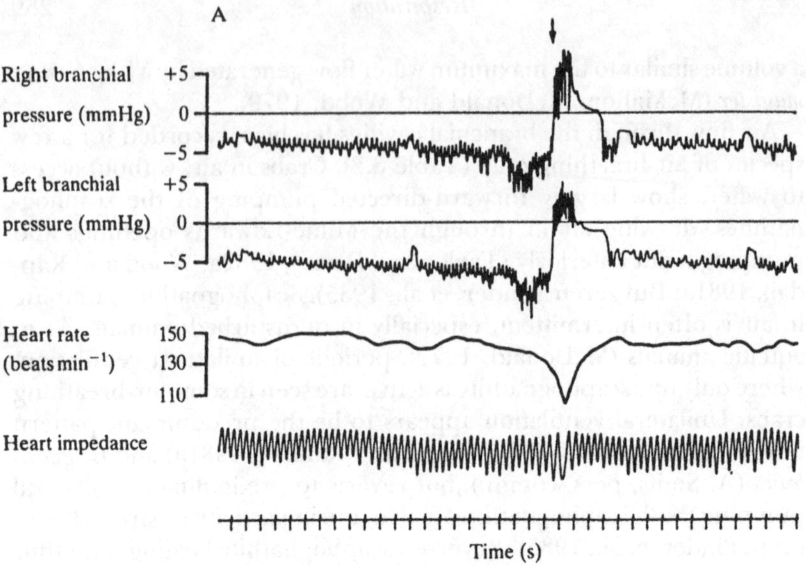

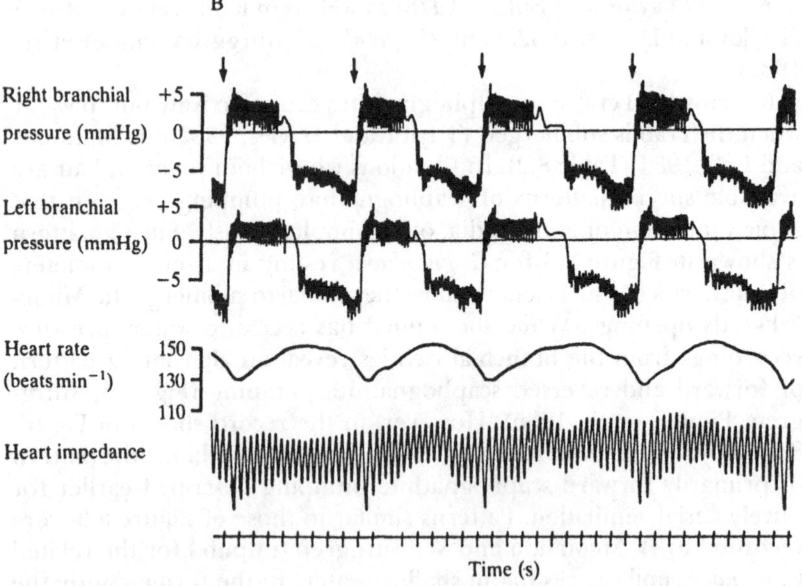

Fig. 8.3. Scaphognathite activity as indicated from pressure recordings from both branchial chambers and heart rate (impedance recording) from *Cardisoma guanhumi*. A. In air without access to standing water. B. After transfer to shallow water, sufficient only to cover the inhalent branchial (Milne-Edwards) openings. Note the change in frequency and patterning of the scaphognathite pumps when water is available. *Arrows* indicate initiation of periods of reversed direction scaphognathite pumping. (From Burggren, Pinder et al. 1985)

merged, allow exhalation of branchial chamber air together with suction of water into the chamber from below. The periods of reversed pumping conversely allow inhalation of fresh air and force water to exit the cavity via the Milne-Edwards openings. The system thus allows simultaneous ventilation of the upper branchial chamber with air and of the lower branchial chamber (and gills) with water. A series of postural adjustments accompany these ventilatory cycles (Burggren, Pinder, et al., 1985; B. McMahon and C. M. Wood, unpub.), but their role is not completely understood. This pattern of simultaneous water and air ventilation would clearly be adaptive in *Cardisoma*, which spends most of the time in a burrow that penetrates to the water table (Chapters 3 and 7), provided both aerial and aquatic phases in the burrow have appropriate gas partial pressures. This pattern of ventilation has not yet been recorded for other genera, but periods of reversed air ventilation have been noted during air exposure in the intertidal portunid crab *C. maenas* (Taylor and Butler, 1978).

The pressure changes involved in water acquisition in *Cardisoma* are relatively small (Fig. 8.3), but the scaphognathite pump can contribute to the generation of much higher forces during water uptake from interstitial water sources. Pressures as low as 40 mmHg subambient have been recorded in *O. quadrata* during suction of interstitial water from the substrate and into the branchial cavity (Wolcott, 1976, and see Chapters 3 and 7).

B. *Other ventilatory mechanisms*

Despite their obvious utility in air, the scaphognathites are not the only ventilatory pump used in air-breathing crabs. Much of the aerial ventilation in the Australian land crab *H. transversa* results from lateral movements of the abdominal mass that act like a piston driving air alternately into and out of the branchial chambers (Greenaway and Taylor 1976; Greenaway, Taylor, and Bonaventura, 1983). Similar movements may possibly contribute to ventilation in Pseudothelphusidae (A. J. Innes and E. W. Taylor, unpub.) and in other land crabs. Sudden large pressure changes, presumably associated with contractions of the epimeral musculature, commonly occur in pressure recordings from the branchial cavities of the land hermit crab *Coenobita clypeatus* (McMahon and Burggren, 1979). Harms (1932) described the position of muscles that could lift the carapace in *B. latro* and assumed that these offered some ventilatory function. Cameron and Mecklenberg (1973) recorded pressure fluctuations of larger magnitude than those associated with the scaphognathite pump in *Birgus*, but could prove no ventilatory function. Such maneuvers are probably not limited to terrestrial forms. Rapid movements of the branchiostegites cause considerable branchial water displacement during reversal of scaphognathite beat in the lobster (Wilkens and McMahon, 1972).

C. Ventilatory flow of water and air

Since air contains approximately 40 times more oxygen than water, the air flow through the branchial chambers is usually considerably less than water flow needed for equivalent O_2 consumption (Table 8.2). Extraction of oxygen from the branchial air flow is also routinely lower in land crabs than in water breathers (Table. 8.2). In part, this too reflects the higher oxygen capacity of air. However, high air flows and low oxygen extraction are also typical of disturbed crabs (McMahon and Wilkens, 1983), and since the majority of experiments have been very short term, higher extraction of O_2 from air may occur in more settled conditions (e.g., see Wood and Randall, 1981a).

D. Patterns of respiratory and circulatory flow

The processes of ventilation and circulation are necessarily closely coupled, since the transport of respiratory gases is a critical function of the circulatory system. Countercurrent flow of water and hemolymph is an important aspect of aquatic gas exchange in decapod crustaceans. This feature is necessarily lost in the adoption of the air-breathing habit, which in most land crabs often depends on oxygen uptake from a large, homogeneous store of air. This concept is similar to that of the "uniform pool" of air in the vertebrate lung outlined by Piiper and Scheid (1975). This situation, however, may not be true of all land crabs. Harms's (1932) interpretation of the circulatory system of *Birgus* suggested to McMahon and Burggren (1981) that the overall flows of air and hemolymph (see Fig. 8.11) may have features of a multicapillary system such as occurs in birds. Additional physiological work is urgently needed on this and almost all other areas of the respiratory physiology of this animal.

The low O_2 capacity of water dictates that heart and scaphognathite pumps be closely coupled in water-breathing crabs, since stoppage of flow in either system at the respiratory membranes would rapidly reduce gas exchange. Such close coupling is in fact typical of aquatic vertebrates and occurs similarly in aquatic crabs such as *Cancer productus* (McMahon and Wilkens, 1977). In this crab, cessation of the scaphognathite beat, which occurs frequently during periods of low O_2 demand, is routinely accompanied by cessation of heart beat. Close coupling between heart and ventilatory pumping is not seen in crabs during strictly air ventilation (Fig. 8.3) and would not be expected. The high O_2 capacity of air encouraged the development of intermittent ventilatory patterns, but continued utilization of O_2 from the large supply in the air trapped in the expanded branchial chambers would necessitate continued perfusion of the branchial circulation and thus continual cardiac pumping.

Details of the perfusion of the gills and accessary respiratory organs are presented in Chapter 9.

4. Gas exchange and metabolic rate

Rates of oxygen uptake ($\dot{M}O_2$) and carbon dioxide elimination ($\dot{M}CO_2$) reported for aquatic crabs do not differ significantly from those reported for other aquatic invertebrates under similar conditions. They are, however, extremely variable (Table 8.2). Previous reviews of respiratory physiology of Crustacea discuss how this variability may result from differences in size, temperature, season, physiological state, and a variety of other factors (E.W. Taylor, 1982; McMahon and Wilkens, 1983; Innes and Taylor, 1986e; and especially Wolvekamp and Waterman, 1960).

If comparison is made between primarily air- and primarily water-breathing crabs, $\dot{M}O_2$ and $\dot{M}CO_2$ from the preferred medium are similar. All "land crabs," however, have gills as well as aerial gas exchange surfaces as discussed earlier. The ratio between $\dot{M}O_2$ and $\dot{M}CO_2$ across these two surfaces (the respiratory quotient) probably varies considerably with the degree of dependence on aerial respiratory structures (Table 8.2) and also with behavioral and ecological differences between animals. In primarily water-breathing crabs that are placed in air, the gills collapse. This limits both the surface area of the gills and possibly also bulk perfusion, and $\dot{M}O_2$ falls (see *Callinectes* in Table 8.2; Fig. 8.4). In some amphibious crabs, $\dot{M}O_2$ in water and air may be approximately equal (e.g., *Cyclograpsus, Cardisoma*; Table 8.2), whereas in the primarily aerial forms, and even in intertidal forms that are active in air, aerial $\dot{M}O_2$ can be considerably enhanced (see *Gecarcinus*, Table 8.2; Fig. 8.4). The pattern in *H. transversa* seems at variance with the aforementioned progression, since this animal spends by far the greater proportion of its time exposed to air but nonetheless has a lower $\dot{M}O_2$ in air than water. This may be correlated with the marked reduction in activity in this animal when it retreats into the burrow in dry conditions. Similar reduction in $\dot{M}O_2$ is seen in estivation in the snail *Otala lactea* (Barnhart, 1986) and during estivation in the lungfish *Protopterus* (Burggren, Johansen, and McMahon, 1986).

Carbon dioxide production has been recorded only rarely in land crabs. In several cases where $\dot{M}CO_2$ and $\dot{M}O_2$ have been recorded simultaneously, $\dot{M}CO_2$ is lower than expected, resulting in gas-exchange ratios (R_E, defined as $\dot{M}CO_2/\dot{M}O_2$) approaching 0.5–0.8 rather than the anticipated 0.8–1.0 (Bliss, 1953; Wood and Randall, 1981b;

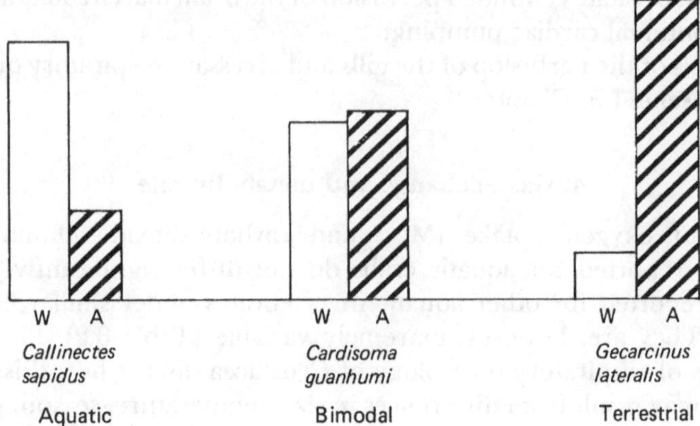

Fig. 8.4. Relative contribution of air- (*A*) and water-breathing (*W*) systems to total oxygen consumption in a range of crabs from sublittoral to supralittoral environments. (From O'Mahoney, 1977.)

Wood and Boutilier, 1985; Wheatly et al., 1986). There are relatively few reliable measurements of $\dot{M}CO_2$ in aquatic crustaceans, but data in Wolvekamp and Waterman (1960) suggest that R_e is higher in aquatic crabs. Very few authors have reported R_E for the same animal in transition from air to water. C. Barnhart and B. McMahon (unpub.) have recently made simultaneous measurements of $\dot{M}O_2$ and $\dot{M}CO_2$ with movement from air to water in the intertidal crab *Hemigrapsus* (Table 8.2). In this animal, similar wide variability in R_E is seen in both media, but standard rates of $\dot{M}O_2$ decreased approximately 25% in air while $\dot{M}CO_2$ decreased less than 10%. R_E thus actually increased significantly on entry into air. Valente (1948) also reported wide variability, but could detect no difference in R_E in air- and water-exposed *Trichodactylus petropolitanus*, a supralittoral brachyuran crab. Unfortunately, no comparable data for other terrestrial crustaceans are known.

Although the very low R_E values reported for land crabs could arise from a portion of the amount of CO_2 excreted being diverted to deposition of carbonates in the shell (Wood and Randall, 1981a; Wood and Boutilier, 1985), it would seem unlikely that this could be a continual activity. Also, a buildup of CO_3^{2-} in the shell also should involve a net production of acidic equivalents that was not observed in Wood and Boutilier's (1985) study. It seems more likely that these R_E values are abnormal, perhaps resulting from an internal buildup of CO_2 (as external PCO_2 rose) in the closed-circuit respirometry systems used for these measurements. This conclusion is supported by preliminary data obtained by flow-through respirometry for the anomuran land crab *Coenobita compressus* (B. McMahon, unpub.). Whereas R_E again varies widely, from 0.6 to 1.6 over a 24-hour period including natural periods of both rest and activity in air, the overall R_E for the 24-hour period was 0.98 ± 0.03 SE, a value similar to those reported for aquatic crustaceans. Partitioning of gas exchange between lungs and gills has not been accurately assessed. Wood and Randall (1981a) measured changes in oxygen and carbon dioxide levels in branchial air and in water trapped in the branchial cavity in the brachyuran *C. carnifex*. These measurements can be used to predict that little O_2 uptake occurs from branchial water, whereas a variable fraction, amounting to as much as 65% of total $\dot{M}CO_2$, could be eliminated by this route. The gills of this land crab may thus still play an important role in CO_2 excretion. This may still be true even of *B. latro* with its highly complex air-breathing structure and very reduced gills. Smatresk and Cameron (1981) show that while ablation of the gills is without obvious effect on CO_2 elimination in resting animals, CO_2 levels are significantly elevated above normal (gills present) following activity. The importance of the gills in CO_2 elimination may be explained by the distribution of the enzyme carbonic anhydrase in gill tissue but not in circulating hemolymph or branchiostegal tissue in crustaceans. Branchial carbonic anhydrase is necessary for efficient CO_2 removal in aquatic forms (McMahon, Burnett, and de Fur, 1984; Burnett and McMahon, 1985) and in the land crab *G. lateralis* (Henry and Cameron, 1983). Since this enzyme is apparently not active in branchiostegal wall tissue in *Birgus*, ablation of the gills may interfere with CO_2 elimination when CO_2 production is high, as in activity.

Rates of $\dot{M}O_2$ and $\dot{M}CO_2$ across the branchial gas exchangers of land crabs (Table 8.2) are apparently similar to those achieved by similar structures in other invertebrate groups.

5. Ventilatory control mechanisms

Respiratory control mechanisms of crustaceans generally have been reviewed by Wilkens (1981) and McMahon and Wilkens (1983). In crabs, ventilatory control, at least of the scaphognathite pump, is resident in the metathoracic ganglion. One of the more predictable differences between control patterns of aquatic and aerial animals, particularly vertebrates, is seen in the switch from the minor role of CO_2 in the primarily O_2- driven ventilatory control system of aquatic forms, to the major role of CO_2 in ventilatory control in air breathers. Since degrees of apparent dependence on aerial respiration within the land crabs have been described, it is of interest to see if these are reflected in equivalent changes in respiratory chemosensory control mechanisms. In aquatic crabs, the presence of a basically oxygen-powered chemosensory drive is demonstrated by the marked increase in scaphognathite pumping seen in response to hypoxic water, and the contrast with a usually weak response to hypercapnic water (see E. W. Taylor, 1982; McMahon and Wilkens, 1983, for reviews). Responses for several land crabs have also been determined. The range of responses seen is predictable, but only with a detailed knowledge of the physiology of these animals shown earlier. Those crabs that are more dependent on air as a primary respiratory medium (*B. latro, G. lateralis, Holthuisana transversa*) show increased CO_2 sensitivity but retain some sensitivity to O_2 (Cameron and Mecklenberg, 1973; Cameron, 1975; Greenaway, Bonaventura et al., 1983; Greenaway, Taylor et al., 1983). The land hermit crab *C. clypeatus* (McMahon and Burggren, 1979) is of particular interest because this apparently very terrestrial animal retains high O_2 and low CO_2 sensitivity. As discussed in more detail later, however, this animal may respire using the water available at most times in the shell and thus might retain respiratory control patterns more typical of aquatic animals. Amphibious animals also tend to show intermediate patterns. Thus, in land crabs we see transitional stages that demonstrate that the importance of the CO_2 drive seems to increase slowly along with greater dependence of the animal on accessory air-breathing organs for gas exchange.

6. Oxygen transport by hemolymph

In decapod crustaceans, as in the majority of complex Metazoa, both oxygen uptake and oxygen transport to tissues are facilitated by oxygen carrier molecules (respiratory pigments). In crabs, whether aquatic or terrestrial, the oxygen carrier is hemocyanin (Hcy). Oxygen binding by the Hcy molecule utilizes change in the oxidation state of copper, rather than of iron as occurs in other blood oxygen carriers. Crab hemocyanin is an aggregate of 6–12 subunits, each with a molecular weight of approximately 70,000 daltons. Oxygen can be bound by individual subunits, but many subunit types may occur and many features of oxygen binding, including oxygen affinity and cooperativity, are affected by subunit composition. Each crab genus thus has at least one specific individual Hcy molecule. Due to the great differences observed earlier between air and water as respiratory sources, we should now examine the functioning of Hcy in aquatic and land crabs to ascertain the extent to which oxygen binding has become modified for uptake and transport in the terrestrial environment.

The functioning of the hemocyanin molecule is best understood, and also best compared, using the classic oxygen equilibrium or oxygen dissociation curve (Fig. 8.5A), which directly portrays the oxygen-carrier function of Hcy by relating the amount of O_2 (Co_2, *Y*-

axis) bound to (i.e., taken up from the environment) or released from the carrier (delivered to the tissues) to the partial pressure of oxygen (Po_2, *x*-axis) available to power its inward diffusion. This relationship is normally sigmoid and allows us to identify two functional zones: the loading zone, which occurs at the high oxygen levels associated with the conditions in the gill/lung, and the unloading zone, which occurs at the lower O_2 levels typical of hemolymph passing through metabolizing tissues. There are two major properties by which O_2 binding of Hcy from air- and water-breathing crabs can be compared: by O_2 capacity and by O_2 affinity. Oxygen capacity is the amount of O_2 carried by the hemolymph at air saturation. Oxygen affinity is the avidity with which oxygen is bound to the Hcy and is quantified by the oxygen partial pressure at which the Hcy becomes 50% saturated with oxygen (P_{50}). A high-affinity Hcy thus has a low P_{50}, and vice versa. Since oxygen affinity is affected by many factors, including both temperature and pH, P_{50} must be compared under equivalent conditions. Oxygen equilibrium curves and variation in both O_2 affinity and O_2 capacity are shown diagrammatically in Fig. 8.5A.

A. Oxygen binding by hemocyanin: aquatic crabs

In the majority of aquatic crabs, the concentration of Hcy in hemolymph is low compared with vertebrate respiratory pigments, and often increases the oxygen capacity above that resulting from dissolved oxygen by only two- to threefold (Mangum, 1980, 1983c; McMahon, 1985). Most reported values of circulating oxygen partial pressures (Po_2) measured from aquatic crabs are very high (McMahon and Wilkens, 1983; McMahon, 1985), and in consequence a substantial fraction of tissue O_2 delivery apparently occurs from O_2 in simple solution (Fig. 8.5B). This has led to the interpretation that Hcy is not well adapted for O_2 transport in decapod crustaceans (Mangum, 1980). However, these high circulating Po_2's may be typical only of the somewhat disturbed conditions common in laboratory experiments. Samples taken from undisturbed crabs in aerated water show that these animals use relatively little of their total capacity to take up O_2 when at rest (McDonald, 1977; McMahon and Wilkens, 1977, 1983). Under these settled conditions circulating O_2 pressures fall to levels at which the respiratory carrier plays a substantial role in O_2 transport (Fig. 8.5B). This optimizes both O_2 uptake and O_2 delivery, with consequent savings to the animal in reduced energy costs for operation of the ventilatory and cardiac pumps.

Additionally, oxygen affinity can be either increased or decreased to optimize O_2 delivery to tissues in various situations. For example, O_2 affinity is decreased (Bohr effect) during activity, that is, when O_2 demand is increased (Fig. 8.5B, C) and can be increased during hy-

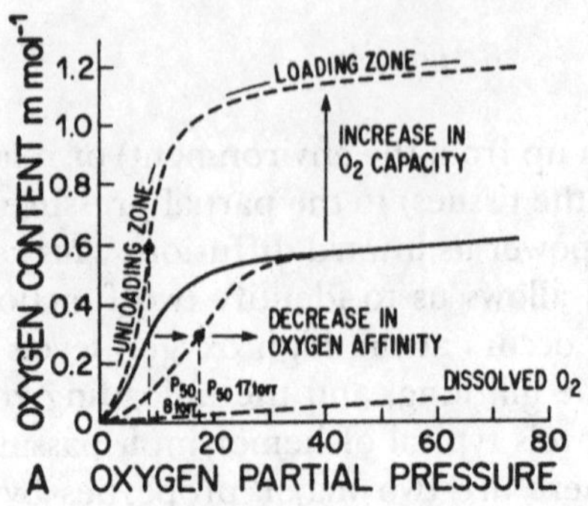

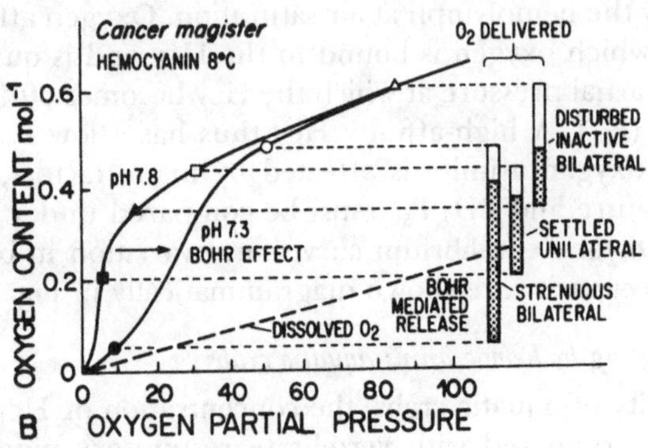

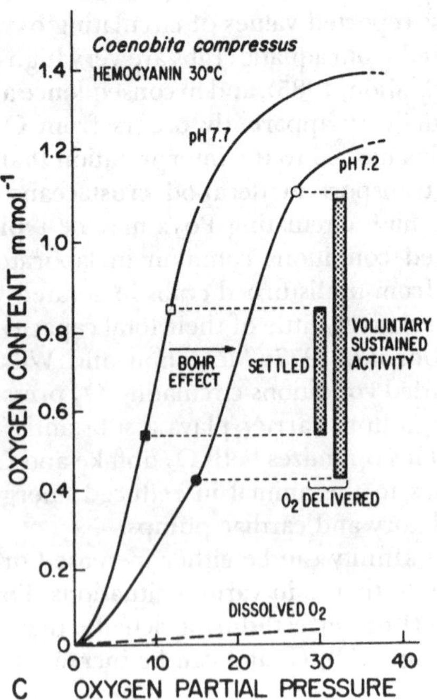

Fig. 8.5.
(Caption on facing page.)

poxia when O_2 supply is diminished (McMahon and Wilkens, 1983; McMahon, 1985, 1986; Bridges and Morris, 1986). These acclimative changes are now known to result almost automatically, since changes in metabolism result in changes in the circulating levels of metabolic end products that themselves act to modify O_2 binding affinity of hemocyanin (McMahon, 1985, 1986). Increases in the hemolymph content of such substances, called modulators, can either increase O_2 affinity (e.g., lactate, urate) or decrease O_2 affinity (e.g., H^+). It is important to emphasize that a single modifier substance (i.e., H^+) may vary O_2 affinity in either direction (i.e., by increase or decrease in concentration). This control of oxygen-binding affinity and hence O_2 uptake and supply to tissues by metabolic end products is a new example of a positive feedback mechanism in physiology. Thus, oxygen-binding characteristics of hemocyanin in aquatic crabs, as exemplified by *C. magister*, are clearly well adapted to function efficiently in gas transport both at rest and in activity.

B. Oxygen binding by hemocyanin: land crabs

The fact that O_2 binding by hemocyanin is so well adjusted for optimum performance in water-breathing crabs leads us to pose two questions: Are the O_2 carrier molecules of the land crabs equally well adjusted to function efficiently in air breathing, and if so, what is the manner by which such modification of oxygen binding might occur? As noted, adjustment could occur in either oxygen capacity or affinity. There is some evidence that change in both aspects may be involved in respiratory adaptation in land crabs, but the situation is complex and each must be considered separately.

Examination of the data compiled in Tables 8.3 and 8.5 reveals that

Fig. 8.5. Oxygen equilibrium curves for hemocyanin from air- and water-breathing crabs at "rest" and during activity. A. Loading and unloading zones and changes in oxygen affinity and oxygen capacity. P_{50} is the oxygen partial pressure for 50% oxygen saturation of hemolymph. B, C. Oxygen transport by hemolymph of an aquatic crab *Cancer magister* (B) and a terrestrial crab *Coenobita compressus* (C). Points (*open symbols* = postbranchial; *closed symbols* = prebranchial) are determined from simultaneous in vivo measures of O_2 pressure and O_2 capacity on hemolymph sampled during the activity stipulated. Histograms show proportion of O_2 delivered to tissues from O_2 carried in solution (*clear areas*) and delivered from hemocyanin-bound oxygen (*stippled areas*). In B and C curves are constructed from "in vitro" data from each species at pH and temperature shown. The pH shown is, in each case, equivalent to that measured during the activity shown. The difference between the curves allows quantification of the Bohr effect. In the case of *C. magister* a moderately large lactate effect has been ignored. In the case of *C. compressus* the magnitude of the lactate effect is negligible. In *C. magister* a 30-min activity period was enforced by disturbance. In *C. compressus* activity (up to 3 h) was voluntary. (Data from McMahon and Wilkens, 1983, and from Wheatly et al., 1986.)

Table 8.3. *Oxygen-binding properties of hemocyanin of crabs from a range of sublittoral to supralittoral habitats*

Species	Air/ water	Temp. (°C)	P_{50}	pH	Bohr factor	O_2 Cap	Reference
ANOMURA							
Pagurus bernhardus	Water	15	110	7.5	−1.55		Jokumsen & Weber (1982)
Coenobita clypeatus	Air	23–25	6	7.84	−0.33	1.84	McMahon & Burggren (1979)
Birgus latro	Air	25–28	20	7.58			McMahon & Burggren (1980)
BRACHYURA							
Callinectes sapidus	Water	20	11	7.5	−1.14	0.66	Booth (1982)
Carcinus maenas	Water	15	7	7.84			Taylor & Butler (1978)
Carcinus maenas	Air	15	10	7.72			Taylor & Butler (1978)
Eurytium albidigitum		25		7.729	−0.45		Burnett (unpub.)
Eurytium albidigitum		35		7.70	−0.30		Burnett (unpub.)
Ocypode quadrata	Air	25	10	8.02			Burnett (1979)
Ocypode saratan	Air	20	19	7.44	−0.67	0.73	Morris & Bridges (1985)
Ocypode saratan	Air	35	20	7.52	−0.67	0.73	Morris & Bridges (1985)
Gecarcinus lateralis	Air	25	17	7.46	−0.43	0.76	Taylor & Davies (1982a)
Pseudothelphusa garmani	Air	25	26–28	P_{CO_2} =6–10	—	—	Innes & Taylor (1986e)

Note: P_{50} in mmHg; O_2 Cap, oxygen capacity in mmol·l^{-1}.

the oxygen capacity of land crab hemolymph tends to be greater than that of a majority of their aquatic relatives. The highest recorded values for O_2 capacity in crabs include the highly "terrestrial" hermit crabs *C. clypeatus* and *C. compressus* with O_2 capacities over 1.6 mmol $O_2 \cdot^{-1}$ (Fig. 8.5C; McMahon and Burggren, 1979; Wheatly et al., 1986). A possible advantage of increased O_2 capacity in the terrestrial habitat is that the higher O_2 capacity allows more O_2 to be picked up and transported per unit hemolymph flow. This in turn could allow reduced hemolymph flow through gas-exchange sites and perhaps reduce respiratory water loss.

The second question, whether adaptive changes in O_2 affinity accompany the transition from an aquatic to an aerial environment, has taxed many authors. The hypothesis formed by early investigators was that the very low O_2 capacitance of water as compared with air would dictate that water-breathing animals should have a carrier of higher oxygen affinity when compared to air-breathing animals. Unfortunately the problem is more complex than this, since many factors other than the O_2 capacity of the respiratory medium must be taken into account. These include activity levels, habitat, and, as seen earlier, degree of development of and dependence on the aerial respiratory surface. Effective comparison of O_2 affinities can perhaps only be made between animals of very close phylogenetic, morphological, and behavioral similarity tested under identical conditions. Since in many parts of the world closely related crustacean genera can be found occupying a range of overlapping habitats from fully aquatic to virtually terrestrial, a comparison of this kind may be legitimately made. In fact, although comparisons have often been made, the question has not been satisfactorily resolved. Two factors may explain this difficulty.

First, in their possession of both lungs and gills, land crabs can potentially take up oxygen from either or both of two completely different gas exchangers. An animal such as the coconut crab *B. latro*, which has evolved a beautifully designed aerial gas exchange organ, probably does not use the gills for oxygen uptake and thus might be expected to possess a hemocyanin of relatively low O_2 affinity. On the other hand, *Coenobita brevimanus*, a very similar and closely related coenobitid crab from the same supralittoral habitat, has no such complex air-breathing structure. *Coenobita* may thus carry out a major proportion of O_2 uptake over the considerably reduced gills, requiring the evolution of a hemocyanin of relatively high O_2 affinity. Both predictions are in fact supported by the data of Fig. 8.6. The point, however, is that animals that are found in the same habitat and that appear similar in structure and even phylogeny may show marked physiological differences. Until we know much more of the biology of a range of land crabs we may not be able to understand the factors that have influenced the evolution of their hemocyanins.

A second possible reason for the difficulty in demonstrating adaptive differences between O_2 binding properties of hemocyanins from aquatic and terrestrial crabs stems from the influence of modulator substances on the Hcy molecule. Although relatively little work has been carried out on modulation of land crab hemocyanins, effects of H^+, Ca^{2+}, and CO_2 are known for some species (Morris and Bridges, 1985, 1986; McMahon, 1986). Since the concentration of many of these modulator substances varies predictably in hemolymph of animals using both water and air breathing (McMahon, 1986), considerable variation in Hcy O_2 affinity could result. Thus, the O_2 affinity of Hcy might vary considerably between individuals depending on the respiratory medium in use at the time of sampling. This could account for much of the variability and confusion in the literature.

Most important, control over the level of a particular modulator substance could allow amphibious species to vary the O_2 affinity of their Hcy from high to low depending on the medium currently available. Empirical verification of this, i.e., of the importance of modulator substances in rapid fine tuning of oxygen affinity to suit the current environmental conditions is, however, technically difficult and has yet to be satisfactorily demonstrated.

Clearly, the oxygen affinity of Hcy of any land crab species in vivo results from several factors. Selection sets oxygen-binding characteristics

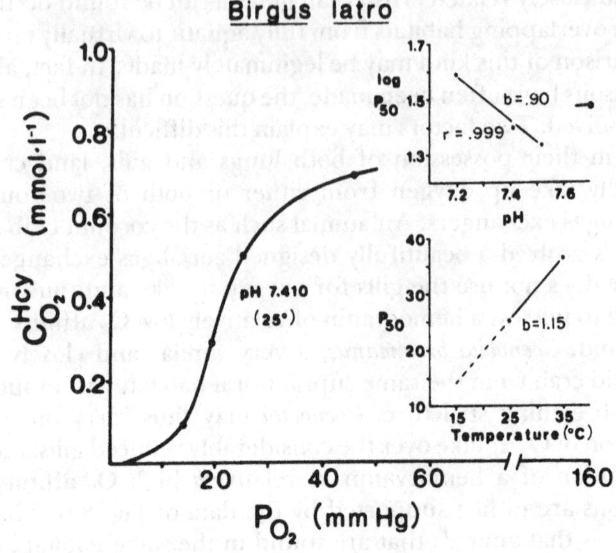

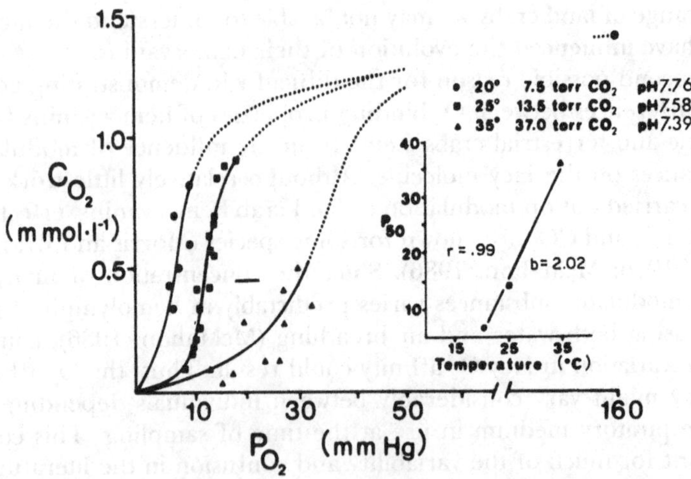

Fig. 8.6. The effect of temperature on hemocyanin oxygen binding in three "land crabs": *Birgus latro, Coenobita brevimanus,* and *Cardisoma carnifex.* The curves in the top panels were constructed in vitro. The curves in the bottom panels were constructed to represent the situation "in vivo" by drawing a curve through plots of oxygen partial pressure and oxygen content determined simultaneously on hemolymph sampled in vivo from animals acclimated to the stated temperature.

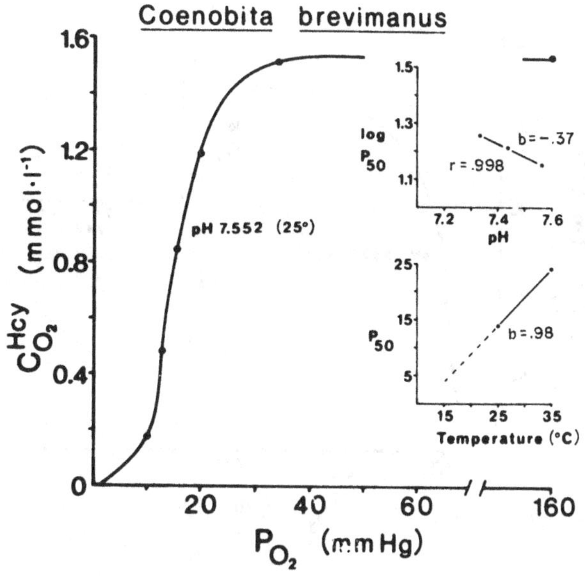

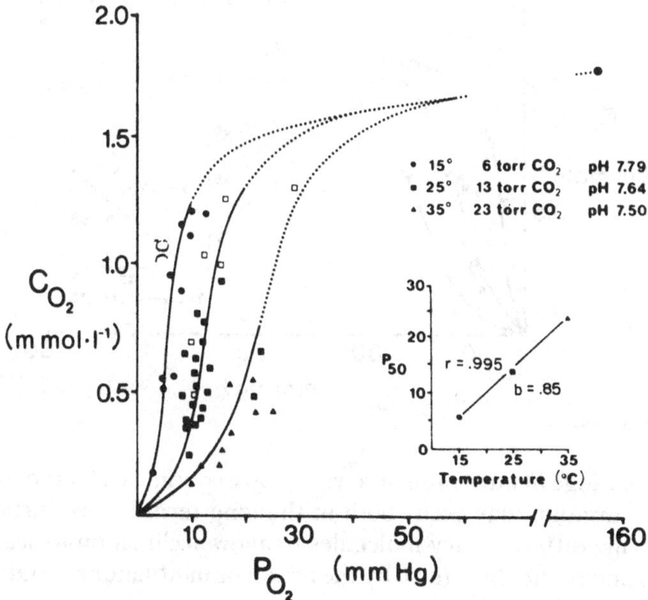

Fig. 8.6 (continued)

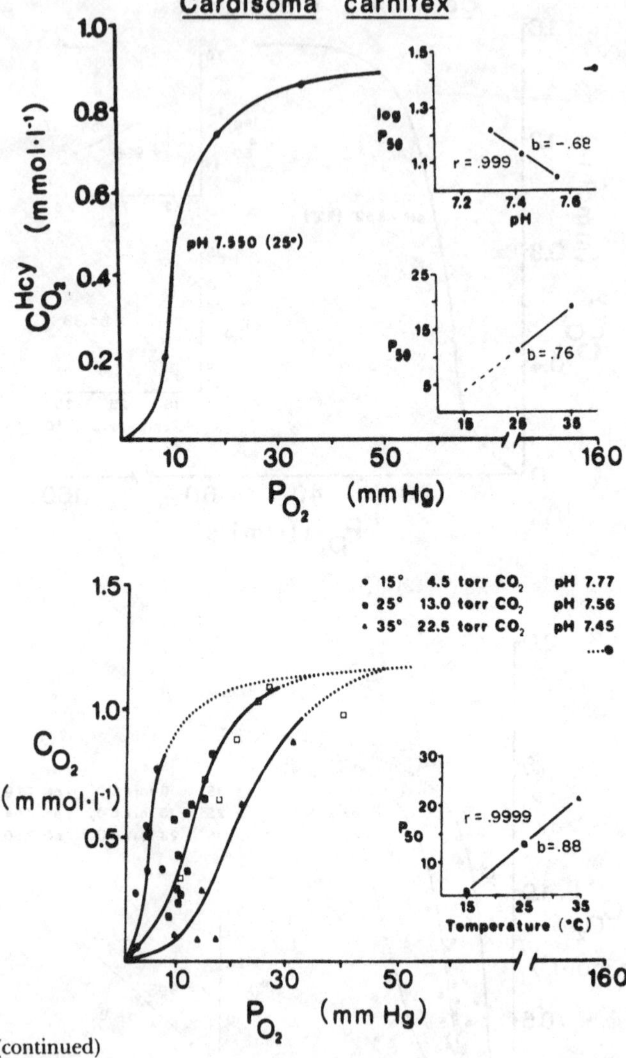

Fig. 8.6 (continued)

to suit the range of environmental variation experienced by the species overall. Variation may occur both in the long term by production of functionally different Hcy molecules to allow acclimation to seasonal changes and in the short term by the action of modulator substances.

7. Carbon dioxide transport by hemolymph

Although the solubility of unbound CO_2 in body fluids is substantially greater than that for oxygen, most CO_2 in crustacean hemolymph is carried in the bound form, either as bicarbonate or possibly as carbamate, i.e., chemically bound to protein (Table 8.4). As for oxygen, CO_2 transport by hemolymph is best demonstrated by CO_2 dissociation curves, which relate changes in hemolymph CO_2 content (amount of CO_2 bound in the tissues) with the CO_2 partial pressure (P_{CO_2}) available to "power" its outward diffusion. Hemolymph dissociation curves for CO_2 for an aquatic crab *Callinectes sapidus* and for the land crab *Coenobita compressus* are reported in Fig. 8.7A and B.

Due to the high solubility of CO_2 in water, adequate CO_2 elimination in aquatic animals has been assumed to occur by simple diffusion of molecular CO_2 across the respiratory surface. This assumption has been questioned for aquatic crabs as a result of recent studies (McMahon et al., 1984; Burnett and McMahon, 1985) that show that the enzyme carbonic anhydrase located in gill tissue is involved in CO_2 elimination across the gills of aquatic crabs. Nonetheless, CO_2 is eliminated readily in aquatic crabs, and P_{CO_2} levels are rarely higher than 3–4 mmHg (Table 8.4).

Much higher circulating CO_2 levels are reported for hemolymph of air-breathing crabs (Table 8.4; see also Fig. 8.10). The increase is due, in part, to the increased O_2 capacitance of air, which allows a relative hypoventilation in air breathers. The buildup of CO_2 in hemolymph and tissues on air exposure is rapid and may cause respiratory acidosis (see the following section), but in many amphibious animals rapid compensation occurs as H^+ is removed by various active and passive buffering mechanisms. Thus, much of the total CO_2 accumulation in the hemolymph occurs as bicarbonate ion.

In addition to its role in facilitating O_2 transport, hemocyanin may also facilitate CO_2 elimination, since binding of O_2 at the lung or gill causes release of protons and hence mobilization of CO_2 from HCO_3^- (Haldane effect). In aquatic crabs a Haldane effect is implicated in CO_2 elimination (Truchot, 1976; Booth, 1982; McMahon, 1985) but is often difficult to quantify since the magnitude of the effect is very small. However, the increased hemocyanin concentration of many terrestrial forms may allow a Haldane effect of correspondingly greater magnitude (Fig. 8.7, Wheatly et al., 1986).

8. Hemolymph acid–base balance

Hemolymph pH (more accurately H^+ concentration) may vary widely in land crabs due to the influence of temperature and other factors in the relatively unstable terrestrial habitat. Nonetheless, regulation of acid–base status (as opposed to pH) is important in land crabs as in all organisms, since this regulation may involve maintenance of the net charge state of the animals' active protein systems and hence

Table 8.4. *Variation in hemolymph acid–base status in air or water for a range of crabs from sublittoral to terrestrial habitats*

Species	Air/water	Temp	pHa	pHv	$P_{CO_2}a$	CO_2v	$C_{CO_2}a$	$C_{CO_2}v$	Reference
ANOMURA									
Coenobita clypeatus	Air	24	7.84	7.66	4.1	6.8	11.7	13	McMahon & Burggren (1979)
Coenobita compressus	Air	30	7.78	7.73	2.5	2.7	5.0	5.3	Wheatly et al. (1986)
Birgus latro	Air	27–30	7.50	7.46	6.2	9.2	14.1	14	McMahon and Burggren (1981)
BRACHYURA									
Callinectes sapidus	Water	20	7.67		2.8		5.7		Booth et al. (1984)
Carcinus maenas	Water	15	7.843		1.9				Taylor & Butler (1978)
Carcinus maenas	Air	15	7.72		3.6				Taylor & Butler (1978)
Cyclograpsus lavauxi	Water	10	7.922		2.31		10.9		Innes, Forester et al. (1986)
Cyclograpsus lavauxi	Air	10	7.931		6.42		10.9		Innes, Forester et al. (1986)
Ocypode quadrata	Air	25	8.02	7.98					Burnett (1979)
Ocypode saratan	Air/water	25	7.837	7.778					A. C. Taylor (pers. comm.)
Cardisoma carnifex	Air/water	25	7.514	7.487	15.4	16.3	19.0	19.5	Wood & Randall (1981b)
Cardisoma carnifex	Water	28	7.66		6.4		10.2		Cameron (1981b)
Cardisoma carnifex	Air/water	28	7.56		10.5		16.3		Cameron (1981b)
Gecarcinus lateralis	Water	25	7.367	7.341	6.47	7.20	6.17	6.43	Taylor & Davies (1981a)
Gecarcinus lateralis	Air	25	7.479	7.414	7.56	9.11	8.7	9.1	Taylor & Davies (1982b)
Gecarcinus ruricola	Air	25	7.487		10.6		14.3		Mileson & Packer (pers. comm.)
Holthuisana transversa	Water	25	7.33	7.36	6.0	6.8	9.8	10.2	Greenaway, Bonaventura et al. (1983)
Holthuisana transversa	Air	25	7.41	7.32	9.6	11.3	13.05	13.17	Greenaway, Taylor et al. (1983)
Pseudothelphusa garmani	Water				3.7–6.9				Innes & Taylor (1986b)
Pseudothelphusa garmani	Air/water				4.6–7.0				Innes & Taylor (1986b)
Pseudothelphusa garmani	Air				8.2–12.5				Innes & Taylor (1986b)

Note: Temperature in °C; PCO_2 in mmHg; CCO_2 in mmol·l^{-1}; a, postbranchial (arterial); v, prebranchial (venous).

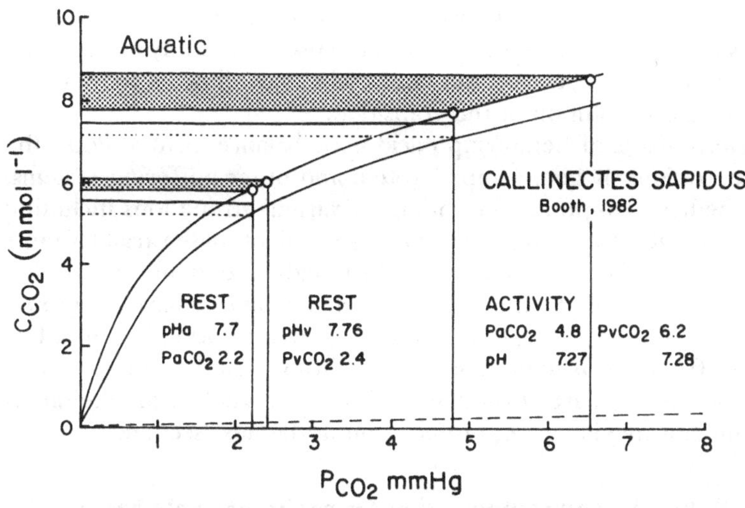

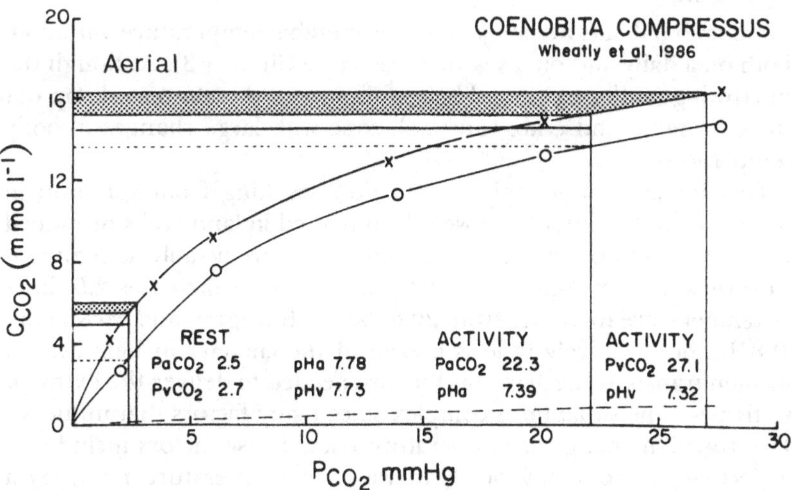

Fig. 8.7. The magnitude and functional significance of the Haldane effect on CO_2 transport in an aquatic crab *Callinectes sapidus* and a land crab *Coenobita compressus*. *Upper curve* is deoxygenated hemolymph; *lower curve*, oxygenated hemolymph in each case.

control of metabolism itself (Reeves, 1977). Regulation of acid–base status is a matter of concern in this chapter, since in aquatic animals, including crabs, this regulation occurs by interaction with the aquatic medium, largely by movement of ions across the gills (Cameron and

Batterton, 1978; McMahon, Sinclair et al., 1978; Cameron, 1985; Wood and Cameron, 1985; Chapter 7). Since with increasing terrestriality animals lose contact with the aquatic medium, these mechanisms become inoperative and must be replaced. A study of the many transitional stages seen in the land crabs may allow us to view the physiological evolution of this transition.

Determinants of hemolymph acid–base balance include P_{CO_2}, the concentrations of hemolymph protein and other buffering systems, and the hemolymph concentrations of various strong ions including Na^+, Cl^-, K^+, Ca^{2+}, Mg^{2+}, etc. Change in all of these variables must be measured if the mechanisms by which acid–base status is controlled are to be understood. Although relatively few studies have attempted this for any land crab species, we have partial information on a few species from which some tentative conclusions can be drawn. These are discussed as part of the mechanisms of physiological adaptation to temperature change and desiccation in the next section.

9. Respiratory compensation for environmental change

A. *Temperature*

Land crab habitats are subject to considerable temperature variation, both on a daily and on a seasonal basis (see Chapter 3). Although the burrowing and/or nocturnal habits of most land crabs may damp out the extremes, land crabs must still cope with large changes in body temperature.

Changes in hemolymph oxygenation resulting from variation in temperature have not been well documented in land crabs or indeed in any crustacean. Oxygen uptake increases predictably as temperature rises. For example, $\dot{M}_{O_2}$ of *Coenobita rugosus* increases 2.5 times as temperature increases from 20 to 30° C (Burggren and McMahon, 1981b), but relatively little is known of the changes in hemolymph oxygen transport mechanisms that are needed to deliver the extra O_2 to tissues. The situation is complex since many factors that influence Hcy function change as temperature rises. These factors include (1) a decrease in solubility of O_2 in water as temperature rises, (2) a decrease in O_2 affinity of hemocyanin due to change in the specific heat of oxygenation, and (3) a decrease in O_2 affinity due to increase in proton concentration. In combination, these factors can have dramatic effects on the ability of the crab's oxygen-transport systems to deliver sufficient O_2 at extremes of a crab's temperature range (Booth, 1982; Mauro and Mangum, 1982; McMahon, 1986).

Solubility of oxygen in hemolymph varies predictably with temperature, but although the magnitude of this contribution to tissue oxygen delivery varies with many factors (including O_2 capacity and the Po_2 difference between arterial and venous hemolymph), temperature-related changes in the amount of O_2 transported in solution are generally insignificant. The magnitude of the change in specific heat of oxygenation also varies between species (Mangum, 1980). In some land crabs hemocyanin appears relatively thermally insensitive, particularly over the preferred temperature range of the animal, but is more variable at extreme temperatures. Examples can be provided both for brachyuran (*Ocypode saratan*; Morris and Bridges, 1985) and anomuran (*Coenobita clypeatus*; Morris and Bridges, 1986) land crabs. Such zones of thermal tolerance would be of clear adaptive importance in the thermally variable environment occupied by many land crabs. The magnitude of the Bohr shift, although variable between species (Table 8.3), is also typically reduced in land crabs as are the effects of the specific modulators lactate and Ca^{2+} (McMahon, 1986; Morris and Bridges, 1986; Wheatly et al., 1986). Thus, although we might predict that the combination of these factors will reduce the effects of temperature change on oxygen binding in land crabs, the blending of these effects is complex and not known accurately for any species. The effects of these in vitro relationships are difficult to predict in vivo and do not allow easy comparison between crabs.

Perhaps the only effective comparison of temperature effects on oxygen binding at present occurs by creating an in vivo O_2 dissociation curve using simultaneous measurements of both oxygen partial pressure and content from crabs acclimated to several temperatures. Unfortunately, few such data are available. Preliminary data, again from *Birgus, Coenobita,* and *Cardisoma* (Fig. 8.6), indicate that the cumulative changes in P_{50} resulting from temperature change in vivo are, indeed, substantially greater than those resulting in vitro (change in specific heat alone). These data also confirm that the overall change in P_{50} observed in land crabs tends to be smaller than similar values calculated for typically aquatic animals (i.e., *Callinectes*; Booth, 1982; McMahon, 1986).

Measurements of circulating O_2 partial pressures and contents of land crabs at different acclimation temperatures are also sparse in the literature. Preliminary data are available for *Birgus, Coenobita,* and *Cardisoma* (McMahon and Burggren, 1980), and for the Australian land crab *Holthuisana* (Greenaway, Burnett, and McMahon, unpub.) In general, circulating oxygen partial pressures increase with increasing temperature in vivo, but O_2 capacity decreases. This is most apparent in venous hemolymph (Fig. 8.8), since in arterial hemolymph these trends are obscured by the oscillations of Co_2 and Po_2 resulting from the intermittent ventilatory patterns typical of these animals (section 3).

Given some decrease in O_2 affinity with increased temperature via

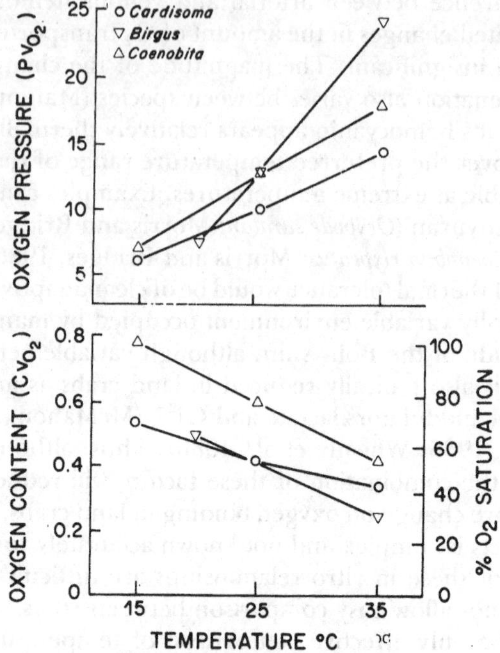

Fig. 8.8. Simultaneous increase in hemolymph oxygen pressure and decrease in hemolymph oxygen content in 3 land crab species as temperature increases.

the mechanisms described here, the increase in Po_2 of arterial hemolymph occurs in an attempt to maintain O_2 saturation of the hemolymph leaving the gills. Increase in Po_2 of venous hemolymph must occur (1) to maintain a Po_2 gradient between hemolymph and tissues sufficient to supply an increased tissue O_2 demand, and (2) to maintain at least a small venous reserve for use in activity, etc. These mechanisms, together with increased ventilation and perfusion, could serve to support increased oxidative metabolism as temperature rises. In fact, the expected increase in metabolic rate may not always occur. In *H. transversa*, for instance, high temperatures occur only in the dry season when the animal is confined largely to the burrow and activity and thus aerial $\dot{M}o_2$ is reduced. At high temperatures, aerial $\dot{M}o_2$ may thus be below that recorded at equivalent temperatures in water (Greenaway and MacMillen, 1978; Greenaway, Taylor, and Bonaventura, 1983).

The reduction in O_2 content of venous hemolymph at higher temperatures, occurring despite an increase in venous O_2 pressure, presumably also results from the decrease in O_2 affinity at a high temperature (Fig. 8.8). A decrease in O_2 content of arterial hemo-

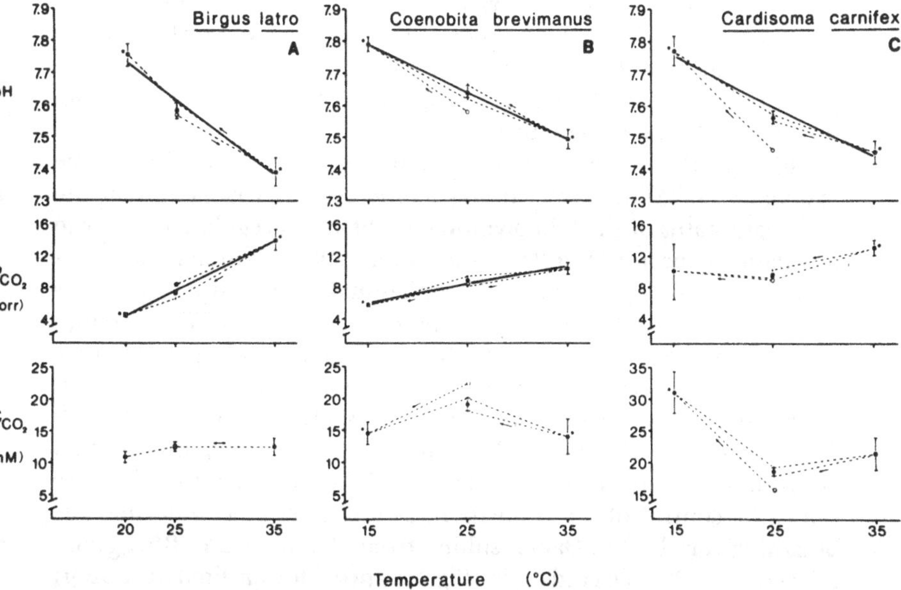

Fig. 8.9. Variation in hemolymph acid–base status after temperature acclimation in three land-crabs. A. *Birgus latro*. B. *Coenobita brevimanus*. C. *Cardisoma carnifex*. All animals were in air with limited access to water. Original acclimation temperature was 25° C. Responses to both 10° C increase, 10° C decrease, and return to 25° C were studied. *Broken lines* connect mean data points at each temperature. *Arrows* show the direction of temperature change. *Solid lines* are regression lines calculated from all data. Bars +/− SE. *Asterisks* denote a significant difference from 25° C. (From McMahon and Burggren, 1981.)

lymph at a high temperature may result from diffusion limitation occurring across the gills.

Changes in temperature of a hemolymph sample in a closed system in vitro result in a change in pH of about 0.018 units per degree temperature change. For the temperature range of 15–35° C (encompassing the normal range for land crabs), hemolymph samples taken in vivo from two anomurans, *Birgus and Coenobita*, and the brachyuran *Cardisoma* acclimated to 15, 25, and 35° C show that the hemolymph pH of these crabs in vivo more or less follows this in vitro relationship (McMahon and Burggren, 1981; Fig. 8.9). However, the magnitude of the change varies both with species and with the temperature range used. It is important to note that change in pH was routinely less for the transition from 25–35° C than for 15–25° C, a range of temperature that rarely occurs in these crabs' tropical habitat (McMahon and Burggren, 1981). This suggests that changes in hemolymph pH (and thus the resulting Bohr effect on O_2 binding) are lower over the animal's normal temperature range as was reported for specific heat of oxygenation. Similarly, no significant depression of pH resulted from a similar 10° C increase (from 20 to 30° C) in acclimation temperature in either *G. lateralis* (Howell et al., 1973) or *G. ruricola* (Mileson and Packer, pers. comm.).

Although the concentration of H^+ in the hemolymph may not be maintained constant over varying temperatures, it is likely that the acid–base status of the hemolymph (and tissue) is regulated, either to maintain a constant H^+/OH^- ratio (Rahn, 1967) or to maintain constant fractional dissociation of the alpha-imidazole groups of proteins and thus the net protein charge state (Reeves, 1972, 1977). Although aquatic crabs regulate hemolymph acid–base status largely by ionic exchange with their aquatic medium (section 8), it seems clear that the mechanisms involved in land crabs must differ, since these animals have limited access to water. It seems reasonable to hypothesize that land crabs regulate hemolymph P_{CO_2}, and thus hemolymph acid–base status, by control of air ventilation, as do many poikilothermic air-breathing vertebrates. Data resulting from McMahon and Burggren's (1981) study of three land crabs (Fig. 8.9) provide only limited support for this hypothesis. However, it seems clear that the highly terrestrial coconut crab *B. latro* is able to regulate hemolymph acid–base status as temperature varies by adjustment of P_{CO_2} at constant HCO_3^-. Both *Coenobita* and *Cardisoma* exhibit similar pH change but show changes in both P_{CO_2} and HCO_3^- (Fig. 8.9). Unlike *Birgus,* both *Cardisoma* and *Coenobita* have water available in or near the branchial apparatus during air exposure. This could allow involvement of ionic-exchange mechanisms in hemolymph acid–base status.

The complexity of the results seen in *C. carnifex* and *C. brevimanus* (Fig. 8.9) may thus reflect a variable degree of dependence on ionic-exchange processes depending both on the availability of water and the degree of activity (and hence O_2 demand). The retention of gills in land crabs can be viewed either as a limitation (i.e., resulting from an inability to fully transfer such functions as acid–base balance to the air-breathing organ) or as an adaptation allowing them the flexibility to adjust the "blend" between two available mechanisms to suit the current demands of their variable amphibious environment.

B. Dehydration

Many of the interpretations of respiration and acid–base balance and how they are affected by temperature assume that the crabs are held in nondehydrating conditions. Respiratory responses to dehydration have been studied in both anomuran and brachyuran crabs. Many land crabs are not particularly resistant to dehydration (see Chapter 7) and may die when they reach 15–25% water loss. Increases in circulating oxygen levels are associated with decreased O_2 consumption during dehydration in *C. carnifex* (Wood, Boutilier, and Randall 1986) and in *P. garmani* (Table 8.5). One possibility is that the decrease in hemolymph volume results in a reduction in the rate of hemolymph movement through the gas exchangers. This in turn would lead to an increased residence time and, as a possible consequence, a more effective O_2 uptake. This could explain an increase in hemolymph–O_2 levels at the same time as reduced Mo_2. Further evidence against linking the fall in metabolism with limitation of O_2 uptake and a switch to anaerobic metabolism is provided by the observation that no increase in lactate or other anaerobic end product occurs. The cause of the reduction in metabolism thus remains unclear.

The maintenance of increased hemolymph CO_2 levels, despite increased residence time for Hcy in the gills, may result from disruption of that fraction of CO_2 elimination normally occurring across the gills by loss of branchial or other water stores.

10. Ecological physiology of land crabs and the evolution of air breathing

Although land crabs share a number of respiratory adaptations, individual species differ both in the degree of reliance on air as a respiratory medium and in the manner and degree of modification for air breathing. Since we have data on the respiratory physiology of a wide range of land crabs, it is of interest to integrate this physiological information into an ecological framework in order to assess the adaptive significance of the various levels of respiratory modification. This assessment may, in turn, allow us to shed some light on the physiological evolution of terrestriality in land crabs and, perhaps, to comment on the evolution of terrestriality generally. In order to avoid a multiplicity of "groupings" or "classifications" of land crabs, the ecological framework used here is based on the scheme of land crab organization presented by Hartnoll in Chapter 2, which was, in turn, based largely on ecological and behavioral criteria.

Hartnoll's first category (T_1) incorporated a range of littoral animals that, although basically aquatic and normally active in water, occasionally showed activity in air. Incorporation of the physiological evidence (Tables 8.1–8.5) would suggest further division into two subgroups based on the animal's ability to utilize the air as a respiratory medium.

The first subgroup would encompass such animals as *Callinectes bellicosus*, which is essentially aquatic but has been observed to make *brief* forays onto land to feed on fiddler crabs (L. E. Burnett, pers.

Table 8.5. *Hemolymph oxygenation in air- and water-breathing crabs from sublittoral to supralittoral habitats*

Species	Air/water	Temp.	PaO_2	PvO_2	CaO_2	CvO_2	Cap	pH	References
ANOMURA									
Coenobita clypeatus	Air	24	14	8	1.34	0.49	1.84	7.84	
Birgus latro	Air	27–30	27	13	0.85	0.45		7.58	McMahon & Wilkens (1983)
BRACHYURA									
Callinectes sapidus	Water	20	64	12	0.32	0.18		7.674	McMahon & Wilkens (1983)
Carcinus maenas	Water	26	75	10	0.43	0.11		7.843	Taylor & Butler (1978)
Carcinus maenas	Air	26	19	7	0.25	0.031		7.72	Taylor & Butler (1978)
Ocypode quadrata	Air	25	20	9				8.02	Burnett (1979)
Ocypode saratan	Air/water	25	72.5	19.9			1.64	7.837	A. C. Taylor (pers. comm.)
Cardisoma carnifex	Air/water	24–26	32					7.485	Wood & Boutilier (1985)
Gecarcinus lateralis	Air	26–27	32	9	0.65	0.27		7.45	Redmond (1968)
Gecarcinus lateralis	Air	25	81	26	0.63	0.38	0.76	7.45	Taylor & Davies (1981)
Gecarcinus lateralis	Water	25	17	9	0.30	0.05			Taylor & Davies (1982)
Pseudothelphusa garmani	Water		22–30						Innes & Taylor (1986)
Pseudothelphusa garmani	Air/water		87–98						Innes & Taylor (1986)
Pseudothelphusa garmani	Air		124–140						Innes & Taylor (1986)
Mictyris longicarpus	Air/water	25	79–95	20–27					Farrelly & Greenaway (in press)

Note: Temp. in °C; PO_2 in mmHg; CO_2 in mmol·l^{-1}; Cap = O_2 capacity at air saturation.

comm.) and the xanthid crab *Eurytium albidigitum*, whose burrows are exposed by the receding tide but which remains largely inactive in the burrows during air exposure. It would also include smaller individuals of the red rock crab *Cancer productus*, which are commonly found intertidally. Although all three species can survive for a few hours in air, none have much capability for aerial oxygen uptake (Table 8.2). In *E. albidigitum*, for instance, $\dot{M}O_2$ is relatively low even in water and is further reduced in air exposure (Table 8.2; Burnett and McMahon, 1987). Additionally, *E. albidigitum* shows a marked increase in P_{CO_2} during air exposure, indicating reduced ability to remove metabolic CO_2. Unlike many forms discussed in this section, *E. albidigitum* is not able to compensate for the respiratory acidosis resulting from both this CO_2 buildup and the possible accumulation of anaerobic products and thus suffers progressive acidosis throughout air exposure (Fig. 8.10). This crab clearly has little ability to carry out aerial gas exchange and, in fact, seldom leaves the burrow, where water is always available. Similar responses are seen in both *Callinectes sapidus* and in *C. productus* (O'Mahoney 1977; deFur, McMahon, and Booth, 1983). Although exposed at low tide, the latter species is normally found buried in the substrate with access to interstitial water. However, some individuals are stranded on hard surfaces and thus exposed to air. Under these conditions they cannot take up sufficient oxygen and rely on anaerobic metabolism (deFur et al., 1983). This condition may be typical of the many intertidal animals that merely survive, rather than exploit, the period of intertidal air exposure.

The second subcategory of intertidal crabs again comprises basically aquatic animals that voluntarily enter air for brief periods, even at high tide. They differ physiologically from the first subcategory in that they can at least maintain O_2 consumption in air. Several grapsid crabs fit into this grouping and, depending on their position in the littoral, may show more or less air-breathing ability. *Hemigrapsus nudus* (a grapsid of the north temperate Pacific intertidal) and *Pachygrapsus crassipes* (from south temperate Pacific shores) leave water voluntarily for short periods but remain within a few centimeters of water. Both crabs show some air-breathing ability, with *H. nudus* able to maintain oxygen consumption during short periods in air (Table 8.2; Burnett and McMahon, 1987). Carbon dioxide elimination ($\dot{M}CO_2$) is not significantly depressed in *H. nudus* in humid air (C. Barnhart and B. McMahon, unpub.), but especially in drier air some retention of CO_2 in hemolymph is seen in this species (Burnett and McMahon, 1987). Both *P. crassipes* and *H. nudus* are, however, better able both to eliminate CO_2 and to compensate for any residual respiratory acidosis, and as a result show only a minor decrease in pH even following several hours in air (Fig. 8.10). Carbon dioxide pressures and contents

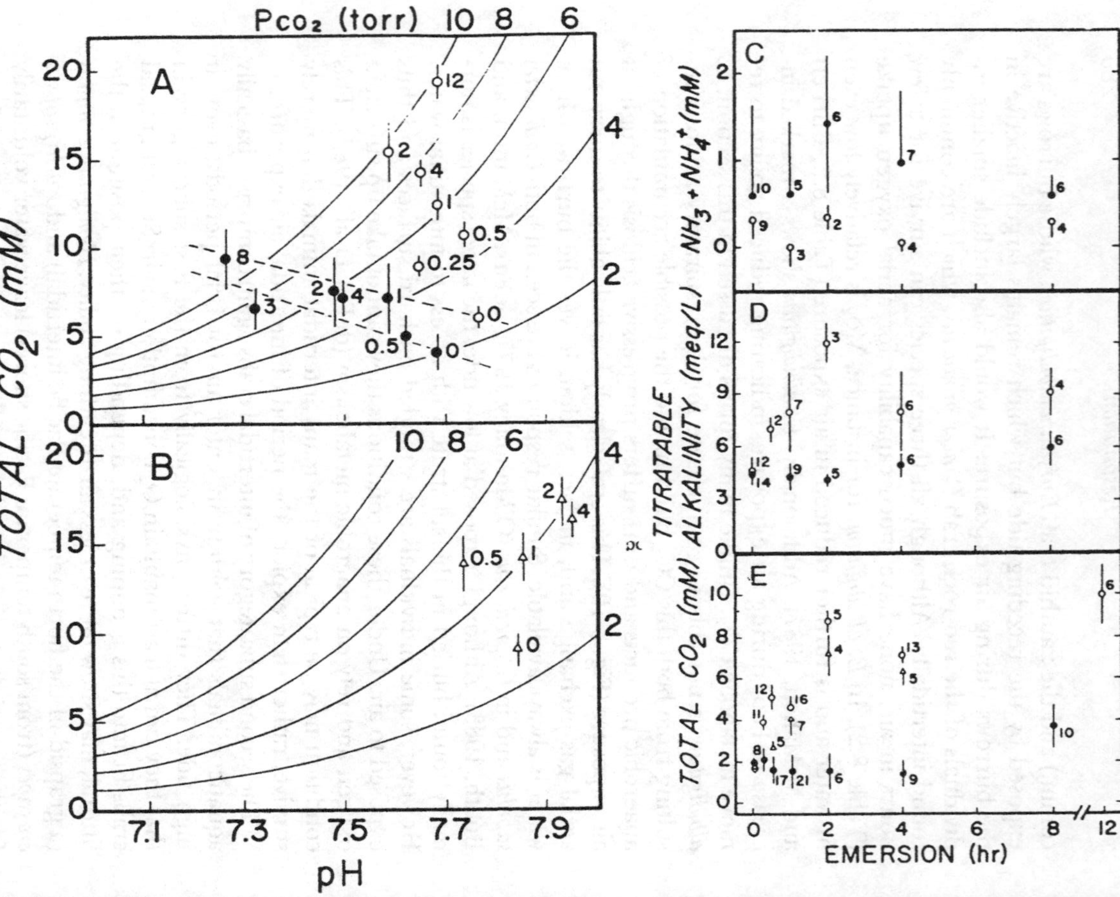

Fig. 8.10. A and B. Changes in hemolymph acid–base status (expressed using a pH–total CO_2 diagram). C, D, and E. Changes in composition of water stored in the branchial cavities, of three littoral crab species both in seawater and during enforced air exposure. –○– *Pachygrapsus crassipes*; –●– *Eurytium albidigitum*; –△– *Hemigrapsus nudus*. In A and B *fine curved lines* are PCO_2 isopleths calculated using the Henderson-Hasselbalch equation. *Dashed lines* drawn through the initial point for both *E. albidigitum* and *P. crassipes* show similar values for the nonbicarbonate buffer value. (From Burnett and McMahon, 1987)

increase on movement into air (Table 8.4) but do not reach levels exhibited by more terrestrial forms.

The most extensively studied example at this level of subcategory may be the green shore crab *Carcinus maenas*, which is air exposed and occasionally active in air at low tide. This species is able to increase O_2 uptake slightly in air (Taylor and Butler, 1978). The gills of both *Carcinus* (Taylor and Butler, 1978) and *H. nudus* (C. Barnhart and B. McMahon, unpub.) are modified to limit collapse in air, perhaps explaining their ability to utilize aerial O_2. Despite this increased air-breathing ability, measurements of respiratory gas pressures in air-exposed *C. maenas* show significant reduction in hemolymph Po_2 and increase in Pco_2 (Tables 8.4 and 8.5). Both *P. crassipes* and *C. maenas* typically show acidosis on initial air exposure (Truchot, 1975a; Taylor and Butler, 1978; Burnett and McMahon, 1987). The acidosis is largely respiratory since *C. maenas* apparently undergoes little or no anaerobic metabolism in short-term air exposure (Taylor and Butler, 1978). As with both *H. nudus* and *P. crassipes*, partial compensation for the acidosis occurs in *C. maenas* (Truchot, 1975a; Burnett and McMahon, 1987), but nonetheless the majority of these animals become moribund after 8–12 hours in dry air. Under natural conditions they rarely spend more than a few hours in the aerial environment.

Fewer data are available for anomurans, but Burggren and McMahon (1981) report similar trends for a range of subtidal to supratidal hermit crabs.

Although capable of air breathing, the crabs comprising this first category cannot really be classified as land crabs. Their adaptive advantage over their littoral relatives (which essentially suffer air exposure passively, e.g., *C. productus*) is the ability for limited aerial O_2 uptake, which often allows exploitation of the habitat during intertidal air exposure (Burggren and McMahon, 1981b). Some of these animals also carry water stored in their branchial chambers (*H. nudus, P. crassipes*; Burnett and McMahon, 1987). This water minimally serves to keep the gills moistened, allows limited O_2 uptake, and thus allows the animals both increased tolerance of air exposure and increased mobility in air. Water stores, whether in the branchial chamber or elsewhere, may have other functions in animals out of water for long periods, as discussed later in this section.

The next category of land crab adaptation (Hartnoll's T_2) includes animals that, although regularly submerged or wetted by the incoming tide, are principally active only when air exposed between high tides. Examples include several upper-level grapsid crabs such as *Cyclograpsus lavauxi* and the fiddler crabs such as *Uca* spp. Sadly, information on the respiratory physiology of such animals is limited. Innes,

Forster, et al. (1986) studied *C. lavauxi* from rocky beaches in New Zealand. This species is found higher up the intertidal zone than the grapsid species discussed earlier, inhabiting the splash zone where it may not always even be wetted by each incoming tide. Not surprisingly, this animal is markedly less dependent on standing water than are *H. nudus* and *P. crassipes*. Individual *C. lavauxi* exhibit similar Mo_2 in either air or seawater O_2 (Table 8.2), but the scope for activity remains higher in water (Innes, Forster, et al., 1986). In *C. lavauxi*, Pco_2 levels also increase in air, but no acidosis is seen even following relatively long-term air exposure. That these animals are basically still aquatic, despite their high littoral position, is suggested by their preferred habit of recycling a small amount of water over the gills. The water exits via the exhalant channel and runs down over a pad of dense hairs under the exhalant opening before reentering the branchial cavity via the inhalant (Milne-Edwards) opening (details in section 3.A). Reaeration probably occurs as the water percolates through the fine mat of hairs. An amount of water sufficient for this type of ventilation apparently can be picked up from a very fine film of water on, or in, the substrate. It is probable that the adoption of this ventilatory mechanism has allowed animals like *C. lavauxi* to colonize the upper regions of the intertidal, while remaining relatively aquatic. Such areas are uninhabitable by crabs that require periodic immersion in standing water to obtain sufficient water for gill ventilation.

At its upper limits, land crabs of category T_2 are exemplified by the fiddler crabs and perhaps some other intertidal Ocypodidae. These are usually intertidal burrowing forms, periodically submerged, but distinguished from the crabs discussed in Hartnoll's group T_1 because they are active only in air. Many species of *Uca* have been very well studied both ecologically and behaviorally (Crane, 1975, and Chapters 3 and 5), and several studies on temperature and salinity adaptation are reviewed by Vernberg (1969). Unfortunately, little information has been gathered on differences in aquatic and aerial respiratory capability in this group. Teal and Carey (1967) found no significant difference in Mo_2 measured from either *U. pugnax* or *U. pugilator* when isolated in either water or air. In fact, many species of *Uca* are still basically aquatic and cannot survive prolonged complete air exposure. They are often described as frequently visiting water to "moisten the gills." Since the branchial chambers are expanded (and the gills reduced) in the more terrestrial species (Crane, 1975), one may imagine that the expanded branchial cavity is used to store both water and air, as in the grapsids discussed earlier. Air may be bubbled through this water to aerate it while the crab is active in air. The actual capability for aerial O_2 uptake in this group might range from relatively low in essentially inactive species such as *U. princeps* (which are seldom active in air and seem to stay close to the burrow area) to high in species such as *U. brevifrons* (which are very active in air and range far from their burrows). Not all *Uca* are confined to the intertidal. *U. brevifrons* may wander many miles inland and may be found in fresh water or some distance from streams (Brusca, 1980) or even climbing trees (Crane, 1975). Greater ability to utilize aerial O_2 and reduced dependence on aquatic exchange has allowed crabs in this category to colonize the upper limits of the littoral. Furthermore, at least the fiddler crabs are able to increase activity in air and to exploit the intertidal air-exposure period to such an extent that most feeding and reproduction occurs only in air. Although these animals thus need frequent access to water, they differ from all the preceding animals in that air has become the preferred medium.

Taking into account some recent physiological studies (Chapter 7 and section 3, this chapter), Hartnoll's third category (T_3) might be reexpressed as crabs resident supratidally, active at night or in humid air, dependent on access to standing water but rarely if ever submerged by the tide. Extensive physiological data have been reported for two species of *Cardisoma*, but rather less for *Ocypode* spp. Individuals of both genera are often seen a considerable distance above the littoral but frequently either visit the sea or retreat to burrows. At least in the case of the *Cardisoma* studied, these burrows routinely penetrate to the water table (Chapter 7). Like several grapsid species mentioned earlier, *Cardisoma* are reported to retain water in the branchial cavity, and thus in contact with the gills, when in air (details in section 3.A). In both *C. carnifex* and *C. guanhumi* (Wood and Randall, 1981a, b; Burggren, Pinder et al., 1985), the gills lie ventrally in a much expanded branchial chamber. In animals removed from water, intermittent forward scaphognathite pumping (Fig. 8.3) ventilates the air in the compartment above the gills. Even during strenuous activity in air, this water is retained tenaciously in the branchial cavities and is kept constantly agitated by the flagellae as described. It is thus possible that the gills remain an important site for gas exchange with the branchial air via this small volume of water (Burggren and McMahon, 1981a; Wood and Randall, 1981a, b). Additional gas exchange occurs across the inner lining of the branchial cavity. Partitioning of O_2 and CO_2 across the gills and branchial chamber lining has not been accurately measured, but both Wood and Randall (1981a, b) and Wood et al., (1986) suggest that CO_2 elimination may occur predominantly across the membranes of the gills and O_2 uptake predominantly across the branchiostegal membranes. Despite their air-breathing capabilities, *C. guanhumi* can tolerate extended periods submerged in water and can extract reasonable amounts of O_2 from this medium (O'Mahoney and Full, 1984). The respiratory interactions of these animals with the water at the base of the burrows have not yet been studied, but in the laboratory crabs with access to shallow water continually circulate water through the branchial cavities.

The water at the base of the burrows of *C. carnifex* may be fresh or seawater. Both in *C. carnifex* burrows in Mooréa (French Polynesia; Wood and Boutilier, 1985) and in *C. guanhumi* burrows (Puerto Rico; A. Pinder and A. Smits, pers. comm.) the water is stagnant and often both extremely hypoxic and hypercapnic, circumstances in which the accessory air-breathing structures are of clear adaptive advantage. Given the wide range of oxygen content (0.4–209 $ml \cdot l^{-1}$) in the media from which *Cardisoma* must acquire oxygen, this animal would be expected to have a Hcy of intermediate O_2 affinity, but one in which O_2-binding affinity remains modifiable in order to allow change of O_2 affinity as the availability of O_2 from the medium varies (see section 8.6). Moderate O_2 binding affinity of the Hcy of *C. carnifex* is demonstrated in Fig. 8.6). Unfortunately, no detailed work on the effect of modulators on O_2-binding affinity of the Hcy of *Cardisoma* has yet been reported.

Many of the animals in this category store water in the branchial cavities while air exposed. Burnett and McMahon (1987) sampled water from the branchial cavities of three amphibious crabs (*Eurytium albidigitum*, *Pachygrapsus crassipes*, and *Hemigrapsus nudus*). Retention of water in the branchial cavity not only keeps the gills moist but also can allow continued excretion of ammonia (*Eurytium*) and CO_2 (*Pachygrapsus*, *Hemigrapsus*; Fig. 8.10). This water may also provide a source of ions for acid–base and other hemolymph ionic adjustments. The retention of water in the branchial chamber allows this category of crabs to appear highly terrestrial. *Cardisoma* spp. are frequently visible in air, and thus are commonly called "land crabs." However, since they have limited tolerance of dehydration, and require periodic access to standing water to replenish their branchial water stores (Burggren and McMahon, 1981a; Wood and Randall, 1981a; Wood and Boutilier, 1985), they are really "amphibious," rather than terrestrial crabs. As such they are active and able to function well in both water and air for short periods and are extremely well adapted for life in the immediate supratidal.

Hartnoll's fourth category (T_4) includes crabs similar to those previously discussed but that are distinguished by not needing to return to standing water to "drink" and by not retaining water in the branchial chamber. However, these crabs still need standing water for release, development, and dispersal of gametes. Knowledge of the respiratory physiology of animals within this category once again suggests the incorporation of several subcategories, each with a different level of modification. The first subcategory might include essentially amphibious crabs such as the Australian freshwater land crab *H. transversa*, which is found under floodwater for part of the year but also is able to survive long periods without access to free water (see section 6 and Chapter 7). Individuals of this species do not routinely maintain water in the branchial cavities.

Recent respiratory studies show that *H. transversa* has a relatively low oxygen consumption when submerged in water (Table 8.2), but can sustain activity and (presumably) undergo mating and most feeding at this time (Greenaway and MacMillen, 1978; Greenaway, Bonaventura, and Taylor, 1983; Greenaway, Taylor, and Bonaventura, 1983). In air, however, oxygen consumption falls to still lower levels and the animals remain essentially inactive, in a state resembling estivation, in the burrow. Estivation may be a common response of air-breathing crabs to drought conditions: Wood et al., (1986) report that *C. carnifex* reduces aerial oxygen consumption by approximately 50% during dehydration. Despite the reduction in metabolism, increase in hemolymph CO_2 levels in air-breathing *H. transversa* are similar to those seen in other air-breathing forms (Greenaway, Taylor, and Bonaventura, 1983).

The land hermit crabs (genus *Coenobita*) have been the focus of several respiratory studies but remain difficult to categorize. These crabs may be found some kilometers from the sea (De Wilde, 1973; see also Chapter 3 and the Appendix). At first sight they appear highly terrestrial. A closer look at the level of respiratory adaptation, however, creates confusion, revealing some features clearly similar to the more terrestrial crabs and some similar to less terrestrial forms. Morphologically, the gills are substantially reduced, indicating a low dependence on water. However, the branchial chambers are not enlarged, nor are the branchiostegal walls particularly modified (section 2). Thus, these crabs are apparently not clearly modified for aerial respiration. *Coenobita* spp. do not voluntarily immerse themselves in water, but a single report of $\dot{M}O_2$ in *C. scaevola* submerged in the laboratory (Achituv and Ziskind, 1985) indicates that aquatic $\dot{M}O_2$ can be increased above that recorded from animals resting in air, at least in the short term. While this increase probably results from the increased activity seen in immersed animals, it does show that reasonable O_2 uptake can occur in submerged animals despite a severely reduced gill area (Table 8.1).

Hemolymph O_2 capacities in this group are among the highest reported for Crustacea, a feature that could be regarded as highly terrestrial. However, the high O_2 capacity is coupled with the lowest O_2 pressures recorded (in air) in any land crab (Fig. 8.6; Table 8.5). This observation, coupled with the poor development of aerial-gas-exchange surfaces, may indicate that these crabs normally conduct a large part of gaseous exchange across the gills despite their apparent terrestriality. The extremely high O_2 capacity of this hemolymph may thus be an adaptation to a very high diffusion limitation occurring across a generally reduced gas-exchange surface. An additional aquatic tendency in the land hermit crabs is the retention of a primarily O_2-driven respiratory control system (section 5).

Some of the retention of apparently aquatic features in *Coenobita* spp. may be associated with the water stores carried in the shell at virtually all times. These crabs replenish these water stores periodically using either standing water (either fresh or brackish) if available, or interstitial water from the substrate. The shell is carried in such a position that the water stores are available to be pumped into the branchial cavity and over the gills (McMahon and Burggren, 1979). This water thus could serve for respiratory gas exchange, as in the branchial water of *Cardisoma* and *Pachygrapsus*, as well as serving for ion and osmoregulation (Chapter 7) and nitrogenous excretion (A. Pinder and B. McMahon, unpub.). The water stored in the hermit crab's "mobile home" apparently serves similar functions to that stored in the brachyuran gill chamber, and has allowed these animals to extend their range into the terrestrial habitat while retaining many features typical of their more aquatic relatives.

A still more terrestrial level within the category T_4 contains crabs that show oxygen consumption drastically reduced or inadequate to support respiration when submerged. Examples are *G. lateralis*, among the Brachyura, and *B. latro*, among the Anomura (Table 8.2).

Although similar in structure to the related species of *Cardisoma* mentioned earlier, *Gecarcinus* spp. differ in terms of degree of apparent terrestriality. Both genera are apparently burrowing forms and both may be found some distance from the sea (see Chapters 3 and 7 and the Appendix). Both genera have the reduced gills and expanded branchial cavity typical of Gecarcinidae. Although *Gecarcinus* can draw water into the branchial cavity from damp substrate (T. G. Wolcott, 1976), it apparently does not commonly carry water in the branchial chamber, does not burrow down to ground water, and thus needs to come to standing water only for reproduction. Although this animal is clearly adapted for terrestrial existence, it can nonetheless survive submergence for several days (A. C. Taylor and Davies, 1982). Unlike *Cardisoma* and *Holthuisana*, however, *G. lateralis* shows reduced O_2 uptake in water (Table 8.2). Circulating Po_2 falls and hemolymph lactate increases, showing that the animal enters anaerobiosis, but the crab does not die, at least not immediately. It is interesting to note here that if we compare hemocyanin O_2 binding between the two gecarcinid crabs, both O_2 affinity and the magnitude of the Bohr factor are lower in the Hcy of *Gecarcinus* than in the hemocyanin of *Cardisoma* (Table 8.3). This would again correlate with increased reliance on air breathing. It is important to indicate that, unfortunately, oxygen binding in Gecarcinids has been relatively poorly studied and modulation and other effects are poorly known.

Birgus latro is a coenobitid crab very closely related to *Coenobita*. Adult *Birgus* do not inhabit molluscan shells, but this feature is found in juveniles (Harms, 1932; Chapter 3). The two coenobitids are often found in the same environment over part of their range, i.e., *C. brevimanus* and *C. perlatus* with *B. latro* in the islands of Palau. Although superficially similar to *Coenobita, Birgus* has important morphological and physiological differences. In contrast with *Coenobita, Birgus* has a very well developed air-breathing organ and the gills not only are further reduced (Cameron, 1981a) but are located in a small fold apart from the area of the branchiostegal chamber containing the aerial respiratory structures (Harms, 1932). Circulating oxygen partial pressures measured in *B. latro* are variable but considerably higher than those found in *Coenobita* spp. (Table 8.5), perhaps implying that the expanded surface of this air-breathing organ allows better oxygen uptake. No data are available for respiratory effects of submergence in this crab, but it actively avoids submergence and is known to die within a few hours if submergence is enforced (Harms, 1932). We can thus assume that the capability of the gas-exchange system to take up oxygen from water is minimal. As is the case for *Gecarcinus*, this animal apparently has reduced Hcy O_2 affinity. For instance, Table 8.3 and Figure 8.6 show substantially higher O_2 affinity in the very closely related *Coenobita*, again suggesting that this is an adaptive response adjusting O_2-binding characteristics of Hcy to suit O_2 uptake in air, now the predominant respiratory medium. However, once again we must caution that the oxygen binding characteristics of Hcy and especially modulator effects for this species are poorly known.

Paradoxically, P_{CO_2} and total CO_2 concentration appear lower in *Birgus* than in *Coenobita* (McMahon and Burggren, 1981) when compared at equivalent temperature, hydration state, etc. A similar correlation may be seen in comparisons of *Cardisoma* and *Gecarcinus*. The reason for this is not clear, but as was noted by McMahon and Burggren (1980), the respiratory and circulatory system of *Birgus* (Fig. 8.11) may show features of a multicapillary exchanger similar to that found in bird lungs. Hemolymph perfusing the elaborate "lung" structure flows basically from anterior to posterior (Harms, 1932), whereas air flow generated by the scaphognathite (Cameron and Mecklenburg, 1973) flows essentially in the reverse direction, providing an overall countercurrent system. The resulting improvement in gas exchange capability could possibly contribute to both the increased O_2 and decreased CO_2 levels observed.

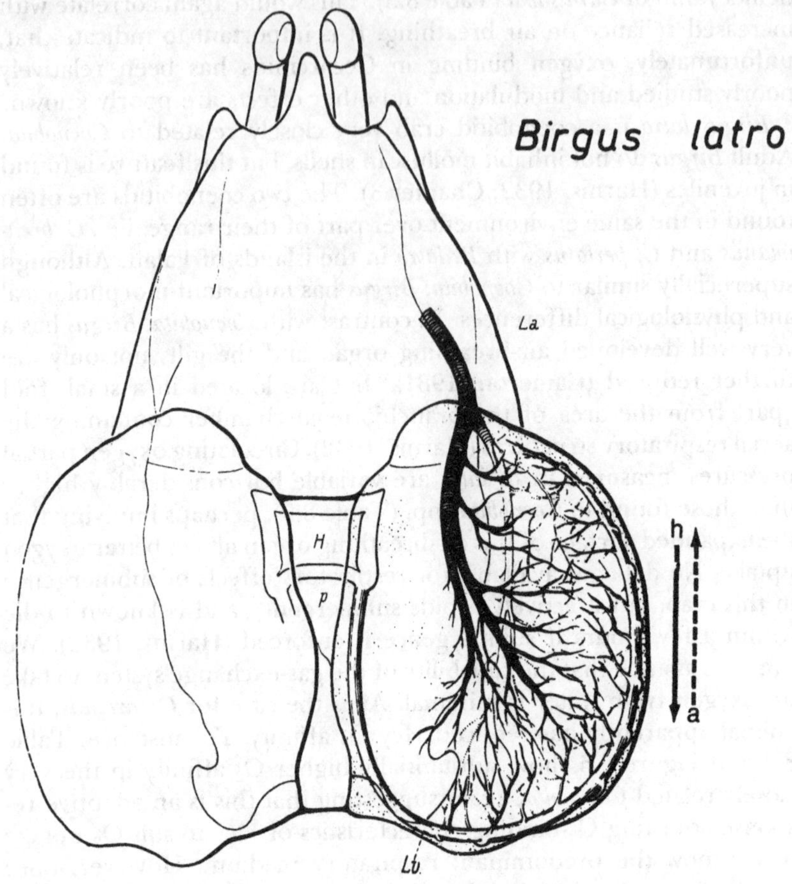

Fig. 8.11. Diagram of hemolymph flow through the "lung" cavity of *Birgus latro* (from Harms, 1932) to show possible overall "countercurrent" flow of air and hemolymph. *a*, airflow; *h*, hemolymph flow; *La*, lung artery; *Lv*, lung vein; *H*, heart; *p*, pericardium.

It is difficult to place the many species of *Ocypode* in an ecophysiological classification partly because very little is known of their respiratory physiology and partly because their terrestrial capabilities vary greatly between species. *O. occidentalis* from the Baja Peninsula of Mexico, for instance, apparently carries water in the branchial cavity (L. Burnett, pers. comm.), but *O. saratan* from a similar dune environment bordering the Red Sea near Jedda apparently is never observed to visit water and occupies burrows that contain no free water (A. C. Taylor, pers. comm.). However, the burrow walls are moist and since we know that other land crabs can easily pick up water from moist sand (T. G. Wolcott, 1976; 1984), the presence of branchial water stores in *O. saratan* cannot be ruled out. Currently respiratory data are available for only two species: *O. quadrata* and *O. saratan* (Table 8.5). *O. quadrata* is primarily an air breather but nonetheless has very similar Po_2 values in air (Burnett, 1979) to those found in *Cardisoma* and *Gecarcinus* (Table 8.5). *O. saratan*, however, has a much higher hemolymph Po_2, possibly associated with a greater reliance on aerial respiration. Clearly, much more physiological work on a range of species of *Ocypode* is needed to understand these crabs.

Very little data have been reported for the respiratory performance of animals from Hartnoll's most terrestrial category (T_5), which is restricted to animals that do not need standing water even for reproduction. Recently Innes and Taylor (1986a, b, c) and El Haj et al. (1986) have published accounts of "lung" structure and function in the Trinidadian mountain crab *P. garmani*, which might be a candidate for this group. Innes and Taylor (1986a, b) describe three respiratory surfaces in these crabs: substantially reduced gills, the branchiostegal lining generally, and an associated structure they term a "lung." This latter is an invagination of the branchial chamber wall formed of a "spongy" tissue consisting of an "elaborate complex of anastomosing respiratory airways which are extensively vascularised and ventilated with a through-flow of air" (Innes et al., 1987; see section 2). This structure may be extremely efficient at oxygen uptake since Pao_2 values of 130–140 mmHg have been reported for this species (Innes et al., 1987). Arterial Pco_2 levels are relatively modest at 8–10 mmHg. These oxygen partial pressures are substantially higher than those reported for other air-breathing crustaceans, with the exception of some early data for *Birgus* (Cameron and Mecklenberg, 1973). Values for *G. lateralis*, *O. saratan*, and *Mictyris longicarpus* (Table 8.5) are also relatively high, suggesting that the specialized "lung" structures seen in these animals may indeed allow enhanced O_2 uptake from the air.

It seems that animals from the last two categories (T_4–T_5) differ more in their reproductive capability, i.e., in the ability to carry out abbreviated development independent of standing water, rather than in respiratory adaptation. For instance, *Birgus* and *Pseudothelphusa* appear equally well equipped to breathe air, and are similarly active in humid air conditions. Comparison of *H. transversa* and *P. garmani* show that the more specialized respiratory structure of the latter may also be associated with increased activity in (humid) air (see Innes and Taylor, 1986e). *H. transversa* shows reduced activity and decreased Mo_2 in air and apparently undertakes the majority of its activities when submerged during the wet season. It is surprising, too, that the more complex air exchanger apparently does not allow *P. garmani* to increase its maximal O_2 uptake from air (Innes, El Haj., et al., 1986) above that of less well equipped land crabs.

The air-breathing behaviors currently observed in land crabs may have evolved from mechanisms similar to those currently used by tidepool animals to supplement aquatic O_2 consumption when trapped in hypoxic water at low tide (McMahon, in press). Recent studies on *C. maenas* may illustrate these early stages. When exposed to hypoxic water, a common occurrence in tidepools where this animal is often found (Truchot and Duhamel-Jouve, 1980; Morris and Taylor, 1983a), individual *C. maenas* rise to the surface and reverse the direction of scaphognathite pumping, causing air to flow backward through the scaphognathite channel (Bohn, 1897; Taylor and Butler, 1973; Taylor, Butler, and Al-Wassia, 1977) and to bubble through the water contained in the branchial cavity (Wheatly and Taylor, 1979). Oxygen from the air thus diffuses first to the branchial water and then to the hemolymph passing through the gills. Although Po_2 of the hemolymph apparently increases only slightly as a result of one of these bouts of aerial ventilation, the high oxygen affinity of the hemocyanin of *Carcinus* during hypoxic exposure (section 5) allows a significant increase in oxygen extracted from the water (Taylor and Butler, 1973; Wheatly and Taylor, 1979).

Perhaps because of this behavior, the gills of *C. maenas* have become strengthened to limit collapse in air (Taylor and Butler, 1978). In fact, *C. maenas* will completely abandon very hypoxic or very warm water (Taylor et al., 1977) and is also often stranded in air at low tide. Under these conditions the branchial cavity is ventilated with air (Taylor and Butler, 1978). *C. maenas* is able to increase O_2 consumption in air above levels recorded in water, presumably using the modified gills. Similar modifications to the gill lamellae occur in a variety of land crabs, suggesting the continued involvement of these structures in gas exchange.

So far we have assumed all gas exchange to occur across the gills in aquatic crustaceans. In fact, some portion of gas exchange probably occurs across the branchial cavity walls even in aquatic forms. Exchange across the inner wall of the branchiostegite may be of special importance since the epithelium is both thin walled and well perfused. Although the magnitude of the contribution of these accessory surfaces to aerial gas exchange is not known either for any water breather or during temporary air breathing in littoral animals like *Carcinus*, we have documented a progressive increase in the air-breathing capability of the branchial cavity lining in land crabs (section 3). This has occurred by the increase in the volume of the branchial chamber and the accompanying increase in its surface area. Eventually these areas are sufficiently specialized that they become the principal respiratory site, as in *B. latro*. It is of interest that this surface appears better suited to O_2 uptake rather than CO_2 elimination. This may arise because the branchial chamber lacks carbonic anhydrase, an enzyme that facilitates CO_2 elimination across crab gills.

Along with the development of a specialized organ for aerial O_2 uptake, we have described a gradual decline in gill area and with this a gradual shift of the complex of nonrespiratory gill functions to other structures. This may be the more complex evolutionary step, since it is apparently not complete even in the most highly accomplished land crabs, adults of which all retain gills. Several possible reasons can be suggested. First, aquatic oxygen consumption may be important periodically in the many amphibious crabs that still spend some time immersed in water. Second, we have seen that CO_2 excretion may occur predominantly over the gills in amphibious forms such as *Cardisoma*. Thus, we may postulate that in the transition from water to air breathing, removal of CO_2 posed more of a problem than O_2 uptake and thus the gills are retained to facilitate CO_2 removal. Some supporting evidence of this is seen even in the obligate air breather *B. latro*, which survives (at least short-term) removal of the gills (Harms, 1932) without apparent ill effect. However, Smatresk and Cameron (1981) showed that *Birgus* with ablated gills were not able to remove all the CO_2 produced in a bout of exercise. It thus seems possible that some portion of CO_2 removal occurs across the gills even in this highly evolved air breather.

Finally, the gills of aquatic animals are involved in many functions other than respiration, including osmotic regulation, ion regulation, regulation of acid–base balance, and nitrogenous excretion. Thus, the retention of the gills in adult land crabs could result from an inability to transfer any of these functions to other systems. Interference with these regulatory systems may not be apparent in the short term, and since long-term experiments have yet to be carried out, the effects or the reason for the persistence of the gills in the land crabs remains obscure.

9: Circulation

WARREN W. BURGGREN and
BRIAN R. McMAHON

CONTENTS

1. Introduction

Effective exchange and transport of respiratory gases, maintenance of tissue acid–base balance, and regulation of water and ion levels are but a few of the many physiological processes that are supported by the cardiovascular system of invertebrates and vertebrates alike. Not only is the morphology of the circulation critical to these processes, but so too are the capabilities for physiological regulation of the circulation, as dictated by changing conditions of the internal and external milieu.

This chapter discusses the cardiovascular system of land crabs, with particular reference to specializations for terrestrial existence. Unfortunately, the study of the cardiovascular system of crustaceans, terrestrial or marine, is still in its infancy compared to our relatively extensive knowledge of their respiratory and acid–base physiology, discussed in the previous chapter. Consequently, this account will be somewhat incomplete and must draw heavily on what is known of marine crabs and other decapods.

2. Cardiovascular morphology

The general morphological pattern of the cardiovascular system of either brachyuran or anomuran land crabs is little changed from that of marine crabs, and other decapods generally (see early reviews by Bouvier, 1891; Pearson, 1908; Maynard, 1960). Important exceptions, particularly relevant to respiratory gas exchange, are the vasculature of the gills and lining of the branchial chambers. Because of the arrangement of the decapod circulation, with its several vascular beds arranged in both series and parallel, the delineation between "arterial" and "venous" circulation is somewhat arbitrary. Consequently, following a discussion of general morphological features of the heart, "arterial" circulation, and tissue vascular bed, a detailed account will be given of the venous circulation and its relationship to the afferent and efferent vasculature of the gills and branchiostegites.*

A. The heart

The heart, composed of striated muscle, is situated within a pericardial sinus lying medially directly under the carapace (Fig. 9.1). Although the heart is generally regarded as consisting of a single chamber, the pericardium acts as a functional antechamber. The heart is usually suspended in the pericardium partly by the arteries but mainly by suspensory ("alary") ligaments, bands of elastic tissue that originate on the wall or ventral septum of the pericardium. In decapods, the suspensory ligaments are flattened horizontally, serving to divide partially the pericardium into dorsal and ventral chambers (Dubuisson, 1928; Burger and Smythe, 1953).

The pericardial sinus receives primarily oxygenated hemolymph from the gills via the branchiopericardial veins. In most decapods the pericardium also receives hemolymph draining the inner wall of the branchiostegites and the thoracic wall of the branchial chamber; in land crabs this circulation is particularly well developed and drains

*As McLaughlin (1983a) indicates in one of the most recent reviews of crustacean morphology, there has been little uniformity in terminology used in previous accounts. In this chapter we adopt the cardiovascular terminology used by McLaughlin (1983a); the reader is referred to her review for a discussion of (numerous) equivalent terms.

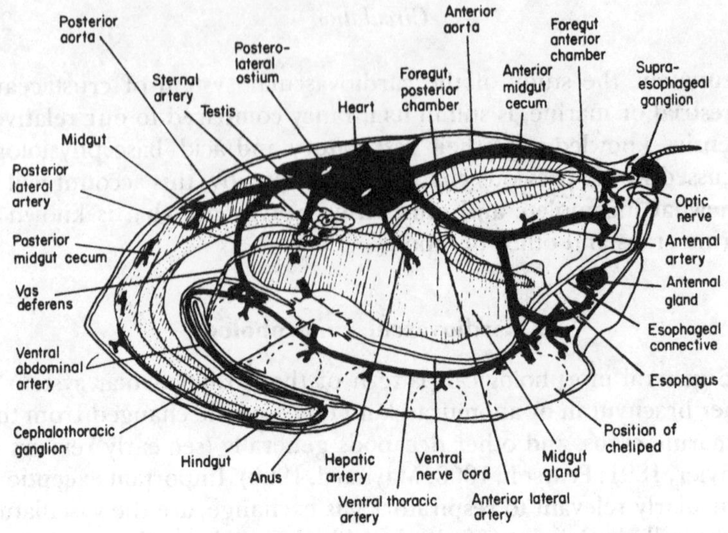

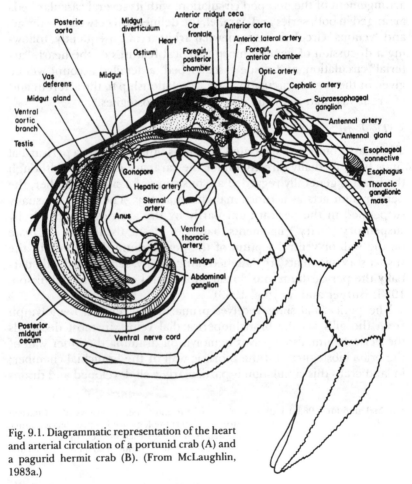

Fig. 9.1. Diagrammatic representation of the heart and arterial circulation of a portunid crab (A) and a pagurid hermit crab (B). (From McLaughlin, 1983a.)

the branchiostegal gas exchange surfaces (see section 2.D). Hemolymph enters the heart from the pericardial sinus via three pairs of lens-shaped ostia; in pagurid and brachyuran crabs one pair is dorsal or anterodorsal and two pairs are lateral (Fig. 9.1). Each ostium is guarded by a set of valves, which prevent reflux of hemolymph from the heart into the pericardial sinus during systole. There is some evidence to suggest that these valves may contain muscle fibers and are innervated, indicating that they may serve as more than simple flap valves (Maynard, 1960).

The decapod heart lacks a "coronary" circulation. Thus cardiac tissue must exchange nutrients, wastes, and respiratory gases via diffusion with the oxygenated hemolymph in both the lumen of the heart and in the surrounding pericardial space.

The innervation of the decapod heart has been the object of intensive investigation (see reviews by Maynard, 1960; Lockwood, 1967; McMahon and Wilkens, 1983). The neurogenic heart is stimulated to contract by rhythmic impulses from neural pacemaker cells in the cardiac ganglion, located on the inner dorsal wall of the heart. The frequency of impulses, which regulates heart rate and strength of contraction, is modified both by changes in concentration of hemolymph-borne hormones (e.g., 5-hydroxytryptamine) and by several cardiomodulatory nerves. These nerves, consisting of two pairs of accelerator fibers and one pair of inhibitory fibers, originate in the metathoracic ganglion (Brachyura) or subesophageal ganglion in Macrura (and possibly Anomura). Although the heart rate and stroke volume of terrestrial crabs clearly can be finely regulated (see section 3.B), it is currently unknown whether the neural/hormonal basis for cardiovascular adjustment typical of aquatic crabs has been modified in any fashion in land crabs.

The reader is referred to Maynard (1960) for details of the cardiac anatomy of aquatic decapods. Unfortunately, the anatomy of the heart and pericardium of land crabs, specifically, has received very little attention.

B. Arterial circulation

The heart of decapods typically ejects hemolymph into five to seven separate arteries emanating from it (Fig. 9.1). One-way semilunar valves at the base of each of these arteries prevent reflux of hemolymph from the artery into the heart during diastole. Like the ostial valves, the arterial valves contain muscle fibers and are thought to be innervated (see Maynard, 1960, for references; also Kuramoto and Ebara, 1984), indicating that valve movement may have active as well as passive elements. The arteries contain elastic tissue but generally are thought to lack smooth muscle, except where arterial valves are located (Maynard, 1960). The posterior aorta of Malacostracans generally contains semilunar valves, though their presence in terrestrial crabs has yet to be confirmed.

The eyes and supraesophageal ganglion are supplied by the anterior aorta. In many decapods, the proximal region of the anterior aorta has a small enlargement, referred to as the *cor frontale* or frontal heart (Fig. 9.1B), first described in malacostracans by von Baumann (1917). This structure, which apparently serves as an auxillary hemolymph pump (Steinacker, 1978, 1979), has been described for several brachyurans and anomurans (Jackson, 1913; Steinacker, 1978; McLaughlin, 1983a), but again its occurrence in terrestrial crabs has yet to be confirmed. Accessory hearts in the form of contractile vessels have been described in the so-called abdominal respiratory lacunae in the terrestrial hermit crab *Coenobita* (Bouvier, 1890).

The cephalic appendages, walls of the foregut, anterior musculature, antennal glands, and much of the carapace are perfused from the anterior lateral artery. The hepatopancreas is supplied by the hepatic artery in all decapods except the anomurans. In hermit crabs the hepatopancreas has been displaced to the abdomen, and the hepatic artery instead perfuses the foregut or midgut (Fig. 9.1B). The sternal artery, which in both brachyurans and anomurans originates from the posterior end of the heart, courses ventrally and either passes directly through or circumvents the ventral nerve cord at the level of the ganglia of the third and fourth pereiopods (McLaughlin, 1983). The sternal artery descends ventrally and eventually bifurcates to form the anterior ventral thoracic artery (supplying the anterior regions of the central neural tissues, the mouthparts, and the first three pairs of pereiopods), and the ventral abdominal artery (supplying the fourth and fifth pairs of pereiopods, the posterior regions of the central nervous system, and the ventral portion of the abdomen). The posterior aorta is the major route for perfusion of the abdomen in many decapods (Pike, 1947). Its course and degree of subdivision depend in part on the general arrangement of the abdomen (Fig. 9.1).

The major perfusion of the gills is via large, central sinuses associated with the "venous" circulation (see section 2.D), but there are also "true" branchial arteries. Hypobranchial and epibranchial arteries that emerge directly from the heart perfuse the gill tissues of *Ocypode cordimanus* and *Holthuisana transversa* (Taylor and Greenaway, 1979, 1984) and are probably present in most crabs. Presumably this arterial circulation supplies nutrients and hormones to the metabolizing tissues of the gills themselves, much as occurs in vertebrate gills (Hughes, 1984).

C. The "capillary bed": general considerations

The circulation of decapods is usually described as an "open" circulation, referring to the fact that hemolymph comes into direct contact with interstitial and lymphatic fluids during its transit of the body tissues. Nonetheless, there is a widespread vascular system for distribution of hemolymph within the tissues. In fact, vessels as small as 2–10 μm in diameter have been described in both marine crabs (Vignal, 1886; Sandeman, 1967) and freshwater and terrestrial crabs (Haeckel, 1857; Taylor and Greenaway, 1979; Greenaway and Farrelly, 1984; see also Figs. 9.3 and 9.5). These vessels are at least as small as the functional exchange vessels of the "closed" circulation of vertebrates, where a lower vessel diameter is set by the size of circulating red blood cells. In some of the tissues of crabs, the walls of these fine vessels retain only an endothelial cell lining (Maynard, 1960), making them structurally as well as functionally comparable to the capillaries in vertebrates.

The density of these fine distribution and exchange vessels in decapods varies greatly between body tissues. Within systemic tissues, the greatest concentrations occur in neural ganglia, where anastomosing capillary networks are found. Dense capillary networks also occur in the excretory organs, the hepatopancreas, and the gastrolith fields of the midgut wall (see Maynard, 1960). Of course, extensive networks of fine vessels also occur in the gills and lining of the branchial chambers, where the vasculature is specialized for respiratory gas exchange (see section 2.D).

The presence in decapods of complex system of fine vessels for distribution of hemolymph within body tissues – i.e., functionally a "capillary bed" – begs further investigation at both the morphological and physiological levels. Of particular importance is whether hemolymph distribution can be altered by peripheral resistance changes at the level of the "capillaries." In the closed circulation of vertebrates, the finer arterioles and the transition zone from arteriole to capillary are potent sites for changing the vascular resistance of a particular tissue, and thus altering the amount of blood that tissue receives. Although redistribution of cardiac output between various body tissues has been documented in the terrestrial crab *H. transversa* (Taylor and Greenaway, 1984), the active role (if any) of the finer vessels in producing redistribution of hemolymph flow in this or any other terrestrial crab is currently unknown.

In all decapods, hemolymph leaving the "capillaries" immediately drains into lacunae, which are relatively amorphous tissue spaces unbounded by an endothelial wall. From the lacunae of the various body tissues, hemolymph enters a complex system of venous sinuses bounded by endothelial tissue, thus separating hemolymph from interstitial fluids.

D. The venous circulation and branchial and branchiostegal vascular beds

i. General plan of the venous circulation. In a closed cardiovascular system like that of vertebrates, the venous circulation is largely a mirror image of the arterial system: in crabs, however, the venous circulation is arranged very differently from the arterial side. Unfortunately, the venous side of the circulation of decapods has received considerably less attention than the arterial circulation. In part this paucity of information exists because of the relative difficulty of making detailed observations of the rather ephemeral venous sinuses, once they have collapsed and drained during dissection. Yet, understanding the arrangement of the venous circulation is crucial in the context of land crabs, since the most striking morphological differences between the circulation of land crabs and aquatic crabs involve the arrangement of the venous circulation, particularly as it relates to the branchial and branchiostegal vascular beds.

Fortunately (and quite atypically when comparing data on aquatic and terrestrial crabs), our knowledge of the anatomy of the venous circulation is most extensive for terrestrial land crabs. Recent application of vascular identification and casting techniques, involving injection into the circulation of methacrylate, latex, and various dyes, has been very useful in elucidating the detailed anatomy of the venous circulation of terrestrial crabs such as *H. transversa* (Taylor and Greenaway, 1984) and *O. cordimanus* (Greenaway and Farrelly, 1984). The following account relies primarily on these studies for information on terrestrial forms, and on the review by Maynard (1960) for decapods in general.

Hemolymph leaving the fine exchange vessels of the various body tissues enters hemocoelic lacunae, which ultimately lead into an interconnected and complex series of sinuses in the abdomen and thorax. Hemolymph from these sinuses can pass only to the gills and branchiostegal membranes for oxygenation and carbon dioxide elimination, after which this hemolymph flows directly to the pericardium and into the heart for recirculation.

The general form of the venous circulation of land crabs may be exemplified by that of *O. cordimanus*, which is represented diagrammatically in Figure 9.2. Large, paired sinuses located medioventrally along the floor of the hemocoelic cavity receive deoxygenated hemolymph flowing ventrally from the dorsal and hepatic sinuses and anteriorly from the abdominal sinuses. The infrabranchial sinuses, as their name implies, lie ventral to the gills and connect with the eye

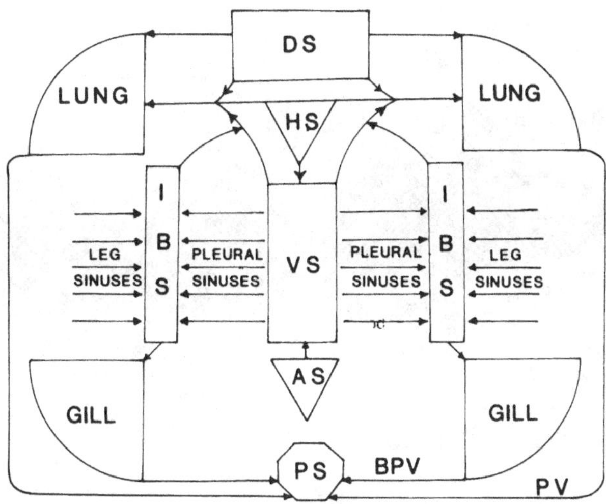

Fig. 9.2. Highly diagrammatic representation of the venous circulation of the terrestrial crab, *Ocypode cordimanus*. *DS*, dorsal sinus; *HS*, hepatic sinus; *VS*, ventral sinus; *IBS*, infrabranchial sinus; *AS*, abdominal sinus; *PS*, pericardial sinus; *BPV*, branchiopericardial vein; *PV*, pulmonary vein. (After Greenaway and Farrelly, 1984.)

sinus anteriorly and the abdominal and ventral sinuses posteriorly. Additionally, in those segments bearing pereiopods, the infrabranchial sinus receives hemolymph directly from the ventral sinus via small hemolymph spaces that Greenaway and Farrelly (1984) term pleural sinuses (Fig. 9.2). These pleural sinuses merge with the small sinuses returning hemolymph from the pereiopods and then course dorsally to join the infrabranchial sinuses. The infrabranchial sinuses, which also receive hemolymph from each leg sinus, thus contain "mixed venous" hemolymph, and have been the sites for hemolymph sampling in numerous studies of gas exchange. Any gill may thus be perfused with hemolymph that may have originated in virtually any of the body tissues. Oxygenated hemolymph draining the gills flows in branchiopericardial vessels directly to the pericardium.

The general venous circulation as outlined earlier for *O. cordimanus* is probably representative of most brachyuran crabs, terrestrial or aquatic. The venous circulation of anomurans has not been so completely described, however, and should be investigated, particularly with respect to the arrangement of the abdominal and hepatic sinuses.

ii. The branchial vascular bed. As has been mentioned on several occasions in this volume, the gills of land crabs show several important modifications for aerial respiration. Primary among these is a reduction in surface area relative to aquatic species (see Table 8.1) and

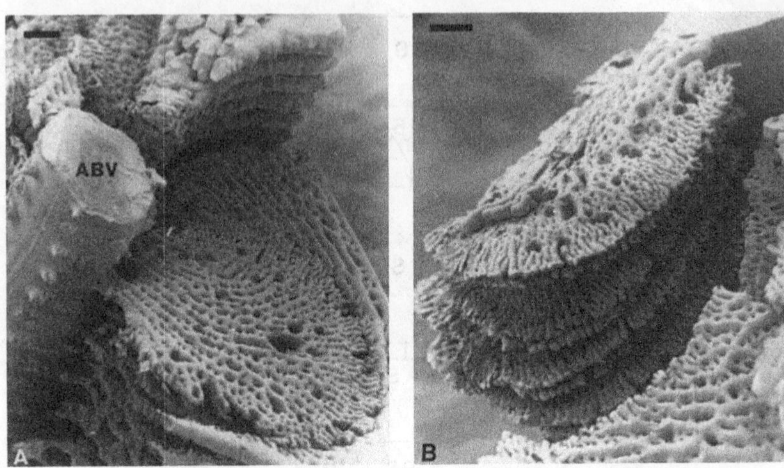

Fig. 9.3. Corrosion cast of the hemolymph spaces of the gills of land crab *Geograpsus crinipes*. A. Prominent afferent branchial vessel (*ABV*) gives rise to afferent hemolymph supply for each secondary gill lamella. Scale bar = 100 μm. B. Secondary lamellae showing two parallel hemolymph spaces formed by septum within the lamella. Efferent hemolymph connections have been broken away to reveal detail of lamellae. Scale bar = 0.5 mm. (Scanning electron micrographs courtesy of Caroline Farrelly.)

structural modifications to prevent gill collapse in air (see also Fig. 8.1).

Internally, the vascular arrangement of the gills of land crabs is not greatly different from that of aquatic crabs, with a few notable exceptions (Taylor and Greenaway, 1979). But for the small nutritive supply of hemolymph from the hypobranchial and epibranchial arteries arising directly from the heart, the branchial vascular supply is from the pleural and leg sinuses. Deoxygenated hemolymph enters each branchial filament by way of a large afferent vessel running the length of the filament. Each gill lamella receives deoxygenated hemolymph from an afferent branchial vessel (Fig. 9.3). Hemolymph from the efferent vessel enters immediately into a marginal channel that, as the name implies, runs around the edge of the gill lamella (Fig. 9.4). Hemolymph leaves the marginal channel and percolates between the pillar cells that hold apart the opposing walls of the lamella. Oxygenated hemolymph drains from the lamella into an efferent vessel, which merges with the major efferent drainage serving the gill filament.

Interestingly, a central septum divides the lamella internally along its long axis into two thin, parallel spaces. This septum is characteristic of aquatic crabs (Drach, 1930; Barra, Pequex, and Humbert, 1983)

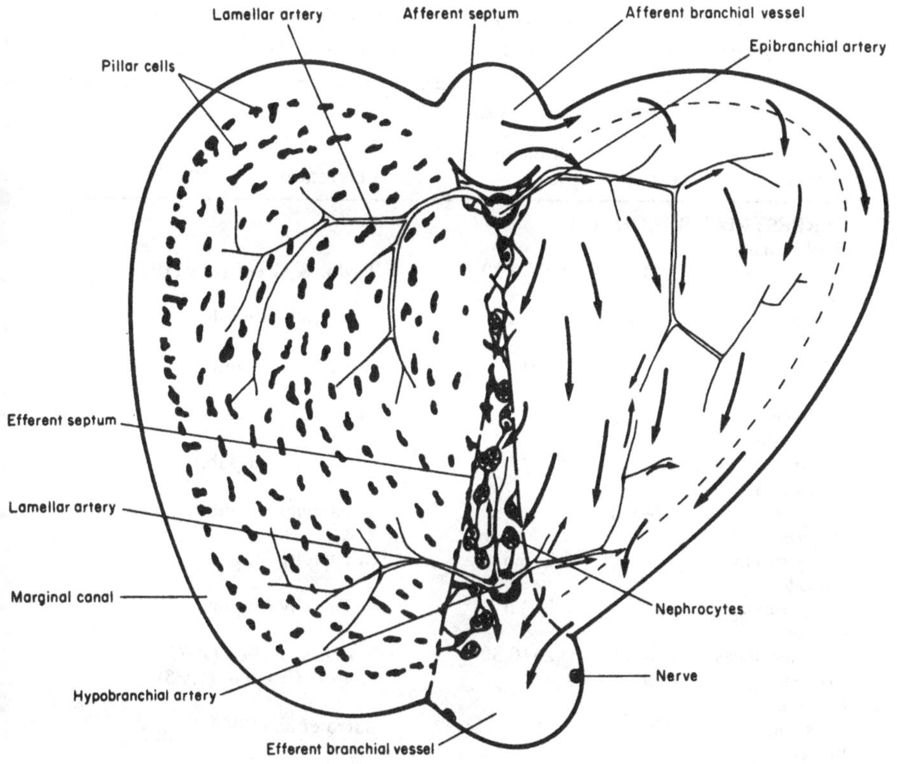

Fig. 9.4. Schematic view of transverse section through gill filament 2 or 3 in the Australian land crab *Holthuisana transversa*. Two opposing lamellae and details of their structure and hemolymph flow within them are represented. See text for details. (From H. H. Taylor and Greenaway, 1979.)

and the land crabs *Geograpsus crinipes, Geograpsus grayi, Gecarcoidea natalis, Cardisoma hirtipes,* and *Ocypode cordimanus,* but is absent in *Holthuisana transversa* (Taylor and Greenaway, 1979; P. Greenaway and C. Farrelly, unpub.) The reader is directed to the account given by Taylor and Greenaway (1979) for further anatomical detail of the microcirculation of the gills of crabs.

It is unknown whether there is active control over the pattern of flow of hemolymph within individual lamella. The many striking structural resemblances between the lamella of the crab and that of fishes (see Hughes, 1984; Laurent, 1984), where intralamellar regulation of blood flow does occur, tempt us to speculate that such redistribution might occur in the lamellae of crabs. Recent experiments by H. H. Taylor and E. W. Taylor (1986) have found valvelike structures within the lamellae of *Carcinus maenas* that apparently maintain directionality of

Table 9.1. *Gas diffusion distances between hemolymph and the respiratory medium in selected terrestrial and aquatic decapod Crustacea*

	Diffusion distance (μm)		
Species	Gills	Branchiostegal "lung"	References
TERRESTRIAL BRACHYURA			
Holthusiana transversa	5–8	0.22–0.30	Taylor & Greenaway (1979)
Gecarcoidea natalis	3.5–10.0	3.2–5.0	C. Farrelly (unpub.)
Cardisoma hirtipes	5–12	2.5–4.5	C. Farrelly (unpub.)
Geograpsus lateralis	6–11	—	Copeland (1968)
Geograpsus grayi	1.5–5.0	2.5–4.0	C. Farrelly (unpub.)
Geograpsus crinipes	3.5–9.0	—	C. Farrelly (unpub.)
Mictyris longicarpus	3.5–7.0	1.2–3.6	C. Farrelly (unpub.)
Ocypode cordimanus	5–10	2.5–4.5	C. Farrelly (unpub.)
Ocypode certophthalmus	2.9–3.5	0.25–0.30	Storch & Welsch (1975)
Uca mordax	6–9	—	Finol & Croghan (1983)
Eriocheir sinensis	2.7–5.0	—	Barra et al. (1983)
Birgus latro	—	0.5–1.2	Storch & Welsch (1984)
Pseudothelphusa garmani	5	0.4–1.0	El Haj et al. (1986)
AQUATIC BRACHYURA AND MACRURA			
Procambarus clarkii	3.15–8.70	NA	Burggren et al. (1974)
Astacus pallipes	1.4–5.7	NA	Fisher (1972)

hemolymph flow from afferent to efferent channels within the lamella. Whether such structures, which are major components of the total branchial resistance, can alter the distribution of hemolymph flow within the lamella remains to be demonstrated.

Diffusion distances between gill hemolymph and the respiratory medium range from 3 to 8 μm in land crabs, which is comparable with that in aquatic decapods (Table 9.1). Interestingly, this distance is considerably greater than the gas diffusion path length in the lining of the branchial membranes of land crabs.

iii. The branchiostegal ("pulmonary") vascular bed. The major feature of cardiovascular morphology that distinguishes terrestrial from aquatic decapods is an elaboration of hemolymph afferent and efferent supply of the branchiostegal membranes lining the branchial chamber, which, as shown earlier, serve as the major or even sole site for gas exchange. That the endothelial lining of the branchial chamber of many different crabs – aquatic as well as terrestrial – receives a venous return of hemolymph from the central sinuses has long been appreciated by morphologists (Jobert, 1876; Bouvier, 1890; von Raben, 1934). In terrestrial crabs, however, the circulation of the branchial membranes is very much elaborated, along with the lining of the branchial chamber itself.

The afferent circulation to the branchial membranes of land crabs is quite complex, and varies substantially between families. The afferent supply to the branchial chambers of ocypodid crabs has been the source of considerable investigation (and misunderstanding). Von Raben (1934), in a detailed study of the venous circulation of *Uca, Ocypode ceratophthalmus*, and *O. cordimanus*, suggested that the afferent circulation to the branchial membranes originated posteriorly directly from the ventral sinuses, thereafter flowing anteriorly through the distributing vessels. Recently Greenaway and Farrelly (1984) have reexamined the circulation of *O. cordimanus* and report it to be considerably more complex than previously reported. In *O. cordimanus*, the primary afferent supply to the branchial lining originates from the eye sinus, which runs ventrally to the orbit before abruptly turning dorsally (Fig. 9.5A). The paired eye sinuses receive hemolymph posteriorly from the hepatic sinus, as well as from dorsal, ventral, and infrabranchial sinuses via a frontal sinus complex. Each eye sinus gives rise to six afferent vessels perfusing the branchial chamber. The thoracic wall of the branchial chamber is perfused by a basal and lateral vessel; the branchiostegal wall and branchiostegal shelf receive three lateral vessels and one dorsal vessel (Fig. 9.5A, B). Thus, although regional variations exist in terms of which regions of the branchial membranes are perfused by hemolymph originating in a particular sinus, the branchial membranes and the gills of *O. cordimanus* will each receive deoxygenated hemolymph from the body tissues.

As mentioned earlier, there is much interspecific variation in the pattern of the afferent circulation of the branchial membranes. In the coconut crab, *Birgus latro*, the lining of the branchial chambers is perfused by deoxygenated hemolymph originating from the ventral region of the head and anterior thorax (Semper, 1878; Harms, 1932). In the Gecarcinidae, deoxygenated venous hemolymph is supplied to the membranes of each branchial chamber via either a single afferent distributing vessel (*Cardisoma*) or a pair of such vessels (*Gecarcoidea*) (C. Farrelly and P. Greenaway, unpub.). These afferent vessels are supplied directly from the eye sinus anteriorly, the ventral sinus pos-

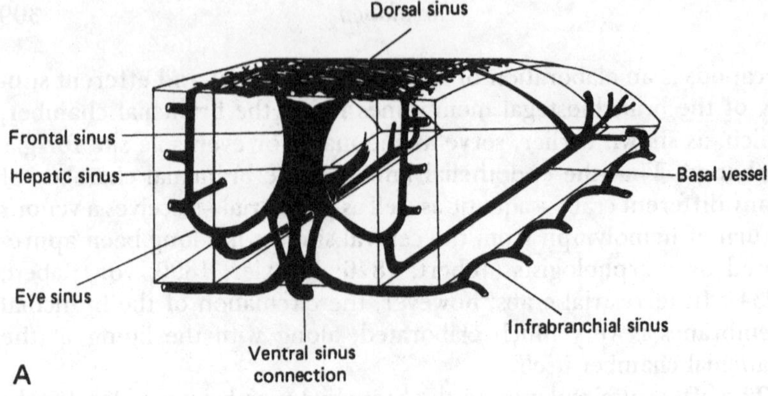

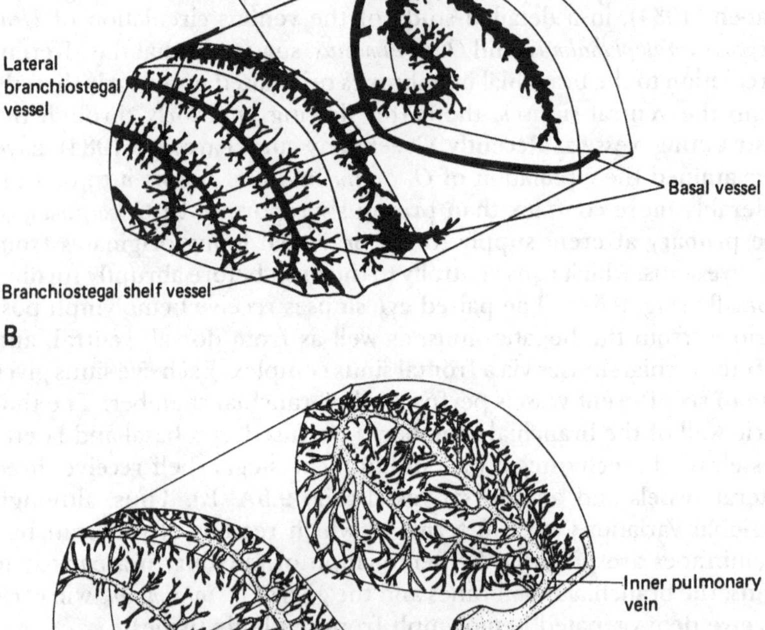

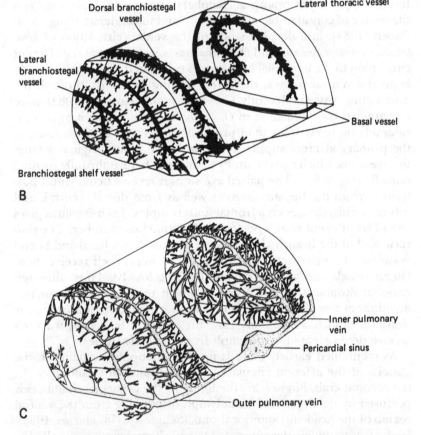

Fig. 9.5. Venous circulation associated with the respiratory organs of *Ocypode quadrata*. A. The distribution of venous hemolymph to the lungs. B. Afferent vasculature supplying deoxygenated hemolymph to the lining of the branchial chamber. C. Efferent vasculature draining oxygenated hemolymph. See text for additional details. (From Greenaway and Farrelly, 1984.)

teriorly, and from the dorsal sinuses. In *Holthuisana*, however, the eye sinus serves as the main afferent supply of deoxygenated hemolymph, whereas in *Geograpsus* the afferent vessels receive hemolymph both anteriorly from the eye sinus and posteriorly from the ventral sinus via a vessel lying anterior to the heart.

In all land crabs that have been examined, the actual exchange sites of the branchial lining consist of lacunar spaces derived from the numerous short, wide trunks given off from the afferent vessels (Fig. 9.6A–C). These short, flat lacunar spaces, which form a very dense network lining the branchial cavity, are covered by a thin layer of epidermal and cuticular tissue. Significantly, these layers are only 0.2–0.5 μm thick, which is about 1/20 as thick as those covering the gill lamellae (Table 9.1). Thus, while the gills of land crabs may have a larger surface area (see also Table 8.1), the diffusion distance for respiratory gases is far less in the "lung" than in the gills.

Interestingly, after breaking down into the flat lacunar system that apparently acts as the site for respiratory exchange in many land crabs, hemolymph gathers into relatively large vessels that again break down into a second set of lacunae (Fig. 9.6D). In the Gecarcinidae this "portal" system is repeated a second time (i.e., three beds of lacunae occur in series) before oxygenated hemolymph enters the efferent vessels directly supplying the pericardium (C. Farrelly and P. Greenaway, unpub.). Two portal systems with three sets of lacuane also typify the branchiostegal membranes of the Grapsidae (*Geograpsus*) and the Sundathelphusidae (*Holthuisana*) (C. Farrelly and P. Greenaway, unpub.).

The efferent drainage of the branchial chambers of *Uca, O. ceratophthalmus*, and *O. cordimanus* consists of two main vessels from each branchial chamber, one draining the thoracic wall and the other the branchiostegal wall (Fig. 9.5C). These two vessels fuse on each side and enter the pericardium quite distinctly from the branchiopericardial vessels. Harms (1932) described the afferent and efferent supply to the branchiostegal membranes of *B. latro*. Generally, the afferent vessel entered the anterior region of the branchiostegal membrane and than broke up into a series of fine distribution vessels. The major efferent vessel appeared to collect hemolymph draining toward the posterior of the branchiostegal membranes. Thus, hemolymph flow generally appeared anterior to posterior, which would be countercurrent to the flow of gas (see Chapter 8, section 3.D). Further information on the patterns of hemolymph flow in *Birgus* is definitely required.

In summarizing the morphology of the venous circulation of land crabs ("typified" by neccessity by relatively few species), it is tempting simply to state that, at least anatomically, hemolymph draining vir-

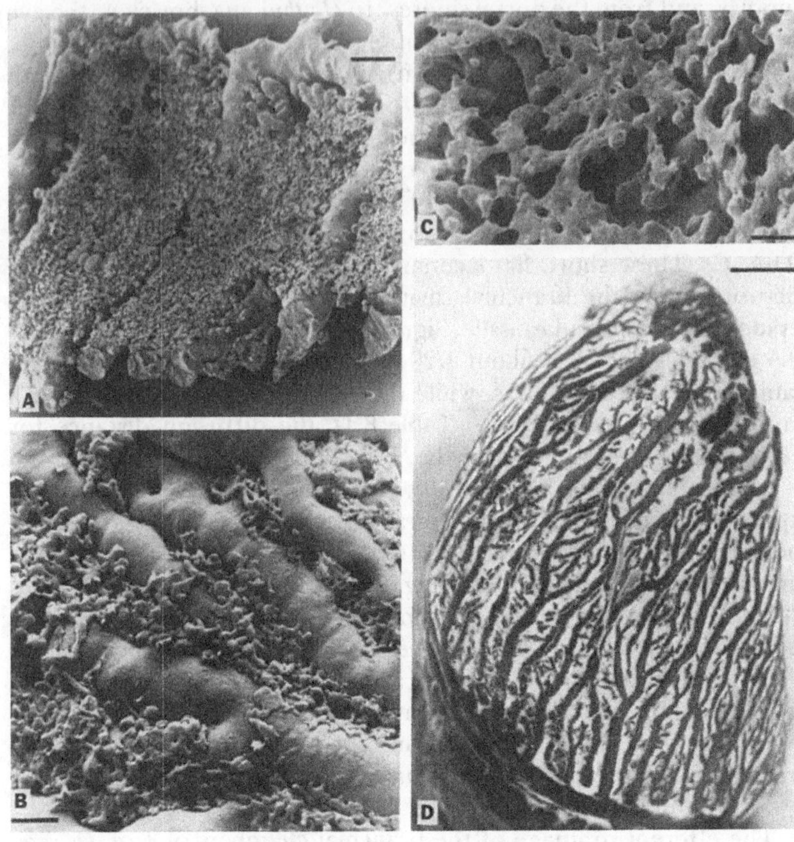

Fig. 9.6. Corrosion casts of the hemolymph vascular spaces of the branchiostegal "lungs" of land crabs. A. Vascular cast of lung of *Geograpsus grayi* viewed from lung lumen. Note large portal systems almost obscured by flat lacunae lining the respiratory epithelium. Scale bar = 400 μm. B. Vascular cast of lung of *Geograpsus grayi* viewed from the carapace side. Note very large interdigitating vessels of the portal systems and the much smaller trunks carrying hemolymph ventrally down to the respiratory lacunae. Scale bar = 100 μm. C. High magnification of respiratory lacunae evident in A. Incomplete filling of lacunae reveals their fingerlike supply vessels. Scale bar = 25 μm. D. Antero-dorsolateral view of corrosion cast of the "lung" of *Gecarcoidea natalis*. Two parallel afferent vessels run along the *lower left side* of the picture and give off lateral branches. The lung continues ventrally, and the pulmonary vein is obscured on the underside of the cast. The cast has been flooded with white dye to enhance contrast. Scale bar = 0.5 mm. (Photographs courtesy of Caroline Farrelly.)

tually any of the body tissues can perfuse either the gills or the respiratory membranes lining the branchial chambers. As to the actual distribution and control of hemolymph flow through these two potential respiratory vascular beds, we have only a few tantalizing findings for the Australian amphibious crab *Holthuisana transversa* (see section 3.B. iii).

It is important to emphasize here that this branchiostegal circulation is not a new development but rather an adaptation of an existing system present in aquatic crabs. P. Greenaway and B. McMahon (unpub.) have found that the branchiostegites of the essentially aquatic crabs *Cancer magister* and *Hemigrapsus nudus* also are perfused, albeit with a much less well developed vascular network. Interestingly, this anatomical network represents a route for functional bypass of the gills, which had hitherto been unsuspected in the Crustacea.

One of the biggest challenges in understanding the circulation of land crabs (or indeed any decapod) is to reconcile the common notion of physiological simplicity in open circulatory systems (purportedly typified by low hemolymph pressures and low and "indiscriminately directed" hemolymph flows) with the obvious complexity of the cardiovascular morphology. In an attempt to understand this morphological complexity, and to determine what specializations for life on land are apparent in land crabs, we now turn to details of their cardiovascular physiology.

3. Cardiovascular physiology

A. *Hemodynamics*

The hemodynamics of land crabs must be inferred largely from morphological features, since direct measurements of hemolymph pressure have been made in only two studies on *C. guanhumi* (Burggren, Pinder, et al., 1985) and *B. latro* (Cameron and Mecklenberg, 1973). In many instances, data on hemolymph pressures and flows in aquatic species (see McMahon and Wilkens, 1983) must be depended on in an attempt to understand circulatory function in land crabs. Yet, even the latter approach is of limited value, since so little information on cardiovascular function in any crustacean is available.*

Decapod circulations are typified by relatively high pressures in the venous sinuses. Infrabranchial pressures from 3 to nearly 20 mmHg have been reported for a variety of aquatic decapods (see McMahon and Wilkens, 1983). These high venous pressures (by the standards of closed vertebrate circulations!) are largely transferred to the pericardial sinus, where hemolymph pressures many mmHg above ambient have usually been recorded (Fig. 9.7A); (Belman, 1975, 1976;

*One of the major hurdles to more invasive studies of the crustacean circulation (and respiratory system) is the development of an effective anticoagulant for their hemolymph, which would permit the use of chronically indwelling catheters for pressure and hemolymph gas measurements.

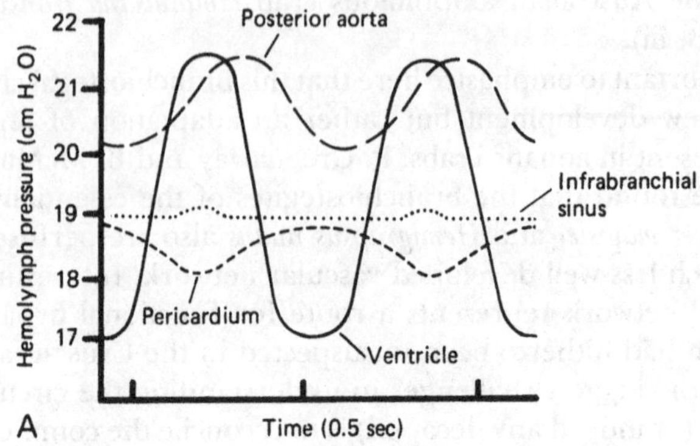

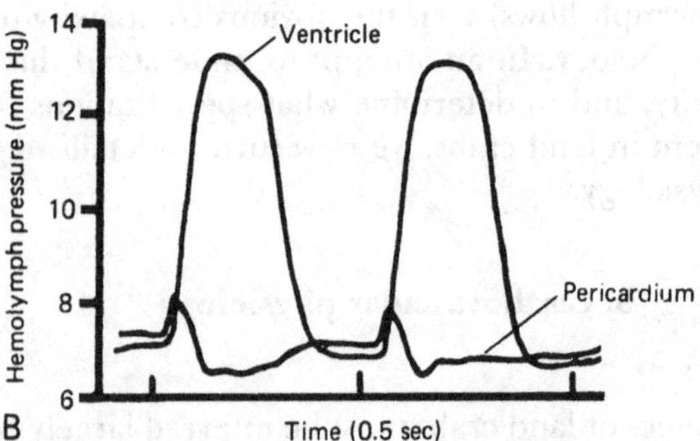

Fig. 9.7. Graphically superimposed hemolymph pressures recorded during two cardiac cycles in the mysid *Gnathophausia ingens* (A) and the brachyuran *Cardisoma guanhumi* (B). (From Belman and Childress, 1976 [*Gnathophausia*] and Burggren et al., 1985 [*Cardisoma*].)

Belman and Childress, 1976; McMahon and Wilkens, 1977; McMahon, McDonald, and Wood, 1979). The land crab *C. guanhumi* complies closely with this pattern (Fig. 9.7B). Mean pericardial pressure is about 6–8 mmHg in restrained but otherwise intact crabs (Burggren, Pinder, et al., 1985). In the anomuran crab *B. latro*, however, mean pressure in the pericardial sinus is substantially lower, oscillating around ambient pressure (Cameron and Mecklenberg, 1973).

Pericardial hemolymph pressure fluctuates slightly during the cardiac cycle in decapods. During diastole in *C. guanhumi*, pressure in the pericardial sinus rises as hemolymph flows in, and, at the end of diastole, pressure in the pericardial sinus may exceed that in the lumen of the heart by as much as 2 mmHg (Fig. 9.7B). This pressure gradient is sufficient to force open the ostial valves and fill the lumen of the expanding heart with oxygenated hemolymph. Diastolic filling must be a relatively rapid process in *Cardisoma* and other land crabs in warm habitats, since at heart rates of about 130 beats·min^{-1} at 30° C the diastolic period for cardiac filling is less than 200 msec (Cameron and Mecklenberg, 1973; Burggren, Pinder, et al., 1985).

With the onset of cardiac contraction, the hemolymph pressure gradient from pericardial sinus to heart lumen is suddenly reversed, resulting in immediate closure of the ostial valves. The heart then enters an initial period of isometric contraction before intracardiac pressure increases above arterial pressure. When this occurs, the valves at the base of the arteries open and arterial ejection commences. Since the heart is completely enclosed within the pericardium, which is relatively noncompliant (Maynard, 1960), the rapid decrease in heart volume accompanying hemolymph ejection into the arteries produces a sharp decrease in pericardial pressure (Fig. 9.7B), and probably helps to aspirate hemolymph from central venous pools.

Hemolymph pressures in the arteries usually rise to peak levels nearly identical with those recorded within the heart (Blatchford, 1971; Belman, 1975; Belman and Childress, 1976), suggesting that the outflow tracts of the arteries from the heart offer little resistance to hemolymph flow. Arterial pressures decline slowly during diastole, with the valves at the bases of the arteries apparently preventing any backflow of hemolymph into the heart. Any elasticity in the central arteries will no doubt contribute to maintaining arterial pressure and flow during diastole.

Intracardiac systolic pressures range from about 5 to nearly 40 mmHg in a wide range of aquatic decapods (see McMahon and Wilkens, 1983). A similarly wide range occurs in the two land crabs in which suah measurements have been made (Table 9.2). In *C. guanhumi* systolic intracardiac pressures average only 14 mmHg. In *B. latro* intracardiac pressures may be as high as 50 mmHg. Cameron and Mecklenberg (1973) propose that the high cardiac systolic pressures (50 mmHg) in *Birgus* may be necessary to keep the complex branchiostegal membranes "inflated" to present maximum surface area to branchial gas.

As implied earlier, diastolic intracardiac pressures in both aquatic crabs and *C. guanhumi* fall to levels below those in the pericardial sinus but still 5–6 mmHg above "zero" (i.e., ambient external) pressure. Once again, *B. latro* proves the exception, with diastolic intracardiac pressures falling to as low as 3 mmHg below ambient (Cameron and Mecklenberg, 1973). Intracardiac and central arterial pressures are influenced greatly by change in air or water pressure within the bran-

Table 9.2. *Mean values (or ranges) for cardiovascular variables for land crabs and representative aquatic decapods under resting,[a] normoxic conditions*

Species	Temp. (°C)	Body mass (g)	Resting heart rate (beats·min^{-1})	Stroke volume (ml·kg^{-1})
TERRESTRIAL SPECIES				
Cardisoma guanhumi	24–27	90–190	70	—
Cardisoma guanhumi	24	128	60	—
Cardisoma guanhumi	25	166	70	—
Cardisoma guanhumi	30	127	134	—
Cardisoma carnifex	25	250–500	85	1.44
Gecarcinus lateralis	25	30	97	1.41
Gecarcinus lateralis	Not stated	28–64	160	—
Ocypode quadrata	25	10–87	Not stated	—
Coenobita clypeatus	23	20–40	120	0.32
Birgus latro	27–30	623–1130	300	—
Birgus latro	25	55–670	105	—
AQUATIC SPECIES				
Cancer productus	16	379	150	—
Cancer productus	12	346	94–100	1.9–2.8
Carcinus maenas	15	50	92	1.3
Carcinus maenas	20	Not stated	108	—
Panulirus interruptus	15	501	64	1.4

[a]Conditions defined as "resting" vary considerably between studies, which no doubt contributes to some of the variations in reported values.

Cardiac output (ml·kg⁻¹·min⁻¹)	Intracardiac pressure – systolic/diastolic (mmHg)	Pericardial sinus pressure (mmHg)	Reference
—	—	—	Shah & Herreid (1978)
—	—	—	Herreid, Lee et al. (1979)
—	—	—	Herreid, Lee et al. (1979)
—	14.1/5.5	8.4 (systole) 6.1 (diastole)	Burggren, Pinder et al. (1985)
122	—	—	Wood & Randall (1981a)
137	—	—	Taylor & Davies (1981)
—	—	—	O'Mahoney-Damon (1984)
70	—	—	Burnett (1979)
38	—	—	McMahon & Burggren (1979)
—	49/3	0	Cameron & Mecklenberg (1973)
—	—	—	Smatresk & Cameron (1981)
—	10.4/6.0	—	Belman (1976)
103–275	—	15–3.0	McMahon & Wilkens (1977)
118	—		Taylor & Butler (1971)
—	10.3/0	0	Blatchford (1971)
128	36/16	17.9	Belman (1976)

chial chambers. These changes in branchial hydrostatic pressure apparently are transmitted across the branchial and branchiostegal membranes and throughout the central venous and arterial circulation. In the land crab *C. guanhumi*, for example, branchial gas pressure increases from a mean of about 5 mmHg below ambient during forward scaphognathite beating to about 2–4 mmHg above ambient during reversed beating (Fig. 9.8). The rise in branchial pressure is reflected in pericardial and intracardiac pressures during both systole and diastole. If the branchial chambers are experimentally opened to air so that reversed scaphognathite beating cannot elevate branchial chamber air pressure, then the transient change in intracardiac hemolymph pressure is eliminated (Burggren, Pinder, et al., 1985).

Hemolymph pressures in the infrabranchial sinuses of land crabs have not yet been measured. However, they must remain elevated above pericardial pressure for a large portion of the cardiac cycle if hemolymph is to flow along a pressure gradient through the gills and into the pericardial sinus. Presumably the pressure in the infrabranchial sinuses is equally (or perhaps most) affected by hydrostatic pressure transients in the branchial chambers. Clearly, hemolymph movement through the venous sinuses will be enhanced by ventilatory movements, particularly when forward scaphognathite beat is interrupted by pauses or reversals (Burggren, Pinder, et al., 1985). Body movements, particularly of the limbs during locomotion, will probably also facilitate hemolymph movement within sinuses, in the same way that limb movement in vertebrates facilitates movement of venous blood toward the heart. Limb movement in *B. latro* causes substantial pressure transients in the pericardium (Cameron and Mecklenberg, 1973), but the effect of such movements on hemolymph flow has yet to be determined. Whether contraction of smooth muscle occurs in any of the sinus walls and whether such muscle could actively propel hemolymph toward the heart are not known.

Comparison of arterial, infrabranchial, and pericardial hemolymph pressures in aquatic decapods indicates that 50–70% of the fall in pressure through the circulatory pathway occurs in transit through the systemic tissues rather than through the gills and membranes of the branchial chamber (see McMahon and Wilkens, 1983). The systemic vascular bed thus appears to be the site of greatest vascular resistance in aquatic forms. While the peripheral resistance of the tissues similarly may be highest in land crabs, this has not been measured. A complicating factor in those terrestrial forms that retain relatively large, well-developed gills is that the vascular resistance of the gills may vary greatly depending on whether the crab is in water (where the gills will be supported and buoyed) or in air (where the gills may partially collapse under their own weight and the capillary

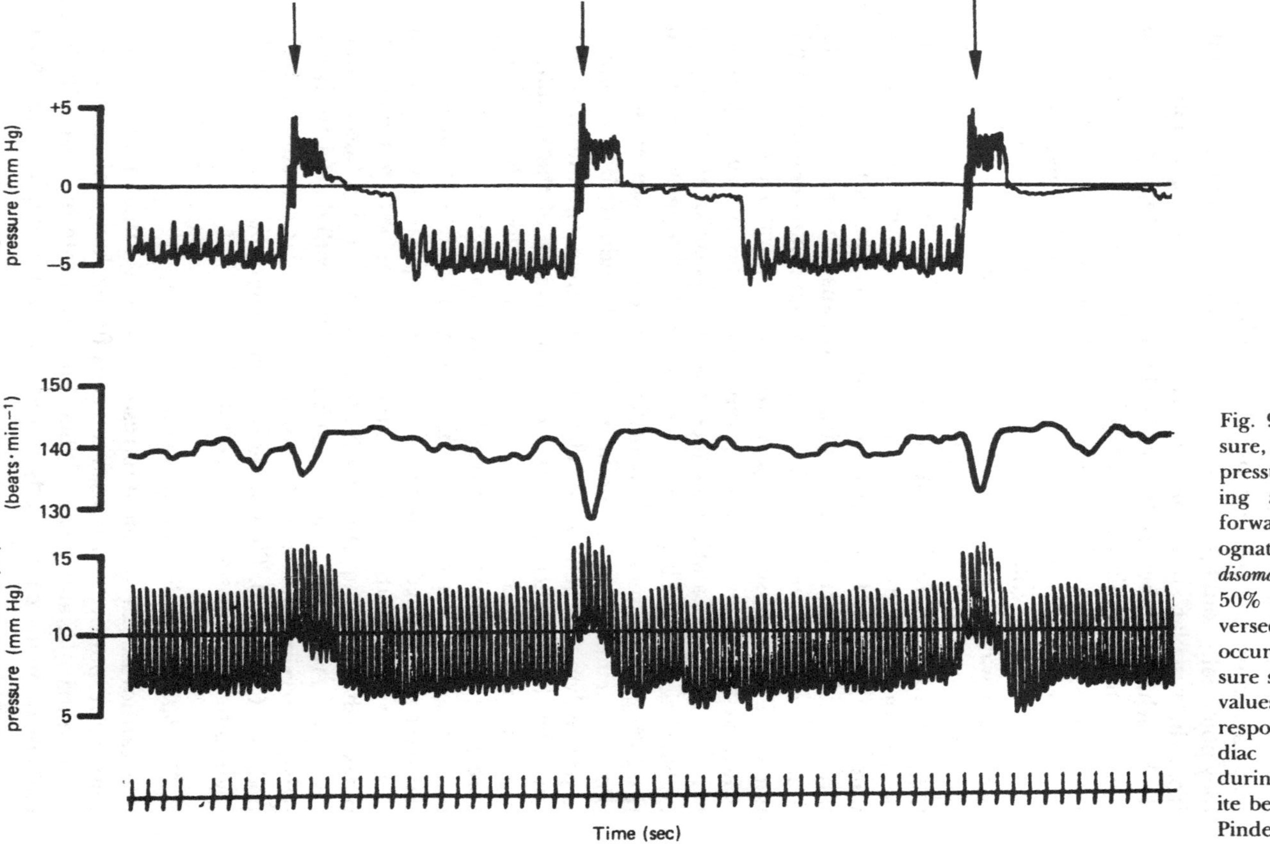

Fig. 9.8. Branchial gas pressure, intracardiac hemolymph pressure, and heart rate during alternating periods of forward and reversed scaphognathite beat in a 142 g *Cardisoma guanhumi* with access to 50% seawater. Periods of reversed scaphognathite beating occur when branchial pressure swings to above ambient values (*arrows*). Note the corresponding rise in intracardiac hemolymph pressure during reversed scaphognathite beating. (From Burggren, Pinder, et al., 1985.)

action of water). Branchial resistance probably will vary with alternating patterns of scaphognathite beat in terrestrial crabs (Taylor and Greenaway, 1984) as has been suggested for aquatic forms (Blatchford, 1971; E. W. Taylor, 1982; H. H. Taylor and E. W. Taylor, 1986).

The relatively high hemolymph pressures in the respiratory lacunae may have important consequences to fluid balance in land crabs, since they potentially could produce a "filtration" or exodus of hemolymph from the vascular spaces of the gills and branchiostegal lining into the branchial chamber. In lower vertebrates, where high pressures in the respiratory capillaries are combined with low blood osmotic pressures and very permeable capillaries, there is a surprisingly high loss of vascular water in the lungs (Burggren, 1982). Because the lungs of vertebrates are completely enclosed within the body wall, this extravascular water can be completely reincorporated into the blood via the lymphatics. In terrestrial crabs, however, fluid filtered across the gill membranes might drain into the branchial chamber, where it could be lost from the body, at least in those species that are unable to retain water in the branchial cavity (see Chapter 7). To our knowledge, filtration of water from the respiratory organs of terrestrial invertebrates (as opposed to evaporative water loss) has not been measured, nor has it been considered a potential route for water loss.

Low systolic arterial pressures (and thus a low pressure gradient between the heart and the smallest vascular spaces of the tissues) in crabs classically have been taken as evidence of a "sluggish" circulation producing low hemolymph velocities and flows. However, it is important to emphasize that the heart of a crab (indeed, of any animal) is primarily a *flow* generator: The pressures developed in the circulation by the contracting heart are a consequence of the peripheral resistance into which the heart attempts to eject blood. A small pressure gradient between the heart and the venous side of the circulation can still generate a high flow of hemolymph to the tissues if the peripheral resistance is low. In fact, the mass-specific cardiac output of resting crabs, both aquatic and terrestrial, may be as much as four times higher than aquatic vertebrates! This surprising finding (surprising to vertebrate physiologists) is associated with the relatively low oxygen-carrying capacity of hemolymph, particularly in aquatic decapods (see Chapter 8). Thus, hemolymph flow must be correspondingly high to maintain tissue oxygenation.

B. Regulation of cardiovascular function

The cardiovascular system of terrestrial crabs, like that of any animal, must be responsive to the tissues' needs for exchange of respiratory gases, nutrients, and wastes. When metabolic rate increases due to exercise or an increase in body temperature, for example, so too must the rate of oxygen transport to the tissues. While adjustments in ventilation and in the respiratory properties of the hemolymph can facilitate oxygen uptake and transport, as described in Chapter 8, it is important to emphasize that an increase in cardiac output (perhaps accompanied by changes in distribution of cardiac output) is fundamental to supporting an increase in metabolic rate. Essentially, cardiac output is the product of the stroke volume of the heart and of heart rate. Largely because measurement of heart rate is far easier than measurement of stroke volume, environmental and physiological factors that affect the rate of heart beat have been studied in some detail, as will now be described.

i. Heart rate. Heart rate has been measured in several different species of land crabs under many different experimental conditions. Table 9.2 presents values for "resting" heart rate, though in many (perhaps most) instances these rates reflect experimental disturbance associated with recent implantation of electrodes, etc.

"Resting" heart rate depends, of course, on body temperature. While heart rate in *Cardisoma guanhumi* at 30° C is higher than that measured at 24–25° C (Table 9.2), a study designed specifically to determine the effects of temperature on heart rate of land crabs has not been published. Heart rate might also be expected to increase with increasing body mass, as occurs in aquatic crabs (Schwartzkopff, 1955; Ahsanullah and Newell, 1971), though not generally in vertebrates. In *G. lateralis*, heart rate increases with body mass (Taylor and Davies, 1982a). Shah and Herreid (1978), however, were unable to observe a relationship between heart rate and body mass over a range from 90 to 190 g in *C. guanhumi*.

At rest, many marine crabs show periods of cardiac inactivity of variable but frequently long duration (McMahon and Wilkens, 1977). However, heart beat is generally continuous in land crabs such as *C. guanhumi* (Herreid, O'Mahoney, and Shah, 1979; Burggren, Pinder, et al., 1985), *G. lateralis* (O'Mahoney-Damon, 1984), *C. clypeatus* (McMahon and Burggren, 1979), and *B. latro* (Cameron and Mecklenberg, 1973; Smatresk and Cameron, 1981). As mentioned in Chapter 8, the high oxygen capacitance of air versus water will allow continued gas exchange even during ventilatory pauses, and consequently the periods of cardiac arrest observed during ventilatory pauses in aquatic crabs might not be anticipated in land crabs. Very brief periods of cardiac arrest have been observed in *C. carnifex* when scaphognathite pumping resumed after a ventilatory pause (Wood and Randall, 1981a), while a transient, moderate bradycardia may occur during intense scaphognathite activity in *C. clypeatus* (McMahon and Burggren, 1979). Interestingly, Shah and Herreid (1978) report

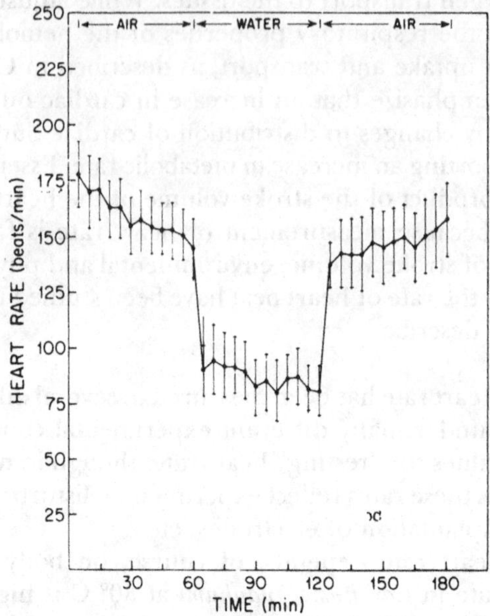

Fig. 9.9. Effect of approximately 1 hour of submersion in 25% seawater on heart rate in the land crab *Gecarcinus lateralis*. Mean values ± 1 SE are presented. (Reprinted with permission of *Comp. Biochem Physiol.* 79A, O'Mahoney-Damon, P., Heart rate of the land crab *Gecarcinus lateralis* during aquatic and aerial respiration, 1984, Pergamon Press, Ltd.)

that resting heart rate in *C. guanhumi* shows a diurnal rhythm, with the highest rates in late afternoon and evening.

Since many "land" crabs may spend some proportion of their time partially or even fully submerged in water, considerable interest has been expressed in the effects of aerial versus aquatic exposure on heart rate. A bradycardia, or reduction in heart rate, occurs on submersion in water in *C. guanhumi* (Shah and Herreid, 1978) and *G. lateralis* (O'Mahoney-Damon, 1984). Figure 9.9 shows the response of *C. guanhumi* to one hour of submergence in 25% seawater. The salinity of water apparently has no influence on the submersion bracycardia in this species (Shah and Herreid, 1978). Aquatic crabs generally show a bradycardia when exposed to air (see deFur and McMahon, 1984a, for references). A bradycardia on air exposure is also seen in those crabs that typically occupy intertidal environments, e.g., *Scylla serrata* (Hill and Koopowitz, 1975), *Hemigrapsus sanguineus* (Depledge, 1984), and *Carcinus maenas* (Ahsanullah and Newell, 1971; Taylor, Butler, and Al-Wassia, 1977; Hume and Berlind, 1976; A. C. Taylor, 1976). Almost all vertebrates show a bradycardia when placed in an environment regarded as unfavorable for respiration (e.g., fish into air, mammals into water). That land crabs show a bradycardia on short-term submersion in water testifies to the extensive adaptations for aerial respiration in these decapods.

Unlike many freshwater aquatic decapods, land crabs may not normally be exposed to a respiratory medium low in oxygen or high in carbon dioxide. Important and numerous exceptions involve those crabs that may seal themselves in burrows during dry seasons or normally inhabit burrows containing stagnant groundwater (see Wood and Boutilier, 1985). Experimental breathing of very hypoxic gas instead of air induces a profound bradycardia in the crabs *Cardisoma* (Herreid et al., 1979) and *Coenobita* (McMahon and Burggren, 1979). At least in *Coenobita*, however, heart rate is unaffected by ambient oxygen partial pressure until Po_2 drops to a very severe level of 20 mmHg. At Po_2's of 20–120, heart rate is not significantly altered, but the infrequent cardiac arrhythmias are eliminated. Exposure to elevated levels of ambient CO_2 (above 7 mmHg) produced a slight elevation of heart rate and elimination of arrhythmias in *C. clypeatus* (McMahon and Burggren, 1979).

Until measurements of stroke volume and cardiac output are made during conditions of hypoxia and hypercapnia, the adaptive benefit, if any of these heart rate responses remains equivocal (see Herreid et al., 1979).

An increased heart rate might reasonably be expected to occur during activity. Unfortunately, heart rate during a specified exercise protocol has been measured in only a few studies on either *C. guanhumi* (Herreid et al., 1979; Wood and Randall, 1981a) or *B. latro* (Smatresk and Cameron, 1981). Paradoxical effects of exercise on heart rate were reported by Herreid et al. (1979) for *C. guanhumi*. Exercise at 150 $cm \cdot min^{-1}$ for 10 minutes had no effect on heart rate. At higher velocities of 300 $cm \cdot min^{-1}$, however, exercise was accompanied by bradycardia and very brief cardiac pauses. Bradycardia and cardiac pauses persisted for up to two hours after exercise. Exercising *C. guanhumi* for 20 minutes rather than 10 minutes made the bradycardia more pronounced and the recovery more protracted. As Herreid et al. (1979) suggest, the relationship between heart rate and stroke volume is poorly understood (see the next section), and whether this bradycardia during exercise represents a real reduction in cardiac output is not clear. Quite the opposite responses to exercise at higher speeds were observed by Wood and Randall (1981a) for *C. carnifex*. In that study, exercise at 270 $cm \cdot min^{-1}$ and 660 $cm \cdot min^{-1}$ for 10 minutes caused heart rate to increase 15% and 30%, respectively (Fig. 9.10). The periods of arrhythmia characteristic of rest disappeared during exercise at either level. Moreover, heart rate remained elevated

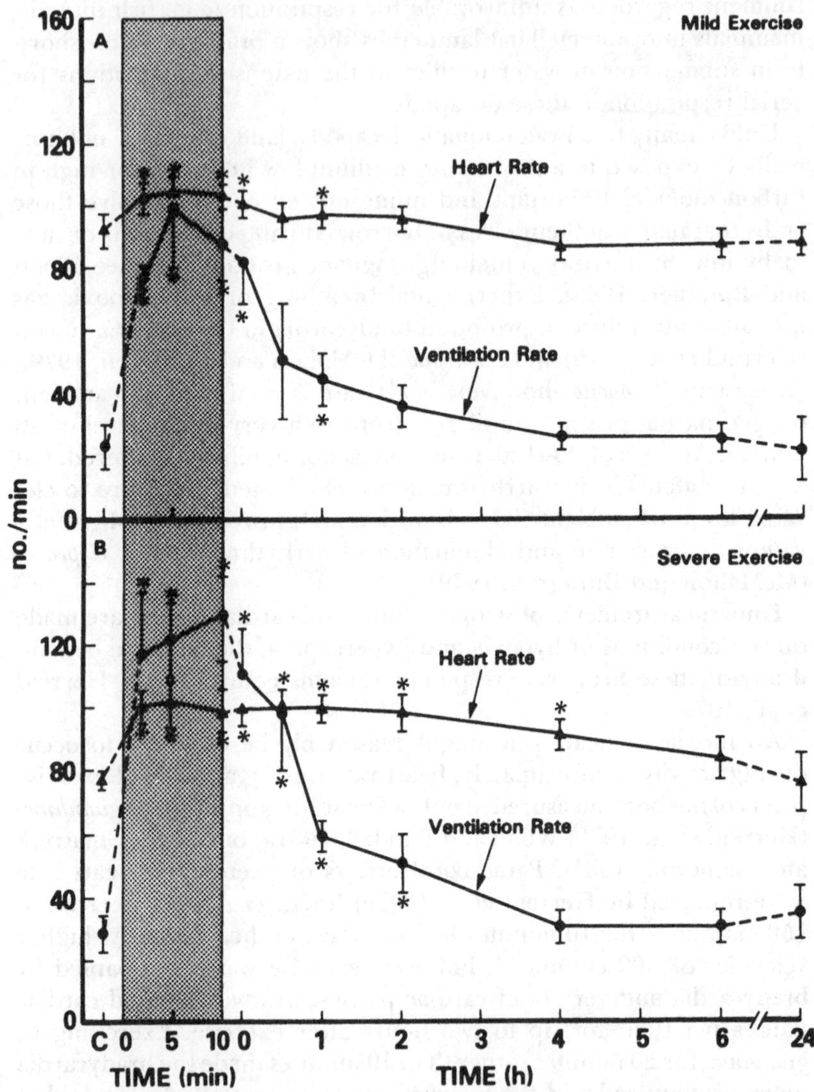

Fig. 9.10. Changes in heart rate and ventilation rate during mild (270 cm·min^{-1}) and severe (660 cm·min^{-1}) exercise in the land crab *Cardisoma carnifex*. The exercise period is indicated by the *stippled bar*. Mean values ± SE are presented. *Asterisk* (*) indicates significant difference from preexercise control value. (From Wood and Randall, 1981a.)

above resting levels for two hours and six hours, respectively, after exercise stopped. Clearly, further experiments are necessary to determine if these represent interspecific differences in the cardiac response to exercise.

Apart from the fairly obvious statement that the heart rate of land crabs is regulated both neurally and hormonally, little is known of the actual regulatory mechanisms for any feature of cardiac performance. Yet, regulation of hemolymph pressure through changes in heart rate may be particularly important in animals like land crabs where the total pressure gradient from heart lumen through the tissues to the pericardial sinus may be only 10 mmHg. Whereas a hemolymph pressure perturbation of a few mmHg is of little significance in most high-pressure vertebrate circulations, a similar change in pressure, if not compensated for, could induce major changes in hemolymph flow in crabs. In *C. guanhumi*, experimental manipulation of hemolymph volume, and thus of hemolymph pressure, has a profound affect on heart rate. Heart rate is inversely proportional to hemolymph pressure, decreasing 1–2% with every mmHg increase in pressure (W. Burggren, A. Pinder, B. McMahon, and M. Wheatly, unpub.). These data are indicative of a "baroreceptor-like" reflex, although the actual mechanism for this pressure-related regulation of heart rate remains unknown. Some cardiorespiratory interactions involving adjustment in heart rate appear to be regulated at the level of the central nervous system, at least in *C. guanhumi* (Burggren, Pinder, et al., 1985). As already discussed, *C. guanhumi* typically shows brief periods of reversed scaphognathite beating. In most crabs examined, heart rate begins to fall simultaneously with the transition from forward to reversed scaphognathite beating (Fig. 9.8). This "reversal bradycardia" apparently does not reflect reflex adjustment in response to the transient rise in central hemolymph pressures: "Reversal bradycardia" persists even when the hemolymph pressure transients are eliminated experimentally by keeping branchial chamber pressure at ambient during reversed scaphognathite beating (Burggren, Pinder, et al., 1985). This continued interaction of heart rate with adjustment in ventilatory pattern probably reflects direct interaction of central neural elements controlling ventilation and circulation, as described for marine crabs (see McMahon and Wilkens, 1983).

ii. Stroke volume. No direct measurements of stroke volume have been made in land crabs, though estimates of resting stroke volume have been made by dividing estimated cardiac output by heart rate (Table 9.2). Values of stroke volume for *C. guanhumi, G. lateralis*, and *O. quadrata* fall within the range reported for aquatic brachyurans; the value for *C. clypeatus* is the lowest stroke volume reported for a decapod (see McMahon and Wilkens, 1983). Whether this represents a difference between Brachyura and Anomura cannot be ascertained at this time. It should be emphasized, however, that all of these data for land crabs were calculated from heart rate and a cardiac output estimated by the Fick principle, and thus are subject to potential error at a number of stages.

Little is known of how variations in stroke volume are used to adjust cardiac output in land crabs. During severe exercise in *C. carnifex* heart rate increases 30% but estimated cardiac output increases 200–300% (Wood and Randall, 1981a). This suggests that adjustment in stroke volume plays the major role in increasing cardiac output during exercise in this species. Cameron and Mecklenberg (1973) report that "disturbance" of *B. latro* causes no change in heart rate but a large increase in minute ventilation. Either the increase in ventilation is not met by a corresponding rise in cardiac output (probably a physiologically inefficient response), or cardiac output is increasing by changes in stroke volume rather than in heart rate in this situation. Recently Burnett, deFur, and Jorgensen (1981) have measured stroke volume in aquatic crabs (*Cancer anthonyi* and *C. magister*) by a relatively direct thermodilution technique. Apparently, much of the variation in cardiac output recorded for these crabs also was due to adjustment in stroke volume rather than heart rate. If stroke volume is equally variable in land crabs, then, coupled with the frequently observed excursions in their heart rate, future experiments may reveal very large changes in cardiac output under various experimental situations.

iii. Cardiac output and its distribution. Cardiac output has been determined in relatively few land crabs (Table 9.2). Although the values for *Gecarcinus, Cardisoma, Ocypode,* and *Coenobita* differ considerably, it is very interesting that in each instance cardiac output is lower than in aquatic crabs (Mangum and Weiland, 1975; A. C. Taylor, 1976; McMahon and Wilkens, 1977; Taylor and Butler, 1978; Burnett, 1979; McMahon et al., 1979; McMahon and Burggren, 1979; Burnett et al., 1981; Taylor and Davies, 1982a), provided that reasonable corrections are made for differences in measurement temperature (e.g., assuming a doubling of cardiac output with a 10° increase in body temperature). Burnett (1979) reported that the resting cardiac output of *O. quadrata* was 1/10 that of the marine spider crab *Libinia emarginata*. Many authors have suggested that the comparatively low cardiac output of terrestrial crabs is a direct reflection of the higher oxygen-carrying capacity frequently observed for the hemolymph of terrestrial crabs (Burnett, 1979; McMahon and Burggren, 1979; Burnett et al., 1981; McMahon and Wilkens, 1983). Figure 9.11 graphically illustrates that cardiac output is correlated with hemolymph oxygen capacity in a variety of terrestrial and aquatic decapods, as well as aquatic vertebrates. What is unclear, however, is what the selection pressures were that led to these differences in land crabs. High hemolymph O_2 capacity and the attendant low hemolymph flow

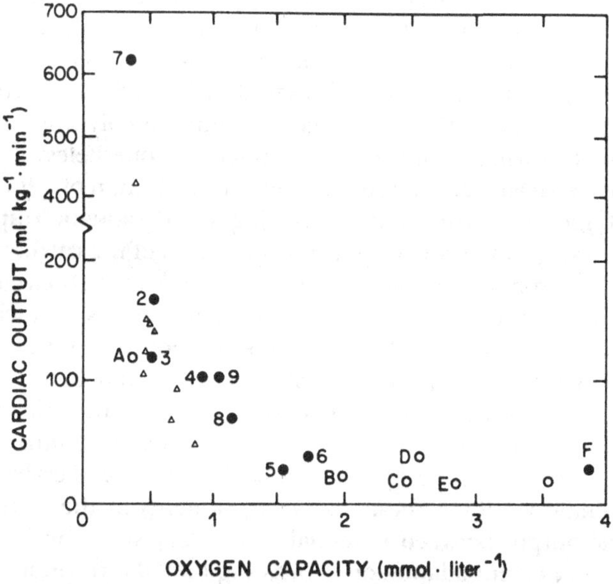

Fig. 9.11. The relationship between cardiac output and hemolymph (blood) oxygen capacity in a variety of decapod crustaceans and aquatic vertebrates. *Open triangles* are values for different individuals of *Cancer magister*. *Closed circles* are mean values for other decapods. 2, *Cancer productus*; 3, *Carcinus maenas*; 4, *Panulirus*; 5, *Cancer magister*; 6, *Coenobita clypeatus*; 7, *Libinia emarginata*; 8, *Ocypode quadrata*; 9, *Cardisoma guanhumi*. *Open circles* are mean values for fishes. *A*, ice fish; *B*, dogfish; *C*, carp; *D*, flounder; *E*, tench; *F*, rainbow trout. (After McMahon and Wilkens, 1983, who quote sources of these data.)

may have been an adaptation to reduce evaporative water loss from the respiratory surfaces via a reduced convective transport of hemolymph to these potential sites of water loss.

Effective exchange of respiratory gases is affected not only by the magnitude of cardiac output but also by the pattern of hemolymph distribution throughout the vascular system. In terrestrial or amphibious crabs, where multiple respiratory sites occur (e.g., gills, branchiostegites, thoracic wall of the branchial chamber; see earlier in this chapter) and either one of two respiratory media may be present in the branchial chamber, the efficacy of gas exchange may pivot on the distribution of cardiac output between three respiratory sites. In crustaceans only a single study has attempted to ascertain whether the distribution of cardiac output can be altered. Taylor and Greenaway (1984) injected small radioactive microspheres into the venous sinus of the amphibious crab *H. transversa*, and observed the distribution of these spheres in the circulation, in an innovative modification of a classic technique used in the investigation of vertebrate circulations. As Taylor and Greenaway (1984) were quick to indicate, this technique gives less than satisfactory results because the relatively slow hemolymph velocities in the venous sinuses allow considerable settling of microspheres close to the site of injection, and hemolymph clots tend eventually to form around the microspheres. Nonetheless, these experiments have yielded a reasonably reliable indication of adjustments in distribution of cardiac output. During strictly aquatic respiration (crabs submerged in water), approximately 80% of the cardiac output was directed from the venous sinuses into the gills, with the remaining 20% perfusing the branchiostegites and membranous thoracic walls. After one day of air breathing, this distribution had reversed such that 87% of cardiac output was distributed to the branchiostegites and thoracic walls, with the small remainder perfusing the gills directly. Since the gills of crabs exposed to air retain relatively little of their functional surface area compared to those of submerged crabs (Taylor and Greenaway, 1984), these major adjustments in the distribution of cardiac output between potential respiratory sites appear to represent an important adaptation facilitating bimodal respiration in *H. transversa*. How these changes in cardiac output are achieved is not clear. It is possible that active physiological regulation of the distribution of cardiac output is being achieved, although Taylor and Greenaway (1984) indicate that, though active regulation should not be discounted, compression of the gills during air exposure could increase branchial vascular resistance sufficiently to divert hemolymph to the branchiostegites, since the branchial and branchiostegal circulations are arranged in parallel (Fig. 9.2).

The finding in decapods of a redistribution of cardiac output between major vascular beds is of great significance, since it portends a similar degree of interest (and subsequent experimental effort) directed toward cardiovascular shunts in vertebrate circulations (see Johansen and Burggren, 1985).

4. Gravity: a circulatory "problem" with terrestrial life

Unlike aquatic crabs, land crabs are subjected to gravitational forces without the offsetting effects of water buoyancy. These gravitational forces could potentially affect cardiovascular function in several ways, both directly and indirectly. For example, the metabolic costs of maintaining posture and of locomotion are much higher in terrestrial animals because they must support their own mass in air (see Chapter 10). Thus the demands made on the circulation with respect to supplying nutrients and oxygen and removing wastes and carbon dioxide may be correspondingly higher than in aquatic crabs.

Gravity also has direct effects on cardiovascular function. The resistance of the branchial circulation of land crabs may be higher than that in aquatic forms, since gill filaments are not supported when in air and collapse under their own weight. However, the gills of both anomuran and brachyuran land crabs are strengthened and supported to maintain adequate spacing in air (see von Raben, 1934; Gray, 1957; Diaz and Rodriguez, 1977; Taylor and Greenaway, 1979; see also Fig. 8.1). The effects on branchial hemodynamics of these morphological modifications to the gills of land crabs beg further investigation.

Yet another potentially disruptive feature of life on land involves the effects on the circulation of a functional reduction in circulating hemolymph volume due to pooling of venous hemolymph in the lower extremities. Such pooling arises when venous pressures necessary to overcome the hydrostatic pressure of the column of hemolymph leading up to the heart cause distension of the veins. Hemolymph is sequestered in the venous circulation below the heart, and cardiac output may be seriously affected. The larger the animal (and the more compliant its venous circulation), the more pronounced this effect will be, since the venous pressure at the lower extremities will be proportional to the height of the venous hemolymph column.

Is venous hemolymph pooling really a problem for land crabs? Most land crabs are relatively small, and the tissues of brachyurans, at least, are confined within a rigid, noncompliant carapace; at first glance it might seem that gravity-induced pooling of hemolymph in ventrally located sinuses may be minimal. Yet, the appearance of life in a rigid box is an illusion. The gills, the lining of the epibranchial chamber, and in anomurans, the abdomen, all represent compliant components of the body wall that could yield in response to elevated venous pressures. In the larger crabs such as *Cardisoma* or *Birgus*, the height of the venous column from lower extremities to the heart can reach 20 cm, particularly in those species that may assume vertical postures during tree climbing or moving in burrows.

Gravity also exerts relatively profound effects on blood flow within individual organs even in quite small animals. For example, the lungs of a variety of mammals and reptiles show quite profound vertical stratification of blood flow, with the most elevated regions of the lung receiving substantially less blood than the lower regions (West, 1977; Seymour, Spragg, and Hartman, 1981). A similar stratification of hemolymph may occur in the branchiostegal respiratory organs of land crabs. Taylor and Greenaway (1984), using injected radioactive microspheres to determine pathways of hemolymph flow in *H. transversa*, reported that the largest numbers of microspheres were trapped in the ventral rather than dorsal regions of the branchiostegites. Whether this results from the effects of gravity on hemolymph flow requires further investigation. Vertical stratification of hemolymph in the branchiostegites of *C. guanhumi* may be prevented by the frequent periods of reversed scaphognathite beating typical of these crabs when they have access to water (Burggren, Pinder, et al., 1985). A bout of reversed beating produces a transient rise in branchial chamber pressure to levels about 10 mmHg above that level maintained during forward scaphognathite beating (Fig. 9.8). A pattern of ventilation with alternate periods above and below ambient pressure may well serve to "wring out" the vascular spaces of the branchiostegite and thus prevent hemolymph pooling and vertical stratification of hemolymph flow.

Interestingly, the typical sustained threat postures of many crabs (see Chapter 4) may have important implications to the circulation of hemolymph. In large crabs such as *Cardisoma*, which when threatened raise and maintain the chelae as much as 15 cm above the level of the heart, such postures may induce several hemodynamic adjustments. At the outset of this posture, venous return to the heart (and thus cardiac output) will be elevated as hemolymph drains from the venous sinuses in the large chelae. If this column of venous hemolymph is interrupted by complete sinus collapse, the "siphon effect" maintaining arterial perfusion of elevated tissues will be interrupted. In this circumstance, increased arterial pressures will be required if the elevated chelae are to receive continued hemolymph flow. If arterial pressure is elevated, those tissues at or below the level of the heart will also be perfused at higher pressure, possibly leading to hemolymph accumulation if the threat posture is maintained for too long.

The gravitational effects that land crabs are exposed to potentially pose many cardiovascular problems – problems that aquatic crabs simply will not experience. While speculation on these problems and their solutions is relatively easy, it is more challenging actually to test hypotheses related to gravitational effects in terrestrial invertebrates. Monitoring of cardiovascular function during experimental tilting, a technique used very effectively in reptiles to elucidate gravitational effects and the morphological and physiological adaptations to them (see Lillywhite, 1986), could be very profitably applied to the study of circulation in land crabs.

5. Conclusions

If measured in terms of species diversity, the decapod Crustacea have not been particularly successful in exploiting the terrestrial habitat (see Chapter 11). Yet, the metabolic, respiratory, and cardiovascular challenges of living in a dry, thermally unstable, and nonbuoyant medium, where sporadically high rates of activity and locomotion are essential to survival, have been very successfully met by many genera of both the brachyuran and anomuran crustaceans.

In order to master these challenges, the respiratory and cardiovascular systems of land crabs have undergone significant modification during the evolutionary transition from aquatic to semiterrestrial or terrestrial habitats. In most cases, however, these structural and functional modifications have not arisen *de novo*, but rather represent changes in existing ancestral structures and processes. Even the magnificent "lung" of the coconut crab, *B. latro*, represents a qualitatively simple elaboration of the surface area and underlying circulation of preexisting branchiostegal linings. Certainly the method of ventilation of this aerial-gas-exchange organ is very similar to that in aquatic anomurans – primarily it is the respiratory medium that differs. As another example, the hermit crabs, *Coenobita* spp., represent an even more specialized case, in which a little of the marine environment is carried about within the adopted molluscan shell. Although these crabs are terrestrial in habitat, they are relatively less adapted to an aerial existence from the perspective of respiration and circulation.

In a sense, then, the land crabs have failed to develop the new respiratory/cardiovascular structures and processes that apparently were directly responsible for the relatively explosive colonization of land by the annelids, mollusks, and the tetrapod vertebrates. Yet, the study of cardiorespiratory adaptations for air breathing in land crabs can tell us much about the primary requisites for living on land that must be met by any animal. For example, like air-breathing fishes (Burggren, Johansen, and McMahon, 1986) and amphibians (Lenfant and Johansen, 1967), the more terrestrially oriented of the crabs tend to have higher hemolymph oxygen contents and lower hemocyanin–oxygen affinities. Similarly, regulation of acid–base balance by ventilatory adjustment of P_{CO_2}, rather than HCO_3^-, becomes increasingly common in more terrestrial forms. As a final example, the diffusion path length between hemolymph and the respiratory medium in the branchiostegal "lung" is much shorter than in the retained but abbreviated gills, as in amphibious vertebrates. When specific cardiorespiratory modifications such as those just mentioned appear repeatedly in a wide range of invertebrate and vertebrate animals undergoing a transition from aquatic to aerial respiration, then these modifications can be assumed to be fundamental to this evolutionary transition.

On the other hand, many adaptations appear to be unique to the land crabs. The elaborate respiratory lacunae and amazingly complex multiple portal systems of the branchiostegal "lung," for example, are not to be found in other transitional air-breathing animals, and apparently represent a solution unique to the land crabs.

Thus, in addition to studying them for the sake of their own interesting adaptations, the land crabs represent an invaluable "calibration point," against which the evolution of respiratory and cardiovascular structures and processes for aerial respiration in other invertebrates and vertebrates can be more easily assessed.

Acknowledgments

During the preparation of this manuscript the authors received financial support from the Canadian National Science and Engineering Research Council (B. R. M) and the American National Science Foundation (W. W. B). We thank Dr. Alan Pinder and Dr. Peter Greenaway for valuable discussions. Dr. Caroline Farrelly and Dr. Peter Greenaway graciously provided much unpublished data on gill ultrastructure for incorporation into this chapter. Robert Infantino provided tremendous assistance in the final stages of manuscript preparation (proofing, indexing) of not only this chapter, but of the entire book, and we are extremely grateful.

10: Energetics and locomotion

CLYDE F. HERREID II and
ROBERT J. FULL

CONTENTS

1. Introduction

The purpose of this chapter is to describe the general way that land crabs consume and use energy. Although techniques are available to measure food intake and its caloric content, and to determine the aerobic and anaerobic metabolism of organisms during activity and rest, a complete energetic evaluation of any animal, much less one for land crabs, has yet to be made. Thus, we are in possession of a few fragments of data – enough to embolden us to speculate but not enough to restrict our imagination.

This chapter presents an overview of the feeding process, digestion, assimilation, metabolism and the energetic cost of a variety of behaviors that crabs perform. Considerable information is available on locomotion, which is easily quantified and permits us to determine the metabolic range for each species. Other behavior patterns have scarcely been considered from an energetic point of view. This paucity of data will become evident in the following pages where we highlight the information that is required before we can work out an energy budget for any species.

2. Feeding

Crustaceans eat an enormous variety of food. Nutrition and feeding are reviewed by Marshall and Orr (1960) and Grahame (1983). Semiterrestrial and terrestrial crustaceans include filter-feeders, scavengers, vegetarians, and predators (also see Chapter 3). Few studies have identified the specific food items taken by a given species, and no caloric studies have been made.

A. Filter-feeding

A large number of crabs living in the intertidal flats consume detritus. For instance, the fiddler crab *Uca signatus* lives in burrows and feeds nearby as the tide recedes. The crab picks up mud with its spoon-shaped chelae and rapidly shovels it into its mouth. Some sorting of organic material from mud occurs by the mouthparts because small mud balls are spit out as fast as new sand is shoveled in. The process is repeated every few seconds, and the burrows are surrounded by rows of little balls. Similar feeding techniques are employed by tropical crabs *Dotilla, Scopimera, Ilyoplax, Metaplax,* and *Paraclei stostoma* (Marshall and Orr, 1960).

The feeding of *Uca pugilator* has been well studied (Altevogt, 1957a; Miller, 1961; Robertson and Newell, 1982). Sand is scooped up by the chelae and passed into the buccal cavity where it is sorted. Heavy inorganic particles fall to the bottom of the cavity and are pushed out of the mouth between the third maxillipeds. The buccal cavity may be flooded with water from the gill chamber, and light organic particles float to the top. On draining, the fine particles are left on the setae of the mouthparts. Food material adhering to sand grains is removed by the setae on the second maxillipeds, which are used to scrape the sand grains across the bristles on the first maxillipeds.

B. Scavengers and vegetarians

Land crabs are largely scavengers. For instance, in Panama the hermit crab *Coenobita compressus* rests above the tide line in trees, and under rocks and drift wood during the day and visits the beach at low tide to search for food. It eats fruit, dead animals, and even human feces on the shore (Herreid and Full, 1986a).

Crabs such as *Cardisoma guanhumi* are normally herbivores. This species' burrows are typically in mangroves or in areas with heavy vegetation. There the crabs harvest the grass as well as fruit and leaves falling from the trees, dragging them to their burrows to feed. Significant competition for food exists because crabs clear the vegetation around their burrows and move immediately to retrieve berries or leaves that fall to the ground. *C. guanhumi* eat a surprising variety of food. Herreid (1963), studying large colonies of crabs in a plant experimental station in Florida, noted them eating 35 families of plants. In addition, they ate filamentous algae, and carrion consisting of crabs, fish, and birds.

Bliss et al. (1978) and D. L. Wolcott and T. G. Wolcott (1984) reported that land crabs, *Gecarcinus lateralis*, in Bermuda are largely herbivorous, but they are also opportunistic carnivores; they are cannibals, scavenge carrion and mammalian feces, and prey on small frogs and insects. Wolcott and Wolcott (1984) argue that nitrogen is a scarce and limiting resource for these land crabs, a problem partially offset by cannibalism (see Chapter 3). Soybean supplements with high nitrogen content markedly reduce the cannibalism of adults on conspecific juveniles.

C. Predators

Predation is common among the large marine decapods and stomatopods. Lobsters and crabs prey on fish, mollusks, worms, and echinoderms as well as other crustaceans. On land, only a few crabs depend on predation to any extent, although it may be a minor component in the diet of many scavengers. Ghost crabs, *Ocypode*, are perhaps the most effective crab predators on land. They are the fastest-running crabs and can easily run down a small *Uca* that strays too far from its burrow. However, even *Ocypode* depend largely on scavenging.

From our comments, one may safely conclude that virtually all land crabs are quite active in seeking food and that their diets are extremely diversified.

D. Caloric intake

There is little direct information on the caloric intake of land crabs. Basic data on oxygen consumption can be used to estimate energy demand, of course. However, there is no information on the efficiency of assimilation, where the caloric content of the following categories are known: Assimilation = Consumption − (Feces + Excreta). Excreta include all losses such as nitrogenous waste, mucus, and shed cuticle (Crisp, 1971). Excreta are often ignored or considered trivial in such calculations. Therefore, authors typically express assimilation as a ratio or efficiency by dividing assimilation by the consumption value. Techniques for its measurement are reviewed by Grahame (1983), who gives some values for aquatic decapods. Spider crabs, *Libinia emarginata*, eating algae, fish, and mussel have an efficiency of 95–99% (Aldrich, 1974). Juvenile lobsters, *Homarus americanus*, eating brine shrimp have an assimilation efficiency of 81% (Logan and Epifanio, 1978). It seems reasonable to expect equally high assimilation in terrestrial species, especially in view of the fact that crabs do possess cellulase and chitinase enzymes to break down plant fibers and exoskeletal material (see Gibson and Barker, 1979, for review).

The caloric content (kcal) of a gram of dry food is surprisingly constant: terrestrial plants, 4.5; algae, 4.9; insects, 5.4; invertebrates except insects, 3.0; vertebrates, 5.6 (see Odum, 1983). However, the water content is so variable that great differences in caloric content exist when wet mass is considered. Nevertheless, most living organisms are two-thirds water and minerals; therefore, a value of 2 kcal·g^{-1} wet mass is the approximate caloric content. Thus, using resting O_2 consumption values from *Ocypode quadrata* (Full, 1984), a ghost crab weighing 2 g would require about 0.048 kcal calories per day. Since the average gram of wet food would have about 2 kcal, the resting crab would only have to eat a fraction of a gram (.02 g) each day to survive. Alternatively, consider a 300 g land crab, *C. guanhumi*, using O_2 consumption data from Herreid, Lee, and Shah (1979). This crab would require 0.865 kcal per day, which could be met by eating only 0.5 g of leaves. This is the equivalent of one small leaf per day. It is no wonder that crab population densities under tropical fig trees may reach 18,500 per hectare (Herreid and Gifford, 1963).

3. Digestive system

Detailed description of the digestive system in crustaceans is available in the recent multivolume work edited by Bliss (1982–5). The internal anatomy is summarized by McLaughlin (1983a); functional aspects of nutrition and digestion are covered by Dall and Moriarty (1983); metabolism and transport of carbohydrates and lipids are reviewed by Chang and O'Connor (1983); control of the mouthparts and gut is described by Wales (1982); and the adaptive aspects of feeding mechanisms are discussed by Grahame (1983).

The digestive system of crustaceans is a simple tube consisting of (1) a mouth with the associated mouthparts; (2) a foregut that consists of a short esophagus and (3) a two-compartment stomach; (4) a midgut; (5) the hepatopancreas and (6) other diverticula that extend from the gastrointestinal tract; and (7) a hindgut that ends at the anus (Fig. 10.1). The digestive system in land crabs appears similar to other crustaceans, but little attention has been given to the group specifically. Since the most careful and detailed work has been done on the mud crab, *Scylla serrata* (Barker and Gibson, 1978), we must of necessity use this species as our descriptive model even though it is not a land crab.

Food is picked up by the chelipeds and passed to the mouthparts. Ingestion consists of an interaction of the third maxillipeds and mandibles before being passed through the mouth into the esophagus and on to the stomach, where mechanical and chemical digestion is initiated. Food leaving the stomach passes either into the midgut or hepatopancreas where chemical digestion continues and absorption occurs. Undigested materials leave the midgut and pass into the hindgut and out the anus.

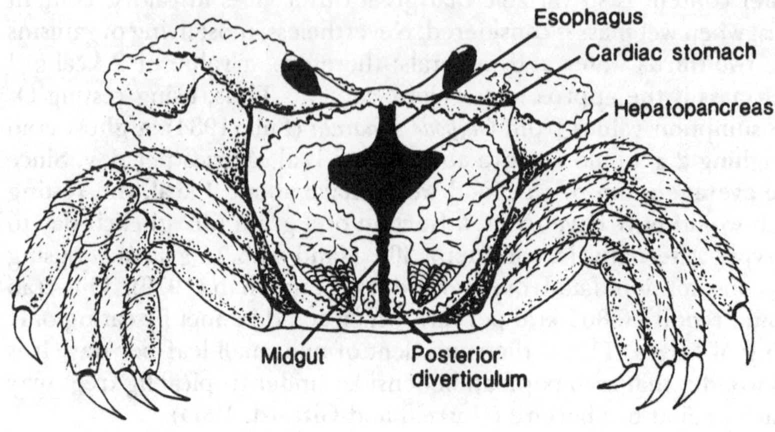

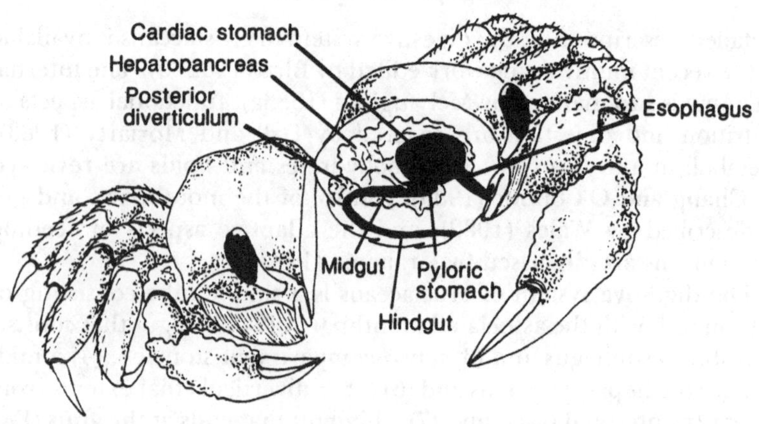

Fig. 10.1. Diagram of the dorsal and lateral views of the digestive tract of the ghost crab *Ocypode quadrata*. The anterior diverticulum of the midgut has been left out of the drawing for clarity. It normally lies in a coiled tangle near the stomach and hepatopancreas.

A. *Esophagus*

The esophagus and stomach comprise the foregut of the crab. The esophagus of an adult mud crab (carapace width 23 cm) is short, about 15 mm, and leads to the first chamber of the stomach.

The lumen of the esophagus is covered by a permeable chitinous cuticle that is shed and replaced during molting. The cuticle is penetrated by ducts that run from the epithelium into the lumen. These ducts connect with tegumental glands of the esophageal wall. The glands produce mucus, which acts as a lubricant for food sliding along the gut. The esophageal epithelium lying under the chitin consists of columnar cells. Connective tissues packed with tegumental gland cells and hemolymph spaces underlie the epithelium. Surrounding this layer is a band of striated circular muscles encircled by a layer of longitudinal muscles irregularly arranged in bundles. Oblique radial muscle strands lead from both the longitudinal and circular layers and terminate at the junction of the epithelium and chitin. Such muscles produce peristaltic waves moving the food along the esophagus. A trilobed valve separates the esophagus and stomach; it is formed from invaginations of the walls of the esophagus, and it prevents a back flow of stomach contents during digestion.

B. Stomach

The stomach of decapod crustaceans has two chambers, the cardiac and pyloric stomachs, separated by a cardiopyloric valve. The stomach produces no digestive enzymes of its own, but does have an active role in chemical as well as mechanical digestion because of digestive juices that are introduced from the hepatopancreas. Small quantities of organic molecules (e.g., 12% of sugars) may be absorbed by the foregut (Dall and Moriarty, 1983).

The first and larger stomach chamber, the cardiac stomach, contains a specialized grinding region called the gastric mill. The gastric mill is least developed in species such as shrimps, where the mouthparts perform an efficient job of mastication or the diet consists of tiny particles. In contrast, in crustaceans such as lobsters, crayfish, and crabs where large pieces of food are quickly eaten, the gastric mill reaches its greatest development (Barnes, 1980). Even within this group there are enormous variations, ranging from crayfish having a few simple chitinous projections that serve as grinding teeth to the spectacular architecture of a ghost crab *(O. quadrata)*, where the cardiac stomach has several elegant cutting plates (Fig. 10.2). The latter have dozens of fine teeth that are part of a marvelously complex movable jaw within the gut. To compare the gastric mills of the crayfish and ghost crab is like comparing the primitive architecture of Stonehenge with the elaborate design of the Taj Mahal.

An extensive muscular system controls the movement of the cardiac stomach (Wales, 1982; McLaughlin, 1983a). These muscles exist as part of the stomach wall, and they extend from the stomach to the body wall. The muscles, teeth, and sclerites act to grind the stomach contents. This action is complemented by enzymes from the hepatopancreas that have been passed forward into the stomach. Together, mechanical and chemical digestion reduces the contents of the cardiac stomach to a fine chyme. In species such as *O. quadrata*, setae line both

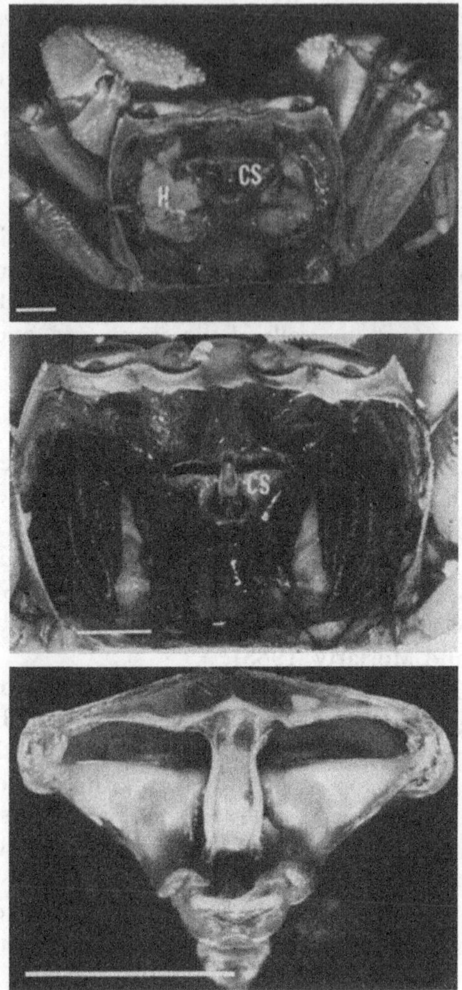

Fig. 10.2. Photographs of the ghost crab digestive system. The *upper picture* shows the interior of a crab with the dorsal carapace removed. The light-colored hepatopancreas (*H*) is readily visible as is the cardiac stomach (*CS*), which has its dorsal wall removed to show the interior sclerites. The *middle picture* gives a more detailed view of the crab's viscera, and the *lower picture* shows the major sclerite architecture of the cardiac stomach. The "T"-shaped structure slides forward and backward articulating with the lower section at the three ends of the "T." Food passes beneath the "T" (posteriorly) toward its base and is chewed up by the underlying teeth, which are barely visible in the channel below leading to the pyloric stomach. Scale bars = 1 cm.

the walls and internal flaps and act to separate large food fragments from small (C. Herreid, unpub.). Only the latter pass into the pyloric stomach where digestion continues. The circulation of the stomach contents is surprisingly complex in thalassinid Crustacea (Powell, 1974), and perhaps this is the general pattern in crustaceans (Dall and Moriarty, 1983).

The pyloric stomach lying ventral to the cardiac stomach has complex folded walls with setae that act as a filtering mechanism (Barker and Gibson, 1978). The folds form distinct channels in the stomach, which direct food toward the two ducts of the hepatopancreas and midgut. Each channel passes through a gland filter laden with fine setae. Only the smallest particles pass through the filter into the ducts of the hepatopancreas; the larger particles are funneled to the midgut. The walls of the pyloric stomach have plates embedded within them and a system of muscles that regulates the size of the channels and assists in squeezing material through the filters and driving it into the midgut or hepatopancreas. The musculature, tooth arrangement, and neuronal control of the gastric mill and the pyloric press are reviewed by Wales (1982) and Wiens (1982). Histologically, the stomach is similar to the esophagus except that tegumental glands are absent.

C. *Hepatopancreas*

The hepatopancreas has been called midgut gland, gastric gland, digestive gland, pancreas, liver, digestive organ, midintestinal gland, and ceca anteriores. It is responsible for virtually all of the free digestive enzymes of the gut and for about 85% of the assimilation. Gibson and Barker (1979) have reviewed the literature and conclude that the term "hepatopancreas" is appropriate; we concur with this decision because the term is distinctive and descriptive of the organ's function without suggesting the organ is identical with the vertebrate liver or pancreas. The following summary is based largely on their review.

The hepatopancreas of crustaceans reaches its greatest elaboration in the decapods, where it makes up 2–6% of the body mass (Brockerhoff and Hoyle, 1967; Stewart et al., 1967; Table 10.1). The organ consists of a large compact cluster of tubules lying on either side of the digestive tract where it fills much of the cephalothorax. In pagurid hermit crabs, the hepatopancreas is located in the abdomen and is developed asymmetrically.

Each half of the hepatopancreas is composed of two or three lobes ensheathed in a connective tissue membrane provided with a delicate network of circular and longitudinal muscle cells. Each side of the hepatopancreas opens from the gut via a primary duct near the pyloric stomach–midgut junction (Gibson and Barker, 1979). The primary duct branches into secondary ductules that enter the lobe masses

Table 10.1. *Estimation of glycogen stores in* Uca pugilator

Organ	Organ mass ($mg \cdot g^{-1}$)	Glycogen concentration ($mg \cdot g^{-1}$)	Glycogen concentration ($mg \cdot organ^{-1}$)	Total glycogen (%)
Hemolymph	250	20	5.0	36
Skeletal muscle	230	32	7.4	53
Hepatopancreas	40	16	0.6	4
Epidermis	4	39	0.2	1
Gills	16	26	0.4	3
Cardiac muscle	4	77	0.3	2

Note: The organ masses are derived from our dissections of 5 *Uca* (mean body mass, 2.5 g). Hemolymph volume is assumed to be 25%. Glycogen levels are from Keller and Andrew (1973).

where they subdivide extensively to form dozens of blind-ending tubules. The tubules are variable in color, often yellow, brown, or greenish.

Muscle cells ensheath each tubule of the hepatopancreas. Thick, striated, mononuclear circular muscle cells have been identified along with thin, nonstriated, longitudinal muscle cells that seem to connect adjacent circular muscles. Contractions of the muscles surrounding the tubules eject digestive enzymes into the gut and draw nutrients into the hepatopancreas (Loizzi, 1971). The nervous system coordinating this activity is not well understood.

The tubules of the hepatopancreas are lined with a layer of epithelial cells, one cell layer thick, except for the blind or distal end where several layers exist. Four types of epithelial cells have been identified as E- (embryonic or undifferentiated), F- (fibrillar), R- (resorptive or absorptive), and B- (secretory) cells (Gibson and Barker, 1979). E-cells are found only at the distal end of the tubules where they undergo mitotic division. These cells seem to specialize into either F- or R-cells as they migrate along the tubule toward the proximal end.

The F-cells have extensively developed rough endoplasmic reticulum, numerous Golgi bodies, and abundant vacuoles and vesicles. These characters suggest F-cells synthesize digestive enzymes. Indeed, as Gibson and Barker (1979) argue, F-cells appear to mature into secretory B-cells, the largest hepatopancreatic cells.

The B-cells contain a single enormous secretory vacuole comprising up to 90% of the cell volume (Barker and Gibson, 1977). B-cells possess a brush border with microvilli at their apex projecting into the tubule lumen. B-Cells are the secreting cells releasing digestive enzymes. Early in digestion as an immediate response to feeding (0.5–1h), they are sloughed off the tubule wall intact in a holocrine secre-

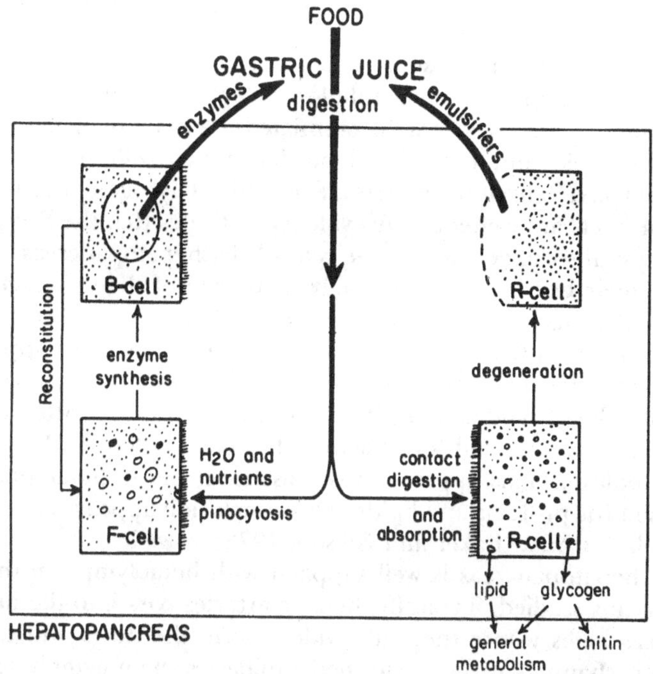

Fig. 10.3. Schematic diagram representing the proposed function of the different cell types within the crab's hepatopancreas. (Modified from Gibson and Barker, 1979.)

1h), they are sloughed off the tubule wall intact in a holocrine secretion. The loose B-cell disintegrates and releases its vacuolar contents. Later after feeding, two other discharges of enzymes have been noted in crabs such as *Scylla serrata*, at three and eight hours after a meal (Barker and Gibson, 1978). These secretion pulses are merocrine or apocrine in nature; B-cells lining the hepatopancreas tubules discharge the digestive enzymes from the vacuole into the lumen. These B-cells and their vacuoles are reconstituted cyclically (Fig. 10.3, Loizzi, 1971; Gibson and Barker, 1979).

Gibson and Barker (1979) and Dall and Moriarty (1983) provide extensive reviews of the specific digestive enzymes that have been discovered in the decapod hepatopancreas. There is an impressive arsenal of enzymes to degrade carbohydrates, lipids, and proteins, along with the more unusual cellulase and chitinase enzymes found in some Crustacea (e.g., Yokoe and Yasumasu, 1964; Muzzarelli, 1977). Only small amounts of gastric juice are produced at any one time (van Weel, 1970). The pH of this fluid usually ranges from 5 to 7. Emulsifying agents (released from R-cells) analogous to bile are an important part of the gastric juice, where they function to reduce particle size in fats and thus enhance hydrolysis (Gibson and Barker, 1979).

The R-cells are the most abundant cell, occurring throughout the length of the hepatopancreas tubules, except at the distal end where the embryonic cells have not differentiated. R-cells contain large numbers of vacuoles and have a brush border of microvilli; they resemble the absorptive cells of the vertebrate intestine. Glycogen granules and lipid droplets are evident. The evidence indicates that the R-cells are the major absorptive and storage cells of the hepatopancreas. In addition, their brush border probably is the site of digestive activities such as the final hydrolysis of lipids (Fig. 10.3).

Within two to five hours after a meal, minute vacuoles appear in the R-cells; these may represent the absorption of short-chain peptides or lipids (Barker and Gibson, 1977). Seven to nine hours after a meal, weak exopeptidase and lipase activity has been noted in *Homarus* R-cell cytoplasm. This appears to represent an intracellular phase of digestion for proteins and lipids. This conclusion appears to hold for the crab *S. serrata* (Baker and Gibson, 1978) as well.

The hepatopancreas is well supplied with hemolymph in the few crustaceans studied. Typically, hepatic arteries pass into the hepatopancreas lobes where they subdivide among the ducts and tubules forming channels variously termed capillaries, hemolymph sinuses, or hemolymph spaces (Gibson and Barker, 1979). Presumably, nutrients pass back and forth between hepatopancreas and hemolymph.

D. Midgut

The midgut has only a minor role in the absorption of nutrients, accounting for perhaps 5–10% of the organic nutrients absorbed during digestion. Unlike the foregut and hindgut, it lacks a chitinous lining (Barker and Gibson, 1978). The midgut varies in length, being quite short in crayfish and many crabs and relatively long in the lobster *Homarus* and snapping shrimp *Alpheus*, where all but the part of the intestine lying in the last abdominal segment is midgut. Diverticula (in addition to the hepatopancreas) are common extensions of the midgut; they appear to have a minor role in digestion and assimilation as well as water balance (Dall and Moriarty, 1983).

In *S. serrata*, the midgut comprises 40–50% of the postgastric intestinal tract and is 2–3 mm in external diameter (Barker and Gibson, 1978). There is a 1 mm posterior diverticulum, which arises from the dorsal wall of the midgut; it is about 100 mm long but is coiled on itself to form a solid disklike mass 10–20 mm in diameter. There is an anterior diverticulum as well, which bifurcates immediately after emerging from the dorsal anterior midgut wall. Each branch is about 1 mm in diameter and 20–30 mm in length, and these extend forward to lie adjacent to the stomach wall where their distal ends lie in a loose tangle (Barker and Gibson, 1978). The origin and histology of the anterior and posterior diverticula are considered by Smith (1978).

In the midgut, enzyme activity (only acid phosphatases and esterases) occurs in the epithelium cytoplasm bordering the lumen; the diverticula show similar phosphatase activity in their epithelial cytoplasm. As with the hepatopancreas, nutrient uptake has occurred two to five hours after a meal, after which intracellular digestion continues. There is evidence that animals with long midguts (e.g., *Homarus*) use passive absorption whereas crabs (e.g., *Scylla*) with short midguts may require active transport to ensure adequate absorption.

E. Hindgut

The hindgut of the crab *Scylla* extends throughout the abdomen, which is flexed beneath the cephalothorax. The hindgut represents 50–60% of the postgastric tract and terminates at the anus, an orifice in the last abdominal segment, the telson. The hindgut is always filled with feces (Barker and Gibson, 1978). The hindgut is infolded with longitudinal ridges, which are covered with a cuticular layer. Underneath lies an epithelium. Located just outside the epithelium are connective tissues containing circular and longitudinal muscles. Tegumental glands, which probably secrete mucus for lubrication of the waste matter, are abundant in the glandular swelling of the anterior hindgut at the junction of the midgut.

To summarize the digestive process: Food is first passed into the mouth by the chelipeds. The mouthparts may cut up the food into smaller bits or help filter it from sand or mud before swallowing. Food is passed via the esophagus into the stomach where it is mechanically chewed and chemically attacked by enzymes added from the hepatopancreas. Small organic particles are drawn into the tubules of the hepatopancreas; larger particles pass into the midgut. In both sites organic molecules are absorbed by epithelial cells that continue digestion intracellularly. The products of digestion and assimilation pass from the epithelial cells into the hemolymph where they are the raw materials for the general metabolism of the body. Undigested matter proceeds from the midgut into the hindgut where water is absorbed and fecal pellets are formed for later expulsion out of the anus.

4. Metabolism

Reviews of crustacean metabolism include Munday and Poat (1970), Vonk (1960), Hohnke and Sheer (1970), and most recently Chang and O'Connor (1983) on carbohydrate metabolism; O'Connor and Gilbert (1968), Gilbert and O'Connor (1970), and Chang and O'Connor (1983) on lipid metabolism; and Schoffeniels and Gilles (1970) and Claybrook (1983) on protein metabolism. From these surveys it is evident that data on the metabolism of land crabs are particularly sparse and fragmentary. Different species have been studied by different techniques under an array of different conditions. Furthermore, even a given species or individual can show extreme variability in metabolic function due to a myriad of processes (e.g., molting stage, reproductive cycle, activity) and environmental factors (e.g., temperature, salinity, food availability) that undergo complex oscillations. As a result, generalizations are of limited value. Here, our intention is to provide a brief overview of crustacean metabolism that will lend structure to future investigations on land crabs.

A brachyuran crab contains about 20 $kJ \cdot g^{-1}$ or 4.8 $kcal \cdot g^{-1}$ in total energy, expressed on an ash-free, dry-mass basis (DuPreez and McLachlan, 1983). The energy can be roughly divided into three pools: 21% carbohydrate, 6% lipid, and 73% protein. Obviously, only a fraction of each pool can serve as an energy source for the animal. This section focuses on (1) the form and location of energy sources, (2) the general route by which they are transported to deliver energy, and (3) the control mechanisms involved. A final energy source, the high-energy phosphate pool, will be mentioned in conclusion.

A. Carbohydrates

i. Location. Most of the carbohydrate in crustaceans is stored as the nitrogenous polysaccharide, chitin. In fact, chitin constitutes 64–74% of all of the organic material in the brachyuran exoskeleton (Drach and Lafon, 1942). However, this source does not appear to be readily available to the carbohydrate pool. Carbohydrate can also be complexed with protein-forming mucopolysaccharides present in the epidermis and hepatopancreas. Glycogen and glucose form a third part of the carbohydrate pool; these forms act as the animal's primary fuel source. Glucose appears to be the major carbohydrate constituent in the hemolymph. Other carbohydrates, such as the oligosaccarides, maltose, maltotriose, and trehalose, have been found mostly in low concentration (Schwoch, 1972; Telford, 1968). Glucose concentrations in the hemolymph of *Carcinus maenas* can vary from 1 to 100 $mg \cdot 100\ ml^{-1}$ hemolymph (Williams and Lutz, 1975). This variation does not appear to be exceptional among crustaceans. A host of parameters have been shown to influence hemolymph sugar levels, including (1) stage of molting, (2) stress, (3) activity, (4) stage of reproduction, (5) starvation, (6) time of sampling after feeding, (7) temperature acclimation, and (8) level of hyperglycemic hormone (Chang and O'Connor, 1983). These diverse effects suggest that the

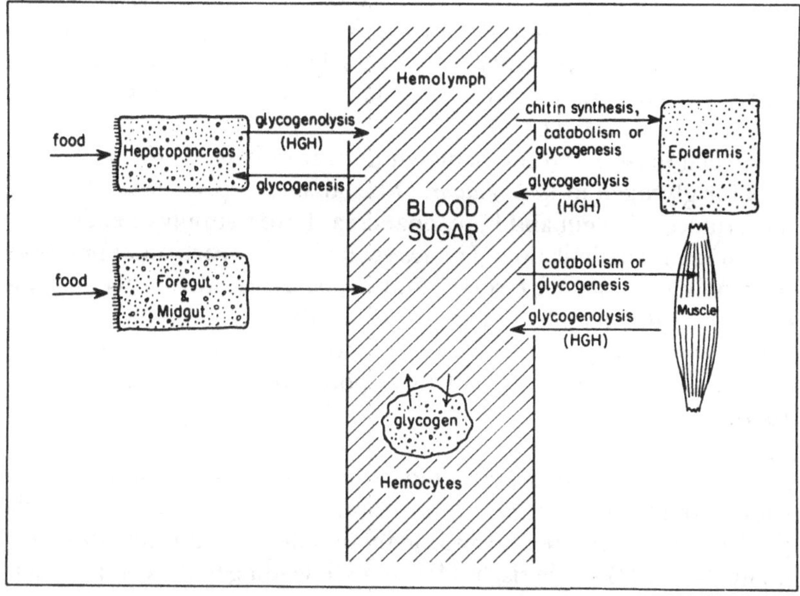

Fig. 10.4. Schematic diagram showing some of the major pathways of carbohydrate flux in crabs. *HGH* (hyperglycemic hormone) promotes glycogenolysis. (Modified from Hohnke and Scheer, 1970.)

hemolymph functions as a transient storage place for glucose removed from cells or tissues (Fig. 10.4).

Of particular interest is the finding that hemolymph itself may store glucose in the form of glycogen. Hemolymph cells called hemocytes may contain large amounts of glycogen and possibly chitin. In fact, they have been argued to be the main repository for glycogen in *C. maenas* (Johnston, Elder, and Davies, 1973). Hemocytes may function as "freely circulating hepatic cells" breaking down and storing carbohydrate as they travel through the circulation. The total contribution of this carbohydrate store could be significant, since crabs have large hemolymph volumes (Table 10.1). Johnston et al. (1973) reported that hemolymph glycogen could be nearly six times the amount found in the hepatopancreas.

Many studies deal with glycogen storage in muscle and hepatopancreas because these organs are often considered the major reservoirs (Parvathy, 1971). Yet, glycogen concentration in the epidermis can exceed that found in either organ (Hohnke and Scheer, 1970). Histological examination of the blue crab, *Callinectes sapidus*, showed that spongy connective-tissue cells (which maintain the position and shape of organs and tissues) were filled with glycogen (Johnson, 1980).

These separate investigations reveal those tissues capable of storing carbohydrate, but a summary of the fuel reserves has not been prepared. Table 10.1 presents such an analysis for *Uca pugilator* based on values from the literature and data from dissection (C. Herreid and D. Sperrazza, unpub.).

Skeletal muscle and the hemolymph function as the major storage sites for glycogen in *U. pugilator*. Other tissues appear to be of minor importance. The gut and hepatopancreas have a supply of energy via digestion and assimilation. Tissues such as gills, heart, and epidermis have only modest fuel reserves and probably withdraw their metabolites from the hemolymph on an ad hoc basis. Skeletal muscle, however, not only has a significant reserve for itself but may act to release carbohydrates into the hemolymph pool when demand arises (Fig. 10.4).

ii. Transport. Carbohydrates removed from storage appear to have at least four major fates: (1) chitin synthesis, (2) mucopolysaccharide production, (3) ribose and nicotinamide adenine dinucleotide phosphate (NADPH) synthesis, or (4) glycolysis leading to the end products L-lactate or carbon dioxide and water.

A new cuticle must be secreted each time an animal molts. Since most of the entire exoskeleton is chitin, its synthesis could represent a substantial drain on a crustacean's organic reserves. De novo synthesis of chitin requires glucose units, which must be aminated, and high-energy phosphate (ATP and UTP) used for its subsequent polymerization. However, since aminated glucose units derived from the breakdown of the old cuticle can serve as a starting point for synthesis (Stevenson, 1985), probably considerable energy is saved by this recycling method.

Two other important routes of carbohydrate utilization are the pentose–phosphate and glucuronic pathways. The pentose–phosphate pathway shunts glucose away from glycolysis and results in the production of NADPH and ribose. NADPH appears to be a required cofactor for lipid synthesis, and ribose acts as a substrate for nucleic acid metabolism. The glucuronic pathway also diverts carbon from glycolysis. Its end products are acid mucopolysaccharides, carbohydrate moieties bound to a polypeptide or protein. Mucopolysaccharides have been found in the digestive tract and hepatopancreas where they function in food lubrication. In addition, they appear to be a component of the cuticle in *Hemigrapsus nudus* (Meenakshi and Scheer, 1959). Unfortunately, no data are available on the amount of carbohydrate diverted into these pathways.

Glycolysis is the major starting point for glucose (or glycogen) breakdown in crustaceans (Hohnke and Scheer, 1970). This pathway is critical, since glycolysis is the only pathway that can operate under anoxic conditions. Crustacean skeletal muscle, in particular, has been reported to be a highly glycolytic tissue (Hochachka, 1976). Despite the many end products described for mullusks, L-lactate apears to be the major anaerobic end product in crustaceans, much the same as in vertebrates (Long, 1976; McMahon, 1981; Zebe, 1982; Full, Herreid, and Assad, 1985). Under conditions of hypoxia or exercise, glycolysis produces a considerable amount of energy as well as lactic acid. *U. pugilator* has been shown to survive 26 hours of anoxia by depleting glycogen stores and tolerating a 20-fold increase in lactic acid (Teal and Carey, 1967). Moreover, 60–70% of the ATP required for locomotion in the fiddler crab may be supplied by accelerated glycolysis (Full and Herreid, 1984).

Glucose can also be completely oxidized to carbon dioxide and water. A functioning tricarboxylic acid cycle and oxidative phosphorylation chain has been demonstrated in mitochondria, and this pathway appears to be the major catabolic route for carbohydrates (Huggins and Munday, 1968). It is important to note, however, that the energy content of oxidized carbohydrate on a per gram basis is about half that of fat.

iii. Control. Carbohydrates must be stored, transported, and utilized at the appropriate time, so that diverse demands, such as growth, activity, or molting, can be met (Fig. 10.3). Given this situation, it is not surprising to find hormonal feedback mechanisms operating to provide the necessary control.

Hyperglycemic hormone (HGH) is probably the best understood crustacean hormone involved in metabolism and can serve as an example of carbohydrate control. Eyestalk removal has been shown to produce hypoglycemia, and injections of eyestalk extract have resulted in elevated hemolymph sugar levels (Keller and Andrew, 1973). HGH is apparently released from the eyestalks and travels in the hemolymph to the site of action. Target tissues include muscle, integument, hepatopancreas, gill, and gonads (Parvathy, 1971; Keller and Andrew, 1973). Various enzymes are then affected causing a mobilization of glucose. For example, glycogen phosphorylase, which catalyzes the breakdown of glycogen to glucose-1-phosphate is activated. By contrast, HGH has also been shown to inhibit the enzyme glycogen synthetase, which would normally favor the formation of glycogen.

HGH is probably not the only hormone involved in the regulation of carbohydrate metabolism. A hypoglycemic factor has been reported by Rangneker, Sabnis, and Nirmal, (1961). Also, Sanders (1983) has found insulin-like peptides in the lobster hemolymph and hepatopancreas that promote glycogenesis in the muscle. However, these peptides do not seem to have a glucostatic role, and their function remains unclear. Recently, Herreid and Mooney (1984) have found that exercise induces hemolymph-borne factors altering the color of land crabs. The color change may be only a secondary effect of a hormonal reaction to exercise, the primary effect being the stimulation of carbohydrate or fat mobilization. Indeed, the neurotransmitter 5-HT is known to trigger the release of hyperglycemic hormone; it also causes pigment dispersion in red chromatophores (e.g., Keller and Beyer, 1968; Rao and Fingerman, 1983).

B. Lipids

i. Location. Lipids have been suggested to be the predominant organic storage reserve in crustaceans (O'Connor and Gilbert, 1968). Total body lipid can exceed levels of glycogen by 10-fold (Vonk, 1960). Moreover, the energy content of fat on a per gram basis is twice that of carbohydrate. However, values of the respiratory quotient (ratio of CO_2 production to O_2 consumption) for the land crabs *C. guanhumi* and *G. lateralis* during rest are about 0.85, which suggests the burning of a mixed carbohydrate, lipid, and protein diet (Herreid, Lee, and Shah, 1979, 1983). Supporting the lipid hypothesis are data from *Cardisoma carnifex* (Wood and Randall, 1981a) showing a resting respiratory quotient of 0.6. Yet in all three cases, respiratory quotient increased over 1.0 during exercise. This probably does not suggest a shift to carbohydrate catabolism so much as an increase in CO_2 release from hemolymph bicarbonate (see Chapter 8). The primary source of fuel remains in doubt.

The hepatopancreas is the major storage site for lipid. Although data for land crabs are not available, in *Cancer pagurus* lipid constitutes 30% of the organ's dry mass (Vonk, 1960). Some of the highest hepatopancreatic lipid levels have been found in the terrestrial anomurans *Birgus latro* and *Coenobita* spp. (Gibson and Barker, 1979). Caloric values of the hepatopancreas, alone, in these species ranged from 2 to 48 kJ. Analysis of the fat content of the hepatopancreas has shown that neutral fats (fatty acids, triglycerides, and sterols) account for 50–60% of the fat, while phospholipid content can range from 12% to 50% (Gibson and Barker, 1979).

ii. Transport. Large variations in lipid concentration occur, possibly as a result of starvation. For example, crayfish have been reported to rely primarily on lipid stores during the first two weeks of starvation (Speck and Urich, 1969). Such variations tend to greatly confound quantitative interpretations of lipid utilization.

Lipid mobilization and transport occur during molting and gametogenesis (Chang and O'Connor, 1983). In late premolt and in postmolt, crabs draw on lipid reserves in the hepatopancreas. The rate of lipid synthesis is decreased, and lipid is secreted into the hemolymph. This lipid can then serve as an energy source for the growth and regeneration of tissues following molting. Hepatopancreatic lipid is also mobilized at the onset of egg maturation. Neutral lipids are transported to the ovary for use by the developing egg (Gibson and Barker, 1979).

One surprising finding is that phospholipids act as the principal circulating lipid in the hemolymph of a number of species (Gilbert and O'Connor, 1970). This is in contrast to that observed in mammals and insects where free fatty acids and diglycerides, respectively, are the transport lipid. In other crustacean species, such as *U. pugilator* and *C. maenas,* hemolymph lipid was shown to exist as a complex lipoprotein moiety (Chang and O'Connor, 1983).

Lipids arriving at the tissues can be converted to triglycerides and stored or used as an energy source. Oxidation is thought to occur by way of beta-oxidation, but the complete crustacean pathway has yet to be described (Chang and O'Connor, 1983). The tricarboxylic acid cycle and respiratory chain appear similar to those described in mammals (Munday and Poat, 1970). Mitochondrial phosphate:oxygen ratios determined in *C. sapidus* and *C. maenas* are consistent with three functioning sites of phosphorylation in the electron-transport chain (Chen and Lehninger, 1973).

iii. Control. Hormones have been implicated in the control of lipid metabolism in crustaceans and in land crabs, in particular. Neutral lipids and phospholipids in the hepatopancreas have been shown to increase one day after eyestalk removal in *G. lateralis* (O'Connor and Gilbert, 1968). This effect appears to be independent of the level of molting hormone, which is also affected by eyestalk removal. Injected eyestalk extract from this land crab resulted in a reduction of the rate of lipid synthesis. Moreover, incubation of *G. lateralis* hepatopancreas with and without eyestalk extract produced similar results.

C. Proteins

i. Location. Proteins and amino acids must not be ignored as a potential energy reserve (Claybrook, 1983). These compounds easily constitute the bulk of crustacean tissues. Muscles appear to be the richest source of protein in crustaceans. In *Cancer magister* over 70% of the muscle dry mass (16% wet mass) is protein (Allen, 1971). Free amino acid levels in the tissues of decapods have been found to be several times higher than in vertebrate tissue; muscle levels range from 80 to 385

$\mu mol \cdot g^{-1}$ muscle mass (Claybrook, 1983). In contrast, hemolymph levels are comparable to those found in the blood plasma of vertebrates: 2–6 $\mu mol \cdot ml^{-1}$. In the green shore crab *Carcinus maenas,* about 50% of the hemolymph free amino acids are located within hemocytes (Evans, 1972). The difference in tissue and hemolymph free amino acid levels probably reflects the cells' role in volume regulation, using free amino acids in osmotic balance (see Chapter 7).

ii. Transport. It is not certain whether protein is the energy source of last resort or whether amino acids serve as a ready substrate for oxidation, as proline does in insects. Protein has been demonstrated to serve as a metabolic fuel during periods of starvation. Speck and Urich (1969) calculated that over 70% of the required energy between two and six weeks of starvation was provided by protein.

Radiolabeled amino acids are oxidized by tissues both in vivo and in vitro (see Claybrook, 1983). Removal of the amino group appears to be accomplished by transamination for a number of the amino acids. However, the pathways of catabolism have not been well defined in Crustacea generally. The pathways are assumed to be conservative and similar to those of mammals. Because the pathways are not yet understood, control cannot be discussed adequately.

D. High-energy phosphates

Carbohydrate, fat, and protein provide the fuel required for prolonged chemical and mechanical work. Many energy demands, however, are of short duration. Muscular contractions during an activity such as escape can be brief, and energy may come directly from the high-energy phosphates stored in the tissue. These stores include ATP and arginine phosphate, the invertebrate version of creatine phosphate. Large stores of ATP have not been found in crustaceans. In the lobster *Homarus vulgaris,* ATP levels are approximately 7 $\mu mol \cdot g^{-1}$, and ADP and AMP levels are considerably lower (Beis and Newsholme, 1975). In general, the adenine nucleotide concentrations are comparable to those of other invertebrates and vertebrates. Low levels of ATP are not surprising, since adenine nucleotides are involved in regulating their own production. On the other hand, arginine phosphate levels in the lobster tail muscle were three to six times the levels found in insects and twice those of creatine phosphate measured in mammals (Beis and Newsholme, 1975). Enzymatic breakdown of arginine phosphate by arginine kinase appeared to be extremely rapid. Since so few data are available, it is difficult to evaluate the significance of this short-term energy source. Yet, high levels of arginine phosphate in leg muscle may very well be critical for burst locomotion on land.

5. Animal energetics

Oxygen consumption ($\dot{V}O_2$) has served as the universal measurement of metabolism for most organisms, and crustaceans are no exception. Anaerobic contributions have been largely neglected. $\dot{V}O_2$ depends on a variety of parameters: body size, temperature, partial pressure of O_2 in the environment, molting cycle, activity, etc. These topics have been reviewed for crustaceans, generally, by Wolvekamp and Waterman (1960), Herreid (1980), and McMahon and Wilkens (1983), and are considered for land crabs in particular in Chapter 8. Our major focus in this section will be to evaluate the normal limits of metabolism in land crabs and to give an estimate of how they use their energy, so we will briefly touch on some of these topics here.

A. Resting metabolism

Many crabs and other crustaceans spend large periods of time, perhaps the majority of their lives, in apparent rest in crevices and burrows. Periods of inactivity may last for hours, as in the case of *U. pugilator* retreating into their burrows during high tide, or extend to weeks and months, as in the case of *C. guanhumi* in southern Florida, which plug their burrows with mud in the winter (C. Herreid, unpub.). We can only speculate on the metabolic state of these animals under natural conditions based on laboratory data.

Prosser (1973) reports standard metabolic rates for crustaceans ranging from 1 to 8 $\mu mol \cdot O_2 \cdot g^{-1} \cdot h^{-1}$. But as McMahon and Wilkens (1983) emphasize, values considerably lower than these are readily measured when animals are allowed to acclimate to experimental conditions resembling their natural habitat. Also, $\dot{V}O_2$, lactic acid levels, and acid–base balance continue to change for 24–48 hours after crustaceans are first introduced into metabolic chambers. Consequently, minimal $\dot{V}O_2$ rates (those values needed for bare maintenance) may vary considerably from the values published as "resting" rates. This clearly has significance when comparing resting and active metabolic rates, i.e., metabolic factorial scope.

B. Body size

Oxygen consumption is directly related to body mass. Thus, $\dot{V}O_2 = aW^b$, where $\dot{V}O_2$ is mass-specific O_2 consumption (i.e., O_2 consumption per gram body mass) and W is the mass of the animal in grams. When this exponential equation is written in logarithmic form ($\log \dot{V}O_2 = \log a + b \log W$), the coefficients a and b represent the intercept on the y-axis and the slope of the function, respectively. Slope values range between 0.67 and 1.0 for crustaceans, with 0.85 apparently representing a reasonable average for interspecific comparisons (see Wolvekamp and Waterman, 1960). These values indicate that the food requirement and mass-specific $\dot{V}O_2$ are considerably higher in small individuals than large. This metabolic principle appears practically universally throughout the animal kingdom in invertebrates and vertebrates and among poikilotherms and homeotherms alike, although the coefficients vary (Zeuthen, 1953). Mammalogists fondly refer to their "mouse to elephant curve" (e.g., Benedict, 1938; Kleiber, 1961), but an analogous situation prevails for carcinologists: We have our "fairy shrimp to spider crab curve" (e.g., Scholander et al., 1953).

In spite of the fundamental nature of this metabolic principle, and the speculations as to its cause (e.g., Wolvekamp and Waterman, 1960), we are largely unable to explain its origin. However, some ecological consequences are clear. An ecosystem can support a larger biomass of animals, *ceteris paribus*, if it is packaged into a few large individuals (with a low metabolic rate per gram) than it can a large number of small individuals (with a high metabolic rate per gram). The maintenance costs for the latter are much higher. Also, as we will see, the mass specific energy cost for activity is greater for small individuals.

C. Temperature

Temperature influences $\dot{V}O_2$ in a predictable fashion in crustaceans. In general, with a 10° C increase of temperature, metabolic rates rise two to three times (i.e., $Q_{10}=2$–3). Most land crabs are limited to a modest temperature range, with thermal limits perhaps between 15 and 40° C, although short-term exposure to higher temperatures is possible. *C. guanhumi* in southern Florida has a normal operating temperature much narrower than this range (C. Herreid, unpub.).

Land crabs use behavioral and physiological methods for regulating temperature. They use their burrows for protection from the more major changes in temperature. In addition, evaporative cooling from their gills and through their shell becomes important at high temperatures (see also Chapter 7). *Cardisoma*, during migration on hot summer days in Florida, dashes from shady spot to shady spot, maintaining body temperature 1–2° C below air temperature (C. Herreid, unpub.). Similar temperature depression has been seen for fiddler crabs and ghost crabs (cf. Cloudsley-Thompson, 1970). Moreover, *C. guanhumi* will urinate over the extensive bristle patch just under its eyes every few minutes when body temperatures approach the lethal temperature of 39° C (C. Herreid, unpub.). The increased evaporation slows the rise in body temperature. This method of temperature regulation under heat stress is reminiscent of storks or vultures urinating on their legs.

Given a Q_{10} of 2–3 and a temperature range of 20° C, temperature may cause a variation in resting $\dot{V}O_2$ by as much as 10-fold. Realistically, a five-fold variation due to temperature might be normal. Obviously, this means that in cool conditions a crab's food reserve could last perhaps five times longer than when the crab is warm. Given that starvation depresses metabolism, as does a lack of disturbance, the food reserves could easily maintain a resting crab 10 times longer in a cold, undisturbed condition than in a warm disturbed state.

D. Air versus water as the respiratory medium

Water contains relatively small amounts of oxygen (generally less than 10 ml $O_2 \cdot l^{-1}$ compared to 210 ml $O_2 \cdot l^{-1}$ of air) and is dense and viscous compared to air. Consequently, one might expect a significantly higher proportion of the metabolic energy to be devoted to breathing in water than in air. Unfortunately, no data exist for crustaceans on this point directly. Some species, such as the land crab *C. guanhumi*, are capable of survival in both water and air, and there is no significant change in $\dot{V}O_2$ in the different media (Herreid and O'Mahoney, 1978; O'Mahoney and Full, 1984). Consistent with our expectations, the ventilation rate is three times higher in water than in air. Since no major change occurs in $\dot{V}O_2$, we must assume that (1) the total cost of ventilation is quite small in both media and thus any differences are difficult to detect or that (2) crabs breathing water compensate for the increased work load by decreasing metabolism in other ways.

The most recent direct measurements of energy involved in crustacean ventilation occur in the shore crab *C. maenas* (Wilkens, Wilkes, and Evans, 1984). The O_2 requirement of ventilation in water was measured by comparing $\dot{V}O_2$ before and after removal of the scaphognathites. The cost of ventilation for resting crabs was 30% of the total $\dot{V}O_2$. Indirect estimates of the cost of aquatic ventilation have been made for *Callinectes sapidus* (0.02% of the total O_2 consumption), *Cancer pagurus* (17–76%) and *Orconectes virilis* (1%), and *Cancer magister* (6%) (Batterton and Cameron, 1978; McDonald et al., 1980; Burnett and Bridges, 1981; Burggren and McMahon, 1983). The relative cost of ventilation in air-breathing species of crabs should be lower, but no direct measurements or calculations have been made.

It would be particularly interesting to determine the energetic cost of ventilation in the Australian land crabs of the genus *Holthuisana*, which have radically different patterns of ventilation in air and water (Greenaway, Taylor, Bonaventura, 1983; Greenaway, Bonaventura, Taylor, 1983; see Chapter 8).

E. Molting, growth, and regeneration

Molting is such an integral part of a crab's existence (see Chapter 6) that it is difficult to separate its energetic cost from other metabolic expenditures. Changes occur in the integument for 70% or more of the time between successive molts, and during the remaining 30% of the intermolt period the crab is storing energy reserves for the new exoskeleton in the hepatopancreas (Passano, 1960a). As Passano (1960a) states, "The normal physiology of the crab is thus continuously and intimately concerned with the successive stages of the intermolt cycle. Hibernation, ovarian maturation, and the carrying of developing eggs by females are the only interruptions in this series of periodically recurring molts." Complicating the problem of energetic measurement is that drastic changes in the activity level of crabs are correlated with the molt cycle. During intermolt (i.e., stage C crabs; see Passano, 1960a) both aquatic and land crabs are fully active, whereas they become inactive at the time of ecdysis (exuviation) when the exoskeleton is shed.

No studies seem to have been made of oxygen consumption in land crabs during molting. However, in the blue crab, *C. sapidus*, molting is apparently associated with relatively minor changes in $\dot{V}o_2$ (Mangum et al., 1985). For example during premolt (stage D_1–D_4 crabs), which occupies about 25% of the duration of the life cycle (Passano, 1960a), blue crabs actually show a slight decline in $\dot{V}o_2$, which is associated with a reduction in activity. No data are given for ecdysis. However, $\dot{V}o_2$ immediately following ecdysis (stage A) is only doubled compared to intermolt (stage C). $\dot{V}o_2$ declines rapidly to intermolt levels by the time late stage B is attained, when feeding begins and activity returns to normal levels. Thus, $\dot{V}o_2$ is elevated for only a brief time in the molt cycle (less than 5%). Moreover, even at its highest level, the aerobic rate is modest.

Hemolymph lactate remained low except prior to molt (stage D_4) whereupon lactate concentrations doubled to 0.53 $mM \cdot l^{-1}$. Immediately after molt (stage A), hemolymph lactate jumped to 2.26 $mM \cdot l^{-1}$ or over six times intermolt conditions, suggesting a significant anaerobic contribution is present. However, this was short-lived and by the B_2 stage, hemolymph lactate had fallen to 0.23 $mM \cdot l^{-1}$. Thus, even though molting and its preparation and aftereffects are profound events in the life of the crab energetically, they do not appear to match the maxima seen during sustained swimming. Booth (1982) noted hemolymph lactate levels for swimming *Callinectes* reached 9.8 $mM \cdot l^{-1}$ with aerobic factorial scopes of 2.6. However, these swimming values are of short duration compared to those elevated values occurring over several days of molting. Finally, we should note that the crab progressively decreases feeding and locomotion until it is immobile at the time of ecdysis (stage E) and then becomes gradually more active until stage B_2 is reached (Passano, 1960a). Thus, energy normally used in digestion and locomotion may be diverted for activities involved with molt.

Limb loss and regeneration is a common phenomenon in Crustacea (Bliss, 1960). The energetics of regeneration are difficult to evaluate and have hardly been considered. One problem is that limb autotomy stimulates precocious molting with limb regeneration typically confined to the premolting stage. As a result, the energetic cost of regeneration is difficult to separate from the general effects of growth and molt itself.

The cost of limb regeneration is clearly distributed over a relatively long period. In Alaskan king crabs (*Paralithodes camtschatica*), a regenerated limb does not reach normal size until after four to seven molts (Skinner, 1985). In the land crab *G. lateralis*, a single lost limb is replaced by a regenerated limb that is only two-thirds the size of the normal limb. If more than six limbs are autotomized, the regenerated limbs are reduced to half normal size (Skinner and Graham, 1972). Crabs are probably restricted in the amount of tissue they can synthesize, perhaps to 12–15% of their metabolically active body mass. Therefore, intensive regeneration reduces the size of the animal itself as well as the regenerating limbs (see Skinner, 1985, for review). It appears that the metabolic resources are mainly redistributed rather than scaled up in a major way; energy normally used for general growth and activity is diverted to regeneration. From these considerations we would not anticipate a major elevation in overall metabolism and O_2 consumption during regeneration.

6. Locomotion

Crabs can have extensive daily and seasonal movement patterns (see Chapter 3). For example, intertidal movements of *Uca, Ocypode,* and *Coenobita* may exceed 100 m (Herrnkind, 1983). Ghost crabs, which are the fastest running crustaceans (up to 3.4 $m \cdot sec^{-1}$, move as much as 300 m during daily activity periods (Wolcott, 1978; Roe, 1980). Seasonal reproductive migrations of *C. guanhumi* (Gifford, 1962a) and *G. lateralis* (Bliss, 1979) may cover several kilometers. The energetic costs of such behaviors have received little attention until recently, when it was discovered that crabs walk well on treadmills. This opened an avenue for studying the metabolism of crabs in motion. Because the intensity of work could be quantified and rigorously controlled, it was possible to explore the limits and range of physical exertion while aerobic metabolism was measured continuously.

Since 1979, when *C. guanhumi* was tested with a respiratory mask while walking at various speeds (Herreid, Lee, and Shah, 1979), metabolic data during locomotion have been accumulated on about a dozen species of land crabs. The use of a mask, applicable in only special circumstances, has yielded ventilation data but tends to encumber the crab. Hence, in most later studies, freely moving crabs have been tested on a treadmill enclosed in an airtight chamber while $\dot{V}O_2$ is measured with an automatic O_2 analyzer (Herreid, 1981). Studying voluntary locomotion, Wheatly et al. (1985) have carried out studies of crabs walking in a low-friction system while metabolism was monitored. Recently, Blickhan and Full (1987) have run ghost crabs across miniature force plates and presented the first analysis of the mechanical energy changes during terrestrial locomotion in an arthropod. Combining metabolic and mechanical data allows estimates of the efficiency of locomotion. Before considering such data, it is helpful to have a general description of the walking and running styles of land crabs.

A. Locomotion patterns in land crabs

Two major groups of crabs walk extensively on land: the brachyurans such as *Uca*, which walk with eight legs, and the anomuran hermit crabs such as *Coenobita*, which walk with six limbs. Both taxa have different styles of walking, and both patterns clearly were established before these groups left the aquatic environment, since their aquatic cousins possess similar patterns.

The brachyurans typically walk sideways, although they do walk forward, backward, and diagonally on occasion (Lochhead, 1960). When walking sideways, the four legs trailing the crab provide the principal thrust, as they push the crab. The chelipeds do not normally participate in locomotion, being held close to the body off the ground, except in slow walking where they may contact the earth intermittently. Walking sideways avoids one of the major problems facing an animal with multiple limbs: The legs cannot overlap or interfere with one another because they swing sideways.

Leg movement can be described by numbering the pairs of legs, anteriorly to posteriorly. The chelipeds on the right and left sides of the body are designated R_1, and L_1, respectively. The first walking legs are R_2 and L_2, etc. When brachyurans such as *U. pugnax* walk, their neighboring legs alternate with one another. They also alternate with their partner on the contralateral side. Limbs L_2, L_4, R_3, and R_5 tend to step in unison, and L_1, L_3, R_2, and R_4 are coupled (Barnes, 1975). This pattern of locomotion results in an alternating tetrapod gait. At any one time the center of mass is always contained within a quadrangle of support and is obviously a very stable gait.

Gait changes have been observed in land crabs (Lochhead, 1960); the most striking is that seen in the ghost crab. *Ocypode ceratophthalmus* uses fewer and fewer legs as the speed of locomotion increases (Burrows and Hoyle, 1973). At low speeds, all four pairs of walking legs are used. The crab can continue this movement indefinitely, although it may change leads, switching the trailing and leading sides, to avoid fatigue. As velocity increases, the crab raises R_5 and L_5 off the ground and uses only three pairs of walking legs. At the highest speeds, the third pair of walking legs (R_4 and L_4) is raised also, leaving only two pairs of legs for locomotion. The latter gait constitutes a true run, since the crab literally leaps off the ground. Virtually all of the power for this sprint comes from the two trailing legs; hence, the crab has been said to have a bipedal gait (Burrows and Hoyle, 1973).

The method of walking used by hermit crabs involves only two pairs of walking legs. The posterior two pairs of appendages not involved in locomotion have been modified to hold a protective snail shell. In contrast, to the brachyurans, hermit crabs normally walk forward, although they can move in other directions. Like insects, the land crab *C. compressus* uses six-legged locomotion (Herreid and Full, 1986a). It uses its chelipeds as support levers, while the two pairs of walking legs are used for thrust. Also like insects, this hermit crab uses an alternating tripod gait. Limbs L_1, L_3, and R_2 alternate with R_1, R_3, and L_2. This is the most stable six-legged gait, since the center of mass always lies within a triangle of support. Moreover, because the crab shell is often dragged, an extra point of support is also gained.

As with all pedestrian animals, energy for locomotion is used in several ways: to move the center of mass up and down with each step, and to accelerate and decelerate the limbs as they provide thrust and support for movement. As will be seen in the following section, land crabs function energetically very much like land vertebrates during locomotion.

B. Metabolic energy patterns during locomotion

The basic procedure in energetic studies during locomotion has been to collect oxygen-consumption data on crabs during a rest period on a stationary treadmill, then to turn the treadmill on for 10–20 min at a particular speed. The bout of exercise is usually followed by a period of recovery. Typically, once modest exercise begins, $\dot{V}O_2$ rapidly rises to a relatively constant rate, which has been called the steady-state ($\dot{V}O_2ss$) and remains there until the exercise is over. The time it takes to arrive at 50% of the $\dot{V}O_2ss$ ($t_{1/2\ \text{on-response}}$) is a convenient measure of the rapid adjustment of the crab to endurance running. Similarly, it is possible to measure the half-time to recovery ($t_{1/2\ \text{off-response}}$) after the treadmill is turned off.

Another method of evaluating the oxygen-consumption pattern involves the calculation of an O_2 deficit and O_2 debt, terms borrowed from vertebrate physiologists (Stainsby and Barclay, 1970). Oxygen deficit represents the lag in an animal's $\dot{V}O_2$ response to steady-state exercise. Thus, O_2 deficit represents the difference between two values: the actual rise in $\dot{V}O_2$ that occurs as the animal begins to run at a constant speed and the theoretical abrupt rise in $\dot{V}O_2$ that should occur if the steady-state $\dot{V}O_2$ were reached the instant that exercise began.

Oxygen consumption does not immediately fall to resting levels when exercise ceases. The delay in the $\dot{V}O_2$ recovery is called O_2 debt or extra postexercise oxygen consumption (EPOC), and is operationally defined as the area under the recovery $\dot{V}O_2$ curve, above resting $\dot{V}O_2$. These terms and concepts are graphically illustrated in Stainsby and Barclay (1970) and Herreid (1981).

i. O_2 consumption during locomotion. Crabs give a range of responses in $\dot{V}O_2$ during locomotion. Three basic patterns have been identified so far.

1. *Aerobic pattern.* Some species such as the Panamanian ghost crab, *Ocypode guadichaudii*, have good endurance and are highly aerobic (Full and Herreid, 1983). For example, the $t_{1/2\ \text{on-response}}$ for submaximal running of *O. guadichaudii* is brief, less than 2 min. The $t_{1/2\ \text{off-response}}$ is only slightly longer with complete recovery usually within 15 min. Consequently, the O_2 deficit and O_2 debt are small. The aerobic factorial scope values are high, reaching a maximum of 12 times the resting rate. Data for *O. quadrata* (shown in Fig. 10.5) and for the hermit crab *C. compressus* (Herreid and Full, 1986b) indicate that similar aerobic patterns exist in these species. The fast on- and off-response in these crustaceans is similar to that in mammals and insects (e.g., Cerretelli et al, 1977a; Herreid and Full, 1984a).

2. *Mixed aerobic and anaerobic pattern.* Anaerobic metabolism plays a prominent role in some crustaceans. Semiterrestrial and terrestrial crabs such as *U. pugilator, G. lateralis, C. guanhumi,* and *C. carnifex* show a mixed aerobic and anaerobic response to exercise (Herreid, Lee, and Shah, 1979; Wood and Randall, 1981a, b; Herreid, O'Mahoney, and Full, 1983; Full and Herreid, 1984). These animals have poor endurance, typically managing only 10–15 min on a treadmill at modest speeds. Even at low speeds, they may not reach steady-state $\dot{V}O_2$ before they fatigue (Fig. 10.6). Their factorial scope values are low; maximum O_2 consumption is about three to five times rest. Recovery from exercise is prolonged, often well over an hour in duration. Also, measurements of whole-body lactate in *U. pugilator* show that such crabs have a significant dependence on anaerobic metabolism (Full

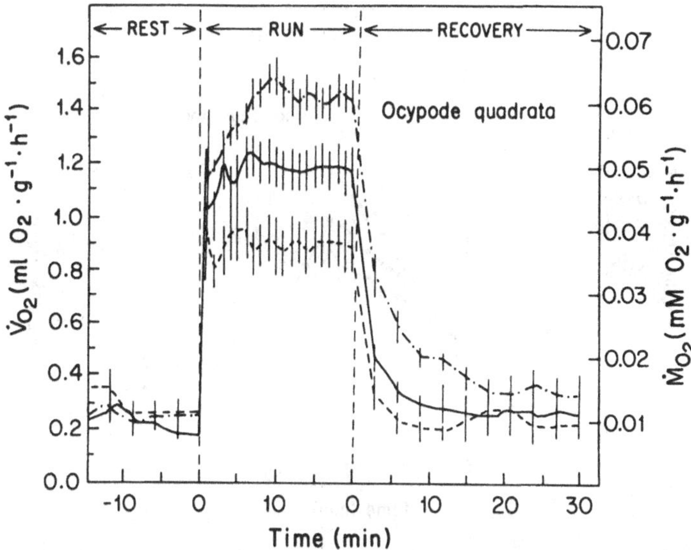

Fig. 10.5. Oxygen consumption ($\dot{V}O_2$) of ghost crabs (2 g) on a treadmill during rest, exercise, and recovery. The *upper, middle,* and *lower curves* represent crabs running at velocities of 0.13, 0.19, and 0.28 km·h^{-1}, respectively. Each curve is the mean of 5 animals, and the baıs represent ± 1 SE. (From Full, 1984.)

and Herreid, 1984). In the fiddler crab *U. pugilator,* whole-body lactate increases in a linear fashion with the speed of locomotion even at the lowest velocity, 0.06 km·h^{-1}. We estimate that even at this speed, anaerobic metabolism (and phosphagen and O_2 stores) accounted for 40% of the total ATP generated. During medium- and fast-velocity experiments (0.11 and 0.16 km·h^{-1}), anaerobic contributions via lactate fermentation rose to 60% and 70% of the total ATP produced.

The mixed metabolic response of exercising land crabs resembles that of exercising mammals running at velocities where O_2 consumption approaches maximum rates; near this point only small increases in $\dot{V}O_2$ can occur, and the rate of lactate production in the blood increases linearly with velocity (Margaria et al., 1963; Seeherman et al., 1981). The mixed response is also similar to that found for humans who are untrained and thus have a poor aerobic response to exercise (Cerretelli et al., 1979).

3. *Nonaerobic patterns.* The wharf crab *Sesarma cinereum,* which lives in intertidal areas, shows an unusual metabolic response to running. Endurance tests reveal that *Sesarma* fatigues sooner than *U. pugilator* and much sooner than *O. quadrata* (Full et al., 1985). Even though *Sesarma* has a resting $\dot{V}O_2$ comparable to *U. pugilator,* running does

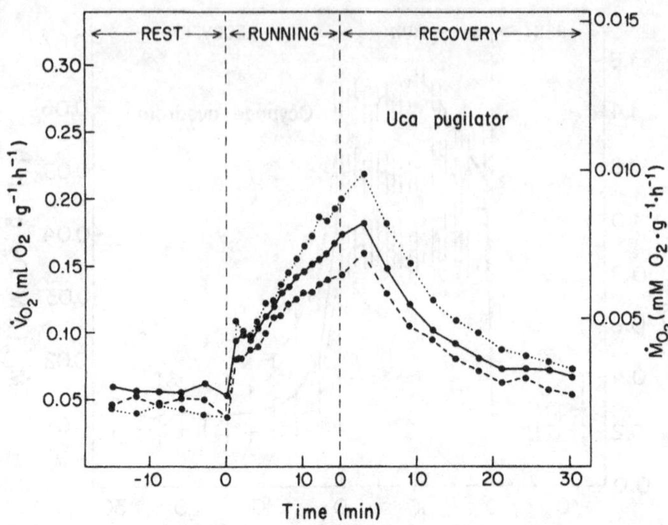

Fig. 10.6. Oxygen consumption ($\dot{V}O_2$) of fiddler crabs on a treadmill during rest, exercise, and recovery. The *upper, middle,* and *lower curves* represent crabs running at velocities of 0.06, 0.11, and 0.16 $km \cdot h^{-1}$, respectively. Each curve is the mean of 7–8 animals. The average standard error is ±0.009. (Modified from Full and Herreid, 1984.)

not stimulate $\dot{V}O_2$ to increase more than 1.6 times the resting rate. This suggests that anaerobic metabolism may play a major role. Surprisingly, whole-body lactate measurements show *Sesarma* to have much lower levels than *U. pugilator*. Thus, the typical aerobic and anaerobic indicators of crustacean metabolism point to a much lower energy demand for the wharf crab running at the same speeds as *U. pugilator*. Preliminary testing for other possible end products of anaerobic metabolism does not show unusual levels of alanine, succinate, D-lactate, octopine, or strombine (Full et al., 1985). Thus, no evidence exists for unusual rates of anaerobic metabolism to compensate for the low aerobic contribution to exercise. This crab's apparent low energy demand could be due to unusual levels of O_2 stored in the hemolymph or to high levels of arginine phosphate in the muscle. This does not seem likely because the crab can walk at least an hour; these energy sources would have to be unrealistically high to produce this result. Moreover, when *Sesarma* was exercised in pure nitrogen gas, it did not show unusual endurance, but it did improve when exposed to a pure oxygen atmosphere, thus emphasizing that the species is not unusual in anaerobic abilities. How this crab accomplishes its locomotion without raising $\dot{V}O_2$ except to a minor degree is unknown.

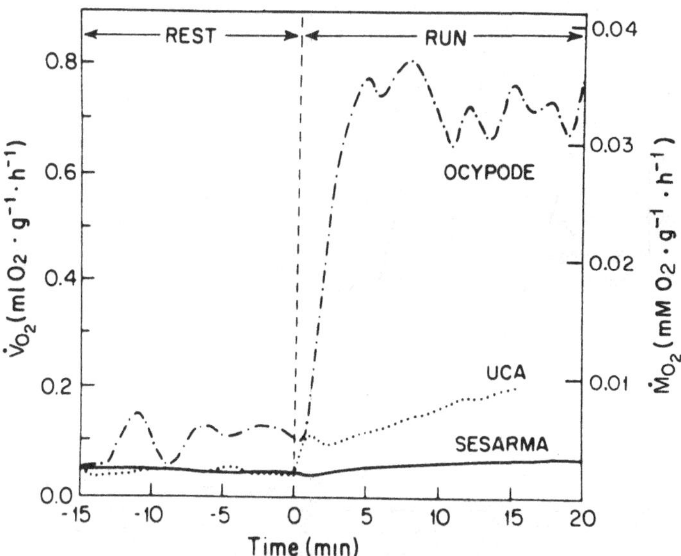

Fig. 10.7. Oxygen consumption of three crabs during exercise and recovery on a treadmill. *Ocypode guadichaudii* and *Sesarma cinereum* were run at 0.13 km·h^{-1} and *Uca pugilator* at 0.11 km·h^{-1}. (From Full and Herreid, 1983, 1984, and Full et al., 1985.)

In summary, at least three different metabolic responses to exercise have been identified. These are exemplified by *Ocypode, Uca,* and *Sesarma* in Figure 10.7. The remarkable feature of this comparison is that the individuals illustrated here are the same size (2–3 g) and are running on a treadmill at similar speeds.

How can such differences be explained? Clearly, anaerobic metabolism, high-energy phosphate reserves, and O_2 stored in the hemolymph may account for some of the variation. But we must entertain the possibility that the crabs have very different energetic efficiencies during locomotion. Moreover, other differences probably exist. First, differences in O_2 conductance and utilization seem likely, but have not yet been identified. It is likely that circulation systems vary, with differences in hemolymph O_2-carrying capacity and O_2 dissociation being important (see Chapter 8). Certainly, the gill surface area is not dramatically different among the species (Full et al., 1985), and the leg muscle mitochondrial density, distribution, and size are similar (C. Privitera and C. Herreid, unpub.).

Second, there are significant differences in endurance (Fig. 10.8). The highly aerobic species such as ghost crabs, with high maximal rates of $\dot{V}O_2$ and low lactate concentration, can sustain walking for well over an hour at modest speeds. It is not surprising that these

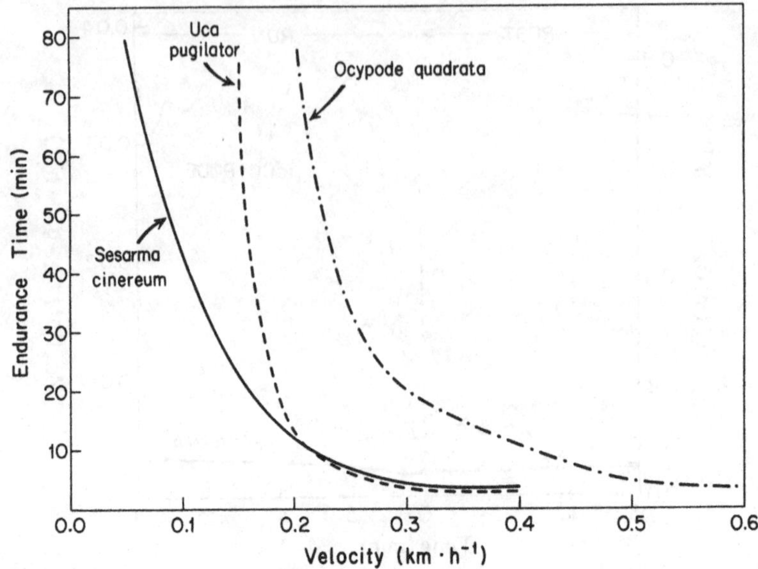

Fig. 10.8. Endurance patterns in three crabs running on a treadmill at different speeds. The *curves* (fitted "by eye") are from 13–26 trials for each species. (From Full, 1984; Full and Herreid, 1984; and Full et al., 1985.)

species are very active foragers in the wild, routinely traveling several hundred meters each day (see Chapter 3). Crabs such as *Uca*, with mixed aerobic and anaerobic patterns, show less stamina. They are less consistent walkers on the treadmill at high speed, have a low maximum $\dot{V}O_2$, and require anaerobic metabolism. *Sesarma*, possessing the poorest aerobic capacity, can sustain only very slow rates of locomotion. Whatever the limitations, it appears that crabs that can consume oxygen at high rates can sustain more intense activity, whereas those with more limited aerobic capacities move at slow speeds or only intermittently.

ii. O_2 consumption versus velocity. In vertebrates (mammals, birds, and reptiles), the $\dot{V}O_2$ for steady-state running tends to rise with an increase in velocity until a maximum ($\dot{V}O_2max$) is reached (e.g., Taylor et al. 1970; Taylor, 1973, 1977). This pattern holds true for aerobic insects, which use progressively more O_2 the faster they run (Herreid and Full, 1984a). A linear rise in $\dot{V}O_2$ with velocity also occurs in the aerobic land crabs such as *Ocypode* (Full and Herreid, 1983; Full, 1987). Figure 10.9 shows the pattern for *O. quadrata.* Even in those less aerobic species that never show a steady-state $\dot{V}O_2$ during exercise (e.g., *C. guanhumi, G. lateralis, U. pugilator*), the sum of the O_2 used

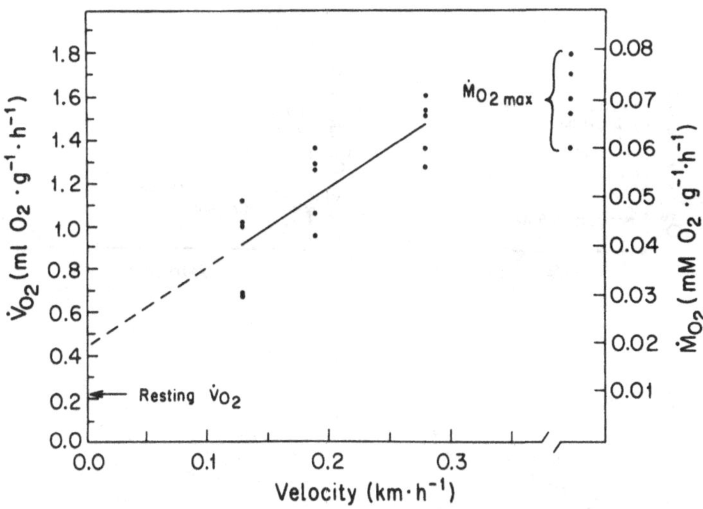

Fig. 10.9. Steady-state oxygen consumption for 5 ghost crabs (*Ocypode quadrata*) as a function of velocity. Maximal O_2 consumption is plotted for comparison on the right. (From Full, 1984.)

during both exercise and recovery is directly related to the velocity of travel (Herreid, Lee, and Shah, 1979; Full and Herreid, 1983).

Oxygen consumption versus velocity shows a curvilinear rise in some animals. Humans walking on a treadmill show this pattern, although $\dot{V}O_2$ during running is linear (Margaria et al., 1963). The curvilinear response also occurs in ponies forced to walk or trot at higher or lower velocities than normal (Hoyt and Taylor, 1981). Such patterns may not be evident in treadmill experiments because animals change gaits. It has been argued that gait changes serve to maintain the most economical travel at a given velocity. Gait changes have been observed in some land crabs (e.g., *Ocypode;* Hafemann and Hubbard, 1969), and they may play a similar role.

iii. Y-intercept problem. When the $\dot{V}O_2$ is plotted as a function of velocity, for many species the regression line passing through the data points may be extrapolated back to zero velocity. This y-intercept is the anticipated $\dot{V}O_2$ value of the animal at rest. In most species tested, the y-intercept is considerably higher than the actual measured resting O_2 consumption (e.g., Taylor et al., 1970). Data for land crabs also show this trend (Herreid, Lee, and Shah, 1979; Full and Herreid, 1983; O'Mahoney, Herreid, and Full, 1983; Fig. 10.9).

Herreid (1981) has reviewed the many possible reasons for the

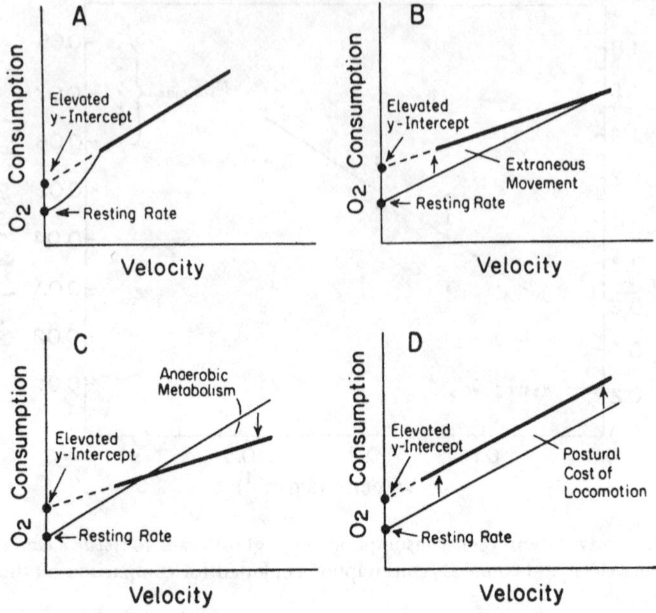

Fig. 10.10. Possible reasons for an elevated y-intercept when Vo_2 is plotted against velocity. A. Case in which Vo_2 would show a curvilinear pattern if measured at slow speeds. Thus, the intercept is an artifact. B. The y-intercept is deflected upward if the animals show extraneous movement at slow speeds. C. Upward deflection results from anaerobic metabolism at high speeds. D. The y-intercept might be elevated if there were a constant postural cost of locomotion.

elevated y-intercept. Perhaps the most likely explanations for the phenomenon in the crabs studied, as illustrated in Fig. 10.10, are the following:

1. At slow speeds, $\dot{V}o_2$ may decrease curvilinearly to resting levels. The elevated y-intercept may be an artifact since very few measurements are made at very slow speeds.

2. Crabs walking at very slow speeds on the treadmill may use more O_2 than predicted because they are apt to wander and have extraneous movements. Consequently, treadmill velocity underestimates their exercise output at slow speeds.

3. Crabs walking at fast speeds may increase their dependency on anaerobic metabolism. Therefore, the $\dot{V}o_2$ at high speeds would not be an accurate measure of the energy used during locomotion; i.e., the measured $\dot{V}o_2$ is less than expected. If either scenario (2) or (3) occurs, it would have the effect of deflecting the $\dot{V}o_2$ versus velocity regression line to elevate the y-intercept (Fig. 10.10).

4. Schmidt-Nielsen (1972) has suggested that the elevation of the y-intercept represents the "postural cost of locomotion." Presumably, this means there is a cost associated with lifting the body into a position for locomotion and maintaining it there. If this is the case, it suggests the entire $\dot{V}O_2$ vs. velocity curve may be displaced upward. At present, except for the hermit crabs discussed later, there are insufficient data on any animal to clearly resolve this problem.

C. Mechanical energy patterns during locomotion

The mechanical energy required for locomotion can be estimated by using a force plate (Cavagna, 1975; Cavagna, Heglund, and Taylor, 1977). As the animal runs or walks over the plates, both horizontal and vertical forces are recorded. Integration of the forces yields the horizontal and vertical velocity changes of the center of mass. Kinetic energy (horizontal and vertical) can then be calculated using these velocities. A second integration of the vertical velocity gives the vertical position changes. By knowing the degree of oscillation in the center of mass, changes in gravitational potential energy can be determined.

i. Mechanical energy changes of the center of mass during locomotion. Changes in mechanical energy during locomotion in land crabs appear to follow at least two different patterns (Blickhan and Full, 1987). During sideways walking in the ghost crab *O. quadrata*, the body moves up and down and the legs are moved in an irregular, alternating tetrapod gait: Adjacent legs on one side of the body move out of phase with one another, as do contralateral legs on opposite sides of the body.

Horizontal kinetic and gravitational potential energy fluctuate out of phase (Fig. 10.11A). As the crab's center of mass rises and falls, energy is alternately transferred from kinetic to potential energy, much like a swinging pendulum or an egg rolling end over end. In this way considerable energy is conserved. In a walking crab, as much as 55% of the energy is recovered during a stride (Blickhan and Full, 1987). This mechanism for conserving energy during walking appears to be very general, as it has been thoroughly described for bipedal and quadrupedal vertebrates (Cavagna et al., 1977; Heglund et al., 1982a).

At high running speeds, photographs show that the ghost crab literally leaps through the air, becoming bipedal and using a running gait (Burrows and Hoyle, 1973). Force plate measurements show the periods of zero vertical force where all eight legs are off the ground (Blicken and Full, 1987). The records also show that at high speed, horizontal, kinetic, and gravitational potential energy are in phase, so little energy exchange is possible (Fig. 10.11B).

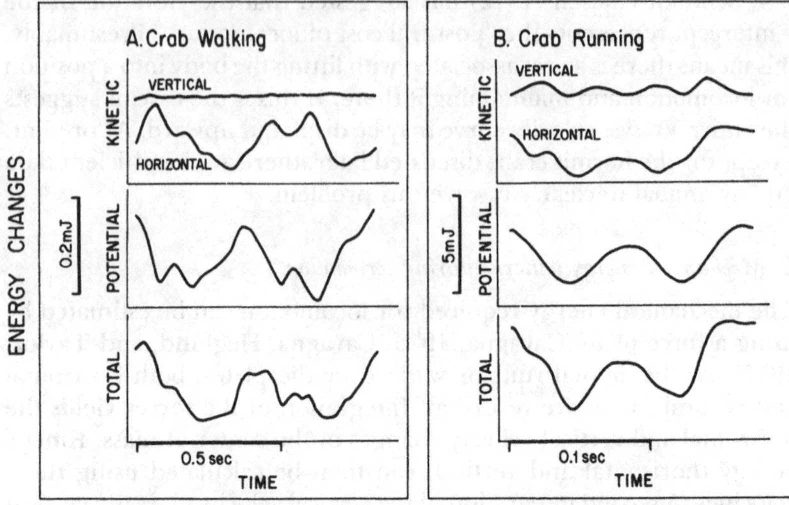

Fig. 10.11. Energy changes of the center of mass of a ghost crab walking and running across force plates. A. Data for a 53 g crab (0.52 N) walking during one stride at 0.25 m·sec^{-1}. B. Data for a 61 g crab (0.60 N) running at 0.60 m·sec^{-1}. The *upper* and *middle traces* represent the changes in vertical and horizontal kinetic energy. The *middle trace* shows the fluctuations in gravitational potential energy. The *lower trace* represents the changes in total mechanical energy of the center of mass. (From Blickhan and Full, 1987.)

The energy fluctuations during running in the crab are very similar to the patterns observed for trotting, hopping, and galloping vertebrates (Heglund, Cameron, and Taylor, 1983). In fact, the frequency at which the ghost crab cycles its legs during fast running is almost identical to that predicted for a galloping mammal of a similar mass, as is the speed at which it switches from a slow to a fast run (Fig. 10.12; Heglund, Taylor, and McMahon, 1974). This could mean that the body of a crab, as well as that of a mammal, functions as a tuned, mechanical spring system (Taylor, 1985).

ii. Mechanical energy changes versus speed. Total mechanical energy is extremely difficult to determine. A major portion of the mechanical power expended for locomotion is required to lift and accelerate the center of mass $\dot{E}_{CM}$, especially for small, slow-moving animals (Heglund et al., 1982b). Additional energy is necessary to accelerate the limbs and the body relative to the center of mass, while energy exchange and elastic storage decrease the energy demanded from muscles. For the ghost crab mechanical energy was estimated from $\dot{E}_{CM}$

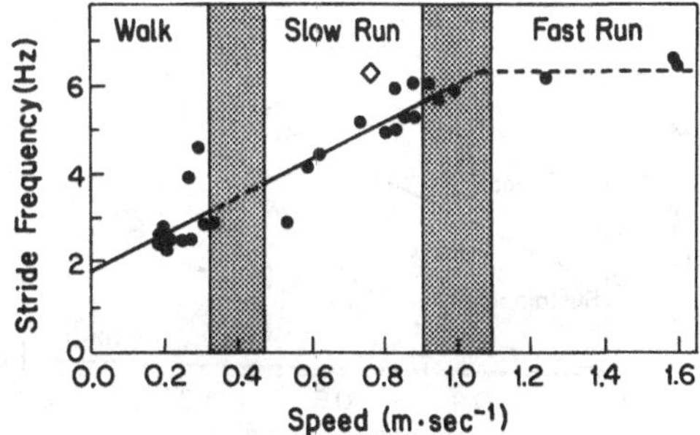

Fig. 10.12. Stride frequency as a function of speed for a 55 g ghost crab. *Shaded regions* represent gait transitions (Blickhan and Full, 1987). The *diamond* indicates the predicted trot–gallop transition for a mammal of the same mass (Heglund et al., 1974).

during walking and running (Blickhan and Full, 1987). $\dot{E}_{CM}$ increased linearly with speed and was directly proportional to body mass.

D. Efficiency of locomotion

How efficient is terrestrial locomotion in land crabs? Since both metabolic and mechanical energy measurements are available, we can estimate efficiency as follows:

$$\text{Efficiency} = \frac{\text{Mechanical energy output}}{\text{Metabolic energy input}} \times 100.$$

For the ghost crab, $\dot{E}_{CM}$ is taken to be the best estimate of the total mechanical energy output, and $\dot{V}o_2ss$ represents the total metabolic input. $\dot{E}_{CM}$ for the ghost crab is not significantly different from that required by a vertebrate of the same mass (Fig. 10.13). As in vertebrates, the mechanical power used for locomotion appears to increase in direct proportion to body mass (Heglund et al., 1982b). This suggests that the mechanical energy required for locomotion is relatively independent of the animal's morphological pattern.

In terms of metabolic energy, endothermic vertebrates demand high rates even at rest. If resting metabolism is removed, the rate of power required during locomotion becomes more similar to that of an ectotherm. Over the range of speeds (0.32–0.50 km·h^{-1}), where both metabolic and mechanical measurements have been obtained, the efficiency of ghost crab locomotion is approximately 5–10% (Blickhan and Full, 1987). These values are remarkably similar to the 2–5% values predicted for a bird or mammal of the same mass (Heglund et al., 1982b). Although more data are certainly needed, the major

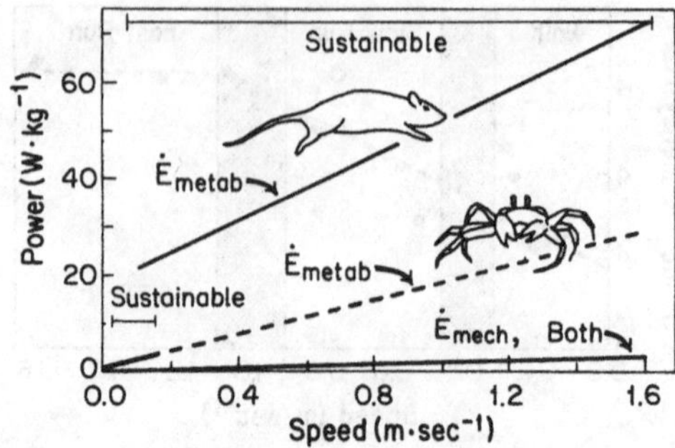

Fig. 10.13. Metabolic ($\dot{E}_{metab}$) and mechanical ($\dot{E}_{mech}$) power required for a ghost crab (25° C) and a mammal of the same mass ($\simeq$ 30g). $\dot{E}_{metab}$ for the mammal was predicted from Taylor et al. (1982). $\dot{E}_{metab}$ for the ghost crab is from Full (1987). The slopes, representing the minimum cost of transport, are not significantly different from that of the mammal. *Horizontal bars* represent the range of sustainable speeds (> 20 min) for each animal.

difference between the crab and the mammal may not involve the economy or efficiency of locomotion, but rather the endurance at high speeds. Mammals can sustain high rates of locomotion for long periods, but the crab is limited to slow walking with occasional brief sprints.

E. Economy of locomotion

The energetic cost of steady-state locomotion can be calculated by dividing the animal's O_2 consumption by its velocity. In aerobic species this yields a relationship indicating that O_2 consumption per distance traveled is highest at slow speeds; it progressively declines with velocity, gradually approaching a minimum cost of transport (C_M). The C_M value is the slope of the regression line for $\dot{V}o_2$ versus velocity (Taylor et al., 1970). Considering only the economy of travel, one might predict that animals traveling long distances should move at relatively fast speeds to conserve energy. Little attention has been given to this point. Certainly, in *C. compressus* traveling over the beach this appears to be the case (see section 6.G). (This hermit crab study is also of interest because it suggests that the type of terrain may alter the energetics of transport; crabs walking on dry sand travel more slowly than those on moist, packed sand.) The total energetic cost of locomotion is greater for large animals than for small (Taylor et al., 1970; Tucker, 1970). However, the relationship is not simple; a doubling of the mass does not necessitate a doubling of the energy output. Less energy is required. Often this relationship is shown by plotting the minimum cost of transport (i.e., ml $O_2 \cdot g^{-1} \cdot km^{-1}$) against body

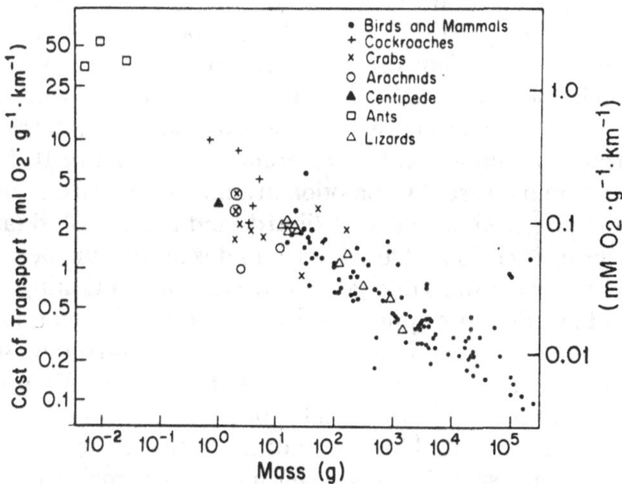

Fig. 10.14. Minimum cost of transport versus body mass on a log scale. Data are from Taylor et al. (1982) for birds and mammals; Bakker (1972) for lizards; Herreid and Full (1984a) for cockroaches; Herreid (1981) for arachnids, some land crabs, and centipede; and Jensen and Holm-Jensen (1980) for ants. Uncircled x = *Ocypode quadrata* from Full (1987). The circled x = *Uca pugilator* from Full and Herreid (1984) calculated by the sum of aerobic and anerobic metabolism.

mass; it is immediately obvious that small animals use relatively large amounts of energy for travel. Land crabs are no exception (Fig. 10.14). What is the explanation for this pattern? Taylor et al. (1980) have argued that small animals must expend more energy than large animals to generate a given force at the same speed. If the intrinsic velocity of muscle shortening is proportional to the rate of myofibral crossbridge cycling, then this rate should increase in direct proportion to the cost of generating force. Since the shortening velocity appears to be inversely related to mass (i.e., small animals have high intrinsic velocities), the cost of generating muscular force should increase for animals of small size.

Because arthropods have a variable number of appendages, it is logical to ask if the number of legs influences the cost of terrestrial locomotion. Consider first the problem of autotomy when land crabs may lose a leg to a predator. Herreid and Full (1986b) have removed limbs from hermit crabs by bilateral amputation, forcing the animals to walk with four limbs rather than the normal six. Of five individuals tested, one showed no increased cost of travel, and the remaining four showed only modest elevation; but the greatest increase in $\dot{V}O_2$ was only 1.8 times the normal rate. At least in this experiment, only modest changes in energy consumption occurred when limb number was reduced from six to four, although the speed and agility of the crab were certainly affected. Possibly, practice would improve performance. Limb autotomy probably has even less effect in species with eight walking legs, especially if it involves only one appendage. The question of leg number may be addressed by comparing the minimum cost of transport among widely different species. Figure 10.14 shows data for minimum cost of locomotion in animals with various numbers of legs ranging from 2 (birds), 4 (lizards and mammals), 6 (ants and cockroaches), 8 (crabs and arachnids), to dozens (centipede). All species tend to follow the same general trend, thus indicating that the number of legs does not appear to affect the cost of locomotion among species. Similarly, there is little correlation with different styles of locomotion or the taxonomic position of the animals. All animals of similar mass appear to have a similar fuel economy when traveling a given distance on land. This does not mean that their total $\dot{V}O_2$ per gram is the same, since homeotherms have higher maintenance costs than poikilotherms, but their minimum costs of transport are similar. This suggests that we are dealing with a fundamental property of muscle energetics.

F. Locomotion on land versus in water

Swimming animals have a much lower minimum cost of transport than pedestrian animals on land (Schmidt-Nielsen, 1972). Aquatic species are buoyed by the water around them and use less energy for support against gravity than terrestrial species. Also, land pedestrians are very uneconomical because their legs tend to slow their progress each time they hit the ground after a recovery stroke. The body must be reaccelerated after each step. To appreciate the effect of this braking and reaccelerating action, consider the relative ease of pedaling a bicycle. Bicycle riders do not have to use their feet to support themselves against gravity. Cyclic accelerations and decelerations are minimized; thus their efficiency of locomotion is very high. Hargreaves (1981) has reviewed the meager evidence available for swimming crustaceans and proposes that a similar reduction in the cost of travel may exist in this case.

Many crustaceans walk along the bottom of the ocean. They are supported by the water and would be expected to use less energy than if walking on land. On the other hand, walking in water should result in greater drag than in air especially in view of the potential turbulence that would be produced from multiple limbs in motion. Lobsters are known to walk in long queues during migrations and in this way may reduce drag (Bill and Herrnkind, 1976).

Attempts have been made to measure O_2 consumption while crabs were walking underwater. Houlihan, Mathers, and El Haj (1984) exercised *C. maenas* in a circular respirometer by prodding them gently. In typical experiments, crabs made 25 circuits of the chamber in 10 min, before they became fatigued. The Po_2 of the seawater was determined at 0-min, 5-min, and 10-min time periods in the run, so it was not possible to determine the O_2 kinetics. The poor endurance and the fact that whole-body lactates were significantly elevated even during the slow speed tests suggest that this crab has a mixed aerobic and anaerobic response to exercise. The authors found $\dot{V}o_2$ increased with velocity in large crabs but not small individuals, which is puzzling. Another surprising finding was that the net minimum cost of transport (including aerobic and anaerobic contributions) for this species walking underwater is similar to that of terrestrial crabs. This finding might have several explanations, including the possibility that the drag in the experiments might be high. However, Houlihan et al. (1984) remark that water currents produced by the crab's movement and the prodding stick seem negligible at low speeds. Yet, the hydrodynamic interactions of a crab in a shallow tank with alterations in speed and periodic reversals in direction are difficult to evaluate.

Houlihan and Innes (1984) compared walking in air and water in a circular respirometer using the amphibious intertidal crab, *Pachygrapsus marmoratus*. The exercise period lasted 6 min or less before fatigue occurred in either water or air. Oxygen kinetics could not be followed, but it was clear that O_2 debts were repaid in 20 min. $\dot{V}o_2$ increased linearly with velocity in both *P. marmoratus* and another crab, *Carcinus mediteraneus* walking in water. Once again Houlihan and Innes (1984) note that the cost of transport for these species walking in water is similar to decapods walking on land, even though the crabs "weigh over 10 times less in water than in air [*sic*]." This problem awaits further study.

G. Energetics of load carrying

Some crabs carry significant loads at some time during their life. A crab may drag food material back to its burrow. Females may carry egg masses. Certain aquatic crustaceans become heavily encrusted with sponges, barnacles, and sea anemones. In each of these cases, we would expect extra energy to be expended to carry the additional mass, less so in the aquatic environment than on land because of the effects of buoyancy. The energetics of load carrying in crustaceans has been recently investigated in the terrestrial hermit crab, *C. compressus* from Panama (Herreid and Full, 1984b, 1986b; Wheatly et al., 1985). Hermit crabs of this species are particularly valuable subjects for investigation because they naturally live in gastropod shells of varying masses. Two crabs of the same size may carry snail shells differing in weight by a factor of 3 (Herreid and Full, 1986a). More-

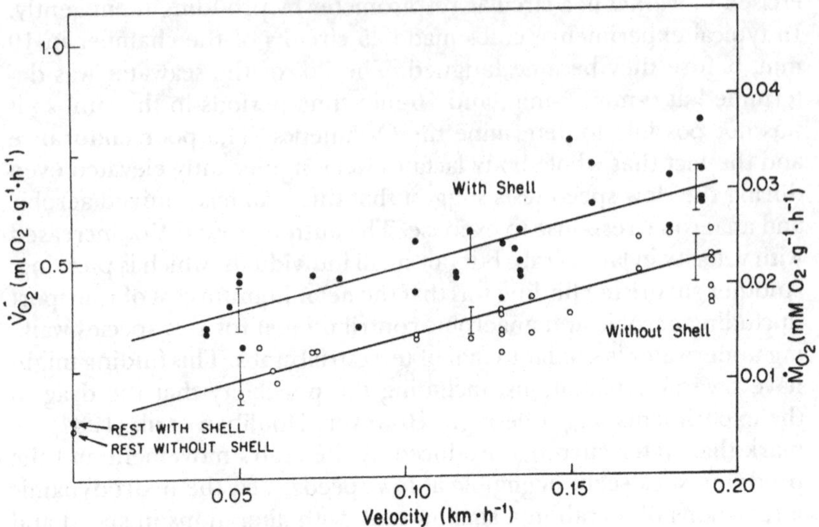

Fig. 10.15. Steady-state O_2 consumption of hermit crabs traveling on a treadmill at different velocities, comparing crabs with and without their snail shell loads. (From Herreid and Full, 1986b.)

over, they can be induced to switch shells or to walk on a treadmill without any snail shell at all. Terrestrial hermit crabs with and without snail shells both show a linear increase in $\dot{V}O_2$ with velocity; both lines are parallel, but the "nude" crabs use significantly less energy than those carrying a shell about the same mass as their body (Fig. 10.15). There is no difference in the resting $\dot{V}O_2$ in crabs with and without shells. Hence, it seems reasonable to conclude that the difference in the two curves represents the cost of carrying a snail shell. It is not the postural cost of locomotion, because crabs lacking shells do not show an elevated y-intercept.

The difference between the cost of transport of loaded and unloaded animals is the extra cost of transport, C_E. This value falls as velocity increases (Fig. 10.16). At slow velocities the extra cost of carrying a gram of shell is similar to carrying a gram of body mass. However, as velocities increase, C_E drops until it is negligible at speeds where the minimum cost of transport occurs. Consequently, another advantage accrues to crabs traveling rapidly across the beach. Perhaps the most interesting finding for the terrestrial hermit crabs is that increasing the load does not cause a corresponding proportional increase in $\dot{V}O_2$. Instead, the $\dot{V}O_2$ss loaded/unloaded ratio rises much less than the loaded/unloaded mass ratio. This is in contrast to data collected for several mammals (Taylor et al., 1980). Terrestrial hermit

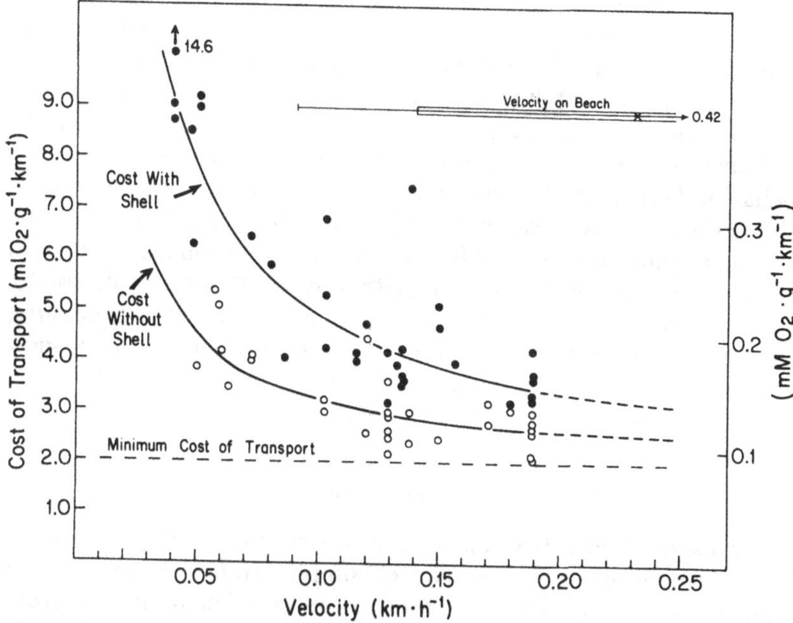

Fig. 10.16. Total cost of transport in hermit crabs as it varies with velocity on a treadmill, comparing crabs with and without snail shell loads. The *upper scale* indicates the speed crabs normally travel on the beach. (Modified from Herreid and Full, 1986b.)

crabs seem especially efficient at carrying loads, especially the heavier shells; there is essentially no increase in $\dot{V}O_2$ as shell sizes increase from two to four times the body mass. The reason for this unusual phenomenon probably has to do with an increase in mechanical efficiency. With greater loads, crabs shift their leg position and tend to drag their shells.

7. Energy budgets

Ideally, one would like to be able to identify each behavior of an organism and measure its energetic cost. This would permit the development of daily and seasonal energy budgets. We are far from this ideal in land crabs. Nevertheless, a few comments can be made. Based on field observations, crabs spend most of their time "resting." There are only brief periods when energy demand may rise severalfold because of behaviors such as an explosive escape response, pursuit of prey, or a fight. These activities are vital to the animal's survival and are energetically expensive, perhaps elevating metabolism to 10 times that at rest. However, these activities are of such short duration and so infrequent as to be of minor importance to the energy budget. Most common activities that are prolonged and frequent are probably energetically inexpensive. Crabs eat, explore, defecate, make courtship displays, and molt with what appears to be modest energy expenditures when we compare these activities to treadmill exercise. Simply stated, most energy is used for maintenance rather than for behavioral activities. Certainly, most daily activities do not even double the resting $\dot{V}O_2$ we observed in the laboratory. Realizing that the resting metabolism is probably much higher in the laboratory than in the undisturbed condition in the field, it seems that the daily energy budget probably is not much higher than the resting rate we measure, certainly no more than double the latter. This may hold true for most animal species. Consequently, the estimate of an energy budget may be fairly approximated using "resting metabolism."

8. Conclusion

In retrospect, energetics in land crabs is reasonably well understood. Their feeding processes are relatively simple and have been observed closely in several species. More information would be useful but probably would not change our perspective. We are, however, grossly ignorant about the caloric intake and assimilation efficiency of crabs in the field. The digestive system of land crabs has not been studied in any detail at all. Even though it is desirable to correct this deficiency, enough crustaceans have been examined that it is unlikely that any major surprises are in store here. Land crabs are probably similar to their aquatic relatives.

Crustacean metabolism appears surprisingly similar to vertebrate metabolism. Moreover, land crabs do not appear unique in any way when compared with aquatic crabs. However, a great deal more attention needs to be focused on intermediary metabolism. For example, we do not even know what is the major metabolic fuel for crustaceans. The role of protein as an energy source and its role in osmotic balance are quite intriguing. The hormonal control of metabolism is undoubtedly complex, but it is poorly understood except for some data on carbohydrate metabolism.

As far as the principles of whole-animal energetics are concerned, crustaceans in general, and land crabs in particular, are reasonably well understood. An impressive array of "resting" rates have been obtained, but it would be useful to have more data on crabs in an undisturbed state in conditions approximating the natural environment.

In contrast to other facets of metabolism, land crabs are better studied during exercise than any other crustacean group. From the dozen species that have been investigated, both "marathoners" and "sprinters" have been studied using treadmills and voluntary walking. Since the techniques are established, the next 10 years should see the number of exercise studies mount. Greater attention will be paid to morphological and physiological differences among species such as *Uca* and *Ocypode* that give rise to the different metabolic responses that have been measured between species. Also, the next decade should bring a clearer analysis of the mechanical energy used in locomotion in pedestrian arthropods.

Although work with treadmills demonstrates the metabolic scope for land crabs and allows a clear indication of the metabolic cost of locomotion, other behavior patterns have not been studied. If we are to evaluate the energy budgets of crabs in the future, we must have good studies of the energetics of feeding, courtship, agonistic displays, egg laying, growth, molting, regeneration, climbing, burrowing, etc. These are now lacking (also see Chapter 4). Oxygen-consumption measurements of these behaviors could be readily obtained in the laboratory. Their overall importance in the energy budget would be appreciated only if there were concomitant field studies where the frequency and duration of the behavior were measured. Collection of this information should be straightforward, and it would be a major advance in our investigations of the energetics of land crabs.

Acknowledgments

We thank Richard Safeer and Daniel Sperrazza for their aid in dissection and James Stamos and Allisa Shulman for their illustrations. Much of the research on locomotion was supported by National Science Foundation grant PCM 79–02890.

11: Epilogue

BRIAN R. McMAHON and
WARREN W. BURGGREN

As intended, this book has gathered a wealth of information concerning many aspects of the biology of land crabs. Most chapters have taken a comparative approach, often starting with the biology of aquatic crabs and showing how the development of morphological, physiological, and behavioral characters have allowed penetration into, and eventual colonization of, the terrestrial habitat. One of the goals of the editors in planning this book was to seek a better understanding of the evolution of terrestriality within decapods. Thus, a reexamination of this evolutionary trend will serve as a suitable epilogue.

From examination of this book's chapters it is clear that, under the descriptor "land crabs," the various contributors have considered many different groups of crabs. These groups vary considerably both in the terrestrial adaptations they exhibit as well as in their ability to move into the terrestrial habitat. The "failure" to achieve a clear definition of what constitutes a "land crab" is, in fact, no disadvantage, for it correctly causes us to focus on a *continuum* of crabs showing progressively greater levels of terrestrial adaptation. An approach that examines a continuum of species not only enhances understanding of the evolution of terrestriality in the land crabs, but may also provide insights into the radiations of protoinsects and vertebrates.

Terrestrial crabs appeared only in relatively recent geologic times, especially when compared with the land colonization by vertebrates or protoinsects. Although fossil evidence is extremely sparse, families of crabs with living species in terrestrial habitats appear in the fossil record no earlier than the Paleocene or late Cretaceous (Chapter 2). Land crabs have thus had relatively little time for radiation into the terrestrial sphere and in many respects might exemplify the early stages of evolution of the complete terrestrial competence evident in other arthropods such as the insects.

It appears that the ancestors of the majority of living species of land crabs entered the terrestrial sphere via the littoral fringe of the sea (Chapter 2). The exceptions were some of the Grapsidae and most, if not all, of the Potamoidea. These groups apparently moved onto land after first colonizing fresh water. This dichotomy in evolutionary pathways to terrestrial habitats is possibly reflected in the current distribution of land crabs (Chapters 2 and 3). Thus, crabs that have entered the terrestrial sphere from seawater are still congregated in "maritime" areas (i.e., around the supralittoral fringe of the oceans), whereas crabs that have originated from fresh water are, perhaps not surprisingly, often found much farther inland. This latter group is not necessarily restricted to areas around permanent or temporary bodies of fresh water, but also occupies areas of high humidity, e.g., the tropical rain forests of New Guinea.

In his excellent treatise on the colonization of land, Little (1983) contended that the selection pressures that led to entry into the terrestrial habitat may have been different for these two groups of crustaceans. Little's (1983) conclusion that freshwater crabs may have evolved mechanisms to utilize aerial oxygen in response to aquatic hypoxia seems reasonable. However, we do not support his contention that aquatic hypoxia would not have been an important selection pressure in marine forms. Hypoxia, is, in fact, also a common stress for intertidal marine forms, and quite severe hypoxia may be routinely associated with nocturnal low tides when respiration exceeds photosynthesis (Chapter 8). Under these conditions crabs such as *Carcinus* either move to the surface and periodically ventilate the branchial chambers with air or leave the water entirely (Chapter 8). Such behavioral modifications may have been a very important step in the evolution of air-breathing structures and processes allowing crabs to invade the supralittoral zone.

Other differences in the evolutionary history of crabs colonizing land from maritime and freshwater environments have probably been at least as influential as respiratory adjustments. Differences in reproductive biology, for instance, are clearly important factors in the difference in distribution between the two groups (Chapters 3 and 5). Crabs that invaded the terrestrial environment directly from the sea must return to the sea for dispersal of the eggs, which consequently develop through at least the early larval stages in water. The distribution of such crabs thus obviously remains essentially maritime. However, crabs that invaded the terrestrial environment via fresh water have circumvented this problem by enclosing the critical early larval stages within the protected environment of the egg. These species lay relatively few, relatively large eggs, which contain sufficient food reserves to delay hatching until the larvae are of a size and level of development to be able to survive on land. These modifications have not evolved specifically as adaptations for a terrestrial life, but were necessary in the initial colonization of the freshwater habitat, where larvae had to be sufficiently large and complex enough at hatching to carry out effective ion and osmotic regulation in fresh water. The freedom from the necessity to visit standing water for reproduction has allowed the freshwater crabs to colonize inland rain forests and other humid microenvironments.

Two other reproductive features that may have aided the ancestral crabs in their colonization of the terrestrial habitat – internal fertilization and "brooding" of the eggs – were both already present in the ancestral aquatic crabs and could be considered preadaptive features.

Several other morphological and physiological features of extant (and presumably ancestral) aquatic crabs have expedited invasion of terrestrial habitats. For example, the arthropod exoskeleton with its low water permeability provides considerable resistance to evaporative water loss and ionic shock (Chapter 7). The exoskeleton as evident in aquatic crabs also allows mobility on land with minimal redesign of the skeletal and other body support systems. Thus, the walking gait of the aquatic brachyuran crabs is hardly changed in transition to land (Chapter 10). The simplicity of this transition is dramatically evident when compared with the complexity involved in the evolution of the pentadactyl limb in the tetrapod vertebrates.

Many aspects of the respiratory system also seem to lend themselves to modification for a terrestrial existence. As only one example, the recently discovered alternative circulatory pathway bypassing the gills of aquatic crabs (see Chapter 9) is clearly the antecedent of the branchiostegal circulation, which has become modified to perfuse the very complex air-breathing organs of the Pseudothelphusid and Myctirid land crabs. The "redesign" of this existing bypass system appears relatively simple when compared with the extensive evolutionary changes in branchial arches necessary for the development of perfusion pathways for the vertebrate lung.

Paradoxically, despite the number of features that are apparently "preadaptive" for success on land, colonization of terrestrial habitats by decapods has been very limited. The most terrestrial of the land crabs remain comparable to the majority of Amphibia, being restricted essentially to the areas around water bodies and to moist or very humid microenvironments. It seems very likely that this persistent limited distribution of land crabs relates to water loss. An important factor in the success of both terrestrial arthropods and vertebrates is the ability to generate hypertonic excretory products. No land crabs have developed this ability, and thus nonurinary water loss becomes even more critical. The land crab *Gecarcinus lateralis* at rest has a water permeability lower than many amphibians, but still 10–20 times greater than occurs in a small desert reptile. No doubt, respiratory water loss increases greatly with activity in land crabs, and specific mechanisms to reduce respiratory water loss, such as the narial countercurrent exchangers of some reptiles, are apparently absent in the land crabs. It is noteworthy that even the most highly specialized of the land crabs essentially become torpid or inactive in dry air.

Ultimately, we can only speculate that the limited distribution of land crabs results from some combination of (1) the problems with respiratory water loss and lack of a fully impermeable integument, (2) the lack of an excretory system capable of producing a"urine" hyperosmotic to hemolymph, and (3) the lack of a "cleidoic" egg that could be left to develop to a sufficiently advanced stage untended on land. Unable to compete effectively in drier terrestrial habitats, the land crabs have made only tentative encroachments on terrestrial ecosystems when compared with many other invertebrates or the vertebrates. However, no one who has witnessed the spectacular mass migration of crabs during their breeding season can deny their overwhelming success in these localized habitats.

The spectrum of decapods adapted for terrestrial life constitutes a fascinating element in the colonization of land by ancestral marine organisms. Yet, our understanding of many aspects of the biology of land crabs remains fragmentary. It is to be hoped that the burgeoning interest in land crabs that we are currently witnessing in biologists from many different fields will generate new insights into these fascinating crustaceans.

Appendix: Natural histories of selected terrestrial crabs

Crabs that are either clearly intertidal or normally found within permanent bodies of fresh water have not been included here, although many of these species can show some limited degree of activity on land. Some supralittoral species are included. Larval development is categorized in one of three ways: PM, larvae are released in a pelagic marine environment; ABB, larvae have an abbreviated development usually associated with a specialized environment; DIR, eggs develop directly into young crabs. Body mass is noted for the largest specimen reported in the literature. Compiled primarily by Sandra L. Gilchrist, with contributions by other authors in this book.

Species	Habitat	Food	Larval development	Body mass	Comments
COENOBITIDAE					
Birgus latro	Unrestricted on islands, found at highest island elevations; sandy areas, coconut groves	Fruit (*Cocos, Pandanus*), dead tortoises, other carrion, crustaceans (e.g., *Cardisoma*), scavenge middens	PM	3,000 g	Largest terrestrial arthropod; found in rock cracks, under rock overhangs, burrows; good climber; one of few terrestrial crabs observed to "drink" water
Coenobita brevimanus	Rarely near shore, found in grassy areas, rain forest	Scavenger; omnivorous	PM	185 g	Fills shell with fresh or brackish water; mostly nocturnal
Coenobita cavipes	Near to the sea, mainly in sandy areas alongside sheltered lagoons	Carrion (dead fish)	PM	160 g	Prefers to enter, and fill shell with, seawater
Coenobita clypeatus	2 population types; one not more than 15 km inland, one close to shore; at elevations up to 1,000 m; dune vegetation; shaded areas with large quantities of leaf litter	Omnivorous; feeds on variety of plants as well as carrion; will feed on feces; cannibalism may occur	PM	110 g	Eats plants toxic to other animals (e.g., cactus fruits and "apples" of the manchineal tree)
Coenobita compressus	Up to 1 km inland, mostly within 100 m of shore; sandy beaches, moist, heavily vegetated	Scavenger; cacao, platano, rice, wood fragments, ciruela, manzillo, fungus, palm copra, carrion	PM	38 g	Agricultural pests; nocturnal in arid areas, active all day in humid areas
Coenobita perlatus	Never far from shore or adjacent dunes, have been found in tidal caves; sandy areas, dune vegetation; usually in humid areas periodically inundated by the sea	Scavenger	PM	80 g	Relatively rare; however, see Holthuis (1953) and Johnson (1965); active nocturnally
Coenobita rubescens	Occurs far inland, to altitudes exceeding 800 m	Omnivorous	PM	48 g	Has been noted using sea urchin tests to cover abdomen
Coenobita rugosus	Occurs up to 300 m from shore; mostly in and around dunes, sandy areas, within dune vegetation	Algae and tortoise feces	PM	67 g	During day, shelter under leaf litter; feed nocturnally

Species	Habitat	Food	Larval development	Body mass	Comments
Coenobita scaevola	Always near sea	Scavenger	PM	74 g	
Coenobita spinosa	Up to 200 m from shore; mostly in dune areas, within dune vegetation	Decaying plants and animals	PM		Mostly nocturnal; burrow under leaf litter during cold months, "hibernate"
DIOGENIDAE					
Clibanarius vittatus	Supralittoral; sandy areas, oyster bars, marshes	Omnivorous; insects, small mollusks, carrion	PM	40 g	Can remain completely exposed for 72 hours, in shaded areas (humid) can remain exposed up to 100 hours; limited activity when exposed
GECARCINIDAE					
Cardisoma armatum	Up to 100 m from shore; moist, sandy areas; mangroves, river mouths; cultivated areas near water	Scavengers; carrion, palm nuts, coconuts, decaying plant materials	PM	64 g	Adults construct deep burrows that intersect; feed nocturnally
Cardisoma carnifex	Moist to wet areas far from shore; habitat use varies, but always near either water or mud	Ingest mud, fallen leaves (fresh or decaying), sedges (*Fimbristylis*), tortoise feces	PM	400 g	Conspicuous sexual dimorphism of chelae (male very much larger than female)
Cardisoma crassum	Beyond dune area into mangrove (up to 7 km from shore); very wet areas with submerged vegetation	Fresh and decaying vegetation; carrion	PM		Diurnally active
Cardisoma guanhumi	Several hundred meters from shore; shaded areas, mangroves, dunes, river banks; saline soils	Wide range of plant materials (fresh and decaying)	PM	400 g	Very sensitive to vibrations; usuall have 2 m feeding radius around burrows
Cardisoma hirtipes	Delimited by fresh water; moist, saline soils; mudbanks near rivers	Herbivorous; will eat carrion	PM	600 g	Crop pests
Cardisoma rotundum	Tends to occur in drier areas than other members of genus	Omnivorous	PM		
Gecarcinus lateralis	Up to 1 km from shore; commonly found 6–9 m above high-tide mark	Largely herbivorous; will ingest carrion, feces, conspecifics	PM		Common; very agile; extends range following humans' activitie

Species	Habitat	Food	Larval development	Body mass	Comments
Gecarcinus malpilensis	At shoreline; rocky substrate free of macrovegetation	Algae scraped from rocks, lizards (*Anolis* spp.), and bird chicks	PM		Endemic to Malpelo Islands
Gecarcinus planatus	Steep slopes up to 120 m from shore; rocky substrates; within beach strand vegetation	Omnivorous	PM		Forage nocturnally
Gecarcinus ruricola	Up to 1.5 km from shore; shaded, woody areas; muddy beaches	Frugivorous (e.g., *Spondias*)	PM		Forage nocturnally; burrows or hides under fallen trees or flat rocks during day
Gecarcoidea lalandei	In burrows within mixed forests, generally in moist soil	Plant material	PM		
Gecarcoidea natalis	Up to 3 km from sea; rain-forest areas where it burrows in clay soil; climbs on rock outcrops	Largely herbivorous; will ingest mollusks, crustaceans, carrion	PM	450 g	Strictly diurnal; drinks water by spooning it with chelae
GECARCINUCIDAE					
Barytelphusa cunicularis	Fresh water; in vicinity of canals, ponds, and rivulets	Omnivorous	DIR		
arytelphusa pulvinata	Fresh water; rice fields (higher elevations)	Omnivorous	DIR		
Gecarcinautes antongilensis	Limestone forests; found near ephemeral pools	Unknown	DIR		Estivates in small caves during dry season
GRAPSIDAE					
ratus elegans	In mangrove trees	Herbivorous?	PM		Very agile, active climbers
ratus pisonii	Mangroves; lives on roots and in canopy; shaded areas on trunks	Fresh and decaying leaves, insects, epizoans on mangroves	PM		Arboreal crabs; leave trees only to reproduce; mate while on tree; very agile and active climbers
pigrapsus politus	In stony areas; close to waterline; under rocks	Unknown	PM		
eograpsus crinipes	Near shore; within 500 m of shoreline; sandy beach to the shrub/grass zone	Active predator on crabs and myriapods attracted to carrion	PM	150 g	Escape into grass, crevices, or *Pandanus*

Species	Habitat	Food	Larval development	Body mass	Comments
Geograpsus grayi	Up to 1 km from shore; more common near shore; sandy areas; within dune vegetation; rarely found directly on shore	Insects, decaying matter (plant and animal)	PM	50 g	Good climbers; have been found using tree holes but most often in rock crevices
Geograpsus lividus	Above high-tide line; (3–5 m from surf); only on rocky or cobbly beaches	Omnivorous?	PM		Found among litter
Geograpsus stormi	High on rocky shores, concealed in caves and under overhangs	Dead crustaceans and other carrion	PM	30 g	Nocturnally active; confused in literature with *G. crinipes*
Geosesarma foxi	Often distant from permanent water	Undetermined	ABB?		Found under stones
Geosesarma gracillimum	In damp forests	Undetermined	ABB?		Nocturnally active; shelters by day in flooded burrow
Geosesarma noduliferum	In streams and in water held in leaf axils of plants	Undetermined	ABB?		Female carries hatched young beneath abdomen
Geosesarma ocypodum	In hilly, wooded areas, which may be very dry	Undetermined	ABB?		
Geosesarma perracae	In and near streams	Undetermined	ABB?		Burrows partially filled with water; larvae hatch into water in base of burrow
Goniopsis cruentata	Marsh areas; within 100 m of seaward mangrove fringe; muddy, marshy areas adjacent to brackish water	Young mangrove shoots, sponges, crustaceans (e.g., *Aratus* and *Uca*); cannibalistic	PM	100 g	Rapid runners on ground; venture only short distance from burrows; will escape into water
Grapsus grapsus	3–4 m above water surface; usually within splash zone	Omnivorous; cannibalistic	PM		Very rapid runners
Grapsus tenuicrustatus	Near shore; within 100 m of shoreline; sandy beaches and abutting lagoons	Omnivorous; attracted to fresh meats and fruits; observed picking at beach wrack	PM	75 g	Mostly crepuscular
Leptograpsus variegatus	Supralittoral; splash zone on rocky coasts	Probably grazes on plant and animal material on rocks	PM	100 g	Very agile, fast runner; will retreat into water or rock crevice

Species	Habitat	Food	Larval development	Body mass	Comments
Metasesarma rousseauxi	Under leaf litter in vegetation bordering beaches	Undetermined	PM	5 g	See Hicks et al. (1984)
Metasesarma rubripes	Mangroves; river banks; mangrove roots and branches (rarely above 6 m)	Roots, leaves, and insects	PM		Active both day and night
Metopaulias depressus	Found at elevations up to 300 m; up to 900 m from shore; in rainwater reservoirs in bromeliads	Herbivorous; decaying plants and animals; insects (?); scavengers	ABB		Phytotelmic; known only from Jamaica
Neosarmatium smithi	High mangrove zone; rarely inundated; burrows in mud	Herbivorous; decaying plant matter, esp. mangrove litter	PM		Nocturnal; spends little time out of burrows; active time spent foraging
Pachygrapsus gracilis	Mangrove roots and river banks; moist, shaded areas; among stones, algae, and sponges at seaward fringe of mangroves; areas of reduced salinity	Omnivorous	PM		Very rapid runner
Pachygrapsus marmoratus	Near shoreline; cliff overhangs	Omnivorous	PM		Found on cliffs of Mediterranean coasts; good climbers?
Sesarma africanum	Near shore; mangrove areas; will follow brackish water wedge; river mouths; salt marshes	Scavengers	PM		
Sesarma angustipes	Under debris; near brackish water; also from water in bromeliads in similar locations	Insects, detritus, bromeliad and other plant materials	PM		Agile climbers
Sesarma bidentatum	In and near freshwater streams; 400–1500 m inland	Primarily plant materials	ABB		
Sesarma cerberus	In caves	Unknown	ABB?		Eyes reduced; legs elongated
Sesarma cookei	Under stones and rubble in damp, shaded sites; far from sea, at elevations of 400–650 m	Plant material; insects?	ABB		

Species	Habitat	Food	Larval development	Body mass	Comments
Sesarma jacksoni	Damp places among cliffs and in caves	Unknown	PM	15 g	See Hicks et al. (1984)
Sesarma jacobsoni	In caves	Unknown	ABB?		Eyes reduced; legs elongated
Sesarma jarvisi	Limestone talus and rubble with damp leaf litter; found at elevations of 300–900 m	Unknown	ABB		Montane
Sesarma meinerti	Within 40 m of saline or brackish water; sandy/clayey areas	Fresh and decaying plant material; scavengers	PM		Construct well-developed burrows; forage nocturnally
Sesarma ortmanni	Burrows above high tide level on sandy and muddy shores	Unknown	PM		
Sesarma ricordi	Up to 40 m from shore; in rocks and debris; among driftwood, leaf litter, stones, and coconut husks	Detritivore; scrape particles from surfaces of stems and branches	PM		Moves from cove[r] at night only
Sesarma roberti	Up to elevations of 400 m; found up to 2 km inland upstream; along freshwater streams and seepage areas; within or near bamboo forests	Herbivorous; will ingest carrion	PM		Retreats into water; may hide under rocks and logs
Sesarma sulcatum	Tidal marshes; stream banks; mangroves; muddy areas; climb marsh plants and mangrove pneumatophores	Fresh and decaying plant materials	PM		Active diurnally; burrow or live under debris
Sesarma verleyi	In caves; in and near fresh water	Unknown	ABB?		Shows cavernicol-ous adaptations; pale color, elongated legs, eyes reduced in size, but functional
OCYPODIDAE					
Ocypode aegyptiaca	Burrows on sandy beaches; essentially supratidal	Unknown	PM		
Ocypode africana	Burrows above high-tide level	Unknown	PM		

Species	Habitat	Food	Larval development	Body mass	Comments
Ocypode cordimanus	Burrows on sandy beaches, greater than 100 m from shore	Insects, small lizards, carrion	PM	48 g	
Ocypode laevis	Burrows above high tide level	Unknown	PM		
Ocypode quadrata	Burrows on sandy beaches from the tide line to 400 m inland	Omnivorous; other crustacea (e.g., hermit crabs), carrion, fruit, feces; also will deposit feed near burrows	PM		Extremely agile runners; retreat into burrows when threatened
Uca annulipes	Close to the tide zones; mud flats, rocky areas, mangroves	Algae, mud, fruit, carrion	PM		
Uca inversa	Burrows at extreme high-tide level	Deposit feeder	PM		
Uca leptodactyla	Burrows in dry flats at extreme high tide level	Deposit feeder	PM		
Uca marionis	Close to tide zone in open, sunny areas; rarely collected among mangroves where tide deposits detritus	Deposit feeder	PM		
Uca minax	Often burrows well above high-tide level	Deposit feeding by sorting substrate	PM		Burrows descend to water table
Uca subcylindrica	Near ephemeral ponds; often many km from tidal waters	Deposit feeder	PM		Burrows to water table; water often hypersaline
Uca tangeri	Above median high-tide water mark; found in well-packed mud and in dense reeds	Deposit feeder	PM		
POTAMIDAE					
Madagopotamon humberti	Limestone forests; found near temporary water pools	Unknown	DIR		Estivates in small caves during dry season
PSEUDOTHELPHUSIDAE					
Guinota dentata	Associated with fresh water; found at altitudes of more than 1 km; near streams, ditches, ponds, lakes, and seepage areas	Omnivorous?	DIR		Constructs burrows or excavates beneath stones

Species	Habitat	Food	Larval development	Body mass	Comments
PARATHELPHUSIDAE[‡]					
Spiralothelphusa hydrodromous	Associated with fresh water; rice fields	Plant material; omnivorous	DIR		
Parathelphusa [*Holthuisana*] *guerini*	Associated with fresh water; seeks higher ground when areas are flooded; banks of streams; dry river beds	Variety of plant and animal materials; has been observed "stocking" burrows before dry season	DIR		In dry season remains burrowed away from water; active in wet season; more than one individual may occupy a single burrow; burrows may be superficially connected near surface; rice pest
SUNDATHELPHUSIDAE[‡]					
Holthuisana agassizi	Pastures; grassy slopes near fresh water; noted at least 4 km from sea; near creek and stream edges in high rainfall	Plant material, especially fruits; scavenger	DIR	20 g	In dry season lives in shallow burrows and under rocks in cree[k] bed; may be with[out] free water
Holthuisana subconvexa	Under moss on mountain slopes	Omnivorous	DIR		
Holthuisana transversa	Associated with fresh water; found in arid areas liable to temporary inundation; burrows into banks of streams, swamps, ground tanks, and water holes	Chiefly herbivorous; will ingest animal material	DIR	40 g	Dry periods spen[t] in deep burrows in clay soil; plug[s] burrows with molded clay to survive long droughts; no fre[e] water available for irregular periods often exceeding several years
Holthuisana valentula	Grassy slopes; forested areas; in high rainfall areas; found in and around vicinity of creeks and rivers	Omnivorous; attracted to cattle and horse feces	DIR	30 g	Adults burrow o[n] slopes well away from river bed; remain in burro[w] in dry season an[d] forage mainly o[n] land during wet season

[‡]See editors' footnote, page 458

References

Abele, L. G. 1970. Semiterrestrial shrimp, *Merguia rhizophorae*. *Nature* 226: 661–662.

Abele, L. G. 1972. A note on the Brazilian bromeliad crabs (Crustacea, Grapsidae). *Arq. Cien. Mar.* 12: 123–126.

Abele, L. G. 1973. Taxonomy, distribution and ecology of the genus *Sesarma* (Crustacea, Decapoda, Grapsidae) in eastern North America, with special reference to Florida. *Amer. Midl. Nat.* 90: 375–386.

Abele, L. G., and D. B. Means. 1977. *Sesarma jarvisi* and *Sesarma cookei*: Montane, terrestrial crabs in Jamaica. *Crustaceana* 32: 91–93.

Abele, L. G., M. H. Robinson, and B. Robinson. 1973. Observations on sound production by two species of crabs from Panama (Decapoda, Gecarcinidae and Pseudothelphusidae). *Crustaceana* 25: 147–152.

Abrams, P. 1978. Shell selection and utilization in a terrestrial hermit crab, *Coenobita compressus* (H. Milne Edwards). *Oecologia* (Berlin) 34: 239–253.

Ache, B. W. 1982. Chemoreception and thermoreception. In *The biology of crustacea*. Vol. 3, *Neurobiology: Structure and function*. Edited by H. L. Atwood and D. C. Snademan, pp. 369–398. New York: Academic Press.

Achituv, Y., and M. Ziskind. 1985. Adaptation of *Coenobita scaevola* (Crustacea: Anomura) to terrestrial life in desert-bordered shore line. *Mar. Ecol. Prog. Ser.* 25: 189–198.

Adiyodi, K. G., and R. G. Adiyodi. 1970. Endocrine control of reproduction in decapod Crustacea. *Biol. Rev.* 45: 121–165.

Adiyodi, K. G., and R. G. Adiyodi. 1975. Morphology and cytology of the accessory sex glands in invertebrates. *Int. Rev. Cytol.* 43: 353–398.

Adiyodi, K. G., and R. G. Adiyodi. 1981. Reproduction vs. growth: Endocrine programming in the Brachyura (Crustacea: Decapoda). In *Comparative endocrinology*. Edited by B. Lofts. Proc. Ninth Int. Symp. Comp. Endocrinol., Hong Kong. Hong Kong: Hong Kong University Press.

Adiyodi, R. G. 1968a. On reproduction and molting in the crab *Parathelphusa hydrodromus*. *Physiol. Zool.* 41: 204–209.

Adiyodi, R. G. 1968b. Protein metabolism in relation to reproduction and molting in the crab *Parathelphusa hydrodromus* (Herbst). Part I. Electrophoretic studies on the mode of utilization of soluble proteins during vitellogenesis. *Indian J. Exp. Biol.* 6: 144–147.

Adiyodi, R. G. 1968c. Protein metabolism in relation to reproduction and molting in the crab *Parathelphusa hydrodromus*. Part II. Fate of conjugated proteins during vitellogenesis. *Indian J. Exp. Biol.* 6: 200–203.

Adiyodi, R. G. 1969a. Protein metabolism in relation to reproduction and molting in the crab *Parathelphusa hydrodromus*. Part III. RNA-activity and protein yolk biosynthesis during normal vitellogenesis and under conditions of acute inanition. *Indian J. Exp. Biol.* 7: 13–16.

Adiyodi, R. G. 1969b. On the storage and mobilization of organic resources in the hepatopancreas of a crab (*Parathelphusa hydrodromus*). *Experientia* 25: 43–44.

Adiyodi, R. G. 1978. Endocrine control of ovarian function in crustaceans. In *Comparative endocrinology*. Edited by P. J. Gaillard and H. H. Boer, pp. 25–28. Amsterdam: Elsevier/North-Holland Biomedical Press.

Adiyodi, R. G. 1984. Seasonal changes and the role of eyestalks in the activity of the androgenic gland of the crab, *Parathelphusa hydrodromus* (Herbst). *Comp. Physiol. Ecol.* 9: 427–431.

Adiyodi, R. G. 1985. Reproduction and its control. In *The biology of crustacea*. Edited by D. E. Bliss and L. H. Mantel, vol. 9, pp. 147–215. New York: Academic Press.

Adiyodi, R. G., and K. G. Adiyodi. 1970a. Lipid metabolism in relation to reproduction and molting in the crab, *Parathelphusa hydrodromus* (Herbst). Phospholipids and reproduction. *Indian J. Exp. Biol.* 8: 222–223.

Adiyodi, R. G., and K. G. Adiyodi, 1970b. Free-sugars in *Parathelphusa hydrodromus* (Herbst), in relation to reproduction and molting. *Physiol. Zool.* 43: 71–77.

Adiyodi, R. G., and K. G. Adiyodi. 1972. Hepatopancreas of *Parathelphusa hydrodromus* (Herbst): Histophysiology and pattern of proteins in relation to reproduction and molt. *Biol. Bull.* 142: 359–369.

Adiyodi, R. G., and K. G. Adiyodi. 1976. Seasonal changes in the activity of the androgenic gland of the crab, *Parathelphusa hydrodromus* (Herbst). *Proc. 4th All-India Symp. Comp. Endocrinol.*, Mysore, p. 1.

Adiyodi, R. G., and T. Subramoniam. 1983. Arthropoda-Crustacea. In *Reproductive Biology of Invertebrates*. Edited by K. G. Adiyodi and R. G. Adiyodi, vol. 1, pp. 443–495. Chichester: John Wiley.

Ahmed, M., and J. Mustaquim. 1974. Population structure of four species of porcellanid crabs (Decapoda: Anomura) occurring on the coast of Karachi. *Mar. Biol.* (Berlin) 26: 173–182.

Ahsanullah, M., and R. C. Newell. 1977. The effects of humidity and temperature on water loss in *Carcinus maenas* (L.) and *Portunus marmoreus* (Leach). *Comp. Biochem. Physiol.* 56A: 593–599.

Aicher, B., M. Markl, W. M. Masters, and M. L. Kirschenlohr. 1983. Vibration transmission through the walking legs of the fiddler crab *Uca pugilator* (Brachyura, Ocypodidae) as measured by laser Doppler vibrometry. *J. Comp. Physiol.* 150A: 483–492.

Alcock, A. 1905. Catalogue of the Indian decapod Crustacea in the collection of the Indian Museum. Part 2. Anomura. Fasc 1. Paguristes. Calcutta: Indian Museum.

Aldrich, J. B., and J. N. Cameron. 1979. CO_2 exchange in the blue crab *Callinectes sapidus* (Rathbun). *J. Exp. Zool.* 207: 321–328.

Aldrich, J. C. 1974. Allometric studies on energy relationships in the spider crab *Libinia emarginata* (Leach). *Biol. Bull.* 147: 257–273.

Alexander, H. G. L. 1976. An ecological study of the terrestrial decapod Crustacea of Aldabra. Ph.D. thesis, University of London.

Alexander, H. G. L. 1979. A preliminary assessment of the role of the terrestrial decapod crustaceans in the Aldabran ecosystem. *Phil. Trans. R. Soc. Lond.* B286: 241–246.

Alexander, R. McN., and T. Goldspink. 1977. *Mechanics and energetics of animal locomotion.* New York: Wiley.

Alexander, S. J., and D. W. Ewer. 1969. A comparative study of some aspects of the biology and ecology of *Sesarma catenata* Ort. and *Cyclograpsus punctatus* M. Edw. with additional observations on *Sesarma meinerti* De Man. *Zool. Afr.* 4: 1–128.

Ali, J. A. 1955. Hunting the land crab *Parathelphusa guerini. J. Bombay Nat. Hist. Soc.* 52: 941–945.

Allen, W. 1971. Amino acid and fatty acid composition of tissues of the Dungeness crab *(Cancer magister). J. Fish. Res. Board. Can.* 28: 1191–1195.

Altevogt, R. 1955. Beobachtungen und Untersuchungen an indischen Winkerkrabben. *Z. Morph. Oekol. Tiere* 43: 501–522.

Altevogt, R. 1957a. Untersuchungen zur Biologie, Oekologie und Physiologie indischer Winkerkrabben. *Z. Morph. Oekol. Tiere* 46: 1–110.

Altevogt, R. 1957b. Beitrage zur Biologie von *Dotilla blandfordi* Alcock und *Dotilla myctiroides* (Milne-Edwards) (Crustacea Decapoda). *Z. Morph. Oekol. Tiere* 46: 369–388.

Altevogt, R. 1965. Lichtokompass- und Landmarken-dressuren bei *Uca tangeri* in Adalusien. *Z. Morph. Oekol. Tiere* 55: 641–655.

Altevogt, R., and H. O. Von Hagen. 1964. Ueber die Orientierung von *Uca tangeri* im Freiland. *Z. Morph. Oekol. Tiere* 53: 636–656.

Alvesmota, M. I. 1975. Sobre a reproducao do caranguejouea, *Ucides cordatus* (Linnaeus), em mangues do estado do Ceara (Brasil). *Arq. Cien. Mar.* 15: 85–91.

Anderson, D. T. 1982. Embryology. In *The biology of crustacea.* Edited by L. G. Abele, vol. 2, pp. 1–41. New York: Academic Press.

Andrews, C. W. 1900. *A Monograph of Christmas Island.* London.

Anger, K., N. Laash, C. Pueschel, and F. Schorn. 1983. Changes in biomass and chemical composition of spider crab (*Hyas araneus*) larvae reared in the laboratory. *Mar. Ecol. Prog. Ser.* 12: 91–102.

Anilkumar, G. 1980. Reproductive physiology of female crustaceans. Ph.D. thesis, University of Calicut, India.

Anilkumar, G., and K. G. Adiyodi. 1980. Ovarian growth, induced by eyestalk ablation during the prebreeding season, is not normal in the crab, *Parathelphusa hydrodromus* (Herbst). *Int. J. Invertebr. Reprod.* 2: 95–105.

Anilkumar, G., and K. G. Adiyodi. 1985. The role of eyestalk hormones in vitellogenesis during the breeding season in the crab, *Parathelphusa hydrodromus. Biol. Bull.* 169: 689–695.

Atema, J. 1977. Functional separation of smell and taste in fish and Crustacea. In *Proceedings of the Sixth International Symposium on Olfaction and Taste.* Edited by J. LeMagnen and P. MacLeod, pp. 165–174. London: Information Retrieval.

Auzou, M. L. 1953. Recherches biologiques et physiologiques sur deux onisciens: *Porcellio scaber* Lat. et *Oniscus asellus* L. *Ann. Sci. Nat. Zool.* 15: 71–98.

Bakker, R. T. 1972. Locomotor energetics of lizards and mammals compared. *Physiologist* 15: 76.

Baldwin, G. F., and L. B. Kirschner. 1976a. Sodium and chloride regulation in *Uca* adapted to 175% seawater. *Physiol. Zool.* 49: 158–171.

Baldwin, G. F., and L. B. Kirschner. 1976b. Sodium and chloride regulation in *Uca* adapted to 10% seawater. *Physiol. Zool.* 49: 172–180.

Ball, E. E. 1972. Observations on the biology of the hermit crab *Coenobita compressus* H. Milne Edwards (Decapoda, Anomura) on the west coast of the Americas. *Rev. Biol. Trop. Biol.* 20: 265–273.

Ball, E. E., and J. Haig. 1974. Hermit crabs from the tropical Eastern Pacific. I. Distribution, color, and natural history of some common shallow-water species. *Bull. Sth. Calif. Acad. Sci.* 73: 95–104.

Banerjee, S. K. 1960. Biological results of the Snellius expedition. 18. The genera *Grapsus*, *Geograpsus* and *Metopograpsus* (Crustacea Brachyura). *Temminckia* 10: 132–199.

Bardach, J. E. 1975. Chemoreception of aquatic animals. In *Olfaction and Taste V*. Edited by D. A. Denton and J. P. Coglan, pp. 121–132. London: Academic Press.

Barker, P. L., and R. Gibson. 1977. Observations on the feeding mechanisms, structure of the gut, and digestive physiology of the European lobster *Homarus gammarus* (L.) (Decapoda: Nephropidae). *J. Exp. Mar. Biol. Ecol.* 26: 297–324.

Barker, P. L., and R. Gibson. 1978. Observations on the structure of the mouth parts, histology of the alimentary tract, and digestive physiology of the mud crab *Scylla serrata* (Forskal) (Decapoda: Portunidae). *J. Exp. Mar. Biol. Ecol.* 32: 177–196.

Barnard, K. H. 1950. Descriptive catalogue of South African decapod Crustacea. *Ann. S. Afr. Mus.* 38: 1–837.

Barnes, R. D. 1980. *Invertebrate Zoology*. 4th ed. pp. 736–740. Philadelphia: Saunders.

Barnes, R. S. K. 1967. The osmotic behaviour of a number of grapsoid crabs with respect to their differential penetration of an estuarine system. *J. Exp. Biol.* 47: 535–552.

Barnes, W. J. 1975. Leg co-ordination during walking in the crab *Uca pugilator*. *J. Comp. Physiol.* 96: 237–256.

Barnhart, M. C. 1986. Respiratory gas tensions and gas exchange in active and dormant land snails *Otala lactea*. *Physiol. Zool.* 59: 733–745.

Barra, J-A., A. Peuquex, and W. Humbert. 1983. A morphological study on gills of a crab acclimated to fresh water. *Tiss. Cell* 15: 583–596.

Barrass, R. 1963. The burrows of *Ocypode ceratopthalmus* (Pallas) on a tidal wave beach at Inhaca Island, Mozambique. *J. Anim. Ecol.* 32: 73–85.

Batie, R. E. 1983. Rhythmic locomotor activity in the intertidal shore crab *Hemigrapsis oregonensis* (Brachyura, Grapsidae) from the Oregon coast. *Northwest Sci.* 57: 49–56.

Batterton, C. V., and J. N. Cameron. 1978. Characteristics of resting ventilation and response to hypoxia, hypercapnia, and emersion in the blue crab *Callinectes sapidus*. *J. Exp. Zool.* 203: 403–418.

Bauchau, A. G. 1961. Regeneration des pereiopodes et croissance chez les Crustaces decapodes Brachyoures. 1. Conditions normales et role des Pedocules oculaires. *Ann. Soc. Roy. Zool. Belgique* 91: 57–84.

Bauer, R. T. 1981. Grooming behavior and morphology in the decapod crustaceans. *J. Crust. Biol.* 1:153–173.

Becker, J. M., and G. W. Hinsch. 1982. Structural differences in male and female burrows of *Uca pugilator*. *Fla. Sci.* 45 (suppl. 1).

Beer, C. G. 1959. Notes on the behaviour of two estuarine crab species. *Trans. R. Soc. N.Z.* 86: 197–203.

Beever, J. W. III, D. Simberloff, and L. L. King. 1979. Herbivory and predation by the mangrove tree crab *Aratus pisonii*. *Oecologia* (Berlin) 43: 317–328.

Begg, G. W. 1981. Cape wagtail *Motacilla capensis* eating fiddler crabs *Uca annulipes*. *Ostrich* 52: 250.

Beis, I., and E. A. Newsholme. 1975. The contents of adenine nucleotides, phosphagens and some glycolytic intermediates in resting muscles from vertebrates and invertebrates. *Biochem. J.* 152: 23–32.

Belman, B. W. 1975. Some aspects of the circulatory physiology of the spiny lobster *Panulirus interruptus*. *Mar. Biol.* 29: 295–305.

Belman, B. W. 1976. New observations on blood pressure in marine Crustacea. *J. Exp. Zool.* 196: 71–78.

Belman, B. W., and J. J. Childress. 1976. Circulatory adaptations to the oxygen minimum layer in the bathypelagic mysid *Gnathophausia ingens*. *Biol. Bull.* 150: 15–37.

Benedict, F. G. 1983. *Vital energetics: A study in comparative basal metabolism.* Washington, D.C.: Carnegie Institute.

Bergin, M. E. 1981. Hatching rhythms in *Uca pugilator* (Decapoda: Brachyura). *Mar. Biol.* 63: 151–158.

Berrill, M., and M. Arsenault. 1982. Mating behavior of the green shore crab *Carcinus maenas*. *Bull. Mar. Sci.* 32: 632–638.

Berreur-Bonnenfant, J., J. J. Meusy, J. P. Ferezou, M. Devys, A. Quesneau-Thierry, and M. Barbier. 1973. Recherches sur la secretion de la glande androgene des Crustaces Malacostruces. Purification et proprietes d'une substance a activites androgenes. *C.R. Acad. Sci. (Paris)*, Ser. D. 274: 971–974.

Bertrand, J.-Y. 1979. Observations sur les crabes *Sudanonautes africanus* (Milne-Edwards 1869) et leurs terriers d'une savane de Lamto (Cote-d'Ivoire). *Bull. Soc. Zool. France* 104: 27–36.

Bill, R. G., and W. F. Herrnkind. 1976. Drag reduction by formation movement in spiny lobsters. *Science* 193: 1146–1148.

Binns, R. 1969. The physiology of the antennal gland of *Carcinus maenas* (L.). II. Urine production rates. *J. Exp. Biol.* 51: 11–16.

Bishop, J. A. 1963. The Australian freshwater crabs of the family Potamonidae (Crustacea: Decapoda). *Aust. J. Mar. Freshwat. Res.* 14: 218–238.

Blatchford, J. G. 1971. Hemodynamics of *Carcinus maenas* (L.). *Comp. Biochem. Physiol.* 39A: 193–202.

Blickhan, R., and R. J. Full. 1987. Ghost crab locomotion: Mechanical energy changes of the center of mass. *Am. Zool.* 25: 55A.

Bliss, D. E. 1953. Endocrine control of metabolism in the land crab, *Gecarcinus lateralis* (Freminville). 1. Differences in the respiratory metabolism of sinusglandless and eyestalkless crabs. *Biol. Bull.* 104: 275–296.

Bliss, D. E. 1960. Autotomy and regeneration. In *The physiology of the Crustacea.* Edited by T. H. Waterman, vol. 1, pp. 561–589. New York: Academic Press.

Bliss, D. E. 1963. The pericardial sacs of terrestrial Brachyura. In *Phylogeny and evolution of Crustacea,* pp. 59–78. Museum Comp. Zool. Harv. Univ., special publication. Cambridge, Mass.: Harvard University Press.

Bliss, D. E. 1968. Transition from water to land in decapod crustaceans. *Am. Zool.* 8: 355–392.

Bliss, D. E. 1979. From sea to tree: Saga of a land crab. *Amer. Zool.*19: 385–410.

Bliss, D., ed. in chief. 1982. *Biology of the Crustacea.* Vols. 1–6. New York: Academic Press.

Bliss, D. E., and J. R. Boyer. 1964. Environmental regulation of growth in the decapod crustacean *Gecarcinus lateralis. Gen. Comp. Endocr.* 4: 15–41.

Bliss, D. E., and L. H. Mantel. 1968. Adaptations of crustaceans to land: A summary and analysis of new findings. *Am. Zool.* 8: 673–685.

Bliss, D. E., J. Van Montfrans, M. Van Montfrans, and J. R. Boyer. 1978. Behavior and growth of the land crab *Gecarcinus lateralis* (Freminville) in southern Florida. *Bull. Am. Mus. Nat. Hist.* 160: 113–151.

Bliss, D. E., S. M. E. Wang, and E. A. Martinez. 1966. Water balance in the land crab, *Gecarcinus lateralis,* during the intermolt cycle. *Am. Zool.* 6: 197–212.

Boas, J. E. V. 1880. Studier over Decapodernes Slaegtskabsforhold. *K. dansk. Vidensk. Selsk. Skr.,* ser. 6, 1: 25–210.

Bohn, G. 1897. Sur le reversement du courant respiratoire chez les Decapodes. *C.R. Acad. Sci.* (Paris) 125: 539–542.

Bohn, G. 1907. Des mechanisms respiratoire chez les crustaces decapodes. *Bull. Sci. Fr. Belg.* 36: 178–551.

Booth, C. E. 1982. Respiration during activity in *Callinectes sapidus.* Ph.D. diss., University of Calgary, Canada.

Booth, C. E., B. R. McMahon, and A. W. Pinder. 1982. Oxygen uptake and the potentiating effects of increased hemolymph lactate on oxygen transport during exercise in the blue crab *Callinectes sapidus. J. Comp. Physiol.* 148: 111–121.

Booth, C. E., B. R. McMahon, P. L. deFur, and P. R. H. Wilkes. 1984. Acid–base regulation during exercise and recovery in the blue crab *Callinectes sapidus. Respir. Physiol.* 58: 359–376.

Borradaile, L. A. 1903a. Land crustaceans. In *The fauna and geography of the Maldive and Laccadive Archipelagoes.* Edited by J. S. Gardiner, vol. 5 pp. 64–100. Cambridge: Cambridge University Press.

Borradaile, L. A. 1903b. Marine crustaceans. 4. Some remarks on the classification of the crabs. In *The fauna and geography of the Maldive and Laccadive Archipelagoes.* Edited by J. S. Gardiner, vol. 1, pp. 424–429.

Borradaile, L. A. 1908. On the classification of the decapod crustaceans. *Ann. Mag. Nat. Hist.* 19: 457–486.

Borradaile, L. A. 1922. On the mouthparts of the shore crab. *J. Linn. Soc.* 35: 115–143.

Borut, A., and Y. Neumann. 1966. Salt and water balance in the ghost crab, *Ocypode cursor. Res. Rep. Hebrew Univ. Jerusalem Sci. Agric.* 1: 369.

Bosche, J. I. 1982. Predation of fiddler crabs *Uca stenodactyla* (Ocypodidae) by the common shore birds in Pangani Beach, Tanzania. *Afr. J. Ecol.* 20: 237–240.

Bott, R. 1955. Susswasser Krabben von Afrika (Crust., Decap.) und ihre Stammesgeschichte. *Ann. Mus. r. Congo Belge* C1: 213–349.

Bott, R. 1970. Die Susswasserkrabben von Europa, Asien, Australien, und ihre Stammesgeschichte. Eine Revision der Potamoidea und der Parathelphusoidea (Crustacea, Decapods). *Abh. Senckenberg. Naturforsch. Ges.* 526: 1–338.

Bouvier, E. L. 1890. Sur le circle circulatoire de la carapace chez les Crustaces Decapodes. *Compt. Rend.* 110: 1211–1213.

Bouvier, E. L. 1891. Recherches anatomiques sur le system arteriel des Crustaces Decapodes. *Ann. Sci. Nat. Zool.* 197–282.

Bouvier, E. L. 1942. Les Crabes de la tribu des "Corystoidea." *Mem. Acad. Sci. Inst. Fr.* 65: 1–52.

Bowman, T. E., and L. G. Abele. 1982. Classification of the recent Crustacea. In *The biology of Crustacea.* Edited by L. G. Abele, vol. 1, pp. 1–27. New York: Academic Press.

Braithwaite. C. J. R., and M. R. Talbot. 1972. Crustacean burrows in the Seychelles, Indian Ocean. *Palaeogeogr. Palaeoclimatol. Palaeoecol.* 11: 265–285.

Bridges, C. R., and S. Morris. 1986. Modulation of haemocyanin oxygen affinity by L-lactate: A role for other cofactors. In *Invertebrate oxygen carriers,* edited by B. Linzen. Berlin: Springer-Verlag.

Bridges, C. R., S. Morris, and M. K. Grieshaber. 1984. Modulation of haemocyanin oxygen affinity in the intertidal prawn *Palaemon elegans* (Rathke). *Respir. Physiol.* 57: 187–200.

Briggs, J. C. 1974. *Marine zoogeography.* New York: McGraw-Hill.

Bright, D. B. 1966. The land crabs of Costa Rica. *Rev. Biol. Trop.* 14: 183–203.

Bright, D. B., and C. L. Hogue. 1972. A synopsis of the burrowing land crabs of the world and list of their arthropod symbionts and burrow associates. Los Angeles County Nat. Hist. Mus., Contribs. in Sci., No. 220.

Britton, J. C., G. C. Kroh, and C. Golightly. 1982. Biometric and ecological relationships in two sympatric Caribbean Gecarcinidae (Crustacea: Decapoda). *J. Crustacean Biol.* 2: 207–222.

Brockerhoff, H., and R. J. Hoyle. 1967. Conversion of dietary triglyceride into depot fat in fish and lobster. *Can. J. Biochem.* 45: 1365–1370.

Brockhuysen, G. J. 1941. The life history of *Cyclograpsus punctatus,* M. Edw.: Breeding and growth. *Trans. Roy. Soc. S. Afr.* 28: 331–336.

Brooke, M. de L. 1981. Size as a factor influencing the ownership of copulation burrows by the ghost crab *Ocypode ceratophthalmus. Z. Tierpsychol.* 55: 63–78.

Brown, G. G. 1966. Ultrastructural studies of sperm morphology and sperm-egg interaction in the decapod *Callinectes sapidus. J. Ultrastr. Res.* 14: 425–440.

Bruce-Chwatt, L. J., and R. A. Fitz-John. 1951. Mosquitoes in crab-burrows on the coast of West Africa and their control. *J. Trop. Med. Hyg.* 54: 116–121.

Brusca, R. C. 1980. *Common intertidal vertebrates of the Gulf of California.* 2nd ed. Tuscon: University of Arizona Press.

Burger, J. W., and C. McC. Smythe. 1953. The general form of circulation in the lobster, *Homarus. J. Cell. Comp. Physiol.* 42: 369–383.

Burggren, W. W. 1982. Pulmonary blood plasma filtration in reptiles: A wet vertebrate lung? *Science* 215: 77–78.

Burggren, W. W., K. Johansen, and B. R. McMahon. 1985. Respiration in phyletically ancient fishes. In *Evolutionary biology of primitive fishes.* Edited by R. E. Foreman, A. Gorbman, J. M. Dodd, and R. Olsson. NATO ISI Series 103, pp. 217–252. New York: Plenum.

Burggren, W. W., and B. R. McMahon. 1981a. Haemolymph oxygen transport, acid-base status and hydromineral regulation during dehydration in three terrestrial crabs, *Cardisoma, Birgus,* and *Coenobita. J. Exp. Zool.* 218: 53–64.

Burggren, W. W., and B. R. McMahon. 1981b. Oxygen uptake during environmental temperature change in hermit crabs: Adaptation to subtidal, intertidal and supratidal habitats. *Physiol. Zool.* 54: 325–333.

Burggren, W. W., and B. R. McMahon. 1983. An analysis of scaphognathite pumping performance in the crayfish *Orconectes virilis*: Compensatory changes to acute and chronic hypoxic exposure. *Physiol. Zool.* 56: 309–318.

Burggren, W. W., B. R. McMahon, and J. W. Costerton. 1974. Branchial water- and blood-flow patterns and the structure of the gill of the crayfish *Preocambarus clarkii. Can. J. Zool.* 52: 1511–1518.

Burggren, W., A. Pinder, B. McMahon, M. Wheatly, and M. Doyle. 1985. Ventilation, circulation and their interactions in the land crab, *Cardisoma guanhumi. J. Exp. Biol.* 117: 133–154.

Burke, W. 1954. An organ for proprioception and vibration sense in *Carcinus maenas. J. Exp. Biol.* 31: 127–138.

Burnett, L. E. 1979. The effects of environmental oxygen levels on the respiratory function of haemocyanin in the crabs, *Libinia emarginata* and *Ocypode quadrata. J. Exp. Zool.* 210: 289–300.

Burnett, L. E., and C. R. Bridges. 1981. The physiological properties and function of ventilatory pauses in the crab *C. pagurus. J. Comp. Physiol.* 45: 81–88.

Burnett, L. E., P. L. deFur, and D. D. Jorgensen. 1981. Application of the thermal dilution technique for measuring cardiac output and assessing stroke volume in crabs. *J. Exp. Zool.* 218: 165–173.

Burnett, L. E., and B. R. McMahon. 1985. Facilitation of CO_2 excretion by carbonic anhydrase located on the surface of the basal membrane of crab gill epithelium. *Respir. Physiol.* 62: 341–348.

Burnett, L. E., and B. R. McMahon. 1987. Gas exchange, hemolymph acid-base status and the role of branchial water stores during air exposure in three littoral crab species. *Physiol. Zool.* 60: 27–36.

Burrows, M., and G. Hoyle. 1973. The mechanism of rapid running in the ghost crab, *Ocypode ceratophthalma. J. Exp. Biol.* 58: 327–349.

Bursey, C. R. 1982. Factors influencing the distribution and co-occurrence of *Sesarma reticulatum* (Decapoda: Grapsidae) and *Uca pugnax* (Decapoda: Ocypodidae) in a Virginia salt marsh. *Amer. Zool.* 22: 950.

Calder, W. A. 1984. *Size, function and life history.* Cambridge, Mass.: Harvard University Press.

Cameron, A. 1966. Some aspects of the behavior of the soldier crab, *Mictyris longicarpus. Pacif. Sci.* 20: 224–234.

Cameron, J. N. 1975. Aerial gas exchange in the terrestrial Brachyura *Gecarcinus lateralis* and *Cardisoma guanhumi. Comp. Biochem. Physiol.* 52A: 129–134.

Cameron, J. N. 1981a. Brief introduction to the land crabs of the Palau Islands: Stages in the transition to air breathing. *J. Exp. Zool.* 218: 1–5.

Cameron, J. N. 1981b. Acid-base responses to changes in CO_2 in two Pacific crabs: *Birgus latro*, and a mangrove crab *Cardisoma carnifex. J. Exp. Zool.* 218: 65–73.

Cameron, J. N. 1985. Compensation of hypercapnic acidosis in the aquatic blue crab *Callinectes sapidus*: The predominance of external sea water over carapace carbonate as the proton sink. *J. Exp. Biol.* 114: 197–206.

Cameron, J. N., and C. V. Batterton. 1978. Temperature and blood acid-base status in the blue crab *Callinectes sapidus. Resp. Physiol.* 35: 101–110.

Cameron, J. N., and T. A. Mecklenberg. 1973. Aerial gas exchange in the coconut crab, *Birgus latro* with some notes on *Gecaroidea lalandii. Respir. Physiol.* 19: 245–261.

Campbell, G. S. 1977. *An introduction to environmental physics.* New York: Springer-Verlag.

Carayon, J. 1941. Morphologie et structure de l'appareil genital femelle chez quelques Pagures. *Bull. Soc. Zool. Fr.* 66: 95–122.

Carlisle, D. B. 1957. On the hormonal inhibition of molting in decapod Crustacea. 2. The terminal anecdysis in crabs. *J. Mar. Biol. Ass. U.K.* 36: 291–307.

Carpenter, A., and W. H. Logan. 1945. Hermits don't always live alone. *Natural History* (American Mus. Nat. Hist.) 54: 286–287.

Carpenter, M. B., and R. De Roos. 1970. Seasonal morphology and histology of the androgenic gland of the crayfish, *Orconectes nais. Gen. Comp. Endocrinol.* 15: 143–157.

Carson, H. L. 1974. Three flies and three islands: Parallel evolution in *Drosophila. Proc. Nat. Acad. Sci.* 71: 3517–3521.

Carson, H. L., and M. R. Wheeler. 1973. A new crab fly from Christmas Island, Indian Ocean. *Pacific Insects* 15: 199–208.

Carter, G. S. 1931. Aquatic and aerial respiration in animals. *Biol. Rev.* 6: 1–220.

Case, J. 1964. Properties of the dactyl chemoreceptors of *Cancer antennarius* Stimpson and *C. productus* Randall. *Biol. Bull.* 127: 428–446.

Case, J., and G. F. Gwilliam. 1961. Amino acid sensitivity of the dactyl chemoreceptors of *Carcinides maenas*. *Biol. Bull.* 121: 449–455.

Cavagna, G. A. 1975. Force platforms as ergometers. *J. Appl. Physiol.* 39: 174–179.

Cavagna, G. A., N. C. Heglund, and C. R. Taylor. 1977. Mechanical work in terrestrial locomotion: Two basic mechanisms for minimizing energy expenditure. *Am. J. Physiol.* 233: R243–R261.

Cerretelli, P., D. Pendergast, W. C., Paganelli, and D. W. Rennie. 1979. Effects of specific muscle training on VO_2 on-response and early blood lactate. *J. Appl. Physiol: Respirat. Environ. Exercise Physiol.* 47: 761–769.

Chace, F. A., and H. H. Hobbs, Jr. 1969. The freshwater and terrestrial crabs of the West Indies with special reference to Dominica. *Smithsonian Inst., U.S. Natl. Mus. Nat. Hist. Bull.* 292.

Chaix, J.-C., and M. De Reggi. 1982. Ecdysteroid levels during ovarian development and embryogenesis in the spider crab, *Acanthonyx lunulatus*. *Gen. Comp. Endocrinol.* 47: 7–14.

Chakrabarti, A. 1981. Burrow patterns of *Ocypode ceratophthalma* and their environmental significance. *J. Paleon.* 55: 431–441.

Chakrabarti, A., and S. Das. 1983. Burrow patterns of *Macrophthalmus telescopicus* and their environmental significance. *Senckeb. Marit.* 15: 43–54.

Chang, E. S., and J. D. O'Connor. 1983. Metabolism and transport of carbohydrates and lipids. In *The biology of Crustacea.* Edited by L. H. Mantel, vol. 5, pp. 263–287. New York: Academic Press.

Charniaux-Cotton, H. 1952. Castration chirurgicale chez un Crustace Amphipode (*Orchestia gammarella*) et determinisme des caracteres sexuels secondaires. Premiers resultats. *C.R. Acad. Sci. Paris.* 234: 2570–2572.

Charniaux-Cotton, H. 1953. Etude du determinisme des caracteres sexuels secondaires par castration chirurgicale et implantation d'ovaire chez un Crustace Amphipode (*Orchestia gammarella*). *C.R. Acad. Sci. Paris.* 236: 141–143.

Charniaux-Cotton, H. 1954. Decouverte chez un Crustace Amphipode (*Orchestia gammarella*) d'une glande endocrine responsable de la differenciation des caracteres sexuels primaires et secondaires males. *C.R. Acad. Sci. Paris.* 239: 780–782.

Charniaux-Cotton, H. 1958. Controle hormonal de la differenciation du sexe et de la reproduction chez les Crustaces superieurs. *Bull. Soc. Zool. Fr.* 83: 314–336.

Charniaux-Cotton, H., and G. Payen. 1985. Sex differentiation. In *The biology of Crustacea.* Edited by D. E. Bliss and L. H. Mantel, vol. 9, pp. 147–215. New York: Academic Press.

Chen, C., and A. L. Lehninger. 1973. Respiration and phosphorylation by mitochondria from the hepatopancreas of the blue crab. (*Callinectes sapidus*). *Arch. Biol. Biophy.* 154: 449–459.

Cheng, L., and C. L. Hogue. 1974. New distribution and habitat records of biting midges and mangrove flies from the coasts of southern Baja California, Mexico (Diptera: Ceratopogonidae, Culicidae, Chironomidae, and Phoridae). *Entom. News* 85: 211–218.

Cheung, T. S. 1966. The inter-relations among three hormonal-control characters in the adult female shore crab, *Carcinus maenas* (L). *Biol. Bull.* 130: 59–66.

Cheung, T. S. 1968. Transmolt retention of sperm in the female stone crab, *Menippe mercenaria* (Say). *Crustaceana* 15: 117–120.

Cheung, T. S. 1969. The environmental and hormonal control of growth and reproduction in the adult female stone crab, *Menippe mercenaria* (Say). *Biol. Bull.* 136: 327–346.

Christy, J. H. 1978. Adaptive significance of reproductive cycles in the fiddler crab *Uca pugilator*: A hypothesis. *Science* 199: 453–455.

Christy, J. H. 1979. Resource-defense polygyny in the sand fiddler crab, *Uca pugilator*. *Amer. Zool.* 19: 933.

Christy, J. H. 1980. The mating system of the sand fiddler crab, *Uca pugilator*. Ph.D. thesis, Cornell University.

Clarac, F. 1982. Decapod crustacean leg coordination during walking. In *Locomotion and energetics in arthropods*. Edited by C. F. Herreid II and C. R. Fortner. New York: Plenum Press.

Clarac, F., and W. J. P. Barnes. 1985. Peripheral influences on the coordination of the legs during walking in decapod crustaceans. In *Coordination of motor behavior*. Edited by B. M. H. Bush and F. Clarac. Society for Experimental Biology Seminar 24. Cambridge: Cambridge University Press.

Claybrook, D. L. 1983. Nitrogen metabolism. In *The biology of Crustacea*. Edited by L. H. Mantel, vol. 5, pp. 163–213. New York: Academic Press.

Clever, F. C. 1949. Preliminary results on the coastal crab, *Cancer magister*. *Washington State Dept. of Fish. Biol. Rep.* A49: 47–82.

Cloudsley-Thompson, J. L. 1970. Terrestrial invertebrates. In *Comparative physiology of thermoregulation*. Edited by G. C. Whitton, vol. 1, pp. 15–77. New York: Academic Press.

Collins, N. M. 1980. The habits and populations of terrestrial crabs (Brachyura: Gecarcinucoidea and Grapsoidea) in the Gunung Mulu National Park, Sarawak. *Zool. Meded. Leiden* 55: 81–85.

Copeland, D. E. 1968. Fine structure of salt and water uptake in the land crab, *Gecarcinus lateralis*. *Am. Zool.* 8: 417–432.

Copeland, D. E., and A. T. Fitzjarrel. 1968. The salt absorbing cells in the gills of the blue crab (*Callinectes sapidus* Rathbun) with notes on modified mitochondria. *Z. Zellforsch. Mikresk. Anat.* 92: 1.

Costlow, J. D., and C. G. Bookhout. 1968. The effect of environmental factors on development of the land-crab, *Cardisoma guanhumi*. *Am. Zool.* 8: 399–410.

Cott, H. B. 1929. Observations on the natural history of the racing crab (*Ocypoda ceratophthalma*) from Beira. *Proc. Zool. Soc. Lond.* for 1929: 755–765.

Crane, J. 1941a. On the growth and ecology of brachyuran crabs of the genus *Ocypode*. *Zoologica* 26: 297–310.

Crane, J. 1941b. Crabs of the genus *Uca* from the west coast of Central America. *Zoologica* 26: 145–208.

Crane, J. 1943. Display, breeding and relationships of fiddler crabs (Brachyura, genus *Uca*) in the northeastern United States. *Zoologica* 28: 217–223.

Crane, J. 1957. Basic patterns of display in fiddler crabs (Ocypodidae, genus *Uca*). *Zoologica* 42: 69–82.

Crane, J. 1975. *Fiddler crabs of the world.* Princeton, N.J.: Princeton University Press.

Creutzberg, F. 1975. Orientation in space: Animals. Invertebrates. In *Marine ecology.* vol. 2, *Physiological Mechanisms,* pp. 555–655. Edited by O. Kinne. New York: John Wiley.

Crisp, D. J. 1971. Energy flow measurements. In *Methods for the study of marine benthos.* Edited by N. A. Holme and A. D. McIntyre. IBP Handbook No. 16, pp. 197–279. Oxford: Blackwell.

Crothers, J. H. 1968. The biology of the shore crab *Carcinus maenas* (L.). 2. The life of the adult crab. *Field Studies* 2: 579–614.

Cundell, A. M., M. S. Brown, R. Stanford, and R. Mitchell. 1979. Microbial degradation of *Rhizophora mangle* leaves immersed in the sea. *Estuar. Coastal Mar. Sci.* 9: 281–286.

Cunningham, P. N., and R. N. Hughes. 1984. Learning predatory skills by shorecrabs *Carcinus maenas* feeding on mussels and dogwhelks. *Mar. Ecol. Prog. Ser.* 16: 21–26.

Dall, W. 1967. Hypo-osmoregulation in Crustacea. *Comp. Biochem. Physiol.* 21: 653–678.

Dall, W., and D. J. W. Moriarty. 1983. Functional aspects of nutrition and digestion. In *The biology of Crustacea.* Edited by L. H. Mantel, vol. 5, pp. 215–261. New York: Academic Press.

Dandy, J. W. T., and D. W. Ewer. 1961. The water economy of three species of the amphibious crab, *Potamon. Trans. Roy. Soc. S. Afr.* 363: 137–162.

Daumer, K., R. Jander, and T. H. Waterman. 1963. Orientation of the ghost crab *Ocypode* in polarized light. *Z. vergl. Physiol.* 47: 56–76.

Davies, N. B. 1978. Ecological questions about territorial behavior. In *Behavioral ecology.* 1st Ed. Edited by J. R. Krebs and N. B. Davies, pp. 317–350. Sunderland, Mass.: Sinauer.

deFur, P. L., and B. R. McMahon. 1984a. Physiological compensation to short-term air exposure in red rock crabs, *Cancer productus* Randall, from littoral and sublittoral habitats. I. Oxygen uptake and transport. *Physiol. Zool.* 57: 137–150.

deFur, P. L., and B. R. McMahon. 1984b. Physiological compensation to short-term air exposure in red rock crabs, *Cancer productus* Randall, from littoral and sublittoral habitats. II. Acid-base balance. *Physiol. Zool.* 57: 151–160.

deFur, P. L., B. R. McMahon, and C. E. Booth. 1983. Analysis of haemolymph oxygen levels and acid-base status during emersion "in situ" in the red rock crab, *Cancer productus. Biol. Bull.* 165: 582–590.

deFur, P. L., P. R. H. Wilkes, and B. R. McMahon. 1980. Non-equilibrium acid-base status in *Cancer productus*: Role of exoskeletal carbonate buffers. *Respir. Physiol.* 42: 247–261.

De Leersnyder, M., A. Dhainaut, and P. Porcheron. 1981. Influence de l'ablation des organes y sur l'ovogenese du crabe *Eriocheir sinensis* dans les condition naturells et apres epedonculation. *Gen. Comp. Endocrinol.* 43: 157–169.

De Leersnyder, M., and H. Hoestlandt. 1963. Premiers donnes sur la regulation osmotique et la regulation ionique du crabe terreste *Cardisoma armatum* Herklots. *Cah. Biol. Mar.* 4: 211–218.

Démeusy, N. 1953. Effets de l'ablation des pedoncules oculaires sur le developpement de l'appareil genital male de *Carcinus maenas* Pennant. *C.R. Acad. Sci.* (Paris) 236: 974–975.

Démeusy, N. 1958. Recherches sur la mue de puberte du Decapode Brachyoure *Carcinus maenas* L. *Arch. Zool. Exp. Gen.* 95: 253–491.

Démeusy, N. 1962. Role de la glande de mue dans l'evolution ovarienne du crabe *Carcinus maenas* Linne. *Cah. Biol. Mar.* 3: 37–56.

Démeusy, N. 1963. Rapports entre mue et vitellogenese chez le crabe *Carcinus maenas* (L). *Proc. XVI Int. Congr. Zool.* 2: 118.

Depledge, M. 1984. Cardiac activity in the intertidal crab *Hemigrapsus sanguineus* (de Haan). *Asian Mar. Biol.* 1: 115–123.

Derby, C. D., and J. Atema. 1982. Chemosensitivity of walking legs of the lobster *Homarus americanus*: Neurophysiological response spectrum and thresholds. *J. Exp. Biol.* 98: 303–315.

de Saint-Laurent, M. 1980a. Sur la classification et la phylogenie des Crustaces Decapodes Brachyoures. I. Podotremata Guinot, 1977, et Eubrachyura sect. nov. *C.R. hebd. Seanc. Acad. Sci.* (Paris) 290: 1265–1268.

de Saint-Laurent, M. 1980b. Sur la classification et la phylogenie des Crustaces Decapodes Brachyoures. II. Heterotremata et Thoracotremata Guinot, 1977. *C.R. hebd. Seanc. Acad. Sci.* (Paris) 290: 1317–1320.

De Wilde, P. A. W. J. 1973. On the ecology of *Coenobita clypeatus* in Curacao with reference to reproduction, water economy and osmoregulation in terrestrial hermit crabs. *Stud. Fauna Curacao* 44: 1–138.

Dhainaut, A., and M. De Leersnyder. 1976. Etude cytochimique et ultrastructurale de l'evolution ovocytaire du crabe *Eriocheir sinesis. Arch. Biol.* (Brussels). 87: 283–302.

Dhillon, B. 1968. Radial processes of decapod sperm. *Microscope* 76: 365–368.

Diaz, H., and J. D. Costlow. 1972. Larval development of *Ocypode quadrata* (Brachyura: Crustacea) under laboratory conditions. *Mar. Biol.* 15: 120–131.

Diaz, H., and G. Rodriguez. 1977. The branchial chamber in terrestrial crabs: A comparative study. *Biol. Bull.* 153: 485–504.

Dingle, H. 1983. Strategies of agonistic behavior in Crustacea. In *Studies in adaptation: The behavior of higher Crustacea.* Edited by S. Rebach and D. Dunham. New York: Wiley.

Diwan, A. D. 1973. Studies on the sexual maturity and growth rate studies of the freshwater crab, *Barytelphusa cunicularis. Marathwada Univ. J. Sci.* 12: 125–145.

Diwan, A. D., and R. Nagabhushanam. 1974. Reproductive cycle and biochemical changes in the gonad of the freshwater crab, *Barytelphusa cunicularis* (Westwood, 1836). *Indian J. Fish.* 21: 164–176.

Doherty, J. A. 1982. Stereotypy and the effects of temperature on some spatiotemporal subcomponents of the "courtship wave" in the fiddler crabs *Uca minax* (Leconte) and *Uca pugnax* (Smith) (Brachyura, Ocypodidae). *Anim. Behav.* 30: 352–363.

Doujak, F. E. 1985. Can a shore crab see a star? *J. Exp. Biol.* 116: 385–393.

Drach, P. 1930. Etude sur le system branchiale des crustaces Decapods. *Archs. Anat. Microsc.* (Paris) 26: 83–133.

Drach, P. 1939. Mue et cycle d'intermue chez les crustaces decapodes. *Ann. Inst. Oceanogr. Monaco* 19: 103–391.

Drach, P, and M. Lafon. 1942. Etudes biochimiques sur le squelette tegumentaire des decapodes brachyoures (Variations au cours du cycle d'intermue). *Arch. Zool. Exp. & Gen.* 82: 100–118.

Drews, G. 1983. NaA +B + KA + B-ATPase activity in the gills and NaA + B concentration in the hemolymph of *Uca tangeri* during osmotic stress. *Abstracts 5th Conference Europ. Soc. Comp. Physiol. Biochem.:* 138–9.

Dubuisson, M. 1928. Recherches sur la circulation du sang chez les Crustaces. 2. Pressions sanguines chez les Decapodes Brachoures. *Arch. Biol.* (Liege) 38: 9–21.

Ducruet, J. 1976. Attraction et reconnaissance sexuelle chez les crustaces. *Bull. Soc. His. Nat. Afr. nord Alger* 67: 57–79.

Dumortier, B. 1963. Morphology of sound emission apparatus in Arthropoda. In *Acoustic behavior of animals.* Edited by R. G. Busnel, pp. 272–345. New York: Elsevier.

Dunham, D. W., and H. Schoene. 1984. Substrate slope and orientation in a land hermit crab, *Coenobita clypeatus* (Decapoda, Coenobitidae). *J. Comp. Physiol.* 154: 511–513.

Dupreez, H. H., and A. McLachlan. 1983. Seasonal changes in biochemical composition of the three-spot swimming crab *Ovalipes punctatus* (De Haan) (Crustacea: Brachyura). *J. Exp. Mar. Bio. Ecol.* 72: 189–198.

Eastman-Reks, S., and M. Fingerman. 1984. Effects of neuroendocrine tissue and cyclic AMP on ovarian growth *in vivo* and *in vitro* in the fiddler crab, *Uca pugilator. Comp. Biochem. Physiol.* 79A: 679–684.

Eastman-Reks, S. B., and M. Fingerman. 1985. *In vitro* synthesis of vitellin by the ovary of the fiddler crab, *Uca pugilator. J. Exp. Zool.* 233: 111–116.

Edney, E. B. 1960. Terrestrial adaptations. In *Physiology of Crustacea* vol. 1, *Metabolism and Growth.* Edited by T. H. Waterman. New York: Academic Press.

Edney, E. B. 1961. The water and heat relationships of fiddler crabs (*Uca* spp.). *Trans. Roy. Soc. S. Afr.* 36: 71–91.

Edney, E. B. 1977. *Water balance in land arthropods.* Berlin: Springer-Verlag.

Ehrhardt, J.-P. 1968. Recensement en 1968 de la population de *Gecarcinus planatus* Stimpson sur l'atoll de Clipperton. *Centre de Recherche du Service de Santee des Armees, Rapport Particulier, Bio.-Eco.* 40: 1–9.

Ehrhardt, J.-P., and P. Niaussat. 1970. Ecologie et physiologie du brachyoure terrestre *Gecarcinus planatus* Stimpson (d'apres les individus de l'atoll de Clipperton). *Bull. Soc. Zool. Fr.* 95: 41–54.

El Haj, A. J., A. J. Innes, and E. W. Taylor. Ultrastructure of the pulmonary, cutaneous and branchial gas exchange organs of the Trinidad mountain crab. *J. Physiol.* 373: 84P.

Elner, R. W. 1978. The mechanics of predation by the shore crab, *Carcinus maenas* (L.), on the edible mussel, *Mytilus edulus* L. *Oecologia* 36: 333–344.

Eurenius, L. 1973. An electron microscope study on the developing oocytes of the crab *Cancer pagurus* L. with reference to yolk formation (Crustacea). *Z. Morph. Tiere.* 75: 243–254.

Evans, P. D. 1972. The free amino acid pool of the hemocytes of *Carcinus maenas* (L.) *J. Exp. Biol.* 56: 501–507.

Evans, D. H., K. Cooper, and M. B. Bogan. 1976. Sodium extrusion by the seawater acclimated fiddler crab *Uca pugilator*: Comparison with other marine Crustacea and marine teleost fish. *J. Exp. Biol.* 64: 203–220.

Evans, S. M., A. Cram, K. Eaton, R. Torrance, and V. Wood. 1976. Foraging and agonistic behavior in the ghost crab *Ocypode kuhli* de Haan. *Mar. Behav. Physiol.* 4: 121–135.

Farrelly, C. A., and P. Greenaway. In press. The morphology and vasculature of the lungs and gills of the soldier crab *Mictyris longicarpus. J. Morphol.*

Feest, J. 1969. Morphophysiologische Untersuchungen zur Ontogense und Fortpflanzungsbiologie von *Uca annulipes* und *Uca triangularis* mit Vergleichsbefunden an *Ilyoplax gangetica. Forma Functio* 1: 159–225.

Felgenhauer, B. E., and L. G. Abele. 1983. Branchial water movement in the grapsid crab *Sesarma reticulatum. J. Crust. Biol.* 3: 187–195.

Feliciano, C. 1962. *Notes on the biology and economic importance of the land crab Cardisoma guanhumi Latreille of Puerto Rico.* Spec. Contr. Inst. of Marine Biol., University of Puerto Rico.

Fellows, D. P. 1973. Behavioral ecology of the ghost crabs *Ocypode ceratopthalmus* and *Ocypode cordimana* at Fanning Atoll, Line Islands. In *Fanning Island expedition.* July and August, 1972, K. E. Chave and E. A. Kay, principal investigators, pp. 219–241. Hawaii Inst. of Geophysics, University of Hawaii, Honolulu.

Fellows, D. P. 1975. On the distribution of the Hawaiian ghost crab, *Ocypode laevis* Dana. *Pacific Sci.* 29: 257–258.

Ferezou, J.-P., J. Berreur-Bonnenfant, A. Tekitek, M. Rojas, M. Barbier, M. Suchy. H. K. Wipf, and J. J. Meusy. 1977. Biologically-active lipids from the androgenic gland of the crab *Carcinus maenas.* In *Marine Natural Products Chemistry.* Edited by D. J. Faulkner and W. H. Fenical, pp. 361–366. New York: Plenum Press.

Fernando, C. H. 1960. The Ceylonese freshwater crabs (Potamonidae). *Ceylon J. Sci. Biol. Sci.* 3: 191–224.

Fielder, D. R. 1970. The feeding behavior of the sand crab *Scopimera inflata* (Decapoda, Ocypodidae). *J. Zool.* 160: 35–49.

Fielder, D. R., K. R. Rao, and M. Fingerman. 1971. A female-limited lipoprotein and the diversity of hemocyanin components in the dimorphic variants of the fiddler crab, *Uca pugilator*, as revealed by disc electrophoresis. *Comp. Biochem. Physiol.* 39B: 291–297.

Fimpel, E. 1975. Phaenomene der Landadaptation bei terrestrischen und semiterrestrischen Brachyura der Brasilianischen Kueste (Malacostraca, Decapoda). *Zool. Jahrb. Syst.* 102: 173–214.

Finol, H. J. and P. C. Croghan. 1983. Ultrastructure of the branchial epithelium of an amphibious brackish-water crab. *Tiss. Cell.* 15: 63–75.

Fioroni, P. 1969. Zum embryonalen und postembryomalen Dotterabbau des Flusskrebses (*Astacus*, Crustaces Malacostraca, Decapoda). *Rev. Suisse Zool.* 76: 919–946.

Fioroni, P. 1970a. Die organogenetische Rolle der Vitellophagen in der Darmentwicklung von *Galathea* (Crustacea, Decapoda, Anomura). *Z. Morphol. Oekol. Tiere.* 67: 263–306.

Fioroni, P. 1970b. Am Dotteraufschluss beteilgte Organ und Zelltypen bei hoheren Krebsen; der Versuch zu einer einheitlichen Terminologie. *Zool. Jahrb. Abt. Anat. Ontog. Tiere.* 87: 481–522.

Fioroni, P. 1981. Zum Auftreten der Geschlechtszellen im Ontogenese-Verlauf der Krebse – mit besonderer Berucksichtigung der Crustacea Malacostraca. *Cah. Biol. Mar.* 22: 388–406.

Fioroni, P., and E. Bandaret. 1971. Mit dem Dotteraufschluss liierte ontogenese Abwandlungen bei einigen Decapoden Krebsen. *Vie Millieu* Ser. A22: 163–188.

Fisher, J. B., and M. J. S. Tevesz. 1979. Within-habitat spatial patterns of *Ocypode quadrata* (Fabricius) (Decapoda Brachyura). *Crustaceana Suppl.* 5: 31–36.

Fisher, J. M. 1972. Fine-structural observations on the gill filaments of the fresh-water crayfish, *Astacus pallipes* Lereboullet. *Tiss. Cell.* 4: 287–299.

Fize, A., and R. Sèrene. 1955. Les pagures du Vietnam. *Not. Inst. Oceanogr. Nhatrang.* 45: 1–228.

Flemister, L. J. 1958. Salt and water anatomy, constancy and regulation in related crabs from marine and terrestrial habitats. *Biol. Bull.* 115: 180–200.

Flower, S. S. 1931. Notes on freshwater crabs in Egypt, Sinai, and the Sudan. *Proc. Zool. Soc. Lond. for 1931*: 729–735.

Forbes, A. T. 1973. An unusual abbreviated larval life in the estuarine burrowing prawn *Callianassa kraussi* (Crustacea, Decapoda, Thalassinidea). *Mar. Biol.* (Berlin) 22: 361–365.

Forest, J., and D. Guinot. 1961. Crustaces Decapodes Brachyoures de Tahiti et des Tuamoto. In *Expedition francais sur les recifs coralliens de la Nouvelle-Caledonie.* vol. 1., pp. 1–195.

Forward, R. B., Jr., and K. J. Lohmann. 1983. Control of egg hatching in the crab, *Rhithropanopeus harrisii. Biol. Bull.* 165: 154–166.

Fox, L. R. 1975. Cannibalism in natural populations. *Ann. Rev. Ecol. Syst.* 6: 87–106.

Freeman, J. A., and J. D. Costlow. 1984. Endocrine control of apolysis in *Rhithropanopeus harrisii* larvae. *J. Crustacean Biol.* 4: 1–6.

Freeman, J. A., T. L. West, and J. D. Costlow. 1983. Postlarval growth in juvenile *Rhithropanopeus harrissi. Biol. Bull.* 165: 409–414.

Frith, D. W., and S. Brunenmeister. 1980. Ecological and population studies of fiddler crabs (Ocypodidae, genus *Uca*) on a mangrove shore at Phuket Island, Western Peninsular Thailand. *Crustaceana* 39: 157–184.

Frith, D. W., and C. B. Frith. 1977. Observations on fiddler crabs (Ocypodidae, genus *Uca*) on Surin Island, Western Peninsula, Thailand, with special reference to *Uca tetragonon* (Herbst). *Bull. Phuket Mar. Biol. Centre* 18.

Full, R. J. 1984. Energetics of invertebrate terrestrial locomotion: A comparison of metabolic response in exercising decapod crustaceans. Ph.D. diss. State University of New York at Buffalo.

Full, R. J., and R. Blickhan. 1985. Ghost crab locomotion: The efficiency of traveling sideways. *Am. Zool.* 25: 55A.

Full, R. J., and C. F. Herreid II. 1983. Aerobic response to exercise of the fastest land crab. *Am. J. Physiol.* 244: R530–R536.

Full, R. J., and C. F. Herreid II. 1984. Fiddler crab exercise: The energetic cost of running sideways. *J. Exp. Biol.* 109: 141–161.

Full, R. J., C. F. Herreid II, and J. A. Assad. 1985. Energetics of the exercising wharf crab *Sesarma cinereum. Physiol. Zool.* 8: 605–615.

Fyffe, W., and J. D. O'Connor. 1974. Characterization and quantification of a crustacean lipovitellin. *Comp. Biochem. Physiol.* 47B: 851–867.

Gangotri, W. S., N. Vasantha, and S. A. T. Venkatachari. 1971. Sexual maturity and breeding behaviour in the freshwater crab *Barytelphusa guerini* Milne Edwards. *Proc. Indian Acad. Sci.* 87B: 195–201.

Garth, S. J. 1948. The Brachyura of the "Askoy" expedition with remarks on carcinological collecting in the Panama Bight. *Bull. Amer. Mus. Nat. Hist.* 82: 1–66.

George, R. W., and D. S. Jones. 1984. Notes on the crab fauna of Mangrove Bay, North West Cape. *Western Aust. Nat.* 15: 169–175.

Ghiradella, H. T., J. F. Case, and J. Cronshaw. 1968. Structure of aesthetascs in selected marine and terrestrial decapods: Chemoreceptor morphology and environment. *Amer. Zool.* 8: 603–621.

Gibbs, P. E. 1974. Notes on *Uca burgesi* Holthuis (Decapoda, Ocypodidae) from Barbuda, Leeward Islands. *Crustaceana* 27: 84–91.

Gibson, R., and P. L. Barker. 1979. The decapod hepatopancreas. *Oceanogr. Mar. Ann. Rev.* 17: 285–346.

Gibson-Hill, C. A. 1947. Field notes on the terrestrial crabs. *Bull. Raffles Mus.* 18: 43–52.

Giddins, R. L., J. S. Lucas, M. J. Nielson, and G. N. Richards, 1986. Feeding ecology of the mangrove crab *Neosarmatium smithi* (Crustacea: Decapoda: Sesarmidae). *Mar. Ecol. Prog. Ser.* 33: 147–155.

Gifford, C. A. 1962a. Some observations on the general biology of the land crab *Cardisoma guanhumi* (Latreille) in South Florida. *Biol. Bull.* 97: 207–223.

Gifford, C. A. 1962b. Some aspects of osmotic and ionic regulation in the blue crab, *Callinectes sapidus*, and the ghost crab, *Ocypode albicans. Pub. Inst. Mar. Sci. Univ. Tex.* 8: 97–125.

Gifford, C. A. and R. F. Johnson. 1962. Distribution of calcium in the land crab *Cardisoma guanhumi* during shell wound recalcification. *Comp. Biochem. Physiol.* 7: 227–231.

Gilbert, L. I., and J. D. O'Connor. 1970. Lipid metabolism and transport in arthropods. In *Chemical zoology.* Edited by M. Florkin and B. T. Scheer, vol. 5, pp. 229–254. New York: Academic Press.

Glaessner, M. F. 1960. The fossil decapod Crustacea of New Zealand and the evolution of the order Decapoda. *Paleont. Bull. Wellington* 31: 1–63.

Goettel, M. S., M. K. Toohey, B. R. Engber, and J. S. Pillai. 1981. A modified garden sprayer for sampling crab hole water. *Mosq. News* 41: 789–790.

Goldspink, G. 1977. Mechanics and energetics of muscle in animals of different sizes, with particular reference to the muscle fiber composition of vertebrate muscle. In *Scale effects in animal locomotion.* Edited by T. J. Pedley, pp. 37–55. New York: Academic Press.

Gomez, L. D. 1977. La mosca del cangrejo terrestre *Cardisoma crassum* Smith (Crustacea: Gecarcinidae) en la Isla del Coco, Costa Rica. *Rev. Biol. Trop.* 25: 59–63.

Gomez, R. 1965. Acceleration of development of gonads by implantation of brain in the crab *Parathelphusa hydrodromus. Naturwissenschaften* 9: 216.

Goodbody, I. 1965. Continuous breeding in populations of two tropical crustaceans, *Mysidium colubiae* (Zimmer) and *Emerita portoriconsis* (Schmitt). *Ecology* 46: 195–197.

Gordan, J. 1956. A bibliography of pagurid crabs, exclusive of Alcock, 1905. *Bull. Amer. Mus. Nat. Hist.* 108: 253–352.

Goshima, S., Y. Ono, and Y. Nakasone. 1978. Daily activity and movement of the land crab *Cardisoma hirtipes* Dana, by radiotelemetry during non-breeding season. *Publ. Amakusa Mar. Biol. Lab., Kyushu Univ.* 4: 175–187.

Goudeau, M. 1982. Fertilization in a crab (*Carcinus maenas*). 1. Early events in the ovary, and cytological aspects of the acrosome reaction and gamete contacts. *Tiss. Cell.* 14: 97–112.

Goudeau, M., and J. Becker. 1982. Fertilization in a crab (*Carcinus maenas*). *Tiss. Cell.* 14: 273–282.

Goudeau, M., and F. Lachaise. 1980a. Fine structure and secretion of the capsule enclosing the embryo in a crab (*Carcinus maenas*). *Tiss. Cell.* 12: 287–308.

Goudeau, M., and F. Lachaise. 1980b. Endogenous yolk as a precursor of a possible fertilization envelope in a crab (*Carcinus maenas*). *Tiss. Cell.* 12: 503–512.

Goudeau,M., and F. Lachaise. 1983. Structure of the egg funiculus and deposition of embryonic envelopes in a crab (*Carcinus maenas*). *Tiss. Cell.* 15: 47–62.

Graf, F. 1968. *Le stockage de calcium avant la mue chez les crustacés amphipodes Orchestia (Talitridé) et Niphargus (Gammaridé hypogé).* Theses Doct. Sci. Nat. Dijon Arch. Orig. Centre Doc. CRNS no. AO 2,690.

Graf, F. 1978. Les sources de calcium pour les crustaces venant de muer. *Arch. Zool. Exp. Gen.* 119: 143–161.

Grahame, J. 1983. Adaptive aspects of feeding mechanisms. In *Biology of Crustacea.* Edited by F. J. Vernberg and W. B. Vernberg, vol. 8, pp. 65–107. New York: Academic Press.

Gray, I. E. 1957. A comparative study of the gill area of crabs. *Biol. Bull.* 112: 34–42.

Green, J. L. 1986. Biology of *Aratus elegans.* M. Sc. thesis. University College of North Wales.

Green, J. W., M. Harsch, L. Barr, and C. L. Prosser. 1959. The regulation of water and salt by the fiddler crabs, *Uca pugnax* and *Uca pugilator. Biol. Bull.* 116: 76–87.

Greenaway, P. 1979. Freshwater invertebrates. In *Comparative physiology of osmoregulation in animals*. Edited by G. M. O. Maloiy, vol. 1, pp. 117–173. London: Academic Press.

Greenaway, P. 1980. Water balance and urine production in the Australian arid zone crab *Holthuisana transversa*. *J. Exp. Biol.* 87: 237–246.

Greenaway, P. 1981. Sodium regulation in the freshwater/land crab *Holthuisana transversa*. *J. Comp. Physiol.* B142: 451–456.

Greenaway, P. 1984a. The relative importance of the gills and lungs in the gas exchange of amphibious crabs of the genus *Holthuisana*. *Aust. J. Zool.* 32: 1–6.

Greenaway, P. 1984b. Survival strategies in desert crabs. In *Arid Australia*. Edited by H. G. Cogger and E. E. Cameron, pp. 145–152. Sydney: Australian Museum.

Greenaway, P. 1985. Calcium balance and moulting in the Crustacea. *Biol. Rev.* 60: 425–454.

Greenaway, P. J., J. Bonaventura, and H. H. Taylor. 1983. Aquatic gas exchange in the freshwater/land crab, *Holthuisana transversa*. *J. Exp. Biol.* 103: 225–236.

Greenaway, P., and C. A. Farrelly. 1984. The venous system of the terrestrial crab *Ocypode cordimanus* (Desmarest 1825) with particular reference to the vasculature of the lungs. *J. Morphol.* 181: 133–142.

Greenaway, P., and R. E. MacMillen. 1978. Salt and water balance in the terrestrial phase of the inland crab *Holthuisana (Austrothelphusa) transversa* von Martens (Parathelphusoidea: Sundathelphusidae). *Physiol. Zool.* 51: 217–229.

Greenaway, P., and H. H. Taylor. 1976. Aerial gas exchange in Australian arid-zone crab *Parathelphusa transversa* Von Martens. *Nature* (London) 262: 711–713.

Greenaway, P., H. H. Taylor, and J. Bonaventura. 1983. Aerial gas exchange in Australian freshwater/land crabs of the genus *Holthuisana*. *J. exp. Biol.* 103: 237–251.

Greenspan, B. N. 1975. Male reproductive strategy in the communal courtship system of the fiddler crab, *Uca rapax*. Ph.D. thesis, Rockefeller University.

Greenspan, B. N. 1980. Male size and reproductive success in the communal courtship system of the fiddler crab, *Uca rapax*. *Anim. Behav.* 28: 387–392.

Greenspan, B. 1982. Semi-monthly reproductive cycles in male and female fiddler crabs, *Uca pugnax*. *Anim. Behav.* 30: 1084–1092.

Griffin, D. J. G. 1968. Social and maintenance behaviour in two Australian ocypodid crabs (Crustacea: Brachyura). *J. Zool.* 156: 291–305.

Gross, W. J. 1955. Aspects of osmotic regulation in crabs showing the terrestrial habit. *Am. Nat.* 89: 205–222.

Gross, W. J. 1957. A behavioral mechanism for osmotic regulation in a semi-terrestrial crab. *Biol. Bull.* 113: 268–274.

Gross, W. J. 1963. Cation and water balance in crabs showing the terrestrial habit. *Physiol. Zool.* 36: 312–324.

Gross, W. J. 1964a. Water balance in anomuran land crabs on a dry atoll. *Biol. Bull.* 126: 54–68.

Gross, W. J. 1964b. Trends in water and salt regulation among aquatic and amphibious crabs. *Biol. Bull.* 127: 447–466.

Gross, W. J., and P. V. Holland. 1960. Water and ionic regulation in a hermit crab. *Physiol. Zool.* 33: 21–28.

Gross, W. J., and L. A. Marshall. 1960. The influence of salinity on the magnesium and water fluxes of a crab. *Biol. Bull.* 119: 440–453.

Gross, W. J., R. C. Lasiewski, M. Dennis, and P. P. Rudy. 1966. Salt and water balance in selected crabs of Madagascar. *Comp. Biochem. Physiol.* 17: 641–660.

Grubb, P. 1971. Ecology of terrestrial decapod crustaceans on Aldabra. *Phil. Trans. Roy. Soc. London, Ser. B* 260: 411–416.

Guinot, D. 1977. Propositions pour une nouvelle classification des Crustaces Decapodes Brachyoures. *C.R. hebd. Seanc. Acad. Sci. Paris* D285: 1049–1052.

Guinot, D. 1978. Principles d'une classification evolutive des Crustaces Decapodes Brachyoures. *Bull. biol. Fr. Belg.* 112: 211–292.

Guinot, D. 1979. Morphologie et phylogenese des Brachyoures. *Mem. Mus. Nat. Hist. Nat. Paris* (A) 112: 3–354.

Guinot-Dumortier, D., and B. Dumortier. 1966. La stridulation chez les crabes. *Crustaceana* 1: 117–155.

Gurney, R. 1937. Notes on some decapod and stomatopod Crustacea from the Red Sea. *Proc. Zool. Soc. London.* 107B: 98–101.

Hadley, N. F. 1981. Cuticular lipids of terrestrial plants and arthropods: A comparison of their structure, composition, and waterproofing function. *Biol. Rev.* 56: 23–47.

Haeckel, E. 1857. Über die Gewebe des Flusskrebses. *Arch. Anat. Physiol. u. wiss. Med.* 469–568.

Hafemann, D. R., and J. O. Hubbard. 1969. On the rapid running of ghost crabs (*Ocypode ceratopthalma*). *J. Exp. Zool.* 170: 25–32.

Hails, A. J., and S. Yaziz. 1982. Abundance breeding and growth of the ocypod crab *Dotilla myctiroides* on a west Malaysian beach. *Estuar. Coast. Mar. Sci.* 15: 229–239.

Hake, S., and C. Teller. 1983. Effects of osmotic stress on the activity of the NaA + B, KA + B-ATPase in gills and NaA + B concentration in the haemolymph of *Carcinus maenas* and *Uca pugilator. Abstracts 5th Conference Europ. Soc. Comp. Physiol. Biochem.*

Haley, S. R. 1969. Relative growth and sexual maturity of the Texas ghost crab, *Ocypode quadrata* (Fabr.) (Brachyura, Ocypodidae). *Crustaceana* 17: 285–297.

Haley, S. R. 1972. Reproductive cycling in the ghost crab, *Ocypode quadrata* (Fabr.) (Brachyura, Ocypodidae). *Crustaceana* 23: 1–11.

Haley, S. R. 1973. On the use of morphometric data as a guide to reproductive maturity in the ghost crab, *Ocypode ceratopthalmus* (Pallas) (Brachyura, Ocypodidae). *Pacific Sci.* 27: 350–362.

Hall, L. A. 1982. Osmotic and ionic regulation in overwintering ghost crabs, *Ocypode quadrata*. Master's thesis, North Carolina State University, Raleigh.

Hamai, I., and E. Hirai. 1940. Relative growth of the crab, *Sesarma (Holometopus) dehaani* M. Edwards. *Sci. Rep. Tohoku Univ. (4) Biol.* 15: 369–384.

Hannan, J. V., and D. H. Evans. 1973. Water permeability in some euryhaline decapods and *Limulus polyphemus*. *Comp. Biochem. Physiol.* A44: 1199–1213.

Harada, E., and H. Kawanabe. 1955. The behavior of the sand-crab, *Scopimera globosa* de Haan, with special reference to the problems of coaction between individuals. *Jpn. J. Ecol.* 4: 162–165.

Hargreaves, B. R. 1981. Energetics of crustacean swimming. In *Locomotion and energetics of arthropods.* Edited by C. F. Herreid II and C. R. Fourtner, pp. 453–490. New York: Pergamon Press.

Harms, J. W. 1932. Die realisation von Genen und die consekutive adaptation. II. *Birgus latro* L. als landkrebs und seine Beziekungen zu den Coenobiten. *Z. wiss. Zool.* 140: 167–290.

Harms, J. W. 1938. Lebensablauf und Stammesgeschichte des *Birgus latro.* L. von der Weihnachtsinsel. *Jena Z. Naturw.* 71: 1–34.

Harris. R. R. 1975. Urine production rate and urinary sodium loss in the freshwater crab *Potamon edulis. J. Comp. Physiol.* 96: 143–153.

Harris, R. R. 1977. Urine production rate and water balance in the terrestrial crabs *Gecarcinus lateralis* and *Cardisoma guanhumi. J. Exp. Biol.* 68: 57–67.

Harris, R. R., and G. A. Kormanik. 1981. Salt and water balance and antennal gland function in three Pacific species of terrestrial crab (*Gecarcoidea lalandii, Cardisoma carnifex, Birgus latro*). II. The effects of desiccation. *J. Exp. Zool.* 218: 107–116.

Harris, R. R., and H. Micallef. 1971. Osmotic and ionic regulation in *Potamon edulis*, a freshwater crab from Malta. *Comp. Biochem. Physiol.* 38A: 769–776.

Hartnoll, R. G. 1964. The freshwater grapsid crabs of Jamaica. *Proc. Linn. Soc. London* 175: 145–169.

Hartnoll, R. G. 1965. Notes on the marine grapsid crabs of Jamaica. *Proc. Linn. Soc. London* 176: 113–147.

Hartnoll, R. G. 1968. Morphology of the genital ducts in female crabs. *J. Linn. Soc. London. Zool.* 47: 279–300.

Hartnoll, R. G. 1969. Mating in the Brachyura. *Crustaceana* 16: 161–181.

Hartnoll, R. G. 1971. *Sesarma cookei*, n. sp., a grapsid crab from Jamaica (Decapoda, Brachyura). *Crustaceana* 20: 257–262.

Hartnoll, R. G. 1973. Factors affecting the distribution and behaviour of the crab *Dotilla fenestrata* on East African shores. *Estuar. Coast. Mar. Sci.* 1: 137–152.

Hartnoll, R. G. 1974. Variation in growth pattern between some secondary sexual characters in crabs. *Crustaceana* 27: 131–136.

Hartnoll, R. G. 1975. The Grapsidae and Ocypodidae (Decapoda: Brachyura) of Tanzania. *J. Zool.* 177: 305–328.

Hartnoll, R. G. 1976. The ecology of some rocky shores in tropical East Africa. *Estuar. Coast. Mar. Sci.* 4: 1–21.

Hartnoll, R. G. 1978a. The effect of salinity and temperature on the postlarval growth of the crab *Rhithropanopeus harrisii.* In *Physiology and behaviour of marine organisms.* Edited by D. S. McLusky and A. J. Berry, pp. 349–358. Oxford: Pergamon Press.

Hartnoll, R. G. 1978b. The determination of relative growth in Crustacea. *Crustaceana* 34: 281–293.

Hartnoll, R. G. 1982. Growth. In *The biology of Crustacea*, vol. 2. Edited by L. G. Abele, pp. 111–196. New York: Academic Press.

Hartnoll, R. G. 1983. Strategies of crustacean growth. *Mem. Austr. Mus.* 18: 121–131.

Hartnoll, R. G. In press. Brachyura. In *Reproduction of marine invertebrates* . Edited by A. C. Giese and J. S. Pearse, vol. 8. New York: Academic Press.

Hashimoto, H. 1965. The spawning of the terrestrial grapsid crab, *Sesarma haematocheir* (de Haan). *Zool Mag. Tokyo* 74: 82–87.

Hawkins, A. J. S., and M. B. Jones. 1982. Gill area and ventilation in two mud crabs *Helice crassa* Dana (Grapsidae) and *Macrophthalmus hirtipes* (Jacquinot) (Ocypodidae). *J. Exp. Mar. Biol. Ecol.* 60: 103–118.

Hazlett, B. A. 1966. Social behavior of the Paguridae andDiogenidae of Curacao. *Stud. Fauna Curacao* 23: 1–143.

Hazlett, B. A. 1968. Stimuli involved in the feeding behavior of the hermit crab *Clibanarius vittatus* (Decapoda, Paguridea). *Crustaceana* 15: 305–311.

Hazlett, B. A. 1971a. Chemical and chemotactic stimulation of feeding behavior in the hermit crab *Petrochirus diogenes. Comp. Biochem. Physiol.* 39A: 665–670.

Hazlett, B. A. 1971b. Antennule sensitivity in marine decapod crustaceans. *J. Anim. Morphol. Physiol.* 18: 1–10.

Healy, A., and J. Yaldwyn. 1970. *Australian crustaceans in colour.* Sydney: A. H. and A. W. Reed.

Heeg, J., and A. J. Cannone. 1966. Osmoregulation by means of a hitherto unsuspected osmoregulatory organ in two grapsid crabs. *Zool. Afr.* 2: 127–129.

Heglund, N. C., G. A. Cavagna, and C. R. Taylor. 1982a. Energetics and mechanics of terrestrial locomotion. III. Energy changes of the center of mass as a function of speed and body size in birds and mammals. *J. Exp. Biol.* 79: 41–56.

Heglund, N. C., M. A. Fedak, C. R. Taylor, and G. A. Cavagna. 1982b. Energetics and mechanics of terrestrial locomotion. IV. Total mechanical energy changes as a function of speed and body size in birds and mammals. *J. Exp. Biol.* 97: 57–66.

Heglund, N. C., C. R. Taylor, and T. A. McMahon. 1974. Scaling stride frequency and gait to animal size: Mice to horse. *Science* 186: 1112–1113.

Held, E. E. 1963. Molting behaviour of *Birgus latro. Nature* 200: 799–800.

Helfman, G. S. 1973. Ecology and behavior of the coconut crab, *Birgus latro* (L.). Master's thesis, University of Hawaii, Honolulu.

Helfman, G. S. 1977a. Agonistic behaviour of the coconut crab, *Birgus latro* (L.). *Tierpsychol.* 43: 425–438.

Helfman, G. S. 1977b. Copulatory behavior of the coconut or robber crab *Birgus latro* (L.) (Decapoda Anomura, Paguridea, Coenobitidae). *Crustaceana* 33: 198–202.

Helfman, G. S. 1979. Coconut crabs and cannibalism. *Nat. Hist.* 88: 76–83.

Henning, H. G. 1975a. Oekologische, ethologische und sinnesphysiologische untersuchungen an der Landkrabbe *Cardisoma guanhumi* Latreille (Decapoda, Brachyura). *Forma et Functio* 8: 253–304.

Henning, H. G. 1975b. Kampf-, Fortpflanzungs- und Hautungsverhalten-

Wachstum und Geschlechtsreife von *Cardisoma guanhumi* Latreille (Crustacea, Brachyura). *Forma et Functio* 8: 463–510.

Henning, H. G., and F. Klaassen. 1973. Dekapode Crustaceen auf der Isla de Salamanca (Atlantik-Kueste, Kolumbien). *Mitt. des Inst. Colombo-Aleman de Invest. Cien.* 7: 63–84.

Henry, R. P., and J. N. Cameron. 1981. A survey of blood and tissue nitrogen compounds in terrestrial decapods of Palau. *J. Exp. Zool.* 218: 83–88.

Henry, R. P., and J. N. Cameron. 1983. The role of carbonic anhydrase in respiration, ion regulation and acid-base balance in the aquatic crab *Callinectes sapidus* and the terrestrial crab *Gecarcinus lateralis*. *J. Exp. Biol.* 103: 205–223.

Herreid, C. F. II. 1963. Observations on the feeding behavior of *Cardisoma guanhumi* (Latreille) in southern Florida. *Crustaceana* 5: 176–180.

Herreid, C. F. II. 1967. Skeletal measurements and growth of the land crab, *Cardisoma guanhumi* Latreille. *Crustaceana* 13: 39–44.

Herreid, C. F. 1969a. Water loss of crabs from different habitats. *Comp. Biochem. Physiol.* 28: 829–839.

Herreid, C. F. 1969b. Integument permeability of crabs and adaptation to land. *Comp. Biochem. Physiol.* 29: 423–429.

Herreid, C. F. II. 1980. Hypoxia in invertebrates. *Comp. Biochem. Physiol.* 20: 333–336.

Herreid, C. F. II. 1981. Energetics of pedestrian arthropods. In *Locomotion and energetics in arthropods*. Edited by C. F. Herreid II and C. R. Fourtner, pp. 491–562. New York: Plenum Press.

Herreid, C. F. II, and R. J. Full. 1984a. Cockroaches on a treadmill: Aerobic running. *J. Insect Physiol.* 30: 395–403.

Herreid, C. F. II, and R. J. Full. 1984b. Hermit crab energetics on treadmill and beach: The cost of shell carrying. *Am. Zool.* 24: 122A.

Herreid, C. F. II, and R. J. Full. 1986a. Locomotion of hermit crabs (*Coenobita compressus*) on beach and treadmill. *J. Exp. Biol.* 120: 283–296.

Herreid, C. F. II, and R. J. Full. 1986b. Energetics of hermit crabs during locomotion: The cost of carrying a shell. *J. Exp. Biol.* 120: 297–308.

Herreid, C. F. II, R. J. Full, R. B. Weinstein, and G. M. Sadlo. 1984. Does the number of legs affect the energetic cost of locomotion? *Amer. Zool.* 24: 122A.

Herreid, C. F. II, R. J. Full, and S. M. Woolley. 1981. Hermit crab locomotion: Energetic cost of carrying a shell. *Physiologist* 24: 57.

Herreid, C. F. II, and C. A. Gifford. 1963. The burrow habitat of the land crab *Cardisoma guanhumi* (Latreille). *Ecol.* 44: 773–775.

Herreid, C. F. II, L. W. Lee, and G. M. Shah. 1979. Respiration and heart rate in exercising crabs. *Resp. Physiol.* 36: 109–120.

Herreid, C. F. II, and S. A. Mooney. 1984. Color change in exercising crabs: Evidence for a hormone. *J. Comp. Physiol.* B154: 207–212.

Herreid, C. F. II, and P. M. O'Mahoney. 1978. Aquatic and aerial respiration of crabs. *Amer. Zool.* 18: 639.

Herreid, C. F. II, P. M. O'Mahoney, and R. J. Full. 1983. Locomotion in land crabs: Respiratory and cardiac response of *Gecarcinus lateralis*. *Comp. Biochem. Physiol.* 74A: 117–124.

Herreid, C. F. II, P. M. O'Mahoney, and G. M. Shah. 1979. Cardiac and respiratory response to hypoxia in the land crab, *Cardisoma guanhumi* (Latreille). *Comp. Biochem. Physiol.* 63A: 145–151.

Herrnkind, W. F. 1968. Adaptive visually-directed orientation in *Uca pugilator. Amer. Zool.* 8: 585–598.

Herrnkind, W. F. 1972. Orientation in shore-living arthropods, especially the sand fiddler crab. In *Behavior of marine animals.* Vol. 1, *Invertebrates.* Edited by H. E. Winn and B. L. Olla, pp. 1–59. New York: Plenum Press.

Herrnkind, W. F. 1983. Movement patterns and orientation. In *The biology of Crustacea*, vol. 7. Edited by F. J. Vernberg and W. B. Vernberg, pp. 41–105. New York: Academic Press.

Hiatt, R. W. 1948. The biology of the lined shore crab, *Pachygrapsus crassipes* Randall. *Pacif. Sci.* 2: 135–213.

Hicks, J. W. 1983. The population, biomass and behaviour of the terrestrial crab *Gecarcoidea natalis* (Decapoda, Brachyura). Unpublished MS, Australian National Parks and Wildlife Service.

Hicks, J. W. 1985. The breeding behaviour and migrations of the terrestrial crab *Gecarcoidea natalis* (Decapoda: Brachyura). *Austral. J. Zool.* 33: 127–42.

Hicks, J., H. Rumpff, and H. Yorkston. 1984. *Christmas crabs.* Christmas Island, Indian Ocean: Christmas Island National History Association.

Hill, B. J., and H. Koopowitz. 1975. Heart-rate of the crab *Scylla serrata* (Forskal) in air and in hypoxic conditions. *Comp. Biochem. Physiol.* 52A: 385–387.

Hindley, J. P. R. 1975. The detection, location and recognition of food by juvenile banana prawns, *Penaeus merguinsis* de Man. *Mar. Behav. Physiol.* 3: 193–210.

Hinsch, G. W. 1968. Reproductive behavior in the spider crab, *Libinia emarginata* (L). *Biol. Bull.* 135: 273–278.

Hinsch, G. W. 1969. Wierotubules in the sperm of the spider crab, *Libinia emarginata* L. *J. Ultrastr. Res.* 29: 525–534.

Hinsch, G. W. 1970. Possible role of intranuclear membranes in nuclear-cytoplasmic exchange in spider crab oocytes. *J. Cell Biol.* 47: 531–535.

Hinsch, G. W. 1972. Some factors controlling reproduction in the spider crab, *Libinia emarginata. Biol. Bull.* 143: 358–366.

Hinsch, G. W., and D. C. Bennett. 1979. Vitellogenesis stimulated by thoracic ganglion implants into destalked immature spider crab, *Libinia emarginata. Tiss. Cell.* 11: 345–351.

Hinsch, G. W., and M. V. Cone. 1969. Ultrastructural observations of vitellogenesis in the spider crab, *Libinia emarginata* L. *J. Cell Biol.* 40: 336–342.

Hochachka, P. W. 1976. Design of metabolic and enzymic machinery to fit lifestyle and environment. *Biochem. Soc. Symp.* 41: 3–31.

Hodge, M. H. 1958. Some aspects of the regeneration of walking legs in the land crab, *Gecarcinus lateralis.* Ph.D. diss., Radcliffe College, Cambridge, Mass.

Hoestlandt, H. 1948. Recherches sur le biologie de l'*Eriocheir sinensis* en France (crustace brachyoure). *Ann. Inst. Oceanogr. Monaco* 24: 1–116.

Hohnke, L., and B. J. Scheer. 1970. Carbohydrate metabolism in crustaceans.

In *Chemical zoology*, Edited by M. Florkin and B. T. Scheer, vol. 5, pp. 147–166. New York: Academic Press.

Holland, C. A., and D. M. Skinner. 1976. Interactions between molting and regeneration in the land crab. *Biol. Bull.* 150: 222–240.

Holliday, C. W. 1985. Salinity-induced changes in gill Na, K-ATPase activity in the mud fiddler crab, *Uca pugnax*. *J. Exp. Zool.* 233: 199–208.

Holmquist, J. (In press). Grooming structure and function in some terrestrial Crustacea. In *Functional morphology of feeding and grooming in selected Crustacea.* Edited by B. E. Felgenhauer and L. Watling. Rotterdam: Balkema.

Holthuis, L. B. 1953. Enumeration of the decapod and stomatopod Crustacea from Pacific coral islands. *Atoll Res. Bull.* 24: 1–66.

Holthuis, L. B. 1954. On a collection of decapod Crustacea from the Republic of El Salvador (Central America). *Zool. Verh. Leiden* 23: 1–43.

Holthuis, L. B. 1964. *Sesarma (Sesarma) cerberus*, a new cavernicolous crab from Amboina. *Zool. Meded. Leiden* 40: 65–72.

Holthuis, L. B. 1974. Notes on the localities, habitats, biology, colour and vernacular names of New Guinea freshwater crabs (Crustacea, Decapoda, Sundathelphusidae). *Zool. Verh. Leiden* 137: 1–47.

Holthuis, L. B. 1977. The Grapsidae, Gecarcinidae and Palicidae (Crustacea: Decapoda: Brachyura) of the Red Sea. *Israel J. Zool.* 26: 141–192.

Holthuis, L. B. 1978. A collection of decapod Crustacea from Sumba, Lesser Sunda Islands, Indonesia. *Zool. Verh. Leiden* 162: 1–55.

Holthuis, L. B. 1979. Cavernicolous and terrestrial decapod Crustacea from Northern Sawawak, Borneo. *Zool. Verh. Leiden* 171: 1–47.

Hopkins, P. M. 1982. Growth and regeneration patterns in the fiddler crab, *Uca pugilator*. *Biol. Bull.* 163: 301–319.

Horch, K. W. 1971. An organ for hearing and vibration sense in the ghost crab *Ocypode*. *Z. Vergl. Physiol.* 73: 1–21.

Horch, K. W. 1975. The acoustic behavior of the ghost crab *Ocypode cordimana* Latreille, 1818 (Decapoda, Brachyura). *Crustaceana* 29: 193–205.

Horch, K. W., and M. Salmon 1969. Production, perception and reception of acoustic stimuli by semiterrestrial crabs (genus *Ocypode* and *Uca*, family Ocypodidae). *Forma et Functio* 1: 1–25.

Hort-Legrand, C., J. Berreur-Bonnenfant, and T. Ginsburger-Vogel. 1974. Etude anatomique et histologique comparee de la differenciation des gonades chez les males et les femelles chez *Orchestia gammarella* Pallas (Crustace Amphipode) pendant le periode post-embryonnaire. *Bull. Soc. Zool. Fr.* 99: 521–524.

Houlihan, D. F., and A. J. Innes. 1984. The cost of walking in crabs: Aerial and aquatic oxygen consumption during activity of two species of intertidal crab. *Comp. Biochem. Physiol.* 77A: 325–334.

Houlihan, D. F., E. Mathers, and A. J. El Haj. 1984. Walking performance and aerobic and anerobic metabolism of *Carcinus maenas* (L.) in sea water at 15° C. *J. Exp. Mar. Biol. Ecol.* 74: 211–230.

Howell, B. J., H. Rahn, D. Goodfellow, and C. Herreid. 1973. Acid-base regulation and temperature in selected invertebrates as a function of temperature. *Ann. Zool.* 13: 557–563.

Hoyt, D. F., and C. R. Taylor. 1981. Gait and the energetics of locomotion in horses. *Nature* 292: 239–240.

Huggins, A. K., and K. A. Munday. 1968. Crustacean metabolism. *Adv. Comp. Physiol. Biochem.* 3: 271–378.

Hughes, D. A. 1966. Behavioral and ecological investigations of *Ocypode ceratopthalmus* (Crustacea: Ocypodidae). *J. Zool.* (London) 150: 129–143.

Hughes, D. A. 1973. On mating and the "copulation burrows" of crabs of the genus *Ocypode* (Decapoda, Brachyura). *Crustaceana* 24: 72–76.

Hughes, G. M. 1984. General anatomy of the gills. In *Fish physiology*. Vol. 10A. Edited by W. S. Hoar and D. J. Randall. New York: Academic Press.

Hughes, G. M., B. Knights, and C. A. Scammel. 1969. The distribution of PO_2 and hydrostatic pressure changes within the branchial chambers of the shore crab *Carcinus maenas*. *J. Exp. Biol.* 51: 203–220.

Hughes, G. M., and M. Morgan. 1973. The structure of fish gills in relation to their respiratory function. *Biol. Rev.* 48: 419–475.

Hume, R. I., and A. Berlind. 1976. Heart and scaphognathite rate changes in a euryhaline crab, *Carcinus maenas*, exposed to dilute environmental medium. *Biol. Bull.* 150: 241–254.

Huxley, J. S. 1924. The variation in the width of the abdomen in immature fiddler-crabs considered in relation to its relative growth rate. *Amer. Nat.* 58: 468–475.

Huxley, J. S. 1932. *Problems of relative growth*. London: Methuen.

Huxley, T. H. 1879. *The crayfish: An introduction to the study of zoology*. International Science Series. London: Paul Trench Trubiner.

Hyatt, G. W. 1975. Physiological and behavioral evidence for colour discrimination by fiddler crabs (Brachyura, Ocypodidae, genus *Uca*). In *Physiological ecology of estuarine organisms*. Edited by F. J. Vernberg. Columbus: University of South Carolina Press.

Hyatt, G. W. 1977a. Field studies of size-dependent changes in waving display and other behavior in the fiddler crab, *Uca pugilator* (Brachyura, Ocypodidae). *Mar. Behav. Physiol.* 4: 283–292.

Hyatt, G. W. 1977b. Quantitative analysis of size-dependent variation in the fiddler crab wave display (*Uca pugilator*, Brachyura, Ocypodidae). *Mar. Behav. Physiol.* 5: 19–36.

Hyatt, G. W. 1983. Qualitative and quantitative dimensions of crustacean aggression. In *Studies in adaptation: The behavior of higher crustacea*. Edited by S. Rebach and D. W. Dunham, pp. 113–139. New York: Wiley.

Hyatt, G. W., and M. Salmon. 1978. Combat in the fiddler crabs *Uca pugilator* and *Uca pugnax*: A quantitative analysis. *Behaviour* 65: 182–211.

Hyatt, G. W., and M. Salmon. 1979. Comparative statistical and information analysis of combat in the fiddler crabs, *Uca pugilator* and *Uca pugnax*. *Behaviour* 68: 1–23.

Hyman, O. W. 1920. The development of *Gelasimus* after hatching. *J. Morph.* 33: 485–525.

Icely, J. D., and D. A. Jones. 1978. Factors affecting the distribution of the genus *Uca* (Crustacea: Ocypodidae) on an East African shore. *Estuar. Coast. Mar. Sci.* 6: 315–325.

Ihle, J. E. W. 1912. Ueber eine klein Brachyuren-Sammlung aus unterirdischen Flussen von Java. *Notes Leyden Mus.* 34: 177–182.

Innes, A. J., A. J. El Haj, and J. F. Gobin. 1986. Scaling of the respiratory, cardiovascular and skeletal muscular systems of the freshwater/terrestrial mountain crab, *Pseudothelphusa garmani garmani* (Rathbun 1898). *J. Zool. Lond.* A209: 595–606.

Innes, A. J., M. E. Forster, M. B. Jones, I. D. Marsden, and H. H. Taylor. 1986. Bimodal respiration water balance and acid-base regulation in a high-shore crab *Cyclograpsus lavaux* H. *J. Exp. Mar. Biol. Ecol.* 100: 127–145.

Innes, A. J., and E. W. Taylor 1986a. An analysis of lung function in the Trinidadian mountain crab. *J. Physiol.* 372: 43P.

Innes, A. J., and E. W. Taylor 1986b. A functional analysis of pulmonary cutaneous and brachial gas exchange in the Trinidad mountain crab: A comparison with other land crabs. *J. Physiol.* 373: 46.

Innes, A. J., and E. W. Taylor 1986c. Lung ventilation in the Trinidad mountain crab. *J. Physiol.* 374: 86.

Innes, A. J., and E. W. Taylor 1986d. Air breathing crabs of Trinidad: Adaptive radiation into the terrestrial habitat. I. Aerobic metabolism and habitat. *Comp. Biochem. Physiol.* 85A: 373–382.

Innes, A. J., and E. W. Taylor 1986e. The evolution of air-breathing in crustaceans: A functional analysis of branchial, cutaneous, and pulmonary gas exchange. *Comp. Biochem. Physiol.* 85A: 621–637.

Innes, A. J., E. W. Taylor, and A. J. El Haj. 1987. Air-breathing in the Trinidad mountain crab: A quantum leap in the evolution of the invertebrate lung. *Comp. Biochem. Physiol.* 87A: 1–9.

Jackson, H. G. 1913. *Eupagurus. Liverpool Mar. Biol. Comm. Mem.* 21: 1–79.

Jacob, V. 1985. Composition of semen of the freshwater crab *Gecarcinucus steniops*: A histochemical study. M. Phil. diss., Calicut University, India.

Jacoby, C. A. 1981. Behavior of the purple shore crab, *Hemigrapsus nudus* Dana, 1851. *J. Crust. Biol.* 1: 531–544.

Jensen, T. F., and I. Holm-Jensen. 1980. Energetic cost of running in workers of ant species, *Formica fuxa* L., *Formica rufa* L. (Hymenoptera, Formicidae). *J. Comp. Physiol.* 137: 151–156.

Jobert, M. 1876. Recherches sur l'appareil respiratoire et le mode de respiration de certaines Crustaces Brachyures (crabes terrestres). *Ann. Sci. Nat. Ser. Zool. Biol. Anim.* 4: 1–5.

Johansen, K., and W. W. Burggren eds. 1985. *Cardiovascular shunts: Phylogenetic, ontogenetic and clinical aspects.* Copenhagen: Munksgaard.

Johnson, D. S. 1965. Land crabs. *J. Malaysian Branch: Roy. Asiatic Soc.* 38: 43–66.

Johnson, P. T. 1980. *Histology of the blue crab, Callinectes sapidus.* New York: Praeger.

Johnston, M. A., H. Y. Elder, and P. S. Davies. 1973. Cytology of *Carcinus* hemocytes and their function in carbohydrate metabolism. *Comp. Biochem. Physiol.* 46A: 569–581.

Jokumsen, A., and R. E. Weber. 1982. Hemocyanin oxygen affinity in hermit crab blood is temperature dependent. *J. Exp. Zool.* 221: 380–394.

Jones, D. A. 1972. Aspects of the ecology and behaviour of *Ocypode ceratopthalmus* (Pallas) and *O. kuhli* de Haan (Crustacea: Ocypodidae). *J. Exp. Mar. Biol. Ecol.* 8: 31–43.

Jones, D. A., and D. Clayton. 1983. The systematics and ecology of crabs belonging to the genera *Cleistostoma* de Haan and *Paracleistostoma* de Man on Kuwait mudflats. *Crustaceana* 45: 183–199.

Jones, M. B., and J. G. Greenwood. 1982. Water loss of a porcelain crab, *Petrolisthes elongatus* (Milne Edwards, 1837) (Decapoda, Anomura) during atmospheric exposure. *Comp. Biochem. Physiol.* 72A: 631–636.

Joshi, P. C., and S. S. Khanna. 1982a. Seasonal changes in the ovary of a freshwater crab, *Potamon koolooense*. *Proc. Indian Acad. Sci. Anim. Sci.* 91: 451–462.

Joshi, P. C., and S. S. Khanna. 1982b. Structure and seasonal changes in the testes of a freshwater crab, *Potamon koolooense*. *Proc. Indian Acad. Sci. Anim. Sci.* 91: 439–450.

Joshi, P. C., and S. S. Khanna. 1984. Neurosecretory system of the thoracic ganglion and its relation to testicular maturation of the crab, *Potamon koolooense*. *Z. Mikrosk. Anat. Forsch.* (Leipzig) 98: 429–442.

Jubb, C. A., R. N. Hughes, and T. A. P. Rheinallt. 1983. Behavioral mechanisms of size-selection by crabs, *Carcinus maenas* (L.) feeding on mussels, *Mytilus edulis* (L.). *J. Exp. Mar. Biol. Ecol.* 66: 81–87.

Juchault, P., J. Maissiat, and J. J. Legrand. 1978. Characterisation chimique d'une substance ayant les effets biologiques de l'hormone androgene chez le Crustace Isopode terrestre *Armadillidium vulgare* Latreille. *C. R. Acad. Sci. Ser D.* 286: 73–76.

Kalber, F. A., and D. E. Costlow. 1968. Osmoregulation in larvae of the land crab, *Cardisoma guanhumi* Latreille. *Amer. Zool.* 8: 411–416.

Katakura, Y., Y. Fujimaki, and K. Unno. 1975. Partial purification and characterization of androgenic gland hormone from the isopod crustacean *Armadillidium vulgare*. *Annot. Zool. Jpn.* 48: 203–209.

Katakura, Y., and Y. Hasegawa. 1983. Masculinization of females of the isopod crustacean *Armadillidium vulgare* following injections of an active extract of the androgenic gland. *Gen. Comp. Endocrin.* 48: 57–62.

Katz, L. C. 1980. Effects of burrows by the fiddler crab *Uca pugnax*. *Estuar. Coast. Mar. Sci.* 11: 233–238.

Kelemec, J. A. 1979. Effect of temperature on the emergence from burrows of the soldier crab, *Mictyris longicarpus* (Latreille). *Aust. J. Mar. Fresh. Res.* 30: 463–468.

Keller, R., and E. M. Andrew. 1973. The site of the crustacean hyperglycemic hormone. *Gen. Comp. Endocrin.* 20: 572–578.

Keller, R., and J. Beyer. 1968. Zur hyperglykamishen Wirkung von Serotonin und Augestielextrakt beim Flusskrebs *Orconectes limosus*. *Z. Vergl. Physiol.* 59: 78–85.

Kerr, M. S. 1968. Protein synthesis by hemocytes of *Callinectes sapidus*: A study in *in vitro* incorporation of ^{14}C-leucine. *J. Cell Biol.* 39: 72a–73a.

Kerr, M. S. 1969. The hemolymph proteins of the blue crab, *Callinectes sapidus*. II. A lipoprotein serologically identical to oocyte lipovitellin. *Develop. Biol.* 20: 1–17.

Kessel, R. G. 1968. Mechanisms of protein yolk synthesis and deposition in crustacean oocytes. *Z. Zellforsch.* 89: 17–38.

King, D. S. 1964. Fine structure of the androgenic gland of the crab, *Pachygrapsus crassipes. Gen. Comp. Endocrinol.* 4: 533–544.

Kinoshita, H., and A. Okajima. 1968. Measuring ability of size and form in shell searching behavior of the land hermit crab. *Zool. Mag.* 77: 233–272.

Kittredge, J. S., M. Terry, and F. T. Takahashi. 1971. Sex pheromone activity of the molting hormone, crustecdysone, on male crabs (*Pachygrapsus crassipes, Cancer antennarius,* and *Cancer anthonyi*). *Fish. Bull. U.S. Fish Wildl. Serv.* 69: 337–343.

Klaassen, F. 1973. Stridulation and communication by substrate vibration in *Gercarcinus lateralis* (Crustacea, Decapoda). *J. Comp. Physiol.* 83: 73–79.

Klaassen, F. 1975. Oekologische und ethologische Untersuchungen zur Fortpflanzungabiologie von *Gecarcinus lateralis* (Decapoda, Brachyura). *Forma et Functio.* 8: 101–174.

Kleiber, M. 1961. *The fire of life: An introduction to animal energetics.* New York: Wiley.

Koba, K. 1936. Preliminary notes on the development of *Geotelphusa dehaani. Per. Imp. Acad. Japan* 12: 105–107.

Koepcke, H. W., and M. Koepcke. 1953. Contribucion al conocimiento de la forma de vida de *Ocypode guadichaudii* Milne Edwards et Lucas (Decapoda, Crustacea). *Publ. Mus. Hist. Hat. "Javier Prado" Ser. A Zool.* 13: 1–46.

Kormanik, G. A., and R. R. Harris. 1981. Salt and water balance and antennal gland function in three Pacific species of terrestrial crab (*Gecarcoidea lalandii, Cardisoma carnifex, Birgus latro*). I. Urine production and salt exchange in hydrated crabs. *J. Exp. Zool.* 218: 97–105.

Krainska, M. K. 1936. On the development of *Eupagurus prideauxi* Leach. *C. R. 12th Int. Cong. Zool.* (1935), pp. 554–565.

Kraus, D. B. 1982. The burrow as a resource for reproduction and molting in the fiddler crab. *Amer. Zool.* 22: 869.

Kraus, H. J., and J. Tautz. 1981. Visual distance-keeping in the soldier crab, *Mictyris platycheles* Latreille (Grapsoidea: Mictyridae). A field study. *Mar. Behav. Physiol.* 6: 123–133.

Krishnakumar, R. 1985. Studies on spermatheca of some decapod crustaceans. Ph.D. thesis, Calicut University, India.

Krishnakumar, R., V. R. Vijayalakshmi, and K. G. Adiyodi. 1979. Uptake of yolk proteins by developing crustacean oocytes as revealed by trypan blue technique. *All-Indian Workshop on Crustacean Reproductive Biology, Calicut,* pp. 29–30.

Kropp, B., and W. J. Crozier. 1928. Geotropic orientation in arthropods. III. The fiddler crab, *Uca. J. Gen. Physiol.* 12: 111–122.

Kuramoto, T., and A. Ebara. 1984. Neurohormonal modulation of the cardiac outflow through the cardioatrial valve in the lobster. *J. Exp. Biol.* 111: 123–130.

Kurata, H. 1962. Studies on the age and growth of Crustacea. *Bull. Hokkaido Reg. Fish Res. Lab.* 24: 1–115.

Kurta, A. 1982. Social facilitation of foraging behavior by the hermit crab, *Coenobita compressus,* in Costa Rica. *Biotropica* 14: 132–136.

Kurup, K. N. P. 1983. Some aspects of limb regeneration in the freshwater crab, *Parathelphusa hydrodromus* (Herbst). Ph.D. thesis, Calicut University, India.

Kurup, K. N. P., and R. G. Adiyodi. 1980. Patterns of regeneration during the breeding season in the crab, *Parathelphusa hydrodromus* (Herbst). *Proc. I. All-India Symp. Inv. Reprod.*, Madras, pp. 25–26.

Kurup, K. N. P., and R. G. Adiyodi. 1981. The programming of somatic growth and reproduction in the crab, *Parathelphusa hydrodromus* (Herbst). *Int. J. Invertebr. Reprod.* 3: 27–39.

Kurup, K. N. P., and R. G. Adiyodi. 1984. Multiple limb autotomy can trigger either ovarian growth or somatic growth in the freshwater crab, *Parathelphusa hydrodromus* (Herbst). *Gen. Comp. Endocrinol.* 56: 433–443.

Lachaise, F., M. Goudeau, C. Hetru, C. Kappler, and J. A. Hoffmann. 1981. Ecdysteroids and ovarian development in the shore crab, *Carcinus maenas*. *Hoppe-Seyler's Z. Physiol. Chem.* 362: 521–530.

Lachaise, F., and J. A. Hoffmann. 1977. Ecdysone et developpement ovarien chez un Decapode, *Carcinus maenas*. *C. R. Acad. Sci., Paris.* 285: 701–704.

Lachaise, F., and J. A. Hoffmann. 1982. Ecdysteroids and embryonic development in the shore crab, *Carcinus maenas*. *Hoppe-Seyler's Z. Physiol. Chem.* 363: 1059–1068.

Lafon, M. 1948. Nouvelles recherches biochimiques et physiologiques sur le squelette tegumentaire des Crustaces. *Bull. Inst. Oceanogr.* 939: 1–28.

Lang, R. 1973. Die Ontogenese von *Maja squinado* (Crustacea, Malacostraca, Decapoda, Brachyura). *Zool. Jahrb., Abt. Anat. Ontog. Tiere* 89: 600–610.

Lang, R., and P. Fioroni. 1971. Darmentwicklung und Dotteraufschluss bei *Macropodia* (Crustacea, Malacostraca, Decapoda, Brachyura). *Zool. Jahrb., Abt. Anat. Ontog. Tiere* 88: 84–137.

Langdon, J. W. 1971. Shape discrimination and learning in the fiddler crab *Uca pugilator*. Ph.D. thesis, Florida State University, Tallahassee. (As cited in Herrnkind, 1983).

Langdon, J. W., and W. F. Herrnkind. 1985. Visual shape discrimination in the fiddler crab, *Uca pugilator*. *Mar. Behav. Physiol.* 11: 315–325.

Langreth, S. G. 1969. Spermiogenesis in *Cancer* crabs. *J. Cell Biol.* 43: 575–603.

Laurent, P. 1984. Gill internal morphology. In *Fish physiology*. Vol. 10A. Edited by W. S. Hoar and D. J. Randall. New York: Academic Press.

Lawrence, J. M. 1970. Lipid content of the organs of the coconut crab, *Birgus latro* (L.) (Decapoda, Paguridae). *Crustaceana* 19: 264–266.

Leber, K. M. 1982. Seasonality of macroinvertebrates on a temperate, high wave energy sandy beach. *Bull. Mar. Sci.* 32: 86–98.

Lee, M. A. B. 1985. The dispersal of *Pandanus tectorius* by the land crab *Cardisoma carnifex*. *Oikos* 45: 169–173.

Lenfant, C., and K. Johansen. 1967. Respiratory adaptations in selected amphibians. *Respir. Physiol.* 2: 247–260.

Leone, C. A. 1950. Serological relationships among common brachyuran crustacea of Europe. *Pubbl. Staz. zool. Napoli* 22: 273–282.

Leone, C. A. 1951. A serological analysis of the systematic relationship of the brachyuran crab *Geryon quinquedens*. *Biol. Bull.* 100: 44–48.

Le Roux, A. 1976. Aspects de la differenciation sexuelle chez *Pisidia longicornis* (Linne) (Crustace, Decapode). *C. R. Acad. Sci., Ser. D.* 283: 959–962.

Lewinsohn, C. 1969. Die Anomuren des roten Meeres (Crustacea Decapoda: Paguridea, Galatheidea, Hippidea). The second Israel South Red Sea expedition, 1965, report no. 6. *Zool. Verh. Leiden* 104: 1–213.

Lewinsohn, C. 1982. Researches on the coast of Somalia. The shore and the dune of Sar Uanle. 33. Diogenidae, Paguridae and Coenobitidae (Crustacea Decapoda Paguridea). *Monitore Zool. Ital. Suppl.* 16: 35–68.

Lighter, F. J. 1974. A note on a behavioral spacing mechanism of the ghost crab *Ocypode ceratopthalmus* (Pallas) (Decapoda, family Ocypodidae). *Crustaceana* 27: 312–314.

Lillywhite, H. B. 1986. Circulatory adaptations of snakes to gravity. *Amer. Zool.* 27: 81–96.

Lindberg, W. J. 1980. Behavior of the Oregon mud crab, *Hemigrapsus oregonensis* (Dana) (Brachyura, Grapsidae). *Crustaceana* 39: 263–281.

Lindqvist, O. V., I. Salminen, and P. W. Winston. 1972. Water content and water activity in the cuticle of terrestrial isopods. *J. Exp. Biol.* 56: 49–55.

Lindsenmaier, E. 1967. Konstruktion und Signalfunktion der Sandpyramide der Reiterkrabbe *Ocypode saraten* Forsk (Decapoda, Brachyura, Ocypodidae). *Z. Tierpsychol.* 24: 403–456.

Lister, J. J. 1888. On the natural history of Christmas Island, in the Indian Ocean. *Proc. Zool. Soc. Lond.* 512–531.

Little, C. 1983. *The colonisation of land.* Cambridge: Cambridge University Press.

Lochhead, J. H. 1960. Locomotion. In *The physiology of the Crustacea.* Edited by T. H. Waterman, vol. 2, pp. 313–364. New York: Academic Press.

Lockwood, A. P. M. 1967. *Aspects of the physiology of Crustacea.* London: Oliver and Boyd.

Logan, D. T., and C. E. Epifanio. 1978. A laboratory energy balance for the larvae and juveniles of the American lobster *Homarus americanus. Mar. Biol.* (Berlin) 47: 381–389.

Loizzi, R. F. 1971. Interpretation of crayfish hepatopancreatic function based on fine structural anatomy of epithelial cell lines and muscle networks. *Z. Zellforsch. Mikrosk. Anat.* 113: 420–440.

Long, G. L. 1976. The stereospecific distribution and evolutionary significance of invertebrate lactate dehydrogenases. *Comp. Biochem. Physiol.* 55B: 77–83.

Losey, G. S., Jr. 1978. Information theory and communication. In *Quantitative ethology.* Edited by P. W. Colgan, pp. 43–78. New York: Wiley-Interscience.

Lui, C. W., and J. D. O'Connor. 1977. Biosynthesis of crustacean lipovitellin. 3. The incorporation of labelled amino acids into the purified lipovitellin of the crab *Pachygrapsus crassipes. J. Exp. Zool.* 199: 105–108.

Lutz, P. L. 1969. Salt and water balance in the West African freshwater/land crab *Sudanonautes africanus africanus* and the effects of desiccation. *Comp. Biochem. Physiol.* 30: 469–480.

McCann, C. 1938. Notes on the common landcrab *Paratelphusa (Barytelphusa) guerini* (M. Edw.) of Salsette Island. *J. Bombay Nat. Hist. Soc.* 39: 531–542.

McCarthy, J. F., and D. M. Skinner. 1977. Proecdysial changes in serum ecdysone titers, gastrolith formation, and limb regeneration following molt induction by limb autotomy and/or eyestalk removal in the land crab, *Gecarcinus lateralis*. *Gen. Comp. Endocrinol.* 33: 278–292.

McDonald, D. G. 1977. Respiratory physiology of the crab *Cancer magister*. Ph.D. thesis, University of Calgary, Alberta, Canada.

McDonald, D. G., B. R. McMahon, and C. M. Wood. 1977. Patterns of heart and scaphognathite activity in the crab *Cancer magister*. *J. Exp. Biol.* 292: 33–44.

McDonald, D. G., B. R. McMahon, and C. M. Wood. 1979. An analysis of acid-base disturbances in the haemolymph following strenuous activity in the Dungeness crab *Cancer magister*. *J. Exp. Biol.* 79: 47–58.

McDonald, D. G., C. M. Wood, and B. R. McMahon. 1980. Ventilation and oxygen consumption in the Dungeness crab, *Cancer magister*. *J. Exp. Zool.* 213: 123–136.

MacDonald, J. D., R. B. Pike, and D. I. Williamson. 1957. Larvae of the British species of *Diogenes, Pagurus, Anapagurus* and *Lithodes* (Crustacea, Dècapoda). *Proc. Zool. Soc. Lond.* 128: 209–257.

McEvoy, W. H., and J. Ayers. 1982. Locomotion and limb movements. In *The biology of Crustacea*. vol. 4, Edited by D. C. Sandeman and H. L. Atwood, pp. 61–105. New York: Academic Press.

Mackay, D. C. G., and F. W. Weymouth. 1935. The growth of the Pacific edible crab, *Cancer magister*. *J. Biol. Board Canada*. 1: 191–212.

McKillup, S., and A. Butler. 1979. Cessation of hole-digging by the crab *Helograpsus haswellianus*: A resource-conserving adaptation. *Mar. Biol.* 50: 157–161.

McLaughlin, P. A. 1980. *Comparative morphology of recent Crustacea*. San Francisco: Freeman.

McLaughlin, P. A. 1983a. Internal anatomy. In *The Biology of Crustacea*, vol. 5, Edited by L. H. Mantel, pp. 1–52. New York: Academic Press.

McLaughlin, P. A. 1983b. Hermit crabs – are they really polyphyletic? *J. Crust. Biol.* 3: 608–621.

McMahon, B. R. 1981. Oxygen uptake and acid-base balance during activity in decapod crustaceans. In *Locomotion and Energetics in Arthropods*. Edited by C. F. Herreid and C. R. Fortner, pp. 299–335. New York: Plenum Press.

McMahon, B. R. 1985. Functions and functioning of crustacean hemocyanin. In *Respiratory pigments in animals: Relation structure-function*. Edited by J. Lamy, J.-P. Truchot, and R. Gilles. Heidelberg: Springer- Verlag.

McMahon, B. R. 1986. Oxygen transport by hemocyanin: Compensation during activity and environmental change. In *Invertebrate oxygen carriers*. Edited by B. Linzen. Heidelberg: Springer-Verlag.

McMahon, B. R. In press. Physiological compensation for oxygen depletion in tidepool animals. *Amer. Zool.*

McMahon, B. R., and W. W. Burggren. 1979. Respiration and adaptation to the terrestrial habitat in the land hermit crab *Coenobita clypeatus*. *J. Exp. Biol.* 79: 265–281.

McMahon, B. R., and W. W. Burggren. 1980. Oxygen uptake and transport in three air breathing crabs. *Physiologist* 23: 928A.

McMahon, B. R., and W. W. Burggren. 1981. Acid-base balance following temperature acclimation in land crabs. *J. Exp. Zool.* 218: 45–52.

McMahon, B. R., L. E. Burnett, and P. L. DeFur. 1984. Carbon dioxide excretion and carbonic anhydrase function in the red rock crab *Cancer productus*. *J. Comp. Physiol.* 154: 371–383.

McMahon, B. R., P. J. Butler, and E. W. Taylor. 1978. Acid-base changes during recovery from disturbance and during long-term hypoxic exposure in the lobster. *Homarus vulgaris*. *J. Exp. Zool.* 205: 361–370.

McMahon, B. R., D. G. McDonald, and C. M. Wood. 1979. Ventilation, oxygen uptake, and haemolymph oxygen transport following enforced exhaustive activity in the Dungeness crab *Cancer magister*. *J. Exp. Biol.* 80: 271–285.

McMahon, B. R., F. Sinclair, C. D. Hassal, P. L. DeFur, and P. R. H. Wilkes. 1978. Ventilation and control of acid-base status during temperature acclimation in the crab *Cancer magister*. *J. Comp. Physiol.* 128B: 109–116.

McMahon, B. R., and J. L. Wilkens. 1975. Respiratory and circulatory responses to hypoxia in the lobster *Homarus americanus*. *J. Exp. Biol.* 62: 637–655.

McMahon, B. R., and J. L. Wilkens. 1977. Periodic respiratory and circulatory performance in the red rock crab *Cancer productus*. *J. Exp. Zool.* 202: 363–374.

McMahon, B. R., and J. L. Wilkens. 1983. Ventilation, perfusion, and oxygen uptake. In *The Biology of Crustacea*, vol. 5. Edited by L. H. Mantel, pp. 289–372. New York: Academic Press.

McMahon, B. R., and P. R. H. Wilkes. 1983. Emergence response and aerial ventilation in normoxic and and hypoxic crayfish *Orconectes rusticus*. *Physiol. Zool.* 56: 133–141.

MacMillen, R. E., and P. Greenaway. 1978. Adjustments of energy and water metabolism to drought in an Australian arid-zone crab. *Physiol. Zool.* 51: 231–240.

Macnae, W. 1968. A general account of the fauna and flora of mangrove swamps and forests in the Indo-West-Pacific region. *Adv. Mar. Biol.* 6: 73–270.

Macnae, W., and M. Kalk. 1962. The ecology of the mangrove swamps at Inhaca Island, Mozambique. *J. Ecol.* 50: 19–34.

McVean, A. 1982. Autotomy. In *The biology of Crustacea. Vol. 4, Neural integration and behavior*. Edited by D. C. Sandeman and H. L. Atwood, pp. 107–132. New York: Academic Press.

Maitland, D. P. 1986. Crabs that breathe air with their legs – *Scopimera* and *Dotilla*. *Nature* 319: 494–495.

Malley, D. F. 1978. Degradation of mangrove litter by the tropical sesarmid crab *Chiromanthes*. *Mar. Biol.* 49: 377–386.

Mangum, C. P. 1980. Respiratory function of the haemocyanins. *Amer. Zool.* 20: 19–38.

Mangum, C. P. 1983a. On the distribution of lactate sensitivity among haemocyanins. *Mar. Biol. Lett.* 4: 139–149.

Mangum, C. P. 1983b. Adaptability and inadaptability among HcO_2 transport systems: An apparent paradox. In *Structure and function of invertebrate respiratory proteins.* Edited by E. J. Wood, pp. 333–352. Life Chemistry Reports Suppl. 1. New York: Harwood.

Mangum, C. P. 1983c. Oxygen transport in the blood. In *Internal anatomy and physiological regulation.* Vol. 5 of *The Biology of Crustacea.* Edited by L. H. Mantel. New York: Academic Press.

Mangum, C. P., B. R. McMahon, P. L. DeFur, and M. G. Wheatly. 1985. Gas exchange, acid-base balance, and the oxygen supply to the tissues during a molt of the blue crab *Callinectes sapidus. J. Crustacean Biol.* 5: 188–206.

Mangum, C. P., and A. L. Weiland. 1975. The function of hemocyanin in respiration of the blue crab *Callinectes sapidus. J. Exp. Zool.* 193: 257–264.

Mann, T. 1964. *Biochemistry of semen and the male reproductive tract.* London: Methuen.

Mantel, L. H. 1968. The foregut of *Gecarcinus lateralis* as an organ of salt and water balance. *Amer. Zool.* 8: 433–442.

Mantel, L. H., D. E. Bliss, S. W. Sheehan, and E. A. Martinez. 1975. Physiology of haemolymph, gut fluid and hepatopancreas of the land crab *Gecarcinus lateralis* (Freminville) in various neuroendocrine states. *Comp. Biochem. Physiol.* 51A: 663–671.

Mantel, L. H., and L. L. Farmer. 1983. Osmotic and ionic regulation. In *The Biology of Crustacea*, vol. 5. Edited by L. H. Mantel, pp. 53–161. New York: Academic Press.

Mantel, L. H., and J. R. Olson. 1976. Studies on the Na + K-activated ATPase of crab gills. *Amer. Zool.* 16: 223.

Manton, S. M. 1977. *The Arthropoda: Habits, functional morphology and evolution.* New York: Oxford University Press.

Margaria, R., P. Cerretelli, P. E. Diprampero, C. Massari, and G. Torelli. 1963. Kinetics and mechanism of oxygen debt contraction in man. *J. App. Physiol.* 18: 371–377.

Markl, H. 1968. Die Verstaendigung durch Stridulationsssignale bei Blattschneiderameisen. II. Erzeugung und Eigenschaften der Signale. *Z. Vergl. Physiol.* 60: 103–115.

Marshall, S. M., and A. P. Orr. 1960. Feeding and nutrition. In *The physiology of Crustacea.* Edited by T. H. Waterman, vol. 1, pp. 227–258. New York: Academic Press.

Mary, R. F., and G. Krishnan. 1974. On the nature and role of protein constituents of the cuticle of crustaceans in relation to permeability of the cuticle. *Mar. Biol.* 25: 299–309.

Mason, C. A. 1970. Function of the pericardial sacs during the molt cycle in the land crab *Gecarcinus lateralis. J. Exp. Zool.* 174: 381–390.

Mathad, S. G. 1983. Biochemistry and physiology of semen in freshwater crabs. Ph.D. thesis, Calicut University, India.

Matsuno, T., W. Yoshiko, and O. Masahiro. 1982. Carotenoids of a river crab, *Potamon dehaani* (in Japanese). *Bull. Jpn. Soc. Sci. Fish.* 48: 661–666.

Mattson, R. A. 1982. Feeding ecology of the mangrove tree crab *Aratus pisonii* (Milne-Edwards): Selection of older leaves for herbivory. *Fl. Scient* (suppl): 26 (Abstract).

Mattson, W. J., Jr. 1980. Herbivory in relation to plant nitrogen content. *Ann. Rev. Ecol. Syst.* 11: 119–161.

Mauro, N. A., and C. P. Mangum. The role of the blood in the temperature dependence of oxidative metabolism in decapod crustaceans. 1. Intraspecific responses to seasonal differences in temperature. *J. Exp. Zool.* 219: 179–188.

Maynard, D. M. 1960. Circulation and heart function. In *The physiology of Crustacea.* Edited by T. H. Waterman. New York: Academic Press.

Meenakshi, V. R., and B. T. Scheer. 1959. Acid mucopolysaccharide of the crustacean cuticle. *Science* 130: 1189–1190.

Messeri, P. 1978. Some observations on a littoral troop of yellow baboons. *Monitore Zool. Ital.* N. S. 12: 69.

Meusy, J. J. 1965a. Modifications ultrastructurales des glandes androgenes de *Carcinus maenas* L. (Crustace Decapode) consecutives a l'ablation des pedoncules oculaires. *C. R. Acad. Sci., Paris* 260: 5901–5903.

Meusy, J. J. 1965b. Contribution de la microscopie electronique a l'etude de la physiologie des glandes androgenes d'*Orchestia gammarella* P. (Crustace Amphipode) et de *Carcinus maenas* L. (Crustace Decapode). *Zool. Jb. Abt. allg. Zool. Physiol. Tiere Dtsch.* 71: 608–623.

Miller, D. C. 1961. The feeding mechanism of fiddler crabs, with ecological considerations of feeding adaptations. *Zoologica* 46: 89–100.

Milne-Edwards, A. 1869. Note sur quelques nouvelles especes du genre *Sesarma. Nouv. Arch. Mus. Hist. Nat. Paris* 4: 25–31.

Milne-Edwards, H. 1834. *Histoire naturelle des Crustaces.* Vol. 1. Paris.

Milne-Edwards, H. 1839. Recherches sur la mechanism de la respiration chez les crustaces. *Ann. de Science Naturelle Zool. & Biol. Animal* 11: 129–142.

Moffett, S. B. 1975. Motor patterns and structural interactions of basi-ischiopodite levator muscles in routine limb elevation and production of autotomy in the land crab, *Cardisoma guanhumi. J. Comp. Physiol.* 96: 285–305.

Monod, T. 1956. Hippidea et Brachyura ouest-africains. *Mem. Inst. franc. Afr. noire* 45: 1–674.

Morgan, D. G., J. W. Goy, and J. D. Costlow, Jr. 1983. Multiple ovipositions from single mating in the mud crab, *Rhithropanopeus harrisii. J. Crustacean Biol.* 3: 543–547.

Morgan, T. H. 1923a. The development of asymmetry in the fiddler crab. *Amer. Nat.* 57: 269–273.

Morgan, T. H. 1923b. Further evidence of variation in the width of the abdomen in immature fiddler crabs. *Amer. Nat.* 57: 274–283.

Morris, S., and C. R. Bridges. 1985. An investigation of haemocyanin oxygen affinity in the semi-terrestrial crab *Ocypode saratan* Forsk. *J. Exp. Biol.* 117: 119–132.

Morris, S., and C. R. Bridges. 1986. Novel non-lactate cofactors of haemocyanin oxygen affinity of crustaceans. In *Invertebrate oxygen carriers.* Edited by B. Linzen. Heidelberg: Springer-Verlag.

Morris, S., C. R. Bridges, and M. K. Grieshaber. 1985. A new role for uric acid: Modulator of haemocyanin oxygen affinity in crustaceans. *J. Exp. Zool.* 234: 151–155.

Morris, S., and A. C. Taylor. 1983a. Diurnal and seasonal variation in physiochemical conditions within intertidal rockpools. *Estuar. Coast. and Shelf Sci.* 17: 339–355.

Morris, S., and A. C. Taylor. 1983b. Heart rate response of the intertidal prawn *Palaemon elegans* to simulated and in situ environmental changes. *Mar. Ecol. Prog. Ser.* 20: 127–136.

Morris, S., A. C. Taylor, C. R. Bridges, and M. K. Grieshaber. 1985. Respiratory properties of the haemolymph of the intertidal prawn *Palaemon elegans* (Rathke). *J. Exp. Zool.* 233: 175–186.

Munday, K. A., and P. C. Poat. 1970. Respiration and energy metabolism in Crustacea. In *Chemical zoology.* Edited by M. Florkin and B. T. Scheer, vol. 6, pp. 191–211. New York: Academic Press.

Murai, M. S., S. Goshima, and Y. Nakasone. 1982. Some behavioral characteristics related to food supply and soil texture of burrowing habitats observed in *Uca vocans vocans* and *Uca lactea perplexa. Mar. Biol.* 66: 191–197.

Murai, M., S. Goshima, and Y. Nakasone. 1983. Adaptive droving behavior observed in the fiddler crab *Uca vocans vocans. Mar. Biol.* 76: 159–164.

Muzzarelli, R. A. 1977. *Chitin.* Oxford: Pergamon.

Mykles, D. L. 1977. The ultrastructure of the posterior midgut caecum of *Pachygrapsus crassipes* (Decapoda: Brachyura) adapted to low salinity. *Tis. and Cell* 9: 681–691.

Nair, K. B. 1941. On the embryology of *Squilla. Proc. Indian Acad. Sci., Sect. B.* 14: 543–576.

Nair, K. B. 1949. The embryology of *Caridina laevis* Weller. *Proc. Indian Acad. Sci.* 29: 211–288.

Nakasone, Y. 1982. Ecology of the fiddler crab *Uca vocans vocans*, Decapoda, Ocypodidae. 1. Daily activities in warm and cold seasons. *Res. Pop. Ecol.* 24: 97–109.

Naylor, E., and B. Williams. 1984. Phase-responsiveness of the circatidal locomotor activity rhythm of *Hemigrapsus eduardsi* (Hilgendorf) to simulated high tide. *J. Mar. Biol. Ass. U.K.* 64: 81–90.

Neil, D. M. 1982. Compensatory eye movements. In *Biology of the Crustacea.* Vol. 4. Edited by H. L. Atwood and D. C. Sandeman. New York: Academic Press.

Niaussat, P., and J. P. Ehrhardt. 1968. Quelques aspects de l'ethologie du crabe terrestre *Gecarcinus planatus* Stimpson de l'atoll de Clipperton. Centre de Recherche du Service de Santee des Armees (Paris), special report no. 39, Bio.-Eco., pp. 1–23.

Niaussat, P., and J.-P. Ehrhardt. 1971. Quelques aspects de l'ethologie du crabe terrestre *Gecarcinus planatus* Stimpson de l'Atoll de Clipperton. *Vie et Milieu* suppl. 22: 167–191.

Nobili, G. 1900. Decapodi e stomatopodi Indo-Malesi. *Annali Mus. Civ. Stor. Nat. Giacomo Doria*, Ser. 2A, 20: 473–523.

Numanoi, H. 1940. Behaviour of blood calcium in the formation of gastrolith in some decapod crustaceans. *Jap. J. Zool.* 8: 357–363.

O'Connor, J. D., and L. I. Gilbert. 1968. Aspecs of lipid metabolism in crustaceans. *Amer. Zool.* 8: 529–539.

Odum, E. P. 1983. *Basic ecology*. Philadelphia: Saunders.

Olmstead, J. M. P., and J. P. Baumberger. 1923. Form and growth of grapsoid crabs. A comparison of the form of three species of grapsoid crabs and their growth at moulting. *J. Morph.* 38: 279–294.

O'Mahoney, P. 1977. Respiration and acid-base balance in brachyuran decapod crustaceans: The transition from water to land. Ph.D. thesis, State University of New York at Buffalo.

O'Mahoney-Damon, P. 1984. Heart rate of the land crab *Gecarcinus lateralis* during aquatic and aerial respiration. *Comp. Biochem. Physiol.* 79A: 621–624.

O'Mahoney, P. M., and R. J. Full. 1984. Respiration of crabs in air and water. *Comp. Biochem. Physiol.* 79A: 275–282.

Ono, Y. 1965. On the ecological distribution of ocypodid crabs in the estuary. *Mem. Fac. Sci. Kyushu Univ.*, Ser. E 4: 1–60.

Ortmann, A. 1896. Das System der Decapoden-Krebse. *Zool. Jb.* 9: 409–453.

Otsu, T. 1963. Bihormonal control of sexual cycle in the freshwater crab, *Potamon dehaani. Embryologia* 8: 1–20.

Pace, F., R. R. Harris, and V. Jaccarini. 1976. The embryonic development of the Mediterranean freshwater crab, *Potamon edulis* (= *P. fluviatile*) (Crustacea, Decapoda, Potamonidae). *J. Zool.* 180: 93–106.

Page, H. M., and S. W. Willason. 1982. Distribution patterns of terrestrial hermit crabs at Enewetak Atoll, Marshall Islands. *Pacific Sci.* 36: 107–117.

Palmer, J. D. 1971. Comparative studies of circadian locomotory rhythms in four species of terrestrial crabs. *Amer. Midl. Nat.* 85: 97–107.

Panning, A. 1939. The Chinese mitten crab. *Smithsonian Report* 1938: 361–375.

Panouse, J. B. 1943. Influence de l'ablation, de pedoncule oculaire sur la croissance de l'ovarie chez la crevette *Leander serratus. C. R. Acad. Sci., Paris* 217: 553–555.

Parvathy, K. 1971. Glycogen storage in relation to the molt cycle in two crustaceans *Emerita asiatica* and *Ligia exotica. Mar. Bio.* 10: 82–86.

Passano, L. M. 1960a. Molting and its control. In *The physiology of Crustacea.* Edited by T. H. Waterman, vol. 1, pp. 473–536. New York: Academic Press.

Passano, L. M. 1960b. Low temperature blockage of molting in *Uca pugnax. Biol. Bull.* 118: 129–136.

Passano, L. M., and S. Jyssum. 1963. The role of Y-organ in crab pro-ecdysis and limb regeneration. *Comp. Biochem. Physiol.* 9: 195–213.

Paul, A. J. 1984. Mating frequency and viability of stored sperm in the Tanner crab *Chionoecetes bairdi* (Decapoda, Majidae). *J. Crustacean Biol.* 4: 375–381.

Paulus, J. E., and H. Laufer. 1987. Vitellogenocytes in the hepatopancreas of *Carcinus maenas* and *Libinia emarginata* (Decapoda Brachyura). *Int. J. Reprod. Dev.* 11: 29–44.

Paulus, J. E. and H. Laufer. 1984. The regulation of yolk production in decapod Crustacea. *Adv. Inv. Reprod.* 3: 623.

Payen, G. 1969. Experiences de greffes de glandes androgenes sur la femelle pubere du crabe *Rhithropanopeus harrisii* (Gould) (Crustace, Decapode). *C. R. Acad. Sci.*, Ser. D 268: 393–396.

Payen, G. 1972. Etude ultrastructurale de la dégénérescence cellulaire dans la glande androgene du crabe *Ocypoda quadrata* (Fabricius). *Z. Zellforsch.* 129: 370–385.

Payen, G. 1974. Morphogenèse sexuelle de quelques Brachyoures (Cyclometopes) au cours du developement embryonnaire, larvaire et postlarvaire. *Bull. Mus. Nat. Hist. Nat.* 3rd series, 209: 201–262.

Payen, G. 1975. Effects masculinisants des glandes androgenes implantées chez la femelle pubere pedonculectomises de *Rhithropanopeus harrisii* (Gould) (Crustacé Décapode Brachyoure). *C. R. Acad. Sci.* Ser. D 280: 1111–1114.

Payen, G. G., and Amato, G. D. 1978. Données actuelles sur le contrôle de la spermatogenèse chez les crustacés Décapodes Reptantia. *Arch. Zool. Exp. Gen.* 119: 447–464.

Payen, G., J. D. Costlow, and M. Charniaux-Cotton. 1971. Etude comparative de l'ultrastructure des glandes androgènes de crabes normaux et pedonculectomisés pendant la vie larvaire ou apres la puberte chez les especes: *Rhithropanopeus harrisii* (Gould) et *Callinectes sapidus* (Rathbun). *Gen. Comp. Endocrinol.* 17: 526–542.

Pearse, A. S. 1929. Observations on certain littoral and terrestrial animals at Tortugas, Florida, with special reference to migrations from marine to terrestrial habitats. *Pap. Tortugas Lab.* 26: 205–223.

Pearson, J. 1908. *Cancer*. Liverpool Marine Biological Committee Memoirs, No. 8. London: Williams and Nordgate.

Pelligrino, C. 1984. The role of desiccation pressures and surface area/volume relationships on seasonal zonation and size distribution of four intertidal decapod crustacea from New Zealand: Implications for adaptations to land. *Crustaceana* 47: 251–268.

Pelligrino, C. R., and J. A. Stoff. 1983. *Darwin's universe: Origins and crises in the history of life.* pp. 106–119. New York: Van Nostrand Reinhold.

Pequeux, A., A. C. Vallota, and R. Gilles. Blood proteins as related to osmoregulation in Crustacea. *Comp. Biochem. Physiol.* 64A: 433–436.

Pesta, O. 1930. Zur Kenntnis der Land- und Susswasserkrabben von Sumatra und Java. *Arch. Hydrobiol. Suppl.* 8: 92–108.

Piiper, J., and P. Scheid. 1975. Gas transport efficiency of lungs, gills and skin: Theory and experimental data. *Resp. Physiol.* 23: 209–221.

Pike, R. B. 1947. *Galathea*. Liverpool Marine Biological Committee Memoirs No. 34. Edited by R. J. Daniel. Liverpool: University Press of Liverpool.

Pillai, C. K., and T. Subramoniam. 1984. Monsoon-dependent breeding in the field crab *Parathelphusa hydrodromus* (Herbst). *Hydrobiologia* 119: 7–14.

Pillai, K. K., and N. B. Nair. 1968. Observations on the reproductive cycles of some crabs from the south-west coast of India. *Mar. Biol. Ass. India* 10: 1–2.

Pillai, K. K., and N. B. Nair, 1971. The reproductive cycle of three decapod crustaceans from the south-west coast of India. *Curr. Sci.* 40: 161–162.

Pochon-Masson, J. 1962. Le chondriofusome des gametes males du Crustace decapods *Carcinus maenas*. *C. R. Acad. Sci., Paris* 254: 4076–4078.

Pochon-Masson, J. 1968a. L'ultrastructure des spermatozoides vesiculaires chez les Crustaces Decapodes avant et au cours de leur devagination experimentale. I. Brachyoures et Anomures. *Ann. Sci. nat. (Zool.)* 10: 1–98.

Pochon-Masson, J. 1968b. L'ultrastructure des spermatozoides vesiculares chez les Crustaces Decapodes avant et au cours de leur devagination experimentale. II. Macroures. Discussion et conclusions. *Ann. Sci. nat. (Zool.)* 10: 367–454.

Pochon-Masson, J. 1983. Arthropoda-Crustacea. In *Reproductive biology of invertebrates*. Edited by K. G. Adiyodi and R. G. Adiyodi, pp. 407–449. Chichester: John Wiley.

Poole, R. L. 1966. A description of laboratory reared zoeae of *Cancer magister* Dana, and megalopae taken under natural conditions (Decapoda, Brachyura). *Crustaceana* 11: 83–97.

Powell, R. R. 1974. The functional morphology of fore-guts of the thalassinid crustaceans, *Calianassa californiensis* and *Upogebia pugettensis*. *Publ. Zool.* (Univ. Of Calif., Berkeley)102: 1–41.

Powers, L. W. 1975. Fiddler crabs in a non-tidal environment. *Cont. Mar. Sci. Univ. Texas* 19: 67–78.

Powers, L. W. 1979. Agonistic behavior in female fiddler crabs, *Uca pugnax* and *U. pugilator*. *Amer. Zool.* 19: 913.

Powers, L. W., and D. E. Bliss. 1983. Terrestrial adaptations. In *Biology of Crustacea*. vol. 8, *Environmental adaptations*. Edited by F. J. Vernberg and W. B. Vernberg, pp. 271–334. New York: Academic Press.

Powers, L. W., and J. F. Cole. 1976. Temperature variation in fiddler crab micro-habitats. *J. Exp. Mar. Biol. Ecol.* 21: 141–157.

Pradeille-Rouquette, M. 1976. Influence de differents facteurs sur la croissance somatique de *Pachygrapsus marmoratus* (Fabricius) Crustace Decapode. *Cah. Biol. Mar.* 17: 77–91.

Prosser, C. L., ed. 1973. *Comparative animal physiology*. 3rd ed. Philadelphia: Saunders.

Provenzano, A. J., Jr. 1962. The larval development of the tropical land hermit *Coenobita clypeatus* (Herbst) in the laboratory. *Crustaceana* 4: 207–228.

Przibram, H. 1905. Die "Heterochelie" bei decapoden Crustaceen. *Arch. Ent. Mech. Org.* 19: 181–247.

Quinn, R. H. 1980. Mechanisms for obtaining water for flotation feeding in the soldier crab, *Mictyris longicarpus* Latreille 1806 (Decapoda, Mictyridae). *J. Exp. Mar. Biol. Ecol.* 43: 49–60.

Rabalais, N. N., and J. N. Cameron. 1983. Abbreviated development of *Uca subcylindrica* (Stimpson, 1859) (Crustacea, Decapoda, Ocypodidae) reared in the laboratory. *J. Crustacean Biol.* 3: 519–541.

Rabalais, N. N., and J. N. Cameron, 1985a. Physiological and morphological adaptations of adult *Uca subcylindrica* to semi-arid environments. *Biol. Bull.* 168: 135–146.

Rabalais, N. N., and J. N. Cameron 1985b. The effects of factors important in semi-arid environments on the early development of *Uca subcylindrica. Biol. Bull.* 168: 147–160.

Rabalais, N. N., and R. H. Gore. 1985. Abbreviated development in decapods. In *Crustacean issues 2: Larval growth.* Edited by A. M. Wenner, pp. 67–126. Rotterdam: Balkema.

Rahn, H. 1967. Gas transport from the external environment to the cell. In *Development of the lung.* A CIBA Foundation Symposium. Edited by A. U. S. de Reuck and R. Porter. London: Churchill.

Ramamurthi, R. 1977. Urine production in the shorecrab, *Hemigrapsus nudus* (Dana). *Indian J. Exp. Biol.* 15: 480–482.

Randall, D. J., and C. M. Wood. 1981. Carbon dioxide excretion in the land crab (*Cardisoma carnifex*). *J. Exp. Zool.* 281: 37–44.

Rangneker, P. V., P. B. Sabnis, and H. B. Nirmal. 1961. The occurrence of a hypoglycemic factor in the eyestalks of a freshwater crab, *Paratelphusa jaquemontii* (Rathbun). *J. Anim. Morph. Physiol.* 8: 259–357.

Rao, K. R. 1966. Studies on the influence of environmental factors on growth in the crab *Ocypode macrocera* H. Milne Edwards. *Crustaceana* 11: 257–276.

Rao, K. R. 1968. The pericardial sacs of *Ocypode* in relation to the conservation of water, molting and behavior. *Amer. Zool.* 8: 561–567.

Rao, K., and M. Fingerman. 1983. Regulation of release and mode of action of crustacean chromatophorotropins. *Amer. Zool.* 23: 517–527.

Rathbun, M. J. 1900. The decapod crustaceans of West Africa. *Proc. U.S. Nat. Mus.* 22: 271–316.

Rathbun, M. J. 1914. New species of crabs of the families Grapsidae and Ocypodidae. In Sci. Res. Philippine Cruise "Albatrosse," no. 31. *Proc. U.S. Nat. Mus.* 47: 69–85.

Rathbun, M. J. 1918. The grapsoid crabs of America. *Bull. U.S. Nat. Mus.* 97: 1–461.

Rebach, S. 1983. Orientation and migration in Crustacea. In *Studies in adaptation: The behavior of higher Crustacea.* Edited by S. Rebach and D. Dunham, pp. 217–264. New York: Wiley Interscience.

Redmond, J. R. 1955. The respiratory function of hemocyanin in crustaceans. *J. Cell. Comp. Physiol.* 46: 209–247.

Redmond, J. R. 1962. The oxygen-haemocyanin relationship in the land crab *Cardisoma guanhumi. Biol. Bull.* 122: 252–262.

Redmond, J. R. 1968. Transport of oxygen by the blood of the land crab *Gecarcinus lateralis. Amer. Zool.* 8: 471–479.

Reese, E. S. 1968. Shell use: An adaptation for emigration from the sea by the coconut crab. *Science* 161: 385–386.

Reese, E. S. 1969. Behavioral adaptations of intertidal hermit crabs. *Amer. Zool.* 9: 343–355.

Reeves, R. B. 1972. An imidazole alphastat hypothesis for vertebrate acid-base regulation: Tissue carbon dioxide content and body temperature in bull frogs. *Resp. Physiol.* 14: 219–236.

Reeves, R. B. 1977. The interaction of body temperature and acid-base balance in ectothermic vertebrates. *Ann. Rev. Physiol.* 39: 559–586.

Reyne, A. 1939. On the food habits of the coconut crab (*Birgus latro* L.), with notes on its distribution. *Arch. Neerland. de Zool.* 3: 283–320.

Rice, A. L. 1981. Crab zoeae and brachyuran classification: A reappraisal. *Bull. Brit. Mus. Nat. Hist. (Zool.)* 40: 287–296.

Riegel, J. A. 1972. *Comparative physiology of renal excretion.* Edinburgh: Oliver and Boyd.

Rittschof, D., L. Barlow, and A. R. Schmidt. 1985. Laboratory studies of olfaction and taste in land hermit crabs *Coenobita rugosus. Amer. Zool.* 25: 113A (Abstr.).

Robertson, J. D. 1937. Some features of the calcium metabolism of the shore crab (*Carcinus maenas* Pennant). *Proc. Roy. Soc. B* 124: 162–182.

Robertson, J. D. 1953. Further studies on ionic regulation in marine invertebrates. *J. Exp. Biol.* 30: 277–291.

Robertson, J. R., K. Bancroft, G. Vermeer, and K. Plaisier. 1980. Experimental studies on the foraging behavior of the sand fiddler crab *Uca pugilator* (Bosc). *J. Exp. Mar. Biol. Ecol.* 44: 67–83.

Robertson, J. R., J. A. Fudge, and G. K. Vermeer. 1981. Chemical and live feeding stimulants of the sand fiddler crab, *Uca pugilator* (Bosc). *J. Exp. Mar. Biol. Ecol.* 53: 47–64.

Robertson, J. R., and S. Y. Newell. 1982. Experimental studies of particle ingestion by the sand fiddler crab, *Uca pugilator. J. Exp. Mar. Biol. Ecol.* 59: 1–21.

Robertson, J. R., and W. J. Pfeiffer. 1982. Deposit feeding by the ghost crab *Ocypode quadrata* (Fabricius). *J. Exp. Mar. Biol. Ecol.* 56: 165–177.

Robinson, M. H., L. G. Abele, and B. Robinson. 1970. Attack autotomy: A defense against predators. *Science* 169: 300–301.

Rodriguez, G. 1982. Les crabes d'eau douce d'Amerique. Familia des pseudothelphusidae. *Office Recher. Scient. Tech. Autre-mer, Faun Trop.* 22: 1–223.

Rodriguez, G., and A. Smalley. 1972. Las cangrejos de aqua dulce de Mexico de la familia Pseudothelphusidae (Crustacea, Brachyura). *Anal. Inst. Biol. Univ. Nat. Auton. Mexico. Ser. Cien. Mar. Lim.* 1: 69–112.

Roe, J. 1980. Zonation and movements of the ghost crab, *Ocypode quadrata.* M.S. thesis, University of Florida, Gainesville.

Roer, R. D. 1980. Mechanisms of resorption and deposition of calcium in the carapace of the crab *Carcinus maenas. J. Exp. Biol.* 88: 205–218.

Rouquette, M. 1970. Etude du tissue ovarien chez le crabe *Pachygrapsus marmoratus* (Fabricius). Premiers resultats concernant les roles de la temperature et des pedoncules oculaires. *Bull. Soc. Zool. Fr.* 95: 233–240.

Ryan, E. P. 1966. Pheromone: Evidence in a decapod crustacean. *Science* 151: 340–341.

Sabourin, T. D., and D. G. Saintsing. 1980. Transport ATPases in the osmoregulating hermit crab, *Clibanarius vittatus. Physiologist* 23: 175.

Saigusa, M. 1978. Ecological distribution of three species of the genus *Sesarma* in winter season. *Zool. Mag.* 87: 142–150.

Saigusa, M. 1981. Adaptive significance of a semilunar rhythm in the terrestrial crab *Sesarma. Biol. Bull.* 160: 311–321.

Salmon, M. 1965. Waving display and sound production in the courtship behavior of *Uca pugilator*, with comparisons to *U. minax* and *U. pugnax*. *Zoologica* 50: 123–150.

Salmon, M. 1983. Courtship, mating systems, and sexual selection in decapods. In *Studies in adaptation: The behavior of higher crustacea.* Edited by S. Rebach and D. Dunham. New York: Wiley.

Salmon, M. 1984. The courtship, aggressive and mating systems of a primitive fiddler crab, *Uca vocans*, Ocypodidae. *Trans. Zool. Soc. Lond.* 37: 1–50.

Salmon, M., and S. P. Atsaides. 1968. Visual and acoustical signalling during courtship by fiddler crabs (genus *Uca*). *Amer. Zool.* 8: 623–639.

Salmon, M., and K. W. Horch. 1972. Acoustic signalling and detection by semiterrestrial crabs of the family Ocypodidae. In *Behavior of marine animals.* Edited by H. E. Winn and B. L. Olla, vol. 1, pp. 60–96. New York: Plenum Press.

Salmon, M., and K. W. Horch. 1973. Vibration reception in the fiddler crab *Uca minax. Comp. Biochem. Physiol.* 44: 527–541.

Salmon, M., K. W. Horch, and G. W. Hyatt. 1977. Barth's myochordotonal organ as a receptor for auditory and vibrational stimuli in fiddler crabs (*Uca pugilator* and *U. minax*). *Mar. Behav. Physiol.* 4: 187–194.

Salmon, M., and G. W. Hyatt. 1979. The development of acoustic display in the fiddler crab *Uca pugilator*, and its hybrids with *U. panacea. Mar. Behav. Physiol.* 6: 197–209.

Salmon, M., and G. W. Hyatt. 1983a. Spatial and temporal aspects of reproduction in North Carolina fiddler crabs (*Uca pugilator* Bosc). *J. Exp. Mar. Biol. Ecol.* 80: 21–43.

Salmon, M., and G. W. Hyatt. 1983b. Communication. In *The Biology of Crustacea*, vol. 7. Edited by F. J. Vernberg and W. B. Vernberg, pp. 1–40. New York: Academic Press.

Salmon, M., G. W. Hyatt, K. McCarthy, and J. D. Costlow, Jr. 1978. Display specificity and reproductive isolation in the fiddler crabs, *Uca panacea* and *U. pugilator. Z. Tierpsychol.* 48: 251–276.

Samuel, K. A. 1984. Reproductive biology of the land crab, *Gecarcinucus steniops*, with special emphasis on the female. M. Phil. diss., Calicut University, India.

Sandeman, D. C. 1967. The vascular circulation of the brain, optic lobes and thoracic ganglion of the crab *Carcinus. Proc. R. Soc. London*, Ser. B 168: 82–90.

Sanders, B. 1983. Insulin-like peptides in the lobster *Homarus americanus*. III. No glucostatic role. *Gen. Comp. Endocrinol.* 50: 378–382.

Sandon, H. 1937. Differential growth in the crab *Ocypoda. Proc. Zool. Soc. Lond.* 107: 397–414.

Santhamma, K. R. 1985. Some aspects of female reproductive physiology of the freshwater crab, *Gecarcinucus steniops*. M. Phil. diss., Calicut University, India.

Santos, M. C. F., M. Engelftein, and M. A. Gabrielli. 1985. Relationships concerning respiratory devices in crabs from different habitats. *Comp. Biochem. Physiol.* 81A: 567–570.

Schaefer, W. 1954. Form und Funktion der Brachyuren-Schere. *Abh. senckenb. naturf. Ges.* 489: 1–65.

Scheidegger, G. 1976. Stadien der Embryonalentwicklung von *Eupagurus prideauxi* Leach (Crustacea, Decapoda, Anomura) unter besonder Berucksichtigung der Darmentwicklung und der am Dotterabbau beteiligten Zelltypen. *Zool. Jahrb., Abt. Anat. Ontog. Tiere* 95: 297–353.

Schembri, P. J. 1982. Feeding behavior of 15 species of hermit crabs (Crustacea, Decapoda, Anomura) from the Otago Region, Southeastern New Zealand. *J. Nat. Hist.* 16: 859–878.

Schenkel, E. 1902. Beitrag zur Kenntnis der Decapoden Fauna von Celebes. *Verh. Nat. Ges. Basel* 13: 485–584.

Schmidt-Nielsen, K. 1972. Locomotion: Energy cost of swimming, flying and running. *Science* 177: 222–228.

Schmitt, B. C., and B. W. Ache. 1979. Olfaction: Responses of a decapod crustacean are enhanced by flicking. *Science* 205: 204–206.

Schoene, H. 1961. Complex behavior. In *The physiology of Crustacea.* Edited by T. H. Waterman, vol. 2, pp. 465–520. New York: Academic Press.

Schoene, H. 1971. Gravity receptors and gravity orientation in Crustacea. In *Gravity and the organism.* Edited by S. A. Gordon and J. J. Cohen. Chicago: University of Chicago Press.

Schoene, H. 1980. *Orientierung im Raum.* Stuttgart: Wissenschaftliche Verlagsgesellschaft.

Schoene, H. 1984. *Spatial orientation.* Princeton, N.J.: Princeton University Press.

Schoene, H., and H. Schoene. 1961. Eyestalk movements induced by polarized light in the ghost crab, *Ocypode quadrata. Science* 134: 675–676.

Schoffeniels, E., and R. Gilles. 1970. Nitrogenous constituents and nitrogen metabolism in arthropods. In *Chemical zoology.* Edited by M. Florkin and B. T. Scheer, vol. 7, pp. 393–420. New York: Academic Press.

Scholander, P. F., W. Flagg, V. Walters, and L. Irving. 1953. Climatic adaptation in arctic and tropical poikilotherms. *Physiol. Zool.* 26: 67–92.

Schwartzkopff, J. 1935. Die Grossenabhangigkeit der Herzfrequenz von Krebsen im Vergleich u. anderen Tiergruppen. *Experentia* 11: 323–324.

Schwoch, G. 1972. Some studies on biosynthesis and function of trehalose in the crayfish *Orconectes limosus* Rafinesque. *Comp. Biochem. Physiol.* 43B: 905–917.

Seeherman, H. J., C. R. Taylor, G. M. O. Maloiy, and R. B. Armstrong. 1981. Design of the mammalian respiratory system. II. Measuring maximum aerobic capacity. *Respir. Physiol.* 44: 11–23.

Seifert, P. 1982. Studies on the pheromone of the shore crab, *Carcinus maenas*, with special regard to ecdysone excretion. *Ophelia* 21: 147–158.

Seiple, W., and M. Salmon. 1982. Comparative social behavior of two grapsid crabs, *Sesarma reticulatum* (Say.) and *S. cinereum* (Bosc). *J. Exp. Mar. Biol. Ecol.* 62: 1–24.

Semper, C. 1878. Ueber die Lunge von *Birgus latro. Z. Wiss. Zool.* 30: 282–287.

Sérene, R. 1968. Note preliminaire sur de nouvelles espéces de *Sesarma. Bull. Mus. natn. Hist. Nat. Paris*, Ser. 2, 39: 1084–1095.

Sérene, R. 1973. Notes on Indo-West Pacific species of *Macrophthalmus* (Crustacea, Brachyura). *Zool. Meded. Leiden* 46: 99–116.

Sérene, R., and C. L. Soh. 1970. New Indo-Pacific genera allied to *Sesarma* Say 1817 (Brachyura, Decapoda, Crustacea). *Treubia* 27: 387–416.

Sérene, R. S., and A. F. Umali. 1972. The family Raninidae and other new and rare species of brachyuran decapods from the Philippines and adjacent regions. *Philipp. J. Sci.* 99: 21–105.

Seurat, L. G. 1904. Observation biologiques sur les cenobites (*Coenobita perlata* Edwards). *Bull. Mus. Hist. Nat. Paris* 10: 238–242.

Seymour, R. S., R. G. Spragg, and M. T. Hartman. 1981. Distribution of ventilation and perfusion in the sea snake, *Pelamis platurus. J. Comp. Physiol.* 145: 109–115.

Shah, G. M., and C. F. Herreid II. 1978. Heart rate of the land crab, *Cardisoma guanhumi* (Latreille), during aquatic and aerial respiration. *Comp. Biochem. Physiol.* 60A: 335–341.

Shaw, J. 1959. Salt and water balance in the East African freshwater crab, *Potamon niloticus* (M. Edw.). *J. Exp. Biol.* 36: 157–176.

Sherfy, J. 1984. The effect of predator avoidance by ghost crabs on the foraging success of black hawks. *Amer. Zool.* 24: 50A (Abstr.).

Shiino, S. M. 1950. Studies on the embryonic development of *Palinurus japonicus* (von Siebold). *J. Fac. Fish. Prefect. Univ. Mie, Otanimachi* 1: 1–168.

Shuchman, E., and M. R. Warburg. 1978. Dispersal, population structure and burrow shape of *Ocypode cursor. Mar. Biol.* 49: 255–263.

Silas, E. G., and G. Sankarankutty. 1967. Field investigations on the shore crabs of the Gulf of Mannar and Palk Bay, with special reference to the ecology and behaviour of the pellet crab *Scopimera proxima* Kemp. *Mar. Biol. India, Symposium Ser.* 2: 1008–1025.

Skinner, D. M. 1962. The structure and metabolism of a crustacean integumentary tissue during a molt cycle. *Biol. Bull.* 123: 635–647.

Skinner, D. M. 1966a. Macromolecular changes associated with the growth of crustacean tissues. *Amer. Zool.* 6: 235–242.

Skinner, D. M. 1966b. Breakdown and reformation of somatic muscle during the molt cycle of the land crab, *Gecarcinus lateralis. J. Exp. Zool.* 163: 115–124.

Skinner, D. M. 1985. Molting and regeneration. In *The biology of Crustacea.* Edited by D. E. Bliss and L. H. Mantel, vol. 9, pp. 44–146. New York: Academic Press.

Skinner, D. M., and D. E. Graham. 1972. Loss of limbs as a stimulus to ecdysis in Brachyura (true crabs). *Biol. Bull.* 143: 222–233.

Skinner, D. M., D. J. Marsh, and J. S. Cook. 1965. Physiological salt solution for the land crab *Gecarcinus lateralis. Biol. Bull.* 129: 355–365.

Smatresk, N. J., and J. N. Cameron. 1981. Post-exercise acid-base balance and ventilatory control in *Birgus latro*, the coconut crab. *J. Exp. Zool.* 218: 75–82.

Smatresk, N. J., A. J. Preslar, and J. N. Cameron. 1979. Post-exercise acid-base disturbance in *Gecarcinus lateralis*, a terrestrial crab. *J. Exp. Zool.* 210: 205–210.

Smith, R. I. 1976. Apparent water-permeability variation and water exchange in crustaceans and annelids. In *Perspectives in experimental biology* Edited by

P. Spencer Davies, vol. 1, pp. 17–24. Oxford and New York: Pergamon Press.

Smith, R. I. 1978. The midgut caeca and the limits of the hindgut of Brachyura: A clarification. *Crustaceana* 35: 195–205.

Smith, W. K., and P. C. Miller. 1973. The thermal ecology of two South Florida fiddler crabs: *Uca rapax* Smith and *U. pugilator* Bosc. *Physiol. Zool.* 46: 186–207.

Soh C. L. 1969. Abbreviated development of a non-marine crab, *Sesarma (Geosesarma) perracae* (Brachyura, Grapsidae) from Singapore. *J. Zool* 158: 357–370.

Soumoff, C., and J. D. O'Connor. 1982. Repression of Y-organ secretory activity by molt inhibiting hormone in the crab, *Pachygrapsus crassipes. Gen. Comp. Endocrinol.* 48: 432–439.

Soumoff, C., and D. M. Skinner. 1983. Ecdysteroid titers during the molt cycle of the blue crab resemble those of other crustaceans. *Biol. Bull.* 165: 321–329.

Spaargaren, D. H. 1975. Notes on the osmotic and ionic regulation of some brachyuran crabs from Curacao. *Neth. J. Sea Res.* 9: 273–286.

Spaargaren, D. H. 1977. On the water and salt economy of some decapod crustaceans from the Gulf of Aqaba (Red Sea). *Neth. J. Sea Res.* 11: 99–106.

Sparkes, S., and P. Greenaway. 1984. The haemolymph as a storage site for cuticular ions during premoult in the freshwater/land crab *Holthuisana transversa. J. Exp. Biol.* 113: 43–54.

Speck, U., and K. Urich. 1969. Der Abbau korpereigener Substanzen in dem Flusskrebs *Orconectes limosus* wärhend des Hungerns. *Z. Vergl. Physiol.* 63: 410–414.

Spencer, A. M., A. H. Fielding, and F. I. Kamemoto. 1979. The relationship between gill Na K-ATPase activity and osmoregulatory capacity in various crabs. *Physiol. Zool.* 52: 1–10.

Sreenarayanan, K. V. 1980. Biochemistry and physiology of semen in some marine decapod crustaceans. Pre-Ph.D. report, Calicut University, India.

Stainsby, W. N., and J. K. Barclay. 1970. O_2 deficit, steady state O_2 uptake and O_2 uptake for recovery. *Med. Sci. Sports* 2: 177–181.

Steinacker, A. 1978. The anatomy of the decapod auxiliary heart. *Biol. Bull.* 154: 497–507.

Steinacker, A. 1979. Neural and neurosecretory control of the decapod crustacean auxiliary heart. *Amer. Zool.* 19: 67–75.

Stevenson, J. R. 1985. Dynamics of the integument. In *The biology of Crustacea.* Edited by D. Bliss and L. H. Mantel, vol. 9, pp. 2–42. New York: Academic Press.

Stewart, J. E., J. W. Cornick, D. M. Foley, M. F. Li, and C. M. Bishop. 1967. Muscle weight relationship to serum proteins, hemocytes, and hepatopancreas in the lobster, *Homarus americanus. J. Fish. Res. Board Can.* 24: 2339–2354.

Storch, V., H. H. Janssen, and E. Cases. 1982. The effects of starvation on the hepatopancreas of the coconut crab, *Birgus latro* (L.) (Crustacea, Decapoda). *Zool. Anz. Jena* 208: 115–123.

Storch, V., and U. Welsch. 1975. Ueber Bau und Funktion der Kiemen and Lungen von *Ocypode ceratophthalma* (Decapoda: Crustacea). *Mar. Biol.* 29: 363–371.

Storch, V., and U. Welsch. 1984. Electron microscopic observations on the lungs of the coconut crab, *Birgus latro* (L.) (Crustacea, Decapoda). *Zool. Anz. Jena* 212: 73–84.

Stutz, A. M. 1978. Tidal and diurnal activity rhythms in the striped shore crab *Pachygrapsus crassipes*. *J. Interdiscipl. Cycle Res.* 9: 41–48.

Subramanian, A. 1984. Burrowing behavior and ecology of the crab-eating Indian snake eel *Pisoodonophis boro*. *Environ. Biol. Fishes* 10: 195–202.

Subramanyam, M. V. V., and R. V. Krishnamoorthy. 1983. Biochemical correlates of the mechanisms of salt uptake and excretion in the gills of freshwater field crabs on adaptation to higher salinity. *Proc. Indian Acad. Sci. Anim. Sci.* 92: 277–284.

Subramoniam, T. 1979. Some aspects of reproductive ecology of a mole crab *Emerita asiatica* Milne-Edwards. *J. Exp. Mar. Biol. Ecol.* 36: 259–268.

Subramoniam, T. 1984. Spermatophore formation in 2 intertidal crabs, *Albunea symnista* and *Emerita asiatica* (Decapoda, Anomura). *Biol. Bull.* 166: 78–95.

Sukumaran, M. 1985. Certain aspects of reproductive biology of the ghost crab, *Ocypode platytarsis* with special reference to the male. Ph.D. thesis, Calicut University, India.

Sundara, R. G., G. Santhanakrishnan, and S. Shyamalanath. 1973. Nature of the sex-attractant pheromone in the crab *Paratelphusa hydrodromus* (Crustacea). *Curr. Sci.* 42: 467–468.

Taissoun, E. 1974a. El cangrejo de tierra *Cardisoma guanhumi* (Latreille) en Venezuela: Distribucion, ecologia, biologia y evaluacion poblacional. *Bol. del Cent. de Invest. Biol.* (Univ. del Zulia) 10: 8–50.

Taissoun, E. 1974b. El cangrejo de tierra *Cardisoma guanhumi* (Latreille) en Venezuela. I. Metodos de captura, comercializacion e industrializacion. II. Medidas y recomendaciones para la conservacion de la especie. *Bol. del Cent. de Invest. Biol.* (Univ. del Zulia) 10: 1–36.

Takeda, M., and T. Yamaguchi. 1973. Occurrences of abnormal males in a fiddler crab, *Uca marionis* (Desmarest), with notes on asymmetry of chelipeds. *Bull. Jap. Soc. Syst. Zool.* 9: 13–20.

Taylor, A. C. 1976. The respiratory responses of *Carcinus maenas* to declining oxygen tension. *J. Exp. Biol.* 65: 309–322.

Taylor, A. C., and P. S. Davies. 1981. Respiration in the land crab, *Gecarcinus lateralis*. *J. Exp. Biol.* 93: 197–208.

Taylor, A. C., and P. S. Davies. 1982. Aquatic respiration in the land crab *Gecarcinus lateralis* (Fréminville). *Comp. Biochem. Physiol.* 72A: 683–688.

Taylor, C. R. 1973. Energy cost of animal locomotion. In *Comparative physiology*. Edited by L. Bolis, K. Schmidt-Nielsen, and S. H. P. Maddrell, pp. 23–41. Amsterdam: North Holland.

Taylor, C. R. 1977. The energetics of terrestrial locomotion and body size in vertebrates. In *Scale effects in animal locomotion*. Edited by T. J. Pedley, pp. 127–141. New York: Academic Press.

Taylor, C. R. 1985. Force development during sustained locomotion: A determinant of gait, speed and metabolic power. *J. Exp. Biol.* 115: 253–262.

Taylor, C. R., N. C. Heglund, and G. M. O. Maloiy. 1982. Energetics and mechanics of terrestrial locomotion. I. Metabolic energy consumption as a function of speed and body size in birds and mammals. *J. Exp. Biol.* 97: 1–21.

Taylor, C. R., N. C. Heglund, T. A. McMahon, and T. R. Looney. 1980. Energetic cost of generating muscular force during running. *J. Exp. Biol.* 86: 9–18.

Taylor, C. R., K. Schmidt-Nielsen, and J. L. Raab. 1970. Scaling of energetic cost of running to body size in mammals. *Amer. J. Physiol.* 219: 1104–1107.

Taylor, E. W. 1982. Control and co-ordination of ventilation and circulation in crustaceans: Responses to hypoxia and exercise. *J. Exp. Biol.* 100: 289–319.

Taylor, E. W., and P. J. Butler. 1973. The behaviour and physiological responses of the shore crab *Carcinus maenas* (L.) during changes in environmental oxygen tension. *Neth. J. Sea Res.* 7: 496–505.

Taylor, E. W., and P. J. Butler. 1978. Aquatic and aerial respiration in the shore crab *Carcinus maenas* (L.), acclimated to 15° C. *J. Comp. Physiol.* 127: 315–323.

Taylor, E. W., P. J. Butler, and A. Al-Wassia. 1977. Some responses of the shore crab, *Carcinus maenas* (L.) to progressive hypoxia at different acclimation temperatures and salinities. *J. Comp. Physiol.* 112: 391–402.

Taylor, E. W., P. J. Butler, and P. J. Sherlock. 1973. The respiratory and cardiovascular changes associated with the emersion response of *Carcinus maenas* (L.) during environmental hypoxia, at three different temperatures. *J. Comp. Physiol.* 86: 95–116.

Taylor, E. W., and M. G. Wheatly. 1979. The behavior and respiratory physiology of the shore crab, *Carcinus maenas* (L.) at moderately high temperatures. *J. Comp. Physiol.* 130: 309–316.

Taylor, H. H., and P. Greenaway. 1979. The structure of the gills and lungs of the arid-zone crab, *Holthuisana (Austrothelphusa) transversa* (Martens) (Sundathelphusidae: Brachyura) including observations on arterial vessels within the gills. *J. Zool., Lond.* 189: 359–384.

Taylor, H. H., and P. Greenaway, 1984. The role of the gills and branchiostegites in gas exchange in a bimodally breathing crab, *Holthuisana transversa*: Evidence for a facultative change in the distribution of the respiratory circulation. *J. Exp. Biol.* 111: 103–122.

Taylor, H. H., and E. W. Taylor. 1986. Observation of valve-like structures and evidence for rectification of flow within the gill lamellae of the crab *Carcinus maenas* (Crustacea, Decapoda). *Zoomorphologie* 106: 1–11.

Taylor, P. 1982. Environmental resistance and the ecology of coexisting hermit crabs: Thermal tolerance. *J. Exp. Mar. Biol. Ecol.* 57: 229–236.

Teal, J. M. 1958. Distribution of fiddler crabs in Georgia salt marshes. *Ecology* 39: 185–193.

Teal, J. M., and F. G. Carey. 1967. The metabolism of marsh crabs under conditions of reduced oxygen pressure. *Physiol. Zool.* 40: 83–91.

Teissier, G. 1960. Relative growth. In *The physiology of Crustacea.* Edited by T. H. Waterman, pp. 537–560. New York: Academic Press.

Telford, M. 1968. The identification and measurement of sugars in the blood of three species of Atlantic crabs. *Bio. Bull.* 135: 574–584.

Terwilliger, N. B., and R. C. Terwilliger. 1982. Changes in the subunit structure of *Cancer magister* hemocyanin during larval development. *J. Exp. Zool.* 221: 181–192.

Thampy, D. M., and P. A. John. 1970. On the androgenic gland of the ghost crab *Ocypode platytarsis* M. Edwards (Crustacea: Brachyura). *Acta Zool.* 51: 203–210.

Thiennemann, A. 1935. Die Tierwelt der tropischen Pflanzengewaesser. *Arch. Hydrobiol. Suppl.* 13: 1–91.

Thompson, L. C. 1970. Osmoregulation of the freshwater crabs *Metapaulius depressus* (Grapsidae) and *Pseudothelphusa jouyi* (Pseudothelphusidae). Ph.D. thesis, University of California at Berkeley.

Thurman, C. L., II. 1984. Ecological notes on fiddler crabs of South Texas, with special reference to *Uca subcylindrica*. *J. Crustacean Biol.* 4: 665–681.

Tinbergen, N. 1951. *The study of instinct.* Oxford: Oxford University Press.

Towle, D. W. 1981. Transport-related ATPases as probes of tissue function in three terrestrial crabs of Palau. *J. Exp. Zool.* 218: 89–95.

Travis, D. F. 1954. The molting cycle of the spiny lobster *Panulirus argus* Latreille. 1. Molting and growth in laboratory-maintained individuals. *Biol. Bull.* 107: 433–450.

Trott, T. J., and J. R. Robertson. 1982. Chemoreception and feeding stimulants of the ghost crab *Ocypode quadrata* (Fabricius). *Amer. Zool.* 22: 853.

Trott, T. J., and J. R. Robertson. 1984. Chemical stimulants of cheliped flexion behavior by western Atlantic ghost crab *Ocypode quadrata* (Fabricius). *J. Exp. Mar. Biol. Ecol.* 78: 237–252.

Truchot, J.-P. 1975a. Blood acid-base changes during experimental emersion and reimmersion of the intertidal crab *Carcinus maenas* (L.). *Respir. Physiol.* 23: 351–360.

Truchot, J.-P. 1975b. Factors controlling the "in vitro" and "in vivo" oxygen affinity of the haemocyanin of the crab *Carcinus maenas* (L.). *Respir. Physiol.* 24: 173–189.

Truchot, J.-P. 1976. Carbon dioxide combining properties of the blood of the shore crab *Carcinus maenas* (L.): CO_2 dissociation curves and the Haldane effect. *J. Comp. Physiol.* 112: 283–293.

Truchot, J.-P. 1980. Lactate increases the oxygen affinity of crab haemocyanin. *J. Exp. Zool.* 214: 205–208.

Truchot, J.-P., and A. Duhamel-Jouve. 1980. Oxygen and carbon dioxide in the marine intertidal environment: Diurnal and tidal changes in rockpools. *Resp. Physiol.* 39: 241–254.

Tucker, V. A. 1970. Energetic cost of locomotion in animals. *Comp. Biochem. Physiol.* 34: 841–846.

Turkay, M. 1970. Die Gecarcinidae Amerikas, mit einem Anhang über *Ucides* Rathbun (Crustacea: Decapoda). *Senckenbergiana Biol.* 51: 333–354.

Turkay, M. 1973a. Bermerkungen zu einigen Landkrabben (Crustacea, Decapoda). *Bull. Mus. Natl. Hist. Nat. Paris* 142: 969–980.

Turkay, M. 1973b. Die Gecarcinidae Afrikas. *Senckenbergiana Biol.* 54: 81–103.

Turkay, M. 1974. Die Gecarcinidae Asiens und Ozeaniens (Crustacea: Decapoda). *Senckenbergiana Biol.* 55: 223–259.

Turkay, M. 1983. The systematic position of an Australian mangrove crab *Heloecius cordiformis* (Crustacea: Decapoda: Brachyura). *Mem. Aust. Mus.* 18: 107–111.

Turkay, M., and K. Sakai. 1976. Die Gecarcinidae von Japan (Crustacea, Decapoda). *Researches Crust.* 7: 11–22.

Tweedie, M. F. W. 1936. On the crabs of the family Grapsidae in the collection of the Raffles Museum. *Bull. Raffles Mus.* 12: 44–71.

Tweedie, M. F. W. 1940. New and interesting Malaysian species of *Sesarma* and *Utica* (Crustacea, Brachyura). *Bull. Raffles Mus.* 16: 88–113.

Tweedie, M. F. W. 1950. Notes on grapsoid crabs from the Raffles Museum. 2. On the habits of three ocypodid crabs. *Bull. Raffles Mus.* 23: 310–324.

Tweedie, M. F. W. 1954. Notes on grapsoid crabs from the Raffles Museum, no. 3. Faunal differentiation in the regions east and west of the Malay Peninsula. *Bull. Raffles Mus.* 25: 118–121.

Uma, K., and T. Subramoniam. 1979. Histochemical characteristics of spermatophore layers of *Scylla serrata* (Forskal) (Decapoda: Portunidae). *Int. J. Inv. Reprod.* 1: 31–40.

Valente, D. 1948. Mechanisms da respiracao de *Trichodactylus petropolitanus* (Goeldi). *Univ. Sao Paulo Fac. filosof. cienc. e letras Zool.* 12: 259–316.

Vannini, M. 1975. Researches on the coast of Somalia. The shore and the dune of Sar Uanle. 5. Description and rhythmicity of digging behaviour in *Coenobita rugosus* H. Milne Edwards. *Monit. Zool. Ital. (N.S.) Suppl.* 6: 233–242.

Vannini, M. 1976a. Researches on the coast of Somalia. The shore and the dune of Sar Uanle. 7. Field observations on the periodical transdune migrations of the hermit crab *Coenobita rugosus* Milne Edwards. *Monit. Zool. Ital. Suppl.* 7: 145–185.

Vannini, M. 1976b. Researches on the coast of Somalia. The shore and the dune of Sar Uanle. 10. Sandy beach decapods. *Monit. Zool. Ital. Suppl. (N.S.)* 8: 255–286.

Vannini, M. 1980a. Notes on the behavior of *Ocypode ryderi* Kingsley (Crustacea, Brachyura). *Mar. Behav. Physiol.* 7: 171–183.

Vannini, M. 1980b. Researches on the coast of Somalia. The shore and dune of Sar Uanle. 27. Burrows and digging behavior in *Ocypode* and other crabs (Crustacea, Brachyura). *Monit. Zool. Ital. Suppl. (N.S.)* 13: 11–44.

Vannini, M., and G. Chelazzi. 1981. Orientation of *Coenobita rugosus* (Crustacea: Anomura): A field study on Aldabra. *Mar. Biol.* 64: 135–140.

Vannini, M., and A. Sardini. 1971. Aggressivity and dominance in the river crab *Potamon fluviatile* (Herbst). *Monitore Zool. Ital.* 5: 173–213.

Vannini, M., and P. Valmori. 1981a. Researches on the coast of Somalia. The shore and the dune of Sar Uanle. 30. Grapsidae (Decapoda, Brachyura). *Monit. Zool. Ital.* 14: 57–101.

Vannini, M., and P. Valmori. 1981b. Researches on the coast of Somalia. The shore and dune of Sar Uanle. 31. Ocypodidae and Gecarcinidae (Decapoda, Brachyura). *Monit. Zool. Ital.* 14: 199–226.

Vantets, G. F. 1956. A study of solar and spatial orientation of *Hemigrapsus oregonensis* (Dana) and *Hemigrapsus nudas* (Dana). Unpubl. B.S. thesis, University of British Columbia, Vancouver, B.C., Canada.

Vanweel, P. B. 1970. Digestion in Crustacea. In *Chemical zoology*. Edited by M. Florkin and B. T. Scheer, vol. 5, pp. 97–115. New York: Academic Press.

Varghese, A. S. 1984. On semen production in the crab, *Gecarcinucus steniops*. M. Phil. diss., Calicut University, India.

Veillet, A. 1945. Recherches sur le parasitisme des crabes et des Galathees par les Rhizociphales et des Epicarides. *Ann. Inst. Oceanogr.* (Paris) 22: 193–341.

Vernberg, F. J. 1969. Acclimation of intertidal crabs. *Amer. Zool.* 9: 332–341.

Vernberg, F. J., and J. D. Costlow. 1966. Handedness in fiddler crabs (genus *Uca*). *Crustaceana* 11: 61–64.

Vernet-Cornubert, G. 1958. Recherches sur la sexualite du crabe *Pachygrapsus marmoratus* (Fabricius). *Arch. Zool. Exp. Gen.* 96: 104–274.

Verwey, J. 1930. Einiges über die Biologie ostindischer Mangrove-krabben. *Treubia* 12: 169–261.

Vignal, W. 1886. Sur l'endothelium de la paroi interne des vaisseaux des invertebres. *Arch. Physiol. Norm. et Pathol.* 8: 1–6.

Vogel, H. H., and J. R. Kent. 1971. A curious case: The coconut crab. *Fauna* 2: 4–11.

Volker, L. 1965. Experimentelle Untersuchungen zur Oekologie des Landeinsiedlerkrebses *Coenobita scaevola* Forskal am Roten Meer. Thesis, Techn. Hochschule, Darmstadt.

von Baumann, H. 1917. Das Cor frontale bei dekapooden Kresben. *Zool. Anz.* 49: 137–144.

von Hagen, H.-O. 1961. Experimentelle Studien zum Winken von *Uca tangeri* in Suedspanien. *Verhandl. deutschen. Zoolog. Gesselsch. Saarbruecken*: 424–432.

von Hagen, H.-O. 1962. Freilandstudien zur Sexual- und Fortpflanzungbiologie von *Uca tangeri* in Andalusien. *Z. Morphol. Oekol. Tiere* 51: 611–725.

von Hagen, H.-O. 1967a. Klopfsignale auch bei Grapsiden. *Naturwissenschaften* 7: 177–178.

von Hagen, H.-O. 1967b. Nachweis einer kinaesthetischen Orientierung bei *Uca rapax. Z. Morphol. Oekol. Tiere* 58: 301–320.

von Hagen, H. O. 1968. Studien an peruanischen Winkerkrabben (Uca). *Zool J. System.* 95: 395–468.

von Hagen, H.-O. 1970. Die Balz von *Uca vocator* (Herbst.) als oeklogisches Problem. *Forma et Functio* 2: 238–253.

von Hagen, H.-O. 1975. Classification and evolution of sound production in ocypodid and grapsid crabs. *Z. Systematik Evol.* 13: 300–316.

von Hagen, H.-O. 1977. The tree-climbing crabs of Trinidad. *Stud. Fauna Curacao* 54: 25–50.

von Hagen, H.-O. 1983. Visual and acoustic display in *Uca mordax* and *U.*

burgersi, sibling species of neotropical fiddler crabs. I. Waving display. *Behaviour* 83: 229–250.

von Hagen, H.-O. 1984. Visual and acoustic display in *Uca mordax* and *U. burgersi*, sibling species of neotropical fiddler crabs. II. Vibration signals. *Behaviour* 91: 204–228.

Vonk, H. J. 1960. Digestion and metabolism. In *The physiology of Crustacea.* Edited by T. H. Waterman, vol. 1, pp. 281–316. New York: Academic Press.

Von Prahl, H. 1983. Cangrejos Gecarcinidos (Crustacea: Gecarcinidae) de la Isla de Providencia, Colombia. *Carib. J. Sci.* 19: 31–34.

Von Raben, K. 1934. Veranderungen im Kiemendeckel und in den Kiemen einiger Brachyuren (Decapoden) im Verlauf der Anpassung an die Feuchtluftatmung. *Z. wiss. Zool.* 145: 425–461.

Vuillemin, S. 1967. La respiration chez les Crustaces Decapoda. *Ann. Biol.* 6: 47–82.

Vuillemin. S. 1970. Observations ecologiques et biologiques sur *Madagapotamon humberti,* Bott 1955 (Crustace, Decapode, Potamonide) du massif de l'Ankarana (Madagascar). *Ann. Univ. Madagascar* 7: 245–266.

Vuillemin, S. 1972. Contribution a l'etude ecologique de la Montagne des Francais (Province de Diego-Suarez): biologie de *Gecarcinautes antongilensis antongilensis* (Rathbun, 1905). *Ann. Univ. Madagascar* 9: 135–167.

Wales, W. 1982. Control of mouthparts and gut. In *The Biology of Crustacea.* Vol. 4. Edited by D. C. Sandeman and H. L. Atwood, pp. 165–191. New York: Academic Press.

Wallace, R. A., S. L. Walker, and P. V. Hauschka. 1967. Crustacean lipovitellin. Isolation and characterization of the major high-density lipoprotein from the eggs of decapods. *Biochemistry* 6: 1582–1590.

Wanson, S. A., A. Pequeux, and R. D. Roer. 1984. NaA + B regulation and (NaA + B + KA + B) ATPase activity in the euryhaline fiddler crab *Uca minax* (Le Conte). *Comp. Biochem. Physiol.* 79A: 673–678.

Warburg, M. R., and S. Goldenburg. 1984. Water loss and haemolymph osmolarity of *Potamon potamios,* an aquatic land crab, under stress of dehydration and salinity. *Comp. Biochem. Physiol.* 79A: 451–455.

Warburg, M. R., S. Goldenberg, and D. Rankevich. 1982. Temperature effect on the behavior and locomotor activity rhythm as related to water balance in the aquatic/land crab *Potamon potamios* Olivier (Crustacea: Brachyura: Potamonidae). *J. Crustacean Biol.* 2: 420–429.

Warburg, W. R., and E. Shuchman. 1979. Experimental studies on burrowing of *Ocypode cursor* (Crustacea: Ocypodidae) in response to sand moisture. *Mar. Behav. Physiol.* 6: 147–156.

Warner, G. F. 1967. The life history of the mangrove tree crab, *Aratus pisoni. J. Zool.* (London) 153: 321–335.

Warner, G. F. 1969. The occurrence and distribution of crabs in a Jamaican mangrove swamp. *J. Anim. Ecol.* 38: 379–389.

Warner, G. F. 1977. *The biology of crabs.* London: Elek Science.

Warner, G. F. 1979. Behaviour of two species of grapsid crabs during intraspecific encounters. *Behaviour* 36: 9–19.

Warren, J. H. 1985. Climbing as an avoidance behaviour in the salt marsh periwinkle *Littorina irrorata* (Say). *J. Exp. Mar. Biol. Ecol.* 89: 11–28.

Watanabe, K., and J. Yamada. 1980. Osmoregulation and the gill Na-K ATPase activity in *Eriocheir japonicus. Bull. Fac. Fish. Hokkaido Univ.* 31: 283–289.

Watt, E. M., D. W. Dunham, and H. Schoene. 1985. Climbing orientation and shell asymmetry in a land hermit crab, *Coenobita clypeatus* (Decapoda, Paguridea). *Crustaceana* 48: 104–105.

Wear, G. F. 1967. Life history studies on New Zealand Brachyura. 1. Embryonic and postembryonic development of *Pilumnus novae-zealandiae* Fihol 1886, and of *P. lumpinus* Bennet, 1964 (Xanthidae, Pilumnae). *N. Z. J. Mar. Freshwater Res.* 1: 482–535.

Weis, J. S. 1976. Effects of environmental factors on regeneration and molting in fiddler crabs. *Biol. Bull.* 150: 152–162.

Weis, J. S. 1977. Limb regeneration in fiddler crabs: Species differences and effects of methylmercury. *Biol. Bull.* 152: 263–274.

Weitzman, M. C. 1963. The biology of the tropical land crab, *Gecarcinus lateralis* (Freminville). Ph.D. diss., Albert Einstein College of Medicine, Yeshiva University, New York.

Weitzman, M. C. 1964. Ovarian development and molting in the tropical land crab, *Gecarcinus lateralis* (Freminville). *Amer. Zool.* 4: 329–330.

Weitzman, M. C. 1966. Oogenesis in the tropical land crab, *Gecarcinus lateralis* (Freminville). *Z. Zellforsch.* 75: 109–119.

Wells, F. E. 1984. Comparative distribution of macromolluscs and macrocrustaceans in a North-western Australian mangrove system. *Aust. J. Mar. Freshw. Res.* 35: 591–596.

Wernick, A. M. 1982. The oxygen consumption of the hermit crab *Clibanarius vittatus* (Bosc.) (Decapoda, Diogenidae) in relation to temperature and size. *Rev. Brasil. Biol.* 42: 267–273.

West. J. B. 1977. *Ventilation/blood flow and gas exchange.* 3rd ed. Oxford: Blackwell.

Weygoldt, P. 1961. Ontogenie der Dekapoden: Embryologische Untersuchungen an *Palaemonetes varians* Leach. *Zool. Jahrb., Abt. Anat. Ontog. Tiere* 79: 223–270.

Wheatly, M. G. 1981. The provision of oxygen to developing eggs by female shore crabs (*Carcinus maenas*). *J. Mar. Biol. Assoc. U.K.* 61: 117–128.

Wheatly, M. G., W. W. Burggren, and B. R. McMahon. 1984. The effects of temperature and water availability on ion and acid-base balance in hemolymph of the land hermit crab *Coenobita clypeatus. Biol. Bull.* 166: 427–445.

Wheatly, M. G., B. R. McMahon, W. W. Burggren, and A. W. Pinder. 1985. A rotating respirometer to monitor voluntary activity and associated exchange of respiratory gases in the land hermit crab (*Coenobita compressus* H. Milne Edwards). *J. Exp. Biol.* 119: 85–102.

Wheatly, M. G., B. R. McMahon, W. W. Burggren, and A. W. Pinder. 1986. Haemolymph acid-base, electrolyte and blood gas status during sustained voluntary activity in the land hermit crab (*Coenobita compressus* H. Milne Edwards). *J. Exp. Biol.* 125: 225–244.

Wheatly, M. G., and E. W. Taylor. 1979. Oxygen levels, acid-base status and heart rate during emersion of the shore crab *Carcinus maenas* (L.)

into air. *J. Comp. Physiol.* 132: 305–311.

Wiens, H. J. 1962. Terrestrial Crabs. In *Atoll environment and ecology.* New Haven, Conn.: Yale University Press.

Wiens, T. J. 1982. Small systems of neurons: Control of rhythmic and reflex activities. In *The biology of the Crustacea*, vol. 4. Edited by D. C. Sandeman and H. L. Atwood, pp. 193–240. New York: Academic Press.

Wilkens, J. L. 1981. Respiratory and circulatory co-ordination in Decapod crustaceans. In *Locomotion and energetics in arthropods.* Edited by C. F. Herreid and C. R. Fortner, pp. 277–298. New York: Plenum.

Wilkens, J. L., and M. Fingerman. 1965. Heat tolerance and temperature relationships of the fiddler crab, *Uca pugilator*, with reference to body coloration. *Biol. Bull.* 128: 133–141.

Wilkens, J. L., and B. R. McMahon. 1972. Aspects of branchial irrigation in the lobster *Homaraus americanus.* I. Functional analysis of scaphognathite beat water pressures and currents. *J. Exp. Biol.* 56: 469–479.

Wilkens, J. L,, P. R. H. Wilkes, and J. Evans. 1984. Analysis of the scaphognathite ventilatory pump in the shore crab *Carcinus maenas.* II. Pumping efficiency and metabolic cost. *J. Exp. Biol.* 113: 55–81.

Williams, A. B. 1965. Marine decapod crustaceans of the Carolinas. *Fishery Bull. Fish. Wildl. Serv. U.S.* 65: 1–298.

Williams, A. J., and P. L. Lutz. 1975. The role of the hemolymph in the carbohydrate metabolism of *Carcinus maenas. J. Mar. Bio. Ass. U.K.* 55: 667–670.

Williams, M. J. 1981. Handedness in males of *Uca vocans* (Linneaus, 1758) (Decapoda, Ocypodidae). *Crustaceana* 40: 215–216.

Wilson, K. A. 1985. Physical and biological interactions that influence habitat use of mangrove crabs. Ph.D. diss., University of Pennsylvana, Philadelphia.

Wolcott, D. L., and T. G. Wolcott, 1984. Food quality and cannibalism in the red land crab, *Gecarcinus lateralis. Physiol. Zool.* 57: 318–324.

Wolcott, D. L., and T. G. Wolcott. 1985. Nitrogen limitation in the herbivorous land crab *Cardisoma guanhumi. Amer. Zool.* 25: 36A.

Wolcott, D. L., and T. G. Wolcott. 1987. Nitrogen limitation in the herbivorous land crab *Cardisoma guanhumi. Physiol. Zool.* 60: 262–268.

Wolcott, T. G. 1976. Uptake of soil capillary water by ghost crabs. *Nature* 264: 756–757.

Wolcott, T. G. 1978. Ecological role of ghost crabs, *Ocypode quadrata* (Fabricius) on an ocean beach: Scavengers or predators? *J. Exp. Mar. Biol. Ecol.* 31: 67–82.

Wolcott, T. G. 1980. Salt availability and range limitation in land crabs. *Amer. Zool.* 20: 295A.

Wolcott, T. G. 1984. Uptake of interstitial water from soil: Mechanisms and ecological significance in the ghost crab *Ocypode quadrata* and two gecarcinid crabs. *Physiol. Zool.* 57: 161–184.

Wolcott, T. G., and D. L. Wolcott. 1982a. Larval loss and spawning behavior in the land crab *Gecarcinus lateralis* (Freminville). *J. Crustacean Biol.* 2: 477–485.

Wolcott, T. G., and D. L. Wolcott. 1982b. Urine reprocessing for salt conservation in terrestrial crabs. *Am. Zool.* 22: 897.

Wolcott, T. G., and D. L. Wolcott. 1984a. Impact of off-road vehicles on macroinvertebrates of a mid-Atlantic beach. *Biol. Conserv.* 29: 217–240.

Wolcott, T. G., and D. L. Wolcott. 1984b. Salt conservation by extrarenal resorption in *Cardisoma guanhumi. Amer. Zool.* 25: 78A. (Abstr.).

Wolcott, T. G., and D. L. Wolcott. 1984c. Extra-renal urine modification for salt conservation in the land crab *Cardisoma guanhumi. Amer. Zool.* 24: 78A.

Wolcott, T. G., and D. L. Wolcott. 1985a. Extrarenal modification of urine for ion conservation in ghost crabs, *Ocypode quadrata* (Fabricius). *J. Exp. Mar. Biol. Ecol.* 91: 93–107.

Wolcott, T. G., and D. L. Wolcott. 1985b. Factors influencing the limits of migratory movements in terrestrial crustaceans. In *Migration: Mechanisms and adaptive significance.* Contr. Mar. Sci. (Univ. Texas Mar. Inst.) Suppl. 27: 257–273.

Wolff, T. 1961. Description of a remarkable deep-sea hermit crab with notes on the evolution of the Paguridea. *Galathea Report* 4: 11–32.

Wolin, E. M., W. Laufer, and D. F. Albertina. 1973. Uptake of the yolk protein, lipovitellin, by developing crustacean oocytes. *Develop. Biol.* 35: 160–170.

Wolvekamp. H. P., and T. H. Waterman. 1960. Respiration. In *The physiology of Crustacea.* Edited by T. H. Waterman, vol. 1, pp. 35–100. New York: Academic Press.

Wood, C. M., and R. G. Boutilier. 1985. Osmoregulation, ionic exchange, blood chemistry, and nitrogenous waste excretion in the land crab *Cardisoma carnifex*: A field and laboratory story. *Biol. Bull.* 169: 267–290.

Wood, C. M., R. G. Boutilier, and D. J. Randall. 1968. The physiology of dehydration stress in the land crab *Cardisoma carnifex*: Respiration, ionoregulation, acid–base balance, and nitrogenous waste excretion. *J. Exp. Biol.* 126: 271–296.

Wood, C. M., and J. N. Cameron. 1985. Temperature and the physiology of intracellular and extracellular acid-base regulation in the blue crab *Callinectes sapidus. J. Exp. Biol.* 114: 151–179.

Wood, C. M., and D. J. Randall. 1981a. Oxygen and carbon dioxide exchange during exercise in the land crab (*Cardisoma carnifex*). *J. Exp. Zool.* 218: 7–16.

Wood, C. M., and D. J. Randall. 1981b. Hemolymph gas transport, acid-base regulation, and anaerobic metabolism during exercise in the land crab (*Cardisoma carnifex*). *J. Exp. Zool.* 218: 23–25.

Wright, H. O. 1968. Visual displays in brachyuran crabs: Field and laboratory studies. *Amer. Zool.* 8: 655–665.

Yaldwyn, J. C., and K. Wodzicki. 1979. Systematics and ecology of the land crabs (Decapoda: Coenobitidae, Grapsidae and Gecarcinidae) of the Tokelau Islands, central Pacific. *Atoll Res. Bull.* 235: 1–58.

Yamaguchi, T. 1977. Studies on the handedness of the fiddler crab *Uca lactea. Biol. Bull.* 152: 424–436.

Yamaguchi, T., Y. Noguchi, and N. Ogawara. 1979. Studies of the courtship behavior and copulation of the sand bubbler crab, *Scopimera globosa. Publ. Amakusa Mar. Biol. Lab.* 5: 31–44.

Yamaguchi, T., and M. Tanaka. 1974. Studies on the ecology of a sand bubbler crab, *Scopimera globosa* de Haan (Decapoda, Ocypodidae). *Jap. J. Ecol.* 24: 165–174.

Yokoe, Y., and I. Yasumasu. 1964. The distribution of cellulase in invertebrates. *Comp. Biochem. Physiol.* 13: 323–338.

Young, A. M. 1978. Desiccation tolerances for three hermit crab species *Clibanarius vittatus* (Bosc), *Pagurus pollicaris* Say and *P. longicarpus* Say (Decapoda, Anomura) in the North Inlet estuary, South Carolina, U.S.A. *Estuarine Coastal Mar. Sci.* 6: 117–122.

Young, A. M. 1979. Osmoregulation in three hermit crab species, *Clibanarius vittatus* (Bosc), *Pagurus pollicaris* Say and *P. longicarpus* Say (Crustacea, Decapoda, Anomura). *Comp. Biochem. Physiol.* 63A: 377–382.

Young, D., and H. Ambrose. 1978. Underwater orientation in the sand fiddler crab, *Uca pugilator. Biol. Bull.* 155: 246–258.

Young, J. E. 1974. Variations in the timing of spermatogenesis in *Uca pugnax* (Smith) and possible effectors (Decapoda, Brachyura, Ocypodidae). *Crustaceana* 27: 68–72.

Young, R. E. 1972. The physiological ecology of haemocyanin in some selected crabs. II. The characteristics of hemocyanin in relation to terrestrialness. *J. Exp. Med. Biol. Ecol.* 10: 193–206.

Young, R. E. 1975. Neuromuscular control of ventilation in the crab *Carcinus maenas. J. Comp. Physiol.* 101: 1–37.

Zagalsky, P. F., D. F. Cheesman, and H. J. Ceccaldi. 1967. Studies on carotenoid-containing lipoproteins isolated from the eggs and ovaries of certain marine invertebrates. *Comp. Biochem. Physiol.* 22: 851–871.

Zanders, I. P. 1978. Ionic regulation in the mangrove crab *Goniopsis cruentata. Comp. Biochem. Physiol.* 60A: 293–302.

Zanders, I. P. 1980. Regulation of blood ions in *Carcinus maenas* (L.). *Comp. Biochem. Physiol.* 65A: 97–108.

Zebe, E. 1982. Anaerobic metabolism in *Upogebia pugettensis* and *Callianassa californiensis* (Crustacea, Thalassinidea). *Comp. Biochem. Physiol.* 72B: 613–617.

Zehnder, H. 1934a. Ueber die Embryonalentwicklung des Flusskrebes. *Acta Zool.* (Stockholm) 15: 261–448.

Zehnder, H. 1934b. Zur Embryologie des Flusskrebses. *Verh. Schweiz. Naturforsch. Ges.* 115: 357–358.

Zeuthen, E. 1953. Oxygen uptake as related to body size in organisms. *Quart. Rev. Biol.* 28: 1–12.

Zimmer-Faust, R. K. 1987. Crustacean chemical perception: Towards a theory on optimal chemoreception. *Biol. Bull.* 172: 10–29.

Zucker, N. 1974. Shelter building as a means of reducing territory size in the fiddler crab, *Uca terpsichores* (Crustacea: Ocypodidae). *Amer. Midl. Nat.* 91: 224–236.

Zucker, N. 1976. Behavioral rhythms in the fiddler crab *Uca terpsichores*. In *Biological rhythms in the marine environment.* Edited by P. J. Decoursey, pp. 145–159. Columbia: University of South Carolina Press.

Zucker, N. 1977. Neighbor dislodgement and burrow-filling activities by male *Uca musica terpsichores*: A spacing mechanism. *Mar. Biol.* 41: 281–286.

Zucker, N. 1978. Monthly reproductive cycles in three sympatric hood-building tropical fiddler crabs (genus *Uca*). *Biol. Bull.* 155: 410–424.

Zucker, N. 1981. The role of hood-building in defining territories and limiting combat in fiddler crabs. *Anim. Behav.* 29: 387–395.

Zucker, N. 1983. Courtship variation in the neo-tropical fiddler crab *Uca deichmanni*: Another example of female incitation of male competition? *Mar. Behav. Physiol.* 10: 57–79.

Zucker, N. 1984. Delayed courtship in the fiddler crab *Uca musica terpsichores*. *Anim. Behav.* 32: 735–742.

Zucker, N., and R. Denny. 1979. Interspecific communication in fiddler crabs: Preliminary report of a female rejection display directed toward courting heterospecific male. *Z. Tierpsychol.* 50: 9–17.

Author index

Systematic index

Notes to systematic index:

* indicates noncrustacean species

[ ] indicates names that have been superseded.

†Note that there is considerable confusion with respect to the classification of the genus Parathelphusa, which has been compounded by frequent misspelling as Paratelphusa and the existence within the Gecarcinucidae of the genus Barytelphusa, which has an overlapping geographical distribution. Additional misunderstanding arises from the precedence of Holthuisana or Parathelphusa, which has yet to be satisfactorily resolved. The validity of the families Parathelphusidae and Sundathelphusidae hangs in the balance.

Subject index